Gondwana Landscapes in southern South America

Springer Earth System Sciences

For further volumes:
http://www.springer.com/series/10178

Jorge Rabassa • Cliff Ollier
Editors

Gondwana Landscapes in southern South America

Argentina, Uruguay and southern Brazil

Springer

Editors
Jorge Rabassa
Laboratorio de Geomorfología y Cuaternario
CADIC-CONICET
Ushuaia, Tierra del Fuego
Argentina

Universidad Nacional de Tierra del Fuego
Ushuaia, Tierra del Fuego
Argentina

Cliff Ollier
School of Earth and Environment
University of Western Australia
Perth, Australia

ISBN 978-94-007-7701-9 ISBN 978-94-007-7702-6 (eBook)
DOI 10.1007/978-94-007-7702-6
Springer Dordrecht Heidelberg New York London

Library of Congress Control Number: 2014935868

Printed on acid-free paper

Springer is part of Springer Science+Business Media (www.springer.com)

Contents

About the Authors

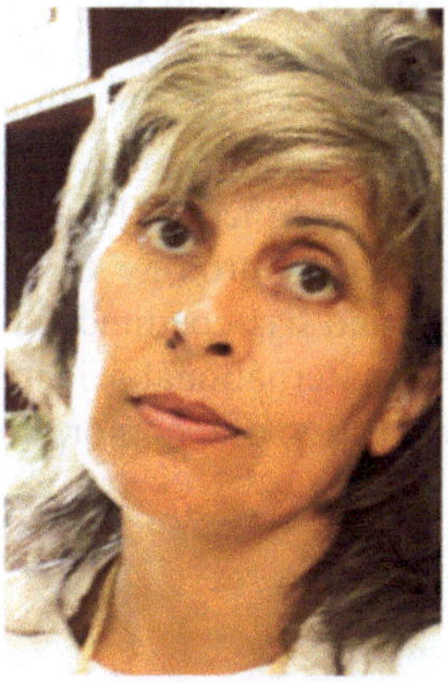

Emilia Y. Aguilera graduated as a geologist (1984) and obtained her doctorate in Remote Sensing and Geomorphology of Paleosurfaces (2006) from the Facultad de Ciencias Naturales y Museo, Universidad Nacional de La Plata, La Plata, Argentina. She is a Professor of Igneous Petrology in the same college since 2009. She is also a senior specialist in Remote Sensing Techniques at the Dirección de Aplicación de Imágenes Satelitarias (DAIS), Ministry of Public Works, La Plata, Buenos Aires province. Her main fields of interest are remote sensing, geomorphology of large areas, and paleosurfaces, especially those developed over igneous rocks. She is active in several projects related to Patagonia and Western Argentina cratonic areas, igneous petrology, volcanology, and paleosurfaces.

María Jimena Andreazzini, M.Sc. graduated as geologist at the Universidad Nacional de Río Cuarto (UNRC, 2002; Río Cuarto, Argentina) and as Master of Sciences in Geology at the Universidade Estadual de Campinas, Brazil (2005). She is presently a doctoral candidate at UNRC, holding a doctoral scholarship from CONICET, the National Research Council of Argentina, and she also teaches in the Department of Geology, UNRC. Since 2009 she takes part in research projects related to the geomorphology and hydrology of mountain and piedmont areas. She is a coauthor of 16 papers and has coedited 2 books. She has participated in the organization of national and international scientific meetings. She is presently secretary of the Argentine Association of Geomorphology and Quaternary Studies.

Eugenio Aragón graduated as a geologist and obtained his doctorate in Petrology (1986) from the Facultad de Ciencias Naturales y Museo, Universidad Nacional de La Plata, La Plata, Argentina.

He is presently a Professor of Petrology at the same university since 1992 and a permanent, senior researcher with CONICET, the National Research Council of Argentina, at the Centro de Investigaciones Geológicas (CIG), Universidad Nacional de La Plata Argentina.

He has participated for more than two decades as a senior researcher in CONICET and also as a Lecturer in the national universities of La Plata, Rio Cuarto and Río Negro, in Argentina. He is responsible for several relevant research projects in the field of Igneous Petrology and Crustal Evolution.

François Bétard is Associate Professor in Physical Geography and Environmental Science at Paris-Diderot University, and a member of the research team CNRS UMR 8586 PRODIG. His main research interests lie in the fields of geomorphology, soil science and rock weathering, with a special focus on tropical and paleotropical environments. Involved in the study of paleolandforms and related paleoweatherings at various geological timescales, he worked extensively in South America, particularly in Northeast Brazil and Southern Patagonia, but also in France (Armorican Massif, Paris and Aquitaine Basins). Ongoing studies include applied research in the fields of conservation and valorization of geological and geomorphological heritage into protected areas (natural parks and reserves, geoparks).

Leda Sánchez Bettucci is the Director of the Institute of Geological Sciences, part of the Facultad de Ciencias at the Universidad de la República in Uruguay. She completed her initial degree in Geology there before completing her Ph.D. in

Geological Sciences at the University of Buenos Aires (UBA) in Argentina. Her academic career began at the Departamento de Geología at the Universidad de la República, where she pursued research on tectonics and paleomagnetism. She has since been working on the tectonic evolution of Uruguay and paleogeographic reconstructions and has also contributed to our geological understanding of aspects related to magnetic paleointensities and environmental magnetism. Over the last 4 years, she has supervised the installation of the first geophysical observatory (for seismology and geomagnetism) in Uruguay. Her professional interests are in tectonics and geophysics.

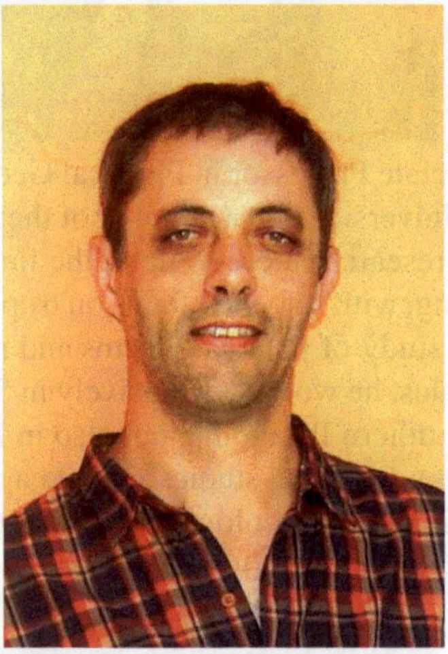

Ernesto Brunetto is a geologist (2003) and doctor in Geological Sciences (2008). He has worked in Quaternary Geology and Neotectonics in cratonic areas and large plains of Argentina. His current position is Assistant Researcher and leads the Laboratory of Neogene-Quaternary Stratigraphy working in collaboration together with paleontology colleagues.

Claudio A. Carignano graduated as a geologist at the School of Geology, Faculty of Physical and Natural Sciences, Universidad Nacional de Córdoba (UNC),

Argentina, in 1990. After completing advanced studies in Geomorphology as Postgraduate Scholarship Holder (1992–1997) at the Argentine National Research Council (CONICET), he obtained in 1997 a Ph.D. (magna cum laude – maximum honors) at UNC. After that, with a CONICET Postdoctoral Scholarship, he completed postdoctoral studies on Quaternary Geology at UNC.

Trained in geomorphology and geomatics, he got several teaching positions: Assistant Professor of Photogeology and Remote Sensing (1990–1997) at the School of Geology UNC; Professor of Geomatics and Environmental Information Systems, Soil Resources Management, and Pollution of Natural Resources (2001–2005) of the Environmental Management Department, Blas Pascal University; Professor of Remote Sensing (2005–2008) of the School of Biology, Sciences and Technology Department of the Universidad Nacional de Chilecito (UNdeC); and Professor of Geomorphology and Remote Sensing (2008) of the Geology Department at the Universidad del Sur (UNS), Bahía Blanca.

He was Director of the Mountain and Arid Regions Environments Research Institute (IAMRA) of UNdeC (2006–2009). Since 2009, he is Assistant Professor of Tectonic Geology, School of Geology of UNC. His main research interests are geomorphological hazards and the interaction between humans and the geological environment.

Silvina Carretero graduated in Geology (2006) and obtained her doctorate in Natural Sciences (Geology, 2011) at the Facultad de Ciencias Naturales y Museo, Universidad Nacional de La Plata, Argentina. She has a permanent research position with CONICET, the National Research Council of Argentina, at the University of La Plata, where she also teaches Mineralogy and Igneous Rocks Petrology.

Her main field of interest is the hydrogeology of coastal aquifers, including modifications in the coastal dune landscape due to urbanization and natural processes. She is also interested in paleosurfaces and pseudokarst landscapes.

Claudia E. Cavarozzi graduated as a geochemist (1990) from the Facultad de Ciencias Naturales y Museo, Universidad Nacional de La Plata, La Plata, Argentina. She is presently a researcher with the Centro de Investigaciones Geológicas (CIG), Universidad Nacional de La Plata, and Full Professor of Geochemistry since 2010. She has participated since 1992 in geochemical and igneous petrology projects as member of CONICET, the National Research Council of Argentina, and for more than 20 years, she has been in charge of geochemistry facilities at the Universidad Nacional de La Plata. She has been part of many research projects in the field of geochemistry of igneous petrology.

Marcela A. Cioccale graduated as geologist at the School of Geology, Faculty of Physical and Natural Sciences, Universidad Nacional de Córdoba (UNC), Argentina, in 1989. After completing advanced studies in geomorphology with an Argentine National Research Council (CONICET) Postgraduate Scholarship

(1993–1998), she got a Ph.D. at the UNC in 1999. After that, she completed postdoctoral studies of Environmental Geology at the UNC on a CONICET postdoctoral fellowship.

Trained in geomorphology and environmental geology, she obtained several teaching positions: Professor of Geomatics and Environmental Information Systems, Soil Resources Management, and Pollution of Natural Resources (2004–2006), in the Environmental Management Department at Blas Pascal University, Córdoba; Professor of Physical Geography (2005–2008) at the School of Biology, in the Sciences and Technology Department of the Universidad Nacional de Chilecito (UNdeC); Professor of Geomorphology and Remote Sensing (2009–2011), at the Geology Department of the Universidad Nacional del Sur (UNS), Bahía Blanca, Argentina; and Professor of Geomorphometry Postgraduate Courses at UNC, UNS, and the National Universities of San Juan (UNSJ) and Río Cuarto (UNRC) (2010–2013). She was also the Dean of Science and Technology of the UNdeC (2007–2009).

Since 1989 she is Assistant Professor of Rock Mechanics at the School of Geology School, UNC. Her main research interests are the relationships between geomorphology and hydrogeochemistry of mountain streams.

Rosa Hilda Compagnucci is currently Full Professor of Climatology at the Department of Atmospheric and Oceans Sciences (DCAO), Faculty of Exact and Natural Sciences (FCEN), Buenos Aires University (UBA), Buenos Aires, Argentina. She holds a *licenciada* (1974) and a doctorate (1989) in Meteorology Sciences. She is also a Principal Investigator with the National Research Council of Argentina (CONICET). Her main fields of interest are synoptic climatology, paleoclimatology, and statistic applications in atmospheric sciences and climate change. She participated as lead author in the Intergovernmental Panel for Climate Change (IPCC), ARA 3, Working Group II. She has recently taken part as a reviewer of the TSU, ARA 5, Working Group I, IPCC.

Susana B. Degiovanni graduated in Geology (1981), has a master's degree in Environmental Geology (1999), and got her doctorate in Geology (2008), all degrees from the Universidad Nacional de Río Cuarto (UNRC, Córdoba, Argentina). She teaches at UNRC since 1981 and she is presently Associated Professor at the Department of Geology, UNRC. Since 1988 she has directed research projects on geomorphology and Quaternary sciences, particularly of surficial hydrological systems and related geoenvironmental topics. She has published four books and 76 scientific papers. She has participated in 46 national and international scientific meetings. Since 2012, she is the President of the Argentine Association of Geomorphology and Quaternary Studies.

Alicia Folguera is a geologist with the SEGEMAR, the Geological Survey of Argentina, in Buenos Aires, Argentina.

She got a Geology degree in 1997 and a doctorate in Geology in 2011, both at the Universidad de Buenos Aires.

She is presently in charge of the regional geological mapping of central Argentina for SEGEMAR. She has been working in the regional geology and the tectonic evolution of the Pampean plains since 2000.

Gabriel Galina is an advanced student of Cartography. He has developed methods for quantitative geomorphology for the fieldworks and mapping laboratory.

Ofelia Gutiérrez completed her degree in Geography and an M.Sc. in Environmental Sciences at the Facultad de Ciencias of the Universidad de la República in Uruguay; she also received a Diploma of Advanced Studies (DEA) from the International University of Andalucía, Spain. She is currently an Assistant Professor at the Instituto de Ecología y Ciencias Ambientales of the Facultad de Ciencias. She is a Visiting Scientist for the Engov Project (Environmental Governance) at the Gino Germani Research Institute, University of Buenos Aires, and has authored or coauthored more than 30 publications on GIS, environmental issues, physical geography, and environmental impacts assessment, including articles in international journals, books, and book chapters.

Martín H. Iriondo is principal investigator of CONICET.

Doctor of Geological Sciences with dedication to regional Quaternary of South America and paleoclimates.

At present, he is developing a program on Stratigraphic Correlation of the Quaternary of the South American countries, which comprises the production of several books (five of them are already published).

He was Vice-President of INQUA and Professor of Geomorphology at UNL (Argentina).

Professor Iriondo is the author of the book "Introducción a la Geología" (4 editions) and some 90 research papers.

Daniela M. Kröhling Associate Researcher of CONICET (National Research Council of Argentina) and Professor of Geomorphology at Facultad de Ingeniería y Ciencias Hídricas (UNL – Universidad Nacional del Litoral), Santa Fe, Argentina. Daniela Kröhling (1967) is a geologist (1992) and a doctor in Geological Sciences (1998).

Her main fields of interest are Quaternary geology, stratigraphy, sedimentology, geomorphology, and palaeoclimatology, with main study areas in Argentina Chaco-Pampa plain and Mesopotamia region, Southern Brazil and Northeastern Argentina basaltic meseta, and Southern Puna (Altiplano) plateau. She is currently the secretary and Webmaster of TERPRO Commission of INQUA, responsible for the Sam-GeoQuat Group (INQUA IFG), and is in charge of national and international research projects.

Francisco Sergio Bernardes Ladeira works in the Department of Geography, Institute of Geosciences, State University of Campinas (UNICAMP), Brazil.

Francisco Ladeira (1965) is a geographer for Universidade Estadual Paulista (UNESP), Rio Claro (1989). He holds a master's degree (1995) and a PhD (2002) in Physical Geography from the University of São Paulo (USP). Minister for 10 years the discipline of Pedology and Structural Geomorphology for the courses of Geography and Geology in UNICAMP and devotes himself to study the relationship between paleosols and paleoenvironments and also old weathering profiles and geomorphological evolution of long duration.

Eduardo Jorge Llambías obtained his doctorate in Geology (1964) from the Universidad de Buenos Aires, Argentina. He is presently an Emeritus Professor at the Centro de Investigaciones Geológicas (CIG), School of Natural Sciences, Universidad Nacional de La Plata. His research interests lie in the transition from shallow plutons to volcanic rocks and the associated epithermal mineral deposits, focused on the Permian silicic magmatism of the Choiyoi plutono-volcanic province and the Miocene volcanic complex of Farallón Negro – La Alumbrera district, northern Argentina. He served as professor at different universities of Argentina and led a research group working in epizonal pluton emplacement at the Universidad Nacional de Río Cuarto, Argentina.

Oscar Alfredo Martínez graduated as a geologist from the School of Natural Sciences, Universidad Nacional de La Plata, La Plata, Argentina, and obtained a doctoral degree in Glacial Geomorphology from the Universidad Nacional de la Patagonia-San Juan Bosco at Comodoro Rivadavia. He is presently Full Professor of Geology and Geomorphology at this same university but in the Esquel campus, at the foot of the Andes. He has done research in different regions of Patagonia and Tierra del Fuego during more than 25 years. His main fields of interest are glacial geology and extra-Andean geomorphology, with special focus in the "Patagonian Gravels" ("Rodados Patagónicos") and the development of endorheic depressions ("Bajos sin salida").

Rodney R. Maud Retired Honorary Professor, School of Geological Sciences, formerly University of Natal, now University of KwaZulu-Natal, Durban, South Africa.

Rodney Maud obtained his B.Sc. (1955) and Ph.D. (1962) degrees at the then University of Natal, concentrating on geology and geomorphology under that very well known, if sometimes contentious, geomorphologist, Lester King. In 1964–1965 he was awarded the Selby Fellowship of the Australian Academy of Science and spent a year researching landscape evolution and soils in the area south of Adelaide (Australia), while at the then CSIRO Division of Soils. Never in full-time academic employment, he became a consultant in the fields of geology, engineering geology, geomorphology, hydrogeology, and soil science, traveling widely for this reason in Africa and elsewhere. He has published extensively in the above-mentioned subjects in the form of numerous refereed journal and conference papers, as well as chapters in books etc., this continuing till the present time. Jointly with his now sadly late, very good friend and colleague, Tim Partridge 1942–2009, for their seminal paper on the evolution of the landscape of Southern Africa (1987), he was awarded the Jubilee Medal of the Geological Society of South Africa (1989). In 2003 he was awarded the Gold Medal Award of the South African Institute for Engineering and Environmental Geologists. Rodney Maud is a Fellow of both the Geological Society (London) and the Geological Society of South Africa.

David M. Mickelson is Emeritus Professor of Geoscience, Geological Engineering, and Water Resources Management at the University of Wisconsin–Madison. He received a B.A. from Clark University, an M.S. from the University of Maine, and a Ph.D. in 1971 from the Ohio State University and taught glacial geology, coastal geomorphology, and other courses in University of Wisconsin from 1971 until his retirement in 2005. He has done research on modern glaciers and glacial deposits in Sweden and Norway, Argentina, China, Alaska, the Rocky Mountains, and other parts of the USA. He is the first author of the book *Geology of the Ice Age National Scenic Trail* published by the University of Wisconsin Press in late 2011.

Cliff Ollier Born in Manchester and educated at Bristol University (B.Sc, M.Sc., D.Sc.), Cliff Ollier spent most of his life in Australia as a geologist and geomorphologist. He is the author of ten books and over 400 scientific papers. He started as a demonstrator in Structural Geology at Bristol and worked as a soil scientist in England and Uganda. He has worked in many universities including Melbourne, Australian National, Oxford, Papua New Guinea, New England, East Anglia, and South Pacific and is currently at the University of Western Australia. His specialty has been applying geomorphology to large-scale tectonics through studies of planation surfaces, drainage, and great escarpments. He has travelled to over a 100 countries and lectured at over a 100 different universities.

Pedro Oyhantçabal is a Professor of Geology at the Universidad de la República (UDELAR, Uruguay) and a researcher at the PEDECIBA (Programme for the Development of Basic Science) and SNI (National Researcher System), part of the ANII (National Research and Innovation Agency). He completed his initial degree in 1982 and his Ph.D. at the University of Göttingen (Germany). He has published over 20 articles in peer-reviewed journals and has acted as coeditor of special issues for the *International Journal of Earth Sciences*. His interests are in geochemistry and the isotope geology of magmas, geochronology, structural geology, and tectonics. At present, his work focuses on the evolution of the Neoproterozoic belts in Uruguay, Southern Brazil, and southwestern Africa.

Daniel Panario is a Professor of Geomorphology, Director of the UNCIEP, Director of the Instituto de Ecología y Ciencias Ambientales, and Coordinator of the Master of Environmental Sciences at the Facultad de Ciencias of the Universidad de la República in Uruguay. He is researcher at the SNI (National Researcher System), part of the ANII (National Research and Innovation Agency). He has also authored or coauthored over 100 scientific publications including articles in international journals, books and book chapters, and papers in national and international conference proceedings. His research interests include fields like environmental issues, physical geography, natural resources, and soil sciences.

Elena Peel obtained her degree in Geology at the Facultad de Ciencias of the Universidad de la República in Uruguay. She started her scientific career at the Departamento de Geología of the Facultad de Ciencias (Uruguay), where she worked in the fields of geochemistry and geochronology. She completed her doctoral studies at the Instituto de Geociências of the Universidade de São Paulo, Brazil, receiving a Ph.D. in Geological Sciences with a specialty in Geochemistry and Geotectonics. Her professional interests are in geochemistry, isotopic geology, and the tectonics of Precambrian rocks. She has been the head of the Departamento de Geología of the Facultad de Ciencias (Uruguay) since 2010.

Jean-Pierre Peulvast is Emeritus Professor of Geomorphology at the University of Paris-Sorbonne (Paris IV). He also teaches in the Master degree and Doctorate program at the Federal University of Ceará (UFC), Fortaleza, Brazil. His principal field of interest is in structural geomorphology and long term landform evolution, but he is also involved in programs of geoarcheology (Egypt) and in studies on natural hazards and risks (Northeast Brazil). He has mainly worked and still works on passive margins around the Atlantic Ocean (Scandinavia, Greenland, Canada, Brazil, Argentina), but also on various regions of France, Mediterranean Europe and Central Asia, as well as on the planet Mars.

Jorge Rabassa obtained his Licenciado and Doctor degrees in Geology at the School of Natural Sciences, Universidad Nacional de La Plata, La Plata, Argentina. He is a researcher with CONICET, the National Research Council of Argentina, at CADIC, Centro Austral de Investigaciones Científicas, Ushuaia, Tierra del Fuego, since 1986. He is presently the director of CADIC, the southernmost, multidisciplinary research center of the World with permanent scientific staff. He is also adjunct Full Professor of Geology at the Universidad Nacional de Tierra del Fuego. He has been a lecturer in several Argentine universities and also the President of Comahue University, one of the Argentine Patagonian universities. He is a member of the Argentine Academy of Sciences and a corresponding member of the Catalonian Academy of Sciences. He has been a Visiting Professor in several universities of United States and Canada, as well as in various Western Europe countries, the Republic of South Africa and China. His main research interests are geomorphology, ancient landscapes, long-term landscape evolution, and glacial and Quaternary studies. He has published over 150 scientific papers and book chapters, and edited 17 books and scientific journal special volumes. He has directed 15 doctoral dissertations and many graduation theses, and is a member of the editorial board of several international journals. He is presently the co-editor for South America of the Springer Brief Monography *Earth System Sciences* series.

Alejandro Ribot is a Professor of Petrology at the Facultad de Ciencias Naturales y Museo, Universidad Nacional de La Plata, Argentina, where he has done research and teaching on Metamorphic Petrology for more than 20 years. He is also a senior geologist at the Laboratorio de Entrenamiento Multidisciplinario para la Investigación Tecnológica (LEMIT), Comisión de Investigaciones Científicas de la Provincia de Buenos Aires (Buenos Aires Province Research Council), where his field of interest is petrography studies applied to the characterization of concrete aggregates.

Ana María Sato graduated as geologist (1981) and obtained her doctorate in Geology (1989) from the Universidad de Buenos Aires, Argentina. She is presently a Professor of Argentine Regional Geology at the School of Natural Sciences, Universidad Nacional de La Plata, Argentina, as well as a permanent researcher with CONICET, the National Scientific and Technological Research Council of Argentina, at the Centro de Investigaciones Geológicas (CIG), of the same university. Her main interests are the stratigraphy and geological events involved in the assemblage and breakup of Gondwana. She has been in charge of a number of research projects on the geological evolution of Sierras Pampeanas and northern Patagonia regions.

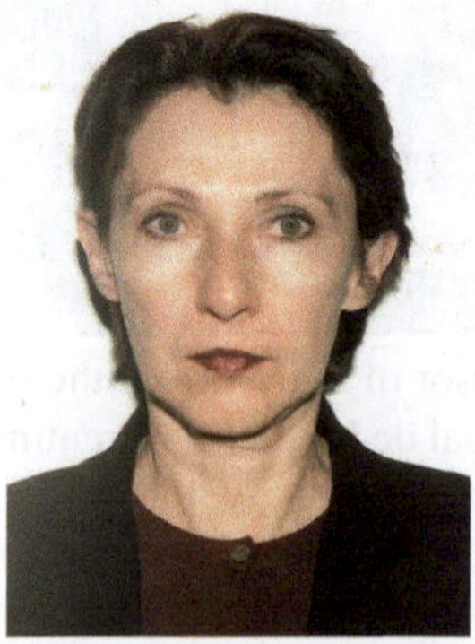

Betty J. Socha is a hydrogeologist and project manager in environmental consulting, working primarily in the investigation and remediation of contaminated soil and groundwater and in geological site characterization for facility development. Her main research interests are glacial sedimentology and stratigraphy and geological mapping. She has master's degrees in Environmental Monitoring and Geology (both 1984) and a Ph.D. in Geology (2007), all from the University of Wisconsin–Madison.

Hugo Tickyj obtained his degree (1994) in Geology at the Universidad Nacional de La Pampa and his doctorate (1999) in Natural Sciences (Geology) at the School of Natural Sciences, Universidad Nacional de La Plata, La Plata, Argentina. He is currently a Professor of Petrology at the Geology Department, Universidad Nacional de La Pampa, Santa Rosa, Argentina. His main interests are related to magmatic events developed in the proto-Andean margin of Gondwana. He participated in several research projects on the geological evolution of the Frontal Andean Cordillera, the San Rafael Block, and the Chadileuvú Block in Central Argentina.

María Cecilia Zalazar (1971) is a mathematician and has worked in techniques of global optimization.

She also specialized in subjects of Geological Mathematics, developing techniques of digital mapping and numerical methods for geophysical and geomorphological applications.

Marcelo Zárate is a geologist who obtained his doctoral degree at the University of La Plata, Argentina. At present he is a researcher of CONICET and Professor at the University of La Pampa at San Rosa, Argentina. He is currently involved in research activities on the geological history, geomorphology, and geoarchaeology of southern South America with the focus on the Pampean plain and the Andes piedmont.

Introduction

Jorge Rabassa and Cliff Ollier

When we started the edition of this volume, we thought that it could serve as a milestone in South American geomorphology, particularly in those topics concerning long-term landscape evolution. We also hoped that it would make a valuable contribution to the geomorphology of the Southern Hemisphere and, even perhaps, to global geomorphology. We do not know if our goals have been fulfilled, but we have tried our best. Firstly, we must express our deepest gratitude to a large group of distinguished colleagues from Argentina, Brazil and Uruguay. Without their participation and much appreciated efforts, our task would have been impossible.

This volume presents extensive and new information on the geomorphology of cratonic areas of southern South America, most of which has never been published before, sometimes with innovative insights into the interpretation of ancient landscapes. It also includes valuable data relating to recurrent controversies in general geomorphology beyond the boundaries of South America. The great need to integrate many different aspects of geomorphology in regional studies is clearly revealed in the varied topics presented in this volume.

Geomorphology sometimes seems to be a rather schizophrenic science. It is studied mainly by geologists and geographers but also by engineers, soil scientists, hydrologists and others, all with rather different points of view. Worse perhaps, sometimes, we have the sad feeling that those specialists do not even listen to each other. Geographers often tend to prepare descriptive accounts of the landscape, often on a short time scale: at the present time, process studies seem to predominate.

J. Rabassa (✉)
Laboratorio de Geomorfología y Cuaternario, CADIC-CONICET, Ushuaia, Tierra del Fuego, Argentina

Universidad Nacional de Tierra del Fuego, Ushuaia, Tierra del Fuego, Argentina
e-mail: jrabassa@gmail.com

C. Ollier
School of Earth and Environment, University of Western Australia, Perth, Australia
e-mail: cliff.ollier@uwa.edu.au

J. Rabassa and C. Ollier (eds.), *Gondwana Landscapes in southern South America*,
Springer Earth System Sciences, DOI 10.1007/978-94-007-7702-6_1,

Geologists bring a longer time scale and often see landforms as just the last contribution to a long geological history – a bit of ornament at the top of the geological column. Many geologists received a rather cursory introduction to geomorphology and the bulk of their professional training is in topics like stratigraphy, petrology and structural geology. But stratigraphy is mainly about ancient seas, not the land; petrology concentrates on rock genesis with little reference to landforms; and structural geologists concentrate on faults, folds and joints rather than the surface features they often control. Engineers focus on particular aspects of applied geomorphology, especially landslides, but often their viewpoints are oriented to the consequences rather than the basic problems.

We are here concerned with ancient landscapes and especially with planation surfaces and find they are very old, dating back to the Mesozoic and, exceptionally, even older. What processes could form such huge features, and how can they be preserved for so long? River systems also often have a very long history, and their evolution is punctuated by tectonic events such as uplift, faulting and – at the greatest scale – the creation of new continental margins as old supercontinents broke up and drifted apart. Tectonic movements are also involved in the evolution of continents and very large landforms such as the Great Escarpments. The step-like succession of planation surfaces that is often recorded is generally assumed to be caused by pulses of uplift but are there other possibilities? A very special feature is the relationship between large landforms and volcanicity. This applies to individual volcanoes and to very large outpourings of lava like the Parana Basalts. We can now get numerical dates for volcanic rocks, so this brings precise time-markers into geomorphic history.

One of the features of many ancient landscapes is the very deep weathering, reported in several chapters in this volume. Where present it has great importance in engineering problems, economic geology, clay mining and hydrology, yet it is scarcely mentioned in most conventional textbooks of geomorphology. In engineering studies, it is often ignored or dismissed as 'residual soil'. Of course there are treasured exceptions, like Lacardi's many studies of landslides in saprolite in Brazil.

Climate affects landscape evolution in various ways, and indeed climatic geomorphology once reigned as a major subdiscipline. The different process can be extreme, as in glacial erosion (and even this finds a place in this volume) or arid climates with dunes. There may be indirect indications of climate, and many writers assume that deep weathering means deep tropical weathering. But climate changes over geological time and many of the chapters here attempt to explain the interplay of climate and landscape formation, and one of the chapters deals specifically with climate modelling. One of the central questions that appear in several of the chapters in this volume is whether the paleoclimates that modelled these Gondwana Landscapes, extreme as they look, have any modern analogues. Perhaps, they existed only in the Late Mesozoic and earliest Tertiary, with bizarre atmospheric conditions unknown in modern times. Perhaps actualism is not the sole available explanation.

A wide range of techniques have been used to determine landscape evolution, some of which are reported in this volume. As in many aspects of geomorphology, mapping is a vital tool, and many of the chapters here include maps of planation surfaces or other geomorphic units. There is a long way to go, but we have here the beginnings of the field-based mapping of the planation surfaces of southern South America. Very little has been done in these fields, since the work of pioneer German and Italian geologists in Argentina and Uruguay at the end of the nineteenth century and the first half of the twentieth century or the investigations of Lester King and several renowned Brazilian geomorphologists between the 1950s and the 1970s.

We hope that this book will be especially valuable in South America, but perhaps it will be useful even outside this continent. The language barrier has usually meant that knowledge of South American science has been generally unavailable to non-Spanish or non-Portuguese readers. We expect that this volume will serve as a starting point for a whole new phase of studies of the fascinating landscape history of southern South America, which provides so many opportunities for landscape research, particularly in long-term landscape evolution.

In this assemblage of chapters, the sum is much more than the parts. The volume collects an up-to-date, state-of-the-art collection of information on South American geomorphology and shows beyond doubt that geomorphology is on the same time scale as global tectonics, biological evolution and major climate change. We do not present here an integrated view of the landscape evolution: facts, theories, speculations and even basic assumptions are not unanimously agreed by all the authors. But this is the state of the art, and we hope this book encourages further work and debate on the entertaining and illuminating art of landscape study on the grand scale.

It should be noted that some of the chapters describe ancient geomorphological features of areas which have never been studied or published before, and some of the chapters describe regions which are totally unknown to the public.

Regarding the geographical areas covered, the scope of this book extends from tropical latitudes beyond the Tropic of Capricorn, down south to freezing Patagonia in the 'roaring fifties', more than 3,500 km in length from north to south. The area described probably covers more than 1.5 million km^2, a nice chunk of territory, poorly known in geomorphological terms until today. This alone suggests that this volume makes a valuable contribution to global geomorphology.

Another important feature of this book is the significant amount of geological and geomorphological literature cited in its various chapters. Over 1,000 contributions have been cited in this volume, many of them written in Spanish or Portuguese and presented in local journals, symposia and congresses, which are not generally available for scientists outside South America.

Although some of the authors may be quite well known internationally due to their previous contributions, many more are younger geomorphologists whose work is perhaps little known outside South America, but who represent the new generation and the future of geomorphological sciences in the southern portion of the continent. They fully deserve a large audience to take note of their efforts.

This book starts with three chapters by the editors to provide a general introduction to the topic and the present state of our knowledge. Jorge Rabassa has presented his own points of view on the nature and characteristics of Gondwana

Landscapes, particularly their genesis, distribution and age in the framework of long-term landscape evolution. This chapter also recovers long-forgotten ideas about North American erosion surfaces and ancient landscapes that North American geomorphologists of today tend to disregard or just ignore their genesis and age. The first contribution by Cliff Ollier exposes the reach and validity of some principles related to the origin and characteristics of planation surfaces, a key aspect of the studied landscapes, which will be a crucial reference to understand the basic rules on which we have structured this volume. Ollier's second contribution goes further into the paleogeographic realm of Gondwana, investigating the planation surfaces found in the different areas of the Gondwana Supercontinent – Africa, Australia, India, Antarctica, Madagascar, Sri Lanka, New Caledonia and South America. This desperately needed review of Gondwana Landscapes outside South America provides an appropriate framework for the contributions of this volume.

Then, we have a precious invited contribution by Rodney Maud, who brings us a review of the Gondwana Landscapes in Africa, as the centrepiece of the huge supercontinent. Having been Lester King's doctoral student and later his professional colleague, with a long and prestigious career in African geomorphology, Maud's discussion of the problem carries us to the very core of the matter on the other side of the Mesozoic rifting ocean. He also brings along the valued memories, ideas and views of his cherished friend and colleague, Tim Partridge, who passed away just when this book was being outlined and whose contributions in the field will enlighten the geomorphology of southern Africa for many decades to come.

The general paleoclimatic conditions of the Late Mesozoic, particularly in the Gondwana mega-continent, have been discussed by Rosa Compagnucci, who has shown that it is possible to model the atmospheric circulation and climatic conditions during Gondwana times, providing a scientific background to understanding those intriguing past climates.

The remaining chapters of this book consider the paleogeomorphological conditions at different regional levels and spatial and time scales.

Firstly, the viewpoint from Brazilian geomorphology is developed by Francisco Ladeira, who discusses the landscape evolution of Southern Brazil and particularly that of the state of Rio Grande do Sul, the southernmost portion of the giant of South American countries. In this case the region that Ladeira studied is the size of Spain and larger than many European countries. This is a region which still has a wet-warm subtropical climate, in which it is certainly hard to unravel the palimpsest of ancient climates and landscapes. Though dealing with only a small part of the huge Brazilian territory, Ladeira provides us with many insights that are applicable to other areas of this core region of Gondwana. Hopefully, his efforts will lead other Brazilian colleagues to publish their regional studies on Gondwana Landscapes.

Daniel Panario and colleagues have thoroughly analysed the remnants of ancient landscapes in Uruguay, but they have gone much further, presenting a general overview of the geomorphology of the entire country which has never been published before: this is a major contribution in itself. When looking at a map of South America, the size of Uruguay may look petty in between Brazil and Argentina but for comparison remember that Uruguay is much larger than England. Bringing geomorphological life to such a large portion of the cratonic world with Gondwana Landscapes is a great achievement by Panario and his colleagues.

The remaining chapters of this volume deal with the Gondwana Landscapes of Argentina, in a variety of structural units, geological provinces and geographical environments.

Jorge Rabassa, Claudio Carignano and Marcela Cioccale have summarized, as an introduction to the problem, the occurrence of these ancient landscapes in the different cratonic areas of Argentina, from those closely related to the Brazilian shield to those scattered in the various hyper-stable areas of the country and from small isolated cratons to the extensive margin of the ancient continent later affected by the Andean tectonics.

Daniela Kröhling and her colleagues present a valuable vision of the ancient landscape of the province of Misiones and surrounding areas, in the northeastern end of Argentina. Here, as in southern Brazil, the regional climate is still wet tropical and deep weathering processes persist even today. The approach, using quantitative geomorphology and morphometric analysis, is precise, interesting and useful. Their findings are described in the framework of their own geomorphological concepts and regional methodology.

Jimena Andreazzini and Susana Degiovanni have brought to our attention the large extent and good preservation of the planation surfaces of the southernmost Sierras Pampeanas in the province of Córdoba, a geographical unit known as Sierra de Comechingones. This is one of the regions of Argentina which has very good geomorphological exposures but very few modern papers dealing with them. Their work is particularly revealing when the different planation surfaces are related to the lithology and structure of the bedrock on which these surfaces were carved.

The chapter by Betty Socha and collaborators is of a different nature. It deals with a Late Paleozoic landscape, an exhumed glacial landscape carved during the Carboniferous Gondwana glaciation. Spectacular, formidable, fossil Alpine troughs and hanging valleys, complex sedimentary units, former moraines, kames and kame deltas, tills, glaciofluvial gravels and glaciolacustrine sedimentary rocks showing an outstanding state of preservation characterize this unique paleolandscape. Moreover, evidence of an Early Carboniferous or even Late Devonian planation surface, with an ancient weathering front and large corestones, has been found underlying the Paleozoic glacial sediments, thus becoming the oldest planation surface described in this book.

Marcelo Zárate and Alicia Folguera discuss the presence, nature and age of planation surfaces in Central Argentina, a very sparsely populated region which has never been studied before in terms of its paleogeomorphology. Their work is solidly based upon stratigraphic, petrological and tectonic reconstructions. Several planation surfaces have been observed, cutting across different rock types and exhibiting ages which may range back even to the Triassic. These ancient surfaces were later covered by a thick and complex Miocene sequence, and from the Late Cenozoic to the present, the surfaces have been exhumed.

Emilia Aguilera and colleagues have carefully studied the erosion surface and morphology of the Sierras de Lihuel Calel, Province of La Pampa, in Central Argentina. In many aspects, this chapter is paired with that of Zárate and Folguera, though much more geographically restricted but with a very comprehensive petrologic study of the ignimbrite rocks which have developed a quite unusual morphology. The discussion basically concerns the influence of petrology in the development of such a characteristic ignimbrite landscape. This area is also very isolated and poorly known, and their contribution will certainly become a benchmark for future investigations.

In another chapter, Emilia Aguilera and collaborators debate the nature and origin of various paleolandscapes in the Northern Patagonian Massif. This region is an isolated craton which merged (or perhaps collided) with South America sometimes during the latest Permian or the earliest Triassic. Extremely stable from a tectonic point of view, this unit was slightly affected by the Andean tectonics but only along its western margin. The crystalline basement occupies extensive areas in this region, and together with outcropping Paleozoic and Triassic-Jurassic rocks, they show the remnants of ample planation surfaces of probably Middle to Late Cretaceous age. This is one of the largest, uninterrupted remains of Gondwana Landscapes found in Argentina, some tens of thousands square kilometres exposed in endless plains.

Also in the Northern Patagonian Massif, Oscar Martínez and Jorge Rabassa have examined the paleosurfaces preserved in the huge 'Rhyolitic' or 'Ignimbritic' plateau, which occupies the southeastern border of the massif, with an extent of over 50,000 km^2. Early to Middle Jurassic ignimbrites show abundant features of an ancient, deep, chemical weathering front. The planation surface appears almost untouched by modern erosion processes, with many endorheic basins occurring between low, rounded hills, bornhardts and corestones, and a poorly integrated drainage network. The authors have also ventured the idea that, in a land of countless closed depressions of many different origins, some of these hollows could be, in fact, shallow, smooth and original irregularities of the paleoweathering front.

François Bétard and colleagues present their first observations and critical findings in the planation surfaces of the Deseado Massif, the southernmost cratonic unit of continental South America and the southernmost one outside Antarctica. This is another rather small cratonic province, a sort of smaller sister of the Northern Patagonian Massif, though itself as large as Portugal. This massif has undergone cold arid to semiarid climates during most of the Tertiary, but it was under permafrost conditions during Quaternary glacial times. In one of the most desolate and bleak environments of the world, very sparsely populated and quite poorly known in geomorphological terms, the authors have presented an integrated study of complex morphostratigraphy, ancient paleosurfaces and deep weathering conditions, which characterize the exposed relief.

Finally, we present two contributions which bring new information about certain aspects of cratonic regions. Emilia Aguilera and colleagues deal with pseudokarst and speleothems in dykes that have undergone weathering under ancient wet-warm climates, using geochemical tools to define some of their past conditions.

Lastly, a short note by Eugenio Aragón and collaborators discusses the process of exhumation of the Gondwana planation surfaces of the Northern Patagonian Massif, after conditions drastically changed during the Middle Tertiary.

Many of the ideas exposed in these chapters were presented or at least suggested by pioneering geologists during the end of the nineteenth century and the first half of the twentieth century, but sadly most of them had been almost forgotten in South America. This volume means to revive the lost word of our predecessors in these fields. Furthermore, we believe that to study the geomorphology of the cratonic areas of South America, it is necessary to adopt a different approach. Since the advent of plate tectonics, the Andean geological viewpoint has dominated geomorphic thinking. As in other Gondwana continents, we now need a passive-margin, cratonic geomorphology interpretation, for this has been active in long-term landscape evolution in southern South America for at least 200 million years.

Some Concepts on Gondwana Landscapes: Long-Term Landscape Evolution, Genesis, Distribution and Age

Jorge Rabassa

"Let the landscape teach me."
Lester C. King, personal letter to Charles Higgins, 1958

"While the geologist may often be in error, the Earth is never wrong."
Lester C. King, 1967

Abstract The concept of "Gondwana Landscape" was defined by Fairbridge (The encyclopedia of geomorphology. Reinhold Book Corporation, New York, p. 483, 1968) as an "ancestral landscape" composed of "series of once-planed remnants" that "record traces of older planation" episodes during the "late Mesozoic (locally Jurassic or Cretaceous)". This has been called the "Gondwana cyclic land surface" in the continents of the southern hemisphere, occurring extensively in Australia, Southern Africa and the cratonic areas of South America. Remnants of these surfaces are found also in India, and it is assumed they have been preserved in Eastern Antarctica, underneath the Antarctic ice sheet which covers that region with an average thickness of 3,000 m. These paleolandscapes were generated when the former Gondwana supercontinent was still in place and similar tectonic conditions in its drifted fragments have allowed their preservation. In Pangaea, remnants of equivalent surfaces, though of very fragmentary condition, have been described in Europe and the United States, south of the Pleistocene glaciation boundary.

These Gondwana planation surfaces are characteristic of cratonic regions, which have survived in the landscape without being covered by marine sediments along extremely long periods, having been exposed to long-term subaerial weathering and

J. Rabassa (✉)
Laboratorio de Geomorfología y Cuaternario, CADIC-CONICET, Ushuaia, Tierra del Fuego, Argentina

Universidad Nacional de Tierra del Fuego, Ushuaia, Tierra del Fuego, Argentina
e-mail: jrabassa@gmail.com

J. Rabassa and C. Ollier (eds.), *Gondwana Landscapes in southern South America*, Springer Earth System Sciences, DOI 10.1007/978-94-007-7702-6_2,

denudation. Their genesis is related to extremely humid and warm paleoclimates of "hyper-tropical" nature, with permanently water saturated soils, or perhaps extreme climates, with seasonal and long-term cyclic fluctuations, from extremely wet to extremely dry. Deep chemical weathering is the dominant geomorphological process, with the development of enormously deep weathering profiles, perhaps of up to many hundreds of metres deep. The weathering products are clays, in some cases kaolinite, pure quartz and other silica types sands, elimination of all other minerals and duricrust formation, such as ferricretes (iron), silcretes (silica) and calcretes (calcium carbonate). Mean annual precipitation in these periods would have been perhaps higher than 10,000 mm, with extremely high, mean annual temperatures, such as 25–30 °C. These deep weathering processes can be achieved only under extremely stable tectonic and climatic conditions. The geomorphological processes continued with fluvial removal of the weathering products in wet climates and with hydro-eolian deflation in the areas with strong climatic seasonality. The final landform products of these deep weathering systems are planation surfaces, inselbergs, bornhardts, duricrust remnants covering tablelands, associated pediments, granite weathered landscape, etc.

Some concepts related of these ancient landform systems were theoretical, developed by Walther Penck in the early twentieth century. The Gondwana Landscapes were studied by Alexander Du Toit and Lester C. King in Africa and more recently, by Timothy Partridge and Rodney Maud in South Africa, C. Rowland Twidale and Cliff Ollier in Australia and Lester C. King and João José Bigarella in Brazil, among others. Both in Australia and Southern Africa, these landform systems have been identified as formed in the Middle to Late Jurassic, throughout the Cretaceous and, in some cases, extending into the Paleogene, when Gondwana was still only partially dismembered.

Keywords Gondwana • Paleosurfaces • Argentina • Planation surfaces • Etchplains

Introduction

This general introduction is a revised version of a previous paper (Rabassa 2010), which then summarized a talk presented at the IV Congreso Argentino de Geomorfología y Cuaternario and the simultaneous Brazilian Quaternary Congress (ABEQUA), La Plata, September 2009, then opening a special symposium on "Paleosurfaces". This was a quite historical event for southern South American Geomorphology since it was the first opportunity that, in recent times, the concepts of "Gondwana Landscapes" and "Long-term Landscape Evolution" of cratonic areas were presented and discussed in Argentina. Several colleagues of Brazil, Uruguay and Argentina got together to analyse the importance of these ideas in the framework of our present knowledge and the availability of modern dating techniques. A renewed overview of the geomorphology of cratonic areas has been growing since then.

The concept of "Gondwana Landscape" was defined by Fairbridge (1968, p. 483) as an "ancestral landscape" composed of "series of once-planed remnants" that "record traces of older planation" episodes during the "late Mesozoic (locally Jurassic or Cretaceous)". This has been called the "Gondwana cyclic land surface" in the continents of the southern hemisphere, occurring extensively in the cratonic areas of Australia, Southern Africa and South America. All fragments of the former Gondwana supercontinent share similar planation conditions because these extensive landforms were all graded to the same base level of a surrounding, common, pre-break-up sea level (Mountain 1968). Remnants of these surfaces are found also in India, and it is assumed they have been preserved in Eastern Antarctica, both in exposed areas and underneath the Antarctic ice sheet which covers that region with an average thickness of 3,000 m (Ollier 2004). These landscapes were generated when the former Gondwana supercontinent was still in place and similar tectonic conditions in its drifted fragments have allowed their preservation. Landscapes of similar ages have been also described in North America and Europe, which are probably related to evolution of Pangaea. Remnants of equivalent surfaces, though of very fragmentary condition, have been described in Europe (for instance, Belgium, France, Germany, Spain (for Sweden, see Lidmar-Bergsson 1988)) and the United States, south of the Pleistocene glaciation boundary. However, there is no clear agreement among the scientists of these continents about the nature and age of these paleosurfaces. These northern hemisphere paleosurfaces are likely to be found in other areas of the world with similar tectonic and paleoclimatic conditions, but they have not been fully described yet. So far, the concept of very old, Mesozoic Landscapes that were never covered by marine sediments or thick continental sequences and that have been part of the landscape since their genesis, is still a matter of study and discussion almost restricted to southern hemisphere geomorphologists.

The Mesozoic paleoclimates and tectonic conditions in the Gondwana supercontinent allowed the formation of ancient landscape systems in Africa, particularly in Southern Africa, Australia, India, Antarctica and South America. In this latter continent, they have been studied in Brazil, Argentina, Uruguay, Venezuela and the Guyana Massif.

The Ideas of Gondwana Landscapes and Long-Term Landscape Evolution: Previous Works

Grove K. Gilbert (1877) was a pioneer of the ideas related to "long-term landscape evolution" when he published his concepts of "dynamic equilibrium". Dynamic equilibrium is a system in which weathering, removal by erosion and further deposition are in a balanced condition and therefore there is no change in form through time. William M. Davis (1899) developed his ideas of the cycle of landscape evolution and the concept of "peneplain" based on the action of fluvial processes and age. Later, Walther Penck (1924) recognized that large planation surfaces were formed by receding headward erosion. He proposed the concepts of "primarrumpf" (initial landscape development phase), "piedmont treppen" (steep erosion terraces developed by headward erosion) and "endrumpf" (final phase, with intersection of wash slopes), which may develop under a variety of paleoclimatic conditions (von Engeln 1948). It should be noted that Walther Penck did not introduce weathering in his discussion of parallel slope retreat (C. Ollier, personal communication, 2011).

The work of Alexander Du Toit (1937, 1954, and other papers cited therein; Du Toit and Reed 1927) defined the ideas of "continental drift" that had been previously suggested by Alfred Wegener (1924) based on paleoclimatic inference. This allowed for the identification of areas that were geographically very closely located in the past and which had split apart since Late Mesozoic times, like Africa and South America. Therefore, those landscape features in both continents that were formed before the rifting would have similar characteristics because they were sharing similar climates and environments. Du Toit (1954) identified that "from the Jurassic onwards, South and Central Africa underwent various cycles of prolonged planation", whose remnants can be identified still today.

The ideas of Du Toit were deeply consolidated by the work of Lester C. King (1949, 1950, 1953, 1956a, 1962, 1963, among other papers) who recognized the long-term action of processes such as "pediplanation", "planation surfaces" and "etchplains", both in South Africa and Brazil. Extensive regional mapping in both continents supported King's ideas and presented for the first time a different overview of these landscapes to the whole world. Lester King's concept was still cyclic, like that of Davis, but he believed instead in parallel slope retreat (an idea that Walther Penck had theoretically developed). Since parallel slope retreat makes pediments, if these grow big and unite to build a larger plain, a pediplain forms. Pediplains are then formed by "backwearing", as opposed to peneplains, which are formed by "downwearing". Some of the landscapes that had been named as peneplains were later reinterpreted as pediplains, particularly in North America (see papers in Melhorn and Flemal 1975).

Extensive regional mapping in both continents supported King's ideas and presented for the first time a different overview of these landscapes to the whole world.

The ideas of King were continued in Brazil by João José Bigarella (Bigarella et al. 1994, and papers cited there; Bigarella and Ab'Sáber 1964), who recognized the existence of these ancient landscapes and by Carlos Schubert and colleagues in Venezuela (Schubert et al. 1986) in his studies about the "tepuis" of the Venezuelan Guyana Shield.

Later on, in Australia, the science of ancient landscapes was deeply developed by C. R. Twidale (2007a, b, and papers cited therein) and Clifford Ollier (1991a; Ollier and Pain 2000, and other papers quoted there). Both authors have an outstanding record of paramount contributions in these fields. Twidale (2007a) recognized the existence of several planation surfaces from the Jurassic, about 200 Ma, characterized by a lateritic surface, and even perhaps from the Triassic.

The importance of weathering under tropical climates was widely recognized by Summerfield and Thomas (1987), who stated that landscape evolution is associated with the formation and removal of deep weathering profiles, which leads to the concept of "etchplanation" as proposed by Wayland (1933). An etchplain is "a form of planation surface associated with crystalline shields and other ancient massifs which do not display tectonic relief and developed under tropical conditions promoting rapid chemical decomposition of susceptible rocks" (M.F. Thomas, in Fairbridge 1968, pp. 331–332). Etchplains depend exclusively on deep weathering processes. Etchplains are formed often under weathering conditions of hundreds of metres, and the new planation surface may be cut entirely across saprolite (Ollier 1960). It is important to note that some planation surfaces are cut across dominant saprolite, but others are cut across fresh, hard rock. Inselbergs rise abruptly from these planation surfaces cut across hard bedrock. All inselbergs are steep, rising like islands, but it should make it clear there are two types – those rising from hard pediments and those rising through a mantle of saprolite, like the ones Ollier (1960), Ollier and Harrop (1959) and Ollier et al. (1969) described from Uganda.

Since there are etchplains developed on weathered volcanic and/or sedimentary rocks, the concept should not be restricted to crystalline shields.

Likewise, one of Lester C. King's doctoral students at the University of Natal, Rodney Maud and his Witwatersrand University colleague, Timothy C. Partridge, developed a similar framework in South Africa, which they later extended to the whole of Southern Africa (Partridge 1998; Partridge and Maud 1987, 1989, 2000).

Gondwana Landscapes: Basic Scientific Concepts Related

The general idea of Gondwana Landscapes is closely related to the concept of long-term landscape evolution. This implies that landscapes may be developed along extreme long time periods, provided that warm/wet climates and tectonic stability are given. These conditions are found along cratonic areas and continental passive margins.

However, it should be taken into consideration that a long-term landscape may evolve under arid conditions; it just happens that the real history for most of these landscapes was of warm-humid conditions.

The conditions of climatic and tectonic stability were active for the last 200 million years only during the Jurassic and Cretaceous, when extremely warm and humid, tropical climates were dominant. Neither orogenic movements nor strong tectonic activity was recorded until the Middle to Late Cretaceous in the Gondwana supercontinent. There were no glaciers on Earth at that time, not even in the polar regions, and sea level was extremely high, overflowing the lowlands of most continents. Therefore, the idea of Gondwana Landscapes is closely associated to the geomorphology of tropical environments. And, indeed, it is from the tropics that most of these concepts are coming from.

Tropical environments are associated with conditions of deep chemical weathering, under wet and warm climates and tectonic stability. This is true also in tectonically active areas like New Guinea. Chemical weathering is developed by percolation of warm waters in heavy rain fall terrains throughout the uppermost levels of the crust. These waters, in large amounts, warm conditions and omnipresent availability, forced the chemical weathering of rocks well beneath the soil, perhaps at hundreds, even up to one thousand metres depth. This altered layer is called the weathering zone. At the bottom of the weathering zone, the weathering front is found, that is, the boundary (often abrupt) in which weathering is active and where it stopped when the active processes were interrupted. The nature and conditions of the weathering front are extremely important to understand the past environments, because in most of the Gondwana Landscape areas, the weathering front is the one of the very few remaining evidences of the existence of an extremely deep weathering zone. Its interpretation is providing much information about the original scenarios. The concept of weathering front is associated with the formation of corestones and etchplains, critical landforms in Gondwana Landscapes.

The concept of tropical soils is clearly linked with the formation of different types of duricrusts, as pedogenetic elements, such as silcretes, ferricretes and calcretes, all of them formed under different environmental conditions. The denudation, partial or complete, of soils and superficial sediments and weathering products is related to processes of pediplanation. The combination of all these circumstances is responsible for the formation of inselbergs and bornhardts (Twidale 2007a, b).

In most of the available scientific texts in geomorphology, particularly those from the northern hemisphere, the consideration of these kinds of landscapes is very rare or absent. Newer books are dealing with these concepts much more carefully. For instance, Thornbury (1954) referred that most of the Earth topography has an age that is not older than the Pleistocene, whereas it is a very rare topography which is older than the Tertiary. Thornbury (1954) expressed his profound doubts about the actual existence of these ancient surfaces. He stated that, if they exist, it is most likely that they are just exhumed erosion surfaces which have not been exposed to degradation through vast periods of geological time. He stated that a vast majority of the present Earth surface has an age younger than the Middle Miocene. He was exposing a vision of geomorphology as seen solely from New England and United States, where everything seems to be interpreted to be of glacial origin and Late Pleistocene in age.

However, during the times of the colonial empires and particularly in the first decades of the twentieth century, the British and French geomorphologists were sent around the world to study the landscape of the colonies. Since both empires extended mostly over tropical regions, they found that the landscapes were very different from what they had observed before in the British Isles and Northern Europe. Many of them settled down in Africa, Asia and Australia, where these landscapes were very obvious and extended. In this sense, the British and French geomorphologists had a much wider view than their contemporaneous American colleagues had, likely because the latter remained at home, mostly devoted to the study of glacial landscapes of Eastern United States, or remained fascinated by the arid-climate landscapes of the Rocky Mountains and Western United States.

The Evolution of the Gondwana Cratonic Areas During the Mesozoic

According to Ronald Blakey (www.nau.edu; http://jan.ucc.nau.edu), in the Late Jurassic, 150 Ma ago, Africa and South America were still united or at least in close contact, stretching over continental areas thousands of km wide. The Atlantic Ocean had already been opened in its northern portion, but the South Atlantic was still unborn. Therefore, the continental mass was enormous and the oceanic circulation was totally different to the present one. It is then expected that there was a continuity of climates, ecosystems and landscapes on the lands located today on both sides of the present Atlantic Ocean. Thus, such landforms of continental scale are extended both in Brazil and in western and southern Africa, with Argentina, Uruguay and the Malvinas/Falklands archipelago as marginal areas. India was located along the eastern side of Africa, but Australia was starting to drift away. Eastern Antarctica was located between India and Australia, but the characteristics of the preglacial Antarctic landscape is still unknown, as most of this continent is totally ice covered by a very thick ice sheet, lacking superficial evidence (Fig. 1).

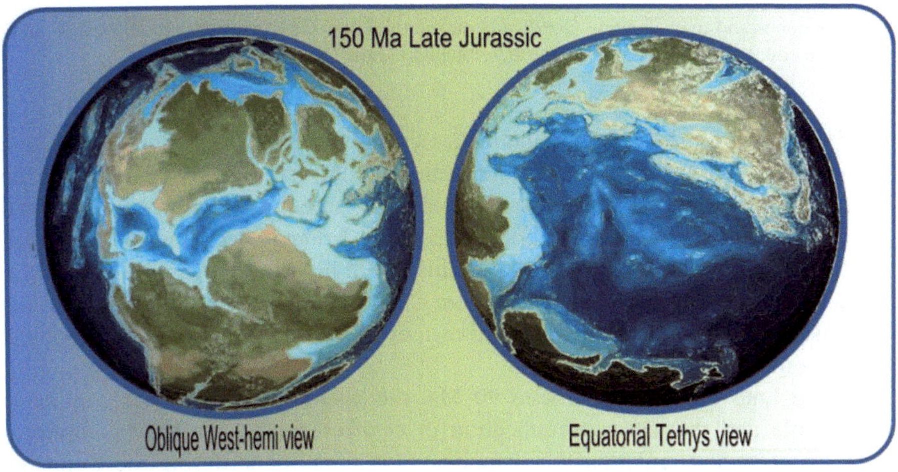

Fig. 1 Gondwana in the Late Jurassic (From Blakey, Ronald: www.nau.edu; http://jan.ucc.nau.edu/~rcb7/globaltext2.html)

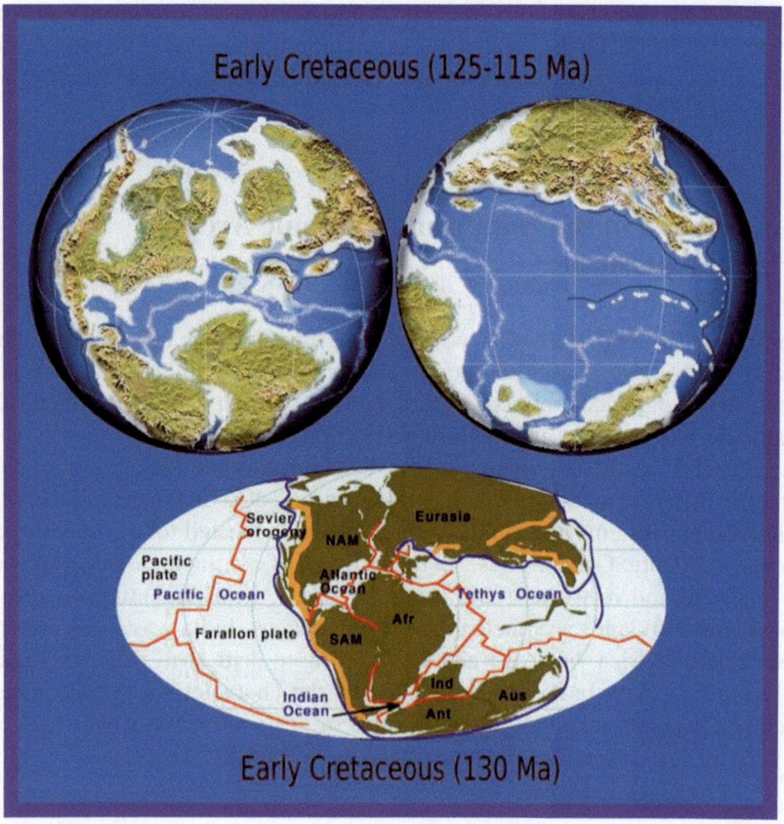

Fig. 2 Globes and map showing Gondwana during the Early Cretaceous (From Blakey, Ronald; www.nau.edu; http://jan.ucc.nau.edu/~rcb7/globaltext2.html)

These conditions were maintained during the Early Cretaceous, around 130–115 Ma, but the rifting and continental drifting had already began, with the opening of ample sectors of the continental platform between both continents, and perhaps the connection between both sides of the Atlantic Ocean had already been established (Fig. 2). It is obvious that in this epoch the environmental conditions in the adjacent areas of both continents were already somewhat different and that the regional climates had already changed as a consequence of the new geographical and tectonic conditions.

In the Late Cretaceous, around 90 Ma, the opening between both continents was ample and the sea communication in between must have been completed, with an oceanic circulation that announces the conditions during the Cenozoic (Fig. 3). As the drifting process continued, it would have generated very different environmental conditions for the landscape evolution on both sides of the Southern Atlantic Ocean.

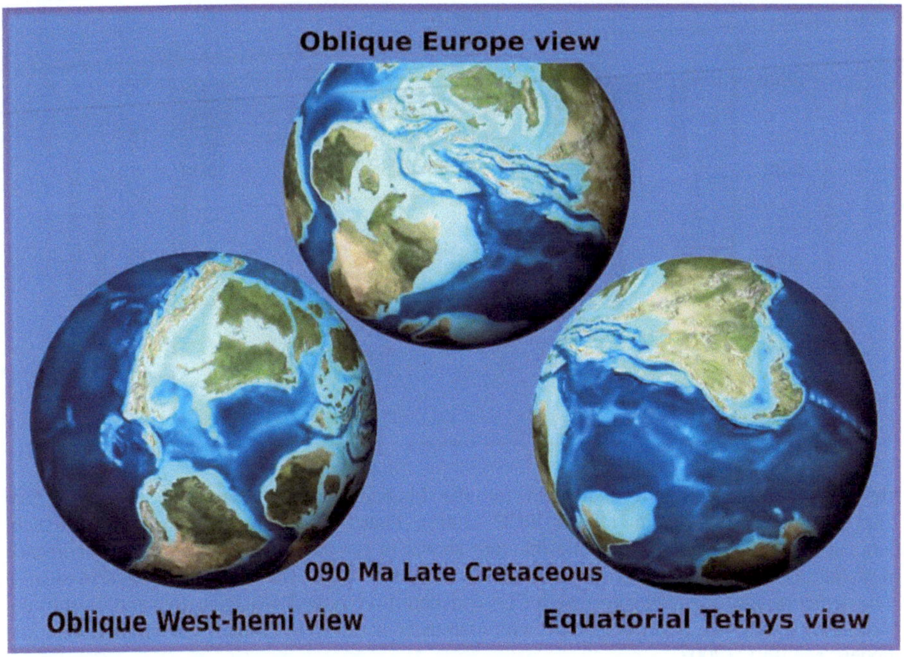

Fig. 3 Globes showing Gondwana fragments distribution in the Late Cretaceous (From Blakey, Ronald. www.nau.edu; http://jan.ucc.nau.edu/~rcb7/globaltext2.html)

Mesozoic and Paleogene Climates

The progressive breaking-up of Pangea, the then global continent at the end of the Triassic, generated much more humid climates, sea level rising and marine transgression on most continents. The relatively larger extension of the global seas reduced the albedo (i.e. reflection of sunlight back to the outer space) and allowed for warmer climates (Uriarte Cantolla 2003, p. 42). Changes in the topography of the ocean floor could also be responsible for the expansion of the shallow seas that forced an increase of evaporation. There is also evidence that methane (CH_4) was released from the bottom of the seas at a large scale during the Jurassic, increasing the atmospheric content of greenhouse gases (Hesselbo et al. 2000).

During the Middle to Late Jurassic (200–150 Ma ago), based on different proxy indicators, the CO_2 content was many times larger than today (presently, around 380 ppm), reaching perhaps above 4,000 ppm, though the final figures are still uncertain (Fig. 4). In the period that is considered in this paper, the CO_2 atmospheric content was extremely high for the Jurassic and the Early to Middle Cretaceous, as shown by paleosoil reconstruction and density of stomas in fossil tree leaves (Royer 2006, in: IPCC 2007) and also by the GEOCARB III model (Berner and Kothavala 2001, in: IPCC 2007). The high CO_2 content was sustained during the Jurassic, more

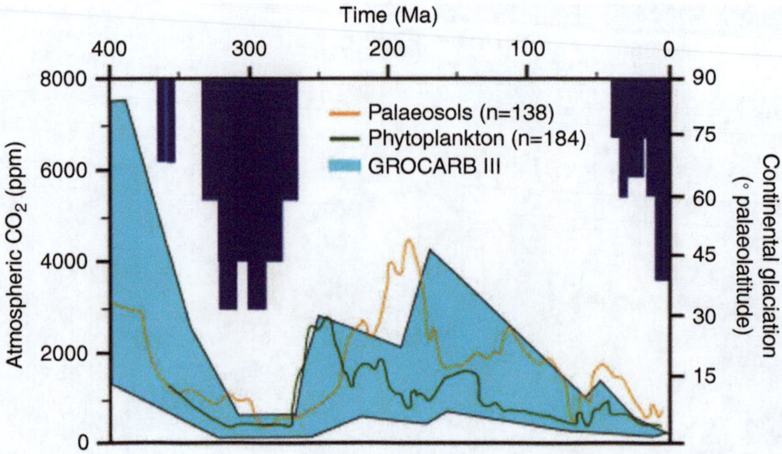

Fig. 4 Global paleoclimatic indicators for the Phanerozoic, with emphasis in the Mesozoic (Simplified and modified from IPCC (2007), Fourth Assessment Report; www.ipcc.ch/publications_and_data/ar4/wg1). The diagram shows in *vertical dark blue bars* the occurrence of global glaciation in the Late Paleozoic and the Late Cenozoic. The *curves* and *greenish-shaded zone* depict proxies indicating very high CO_2 concentration in the Early Paleozoic and most of the Mesozoic, which would be compatible with extremely warm/wet climates. For original data and models, see IPCC (2007)

than 50 million years long, above 4,000 ppm, up to 10 times larger than today, and certainly above 2,000 ppm during the Late Cretaceous, according to paleosoil data. The increased CO_2 content was also fed by huge volcanic eruptions along the rifting areas. Maximum photosynthetic capability in flowering plants should be achieved between 1,500 and 1,000 ppm, suggesting a huge expansion of rain forests and savannas in those times (Uriarte Cantolla 2003). Even beyond the K/T boundary, the ocean bottom temperatures were still very high until the Late Paleocene and Early Eocene. Evidence was also deduced from oxygen isotopes in Late Jurassic belemnites, indicating maximum temperatures of surface sea water of about 14 °C at 75° S latitude, which would be at least 7 °C warmer than present-day temperatures, indicating that this was a period of very warm Earth (Frakes 1979, 1986).

The Cretaceous was also an extended, quite homogeneous period, more than 80 million years long, with most of it under very warm and wet climate. Sea expansion continued in the Cretaceous, when enormous portions of the continents were submerged. Bottom water temperatures for the Early Cretaceous were at least 5–7 °C warmer than today. In the Middle Cretaceous, around 100 Ma, global mean temperature at the surface was between 6 and 12 °C higher than today (Uriarte Cantolla 2003). Jurassic coal and bauxite deposits around the world are related with these warm/wet climates, probably with high rainfall seasonality, at least regionally. The global climate was probably uniformly very warm. The Albian stage was the warmest part of the Cretaceous according to the sea-surface temperatures, around 28 °C. The Albian-Santonian time lapse was the summit for global temperatures in the Late Mesozoic, before the rapid cooling of about 10 °C in the Maastrichtian (Frakes 1979, p. 171). Tropical to subtropical conditions extended perhaps as far south as 70° S due to unique ocean current circulation (Frakes 1979), with large transfer of heat from the equatorial zones to the poles.

Thus, the entire Gondwana supercontinent was undoubtedly under extremely wet/warm conditions in the Cretaceous. The equatorial zone would have been heated much more intensely than today. Besides, the huge extension of the Pacific Ocean at low latitudes would have lowered the total albedo of the globe, strongly increasing the heat capacity of the oceans and, therefore, the influence of the largest heat reservoir on Earth. Frakes (1979, p. 185) stated that, between the Middle Triassic and the Middle Cretaceous, climates were characterized by mean annual temperatures possibly as much as 10 °C higher than today at the global scale, forcing unheard geographical scenarios in present times.

These conditions of very high CO_2 content in the atmosphere would have enhanced the magnitude of the greenhouse effect during this period, compared to present conditions. But temperature at the ground level could not rise indefinitely, because living beings would not bear it. It should be taken into consideration that life forms for these periods were essentially identical to those living today, because all groups that survive today in both in the continents and the oceans were already on Earth, including hair-bearing mammals and feathered birds. According to this, the resolution of this enormous greenhouse effect would have taken place in an immense evaporation rate from the oceans compared with today's conditions. Higher temperatures forced higher evaporation rates, increasing the water content of the atmosphere and the global greenhouse effect, and therefore, a much higher precipitation rate over the continents, under very warm climates. Thus, the precipitation during these times would have been enormous, several times the largest present rates, without glaciers growing on the continents and higher sea levels, forcing major transgressions. This very high precipitation would have generated huge water volumes as surface runoff and soil infiltration, perhaps down to very deep levels, many hundreds of metres, as a consequence of hydrostatic pressure of the hyper-saturated soils, all year-around.

This would have generated extremely intense weathering processes and very thick weathering mantles, with huge weathering profiles. Weathering profiles in the order of 100–200 m are found today in the very wet tropical zones, such as Indonesia or some areas of Brazil (Small 1978; Leopold et al. 1964). Thus, it may be assumed that in those CO_2-rich epochs, the thickness of the weathered zone would have been much higher, perhaps of 700–1,000 m. This would be proven by the presence of bornhardts, inselbergs and other deep weathering, residual landforms which, due to their local relief between their summits and the surrounding denudated surfaces, suggest that the weathered material thicknesses could have been around these values.

The present climatic zonation shows the close relationship between the subsuperficial weathering thickness and the mineralogy of the weathering products (Lisitzin 1972; Strakhov 1967; Fig. 5). As seen in the present conditions, the thickness of the weathering layer reaches a maximum along the Equatorial wet zone, with

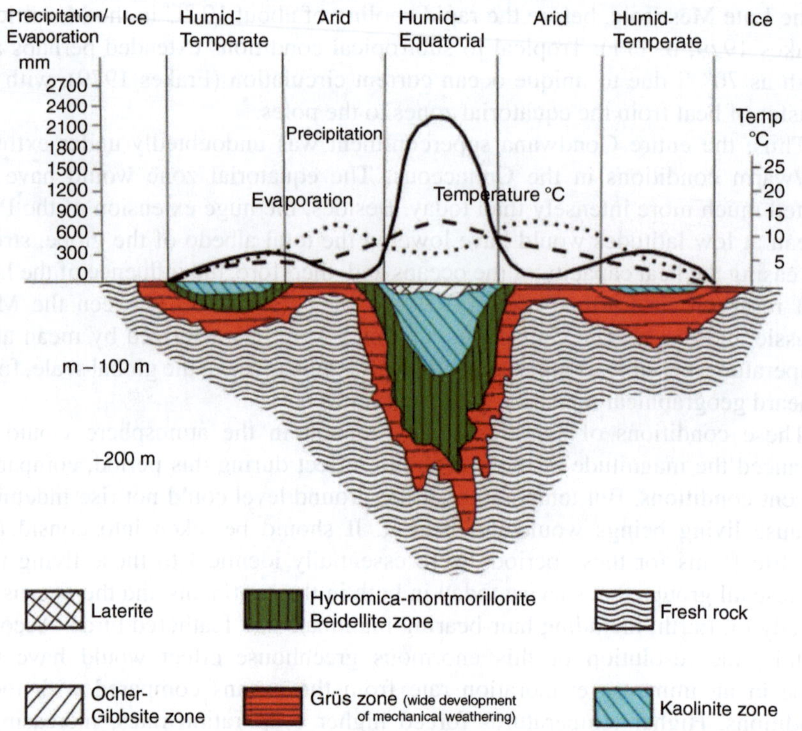

Fig. 5 Global climatic zonation in present times, showing the relationship between latitude, depth of the weathering profile and the mineralogy of the subsurface weathering products (Simplified from Strakhov 1967)

depth values of up to 200 m, with precipitations in the order of 2,100 mm/year and the minimum mean annual temperatures above 15 °C. In this case, there is a development of a thickness of 100 m of the kaolinite zone, with another 100 m of maximum thickness of montmorillonite-beidellite-hydromica and, underneath, up to additional 50 m of gruss, also with development of mechanical weathering. It could also be expected in the presence of laterites and ochre-gibbsite in the superficial zone. The expected weathered zone thicknesses would be much larger in a hyper-tropical climate than in the present conditions, a fact that supports the interpretation of thicknesses of more than 700 m. If the weathering front is today at around -200 m in selected tropical zones, how deep could it have reached during Mesozoic times in Gondwana? Probably up to 4–5 times the present figures, at least, perhaps up to 1,000 m.

According to these circumstances, the Jurassic-Early Cretaceous climates that would have dominated the Gondwana regions could be considered as hyper-tropical, with no present analogues, which would be responsible for the genesis of these noted paleolandscapes.

These climates generated immense weathered debris which remained stable for a very long time, as denudation was slow due to tectonic quietness. A long-term equilibrium was achieved between weathering and denudation, allowing for the development of the intriguing landforms that are found today in the Gondwana Landscapes. This weathered debris cover was denudated during the Cenozoic, particularly since the Middle Eocene, when the world climate changed, until the ancient weathering fronts were exposed as the weathered debris was removed. Thus, usually only the roots of the weathering profile are preserved and observed at the surface, with occasional presence of clays, most frequently kaolinite, or lateritic materials, until only the fresh, unweathered rock is exposed. These rocks cannot be further altered after denudation, because the climate is not hyper-tropical anymore and there is not enough heat and water available. The new conditions do not allow the return of these weathering processes during the entire Cenozoic, not even in tropical areas, as the greenhouse effect diminished due to the reduction in the atmospheric CO_2 content. Therefore, the cited landforms are paleoclimate indicators, and the Gondwana Landscapes were unrepeatable because they could not fully develop today anywhere in the planet.

All these conditions were accompanied by high tectonic stability in the Gondwana cratonic regions, which allowed for deep weathering without debris removal, until the Alpine-Andean tectonic reactivation in the Tertiary, particularly since the end of the Eocene, triggered worldwide denudation.

In spite of a relative cooling at the end of the Cretaceous which has been referred to several causes, the warm climates continued during the Paleocene. Moreover, around 55 Ma, at the end of the Paleocene and beginning of the Eocene, there was an abrupt and short warm peak, with mean annual global temperatures of 5–7 °C above the temperatures at the K/T boundary. This unusual warm event is probably linked to methane released from the bottom of the oceans. Global environmental conditions inherited from Cretaceous times were then sustained until the end of the Paleocene and perhaps even into the Eocene (Uriarte Cantolla 2003).

Granite Deep Weathering

One of the geomorphological processes, which are particularly significant in terms of ancient landscape interpretation, is granite deep weathering. Since one of the conditions required for the development of Gondwana Landscapes is tectonic stability, the occurrence of granites and similar intrusive and/or metamorphic rocks in shields and other cratonic regions is common in such landscapes. Since the other prerequisite is warm/wet climate, granitic rocks are highly sensitive to deep weathering under these conditions. The relatively homogenous and isotropic nature of granites enhances the development of these processes. Granite landscapes are excellent examples of deep weathering paleoclimates and usually diagnostic features for Gondwana Landscapes (Twidale 1982; Vidal Romaní and Twidale 1998).

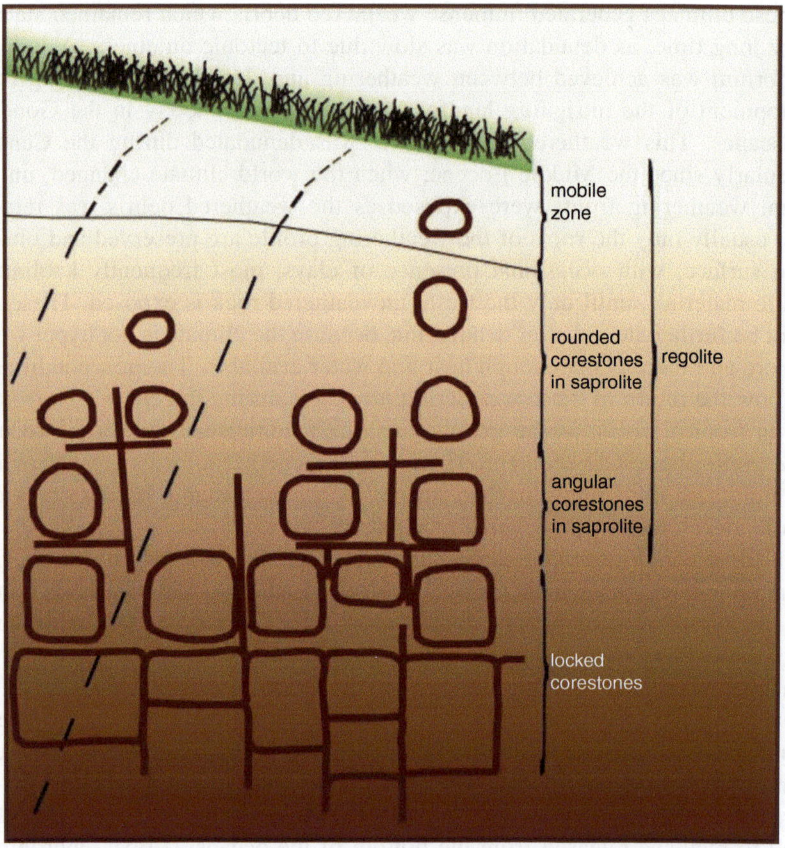

Fig. 6 Granite weathering. A typical section of deep chemical weathering in granites. Granite blocks are bounded by joints. Saprolite is an in situ weathered rock, as indicated by the unremoved quartz veins. The alteration is isovolumetric. Regolith is a term that includes all unconsolidated materials near the surface of the Earth, including saprolite (Redrawn from Ollier 1990)

Ollier (1984, 1990) discusses a typical weathering profile in granitic rocks (Fig. 6). Saprolite, or "rotten rock", is an in situ deeply weathered rock, which is usually indicated by the nonmobilized quartz veins, being quartz almost totally immune to weathering, unless extreme conditions are present. The alteration forced by weathering is isovolumetric, that is, no changes in the volume of the original minerals after being weathered are recorded. Regolith is a term that covers all unconsolidated materials at or near the Earth surface, and it includes saprolite (Fairbridge 1968, p. 933).

In tropical climates, warm, acid-rich (HCO_3) waters penetrate the granite outcrops following joints and other fractures. They react chemically with the poorly resistant minerals of the granites, such as amphiboles, micas, feldspars and other minor components. Quartz is not affected, except perhaps as surface etching, and

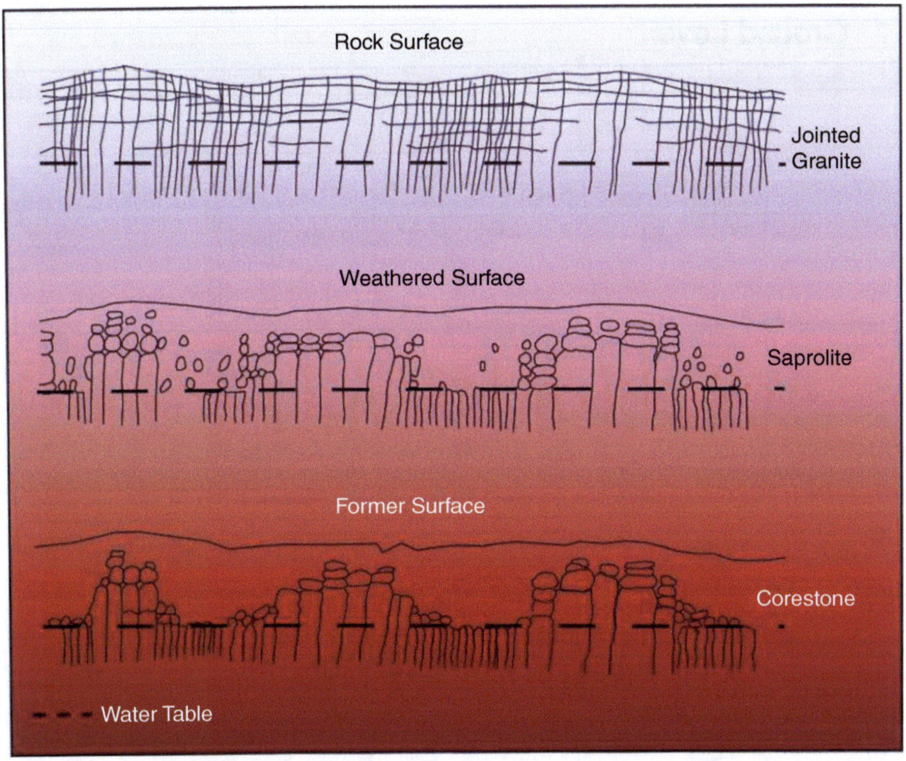

Fig. 7 Corestone and tor development due to deep weathering and subsequent denudation, with finer, weathered material removal (From: Linton 1955; see Fairbridge 1968)

it becomes the main residual material in the gruss, together with clays, usually kaolin. Weathering advances from the surface downwards and from the fracture or exfoliation planes to the inner part of the resulting blocks. This leads to the formation of unweathered cores in the blocks, surrounded by a saprolitic material. These unweathered nuclear remnants are called "corestones" (Fig. 7; Linton 1955), which become roughly rounded in situ as spheroidal weathering makes progress. The process continues indefinitely as the prevailing conditions are maintained. But, when climate changes moving towards drier conditions, denudation starts and the corestones are progressively dismantled. With time, all residual materials are removed and the corestones pile up on the surface. As the climate has changed, the corestones cannot be further weathered, and they remain as unchallenged testimonies of past climates. The corestone accumulations on unweathered bedrock are called "tors" (Fig. 8) and defined as "a bare rock mass surmounted and surrounded by blocks and boulders" (Linton 1955). The equivalent term "kopje", of Afrikaans origin, is used in Southern Africa. Many tors occupy the summit of "bornhardts" (Twidale 2007b) and their evolution ends with the total dismantling (Fig. 9).

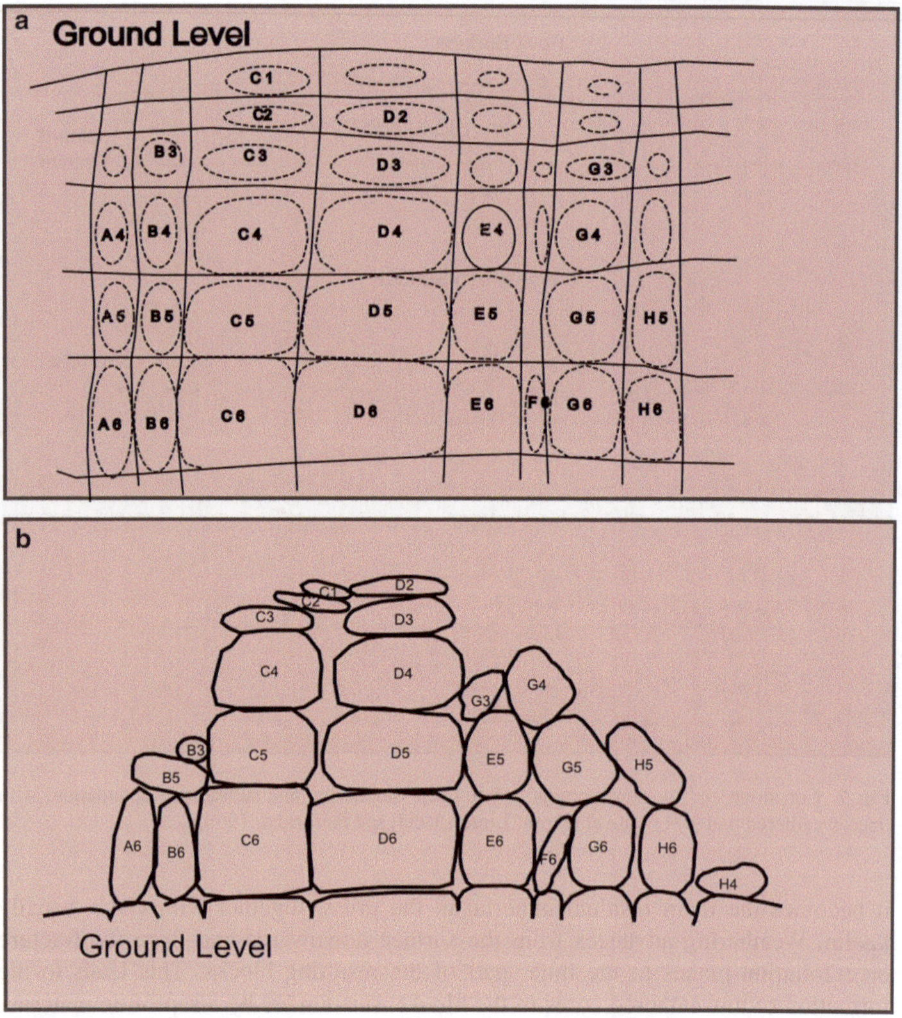

Fig. 8 Stages in the evolution of a tor by subsurface weathering (From Linton 1955). (**a**) Subsurface corestone formation. (**b**) Dismantling by denudation of the weathered materials

Thus, corestones and tors are formed by a two-stage process, involving firstly a period of prolonged subsurface groundwater weathering, under wet tropical climates and tectonic stability, followed by a period of erosion stripping with no further significant weathering. Therefore, corestones and tors are very common in Gondwana Landscapes and important features in the interpretation of ancient climates, no longer active in the study region.

Etchplains are landforms developed essentially by these deep weathering processes, and they are characterized by the abundance of corestones, domes, bornhardts and inselbergs. As denudation proceeds, etchplains are stripped off their

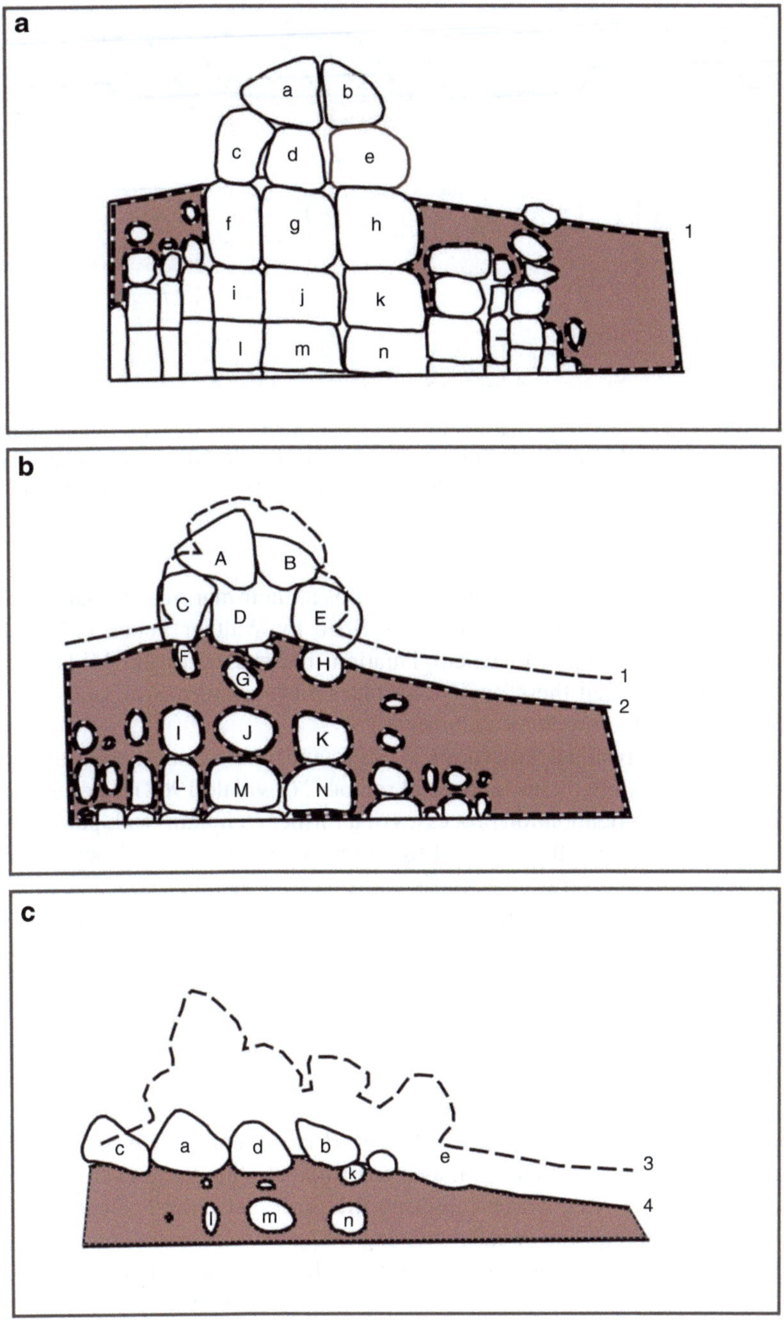

Fig. 9 Stages in the collapse of a domical tor. (**a**) Initial phase in the dismantling of the tor group, with partial removal of the weathered debris; (**b**) progressive collapse of the tor as a result of the washing out of the weathered materials; and (**c**) superficial distribution of the remaining corestones (From Thomas 1965)

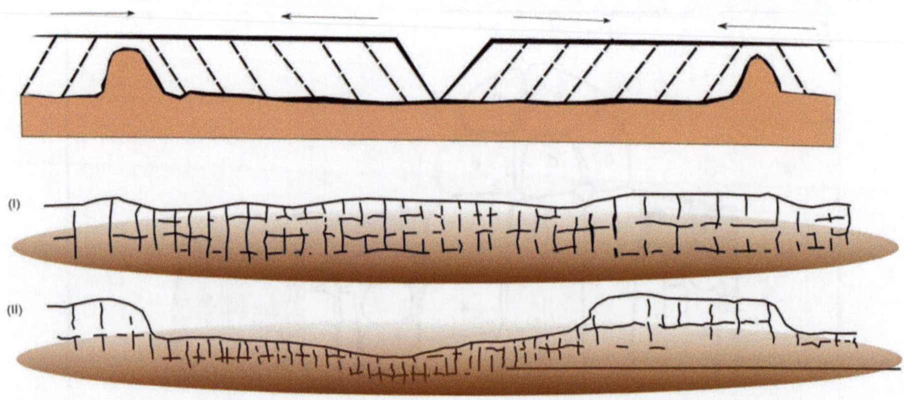

Fig. 10 Formation of inselbergs and bornhardts. *Upper figure*, inselbergs resulting from long-distance scarp retreat; *lower figure*, (*i*) initial stage of etching of differentially jointed bedrock and (*ii*) final stage after removal of the weathered cover and genesis of bornhardts (From Twidale 2007b)

weathering residues and pediplanation develops at their margins, by parallel retreat of the slope, following the processes described by Walther Penck (1924, 1953). Thus, it is highly probable that pediplanation that started in the Middle to Late Cretaceous had affected these ancient surfaces during millions of years while the regions conserved their tectonic stability.

Inselbergs are "residual landforms which stand in isolation above the general level of the surrounding plains in tropical regions" (Twidale 1968). They are formed by a combination of denudation processes of a former etchplain, with parallel retreat of the slope under the influence of differential weathering of bedrock, either due to lithological or structural characteristics (Figs. 10 and 11). The bedrock areas with scarce, closed or no jointing remain unweathered and will become the relict positive features after denudation. These processes are developed during the final phases of the evolution of the Gondwana Landscapes. Inselbergs may be formed on many different rock types. One of the better known inselbergs in the world, the Ayers Rock, in Central Australia, is made of arkose sandstones. Another one, the "Sugarloaf" (Pão de Açucar) in Rio de Janeiro, Brazil, is composed of granitic rocks. The mineralogical and structural characteristics of granitic rocks are particularly appropriate for inselberg formation. Steepened basal slopes of granite inselbergs are due to subsurface weathering, under wet tropical climates, and they are called "flared slopes" (Fig. 12), which represent the position of the ancient weathering front. Granite inselbergs are usually showing frequent caves or smaller holes named as "tafoni". These types of inselbergs should not be confused with hills developed in dry, arid climates, also by parallel retreat but without previous regional weathering. Similar processes are responsible for the formation of "ruwars" and "low domes", massive bedrock features due to differential weathering in tropical environments (Fig. 13). Landscape evolution includes several phases, from the development of the

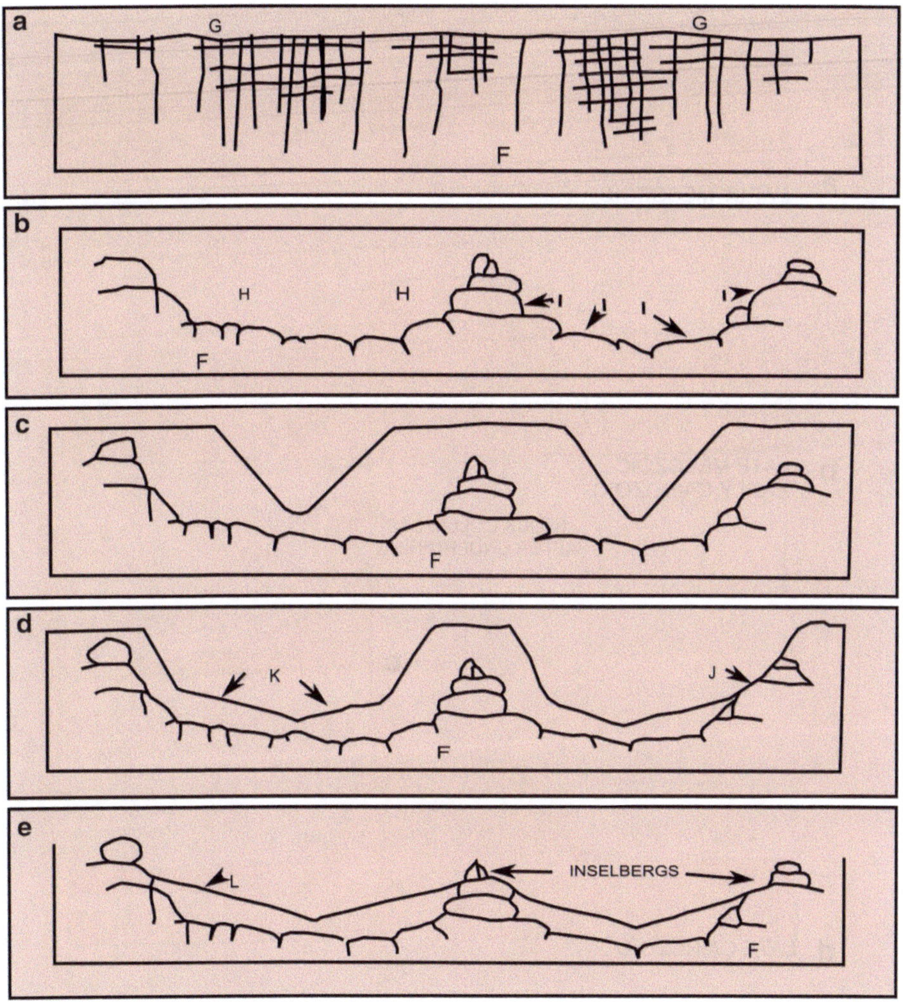

Fig. 11 Formation of inselbergs. (**a**) Gondwana erosion surface (g) developed by etching on differentially jointed, fresh rocks (f). (**b**) Development of an irregular weathering front (i) with uneven weathered debris thickness (h). (**c**) Partial incision of the weathered mantle that develops in the deeper portions of the debris cover. (**d**) Pediment (k) development on the weathered mantle and partial outcropping of the unweathered rocks (j). (**e**) Further removal of most of the weathered mantle and development of a regional surface (l) (From Ollier 1960. Reproduced also in Bigarella et al. 1994)

etchplain, the denudation of the former weathering front, removal of the weathered cover surrounding the fresher, core areas and the preservation of more resistant portions of the bedrock.

Bornhardts (Fig. 14) are "bare surfaces, dome-like summits, precipitous sides becoming steeper towards the base, an absence of talus, alluvial cones or soils, with

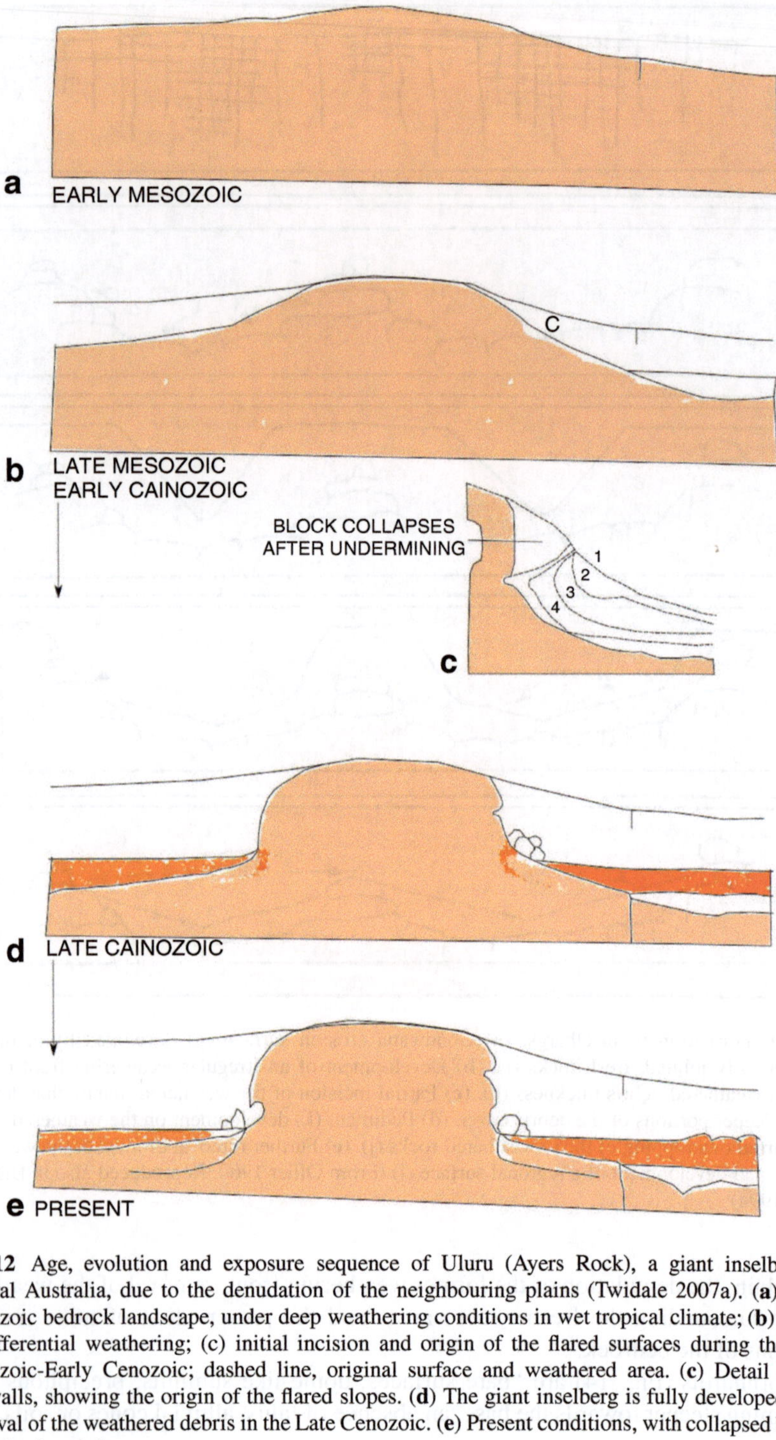

Fig. 12 Age, evolution and exposure sequence of Uluru (Ayers Rock), a giant inselberg in Central Australia, due to the denudation of the neighbouring plains (Twidale 2007a). (**a**) Early Mesozoic bedrock landscape, under deep weathering conditions in wet tropical climate; (**b**) Effect of differential weathering; (c) initial incision and origin of the flared surfaces during the Late Mesozoic-Early Cenozoic; dashed line, original surface and weathered area. (**c**) Detail of the sidewalls, showing the origin of the flared slopes. (**d**) The giant inselberg is fully developed after removal of the weathered debris in the Late Cenozoic. (**e**) Present conditions, with collapsed blocks from the flared slopes

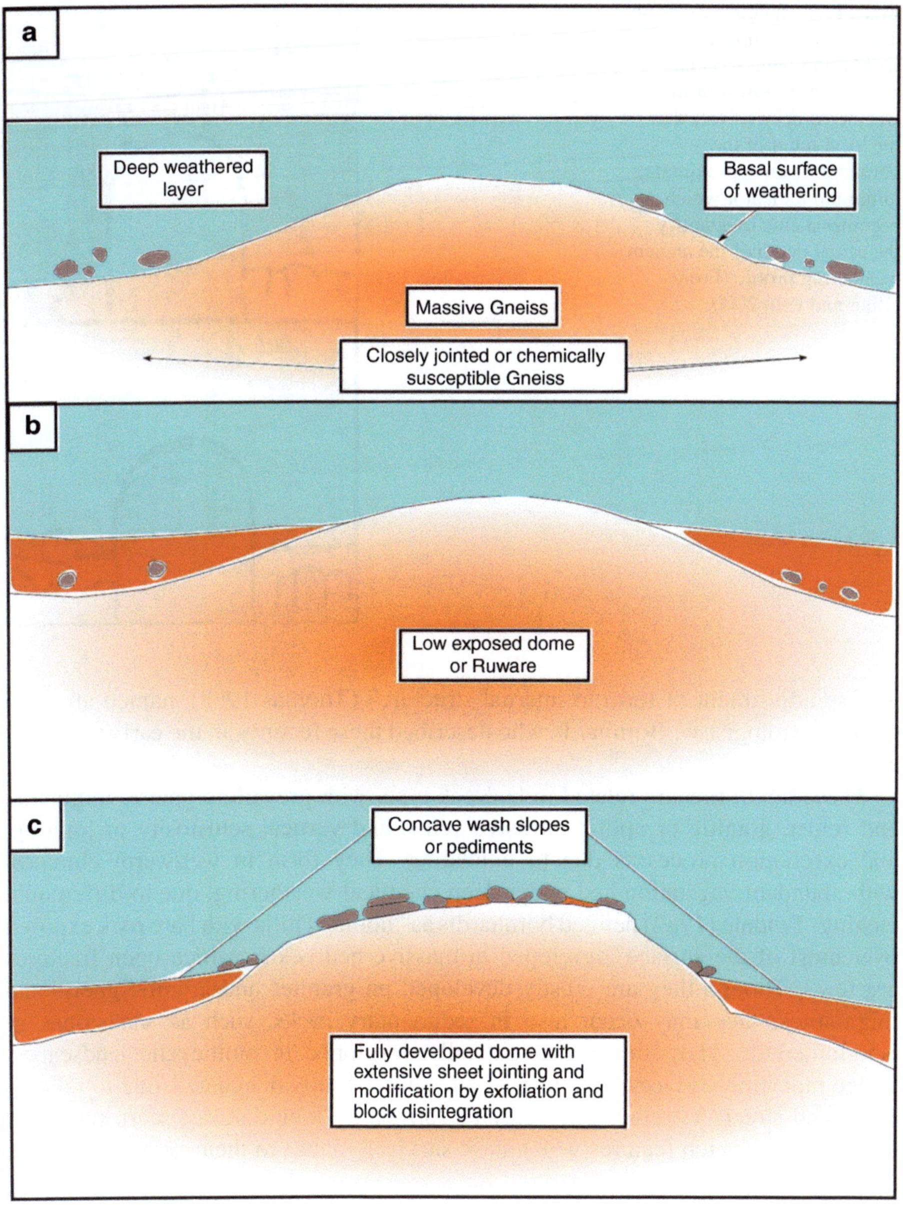

Fig. 13 The development of ruwares and low domes by differential deep weathering and subsurface stripping of the deep weathered layers. (**a**) Differential deep chemical weathering, affecting upon two bedrock types, "massive gneiss" and "closely jointed or chemically susceptible gneiss"; (**b**) after environmental and climatic conditions changed, partial removal of the weathered layer, exposing the top of the ruware; (**c**) formation of the low dome by prolonged denudation of the weathered layer, with modification of the top of the dome, due to exfoliation and block disintegration, and formation of concave wash slopes or pediments (From Small 1978)

Fig. 14 Bornhardt formation due to deep chemical weathering (etching) and subsequent removal of the weathered debris. Note that the local relief of the bornhardt is an approximate, minimum indicator which becomes a clue to quantify the actual depth of the ancient weathering profile (From Ollier and Pain 2000)

a close adjustment of form to internal structure" (Thomas 1968), named after the German geologist W. Bornhardt, who described these features in the early twentieth century.

These landforms are related to bedrock type, with prevailing gneiss, migmatite and schist, granitic or aplitic intrusive veins, and vertical schistosity or jointing, and exfoliation processes due to unloading. They form in wet/warm climates, with abundant vegetation and under deep chemical weathering, due to differential etching. Twidale (1982) defined bornhardts as "domical hills with bare rock exposed over most of the surface, developed in massive bedrock in which open fractures are few". Though they are mostly developed on granites and granitic gneiss and migmatites, they may occur also in sedimentary rocks, such as sandstones or conglomerates. They are characteristically developed in multicyclic landscapes, where planation surfaces were formed and subsequently denudated, due to relative uplift and stream incision. They may be formed by long-distance scarp retreat or as two-stage or etch features which have survived thanks to their massive structure (Twidale 2007b).

Passive-Margin Geomorphology

At passive continental margins (Fig. 15), etchplains were developed during periods of tropical climate and long-term tectonic stability. Other types of paleoplains may also be present. A general slope recession took place as the tectonic conditions

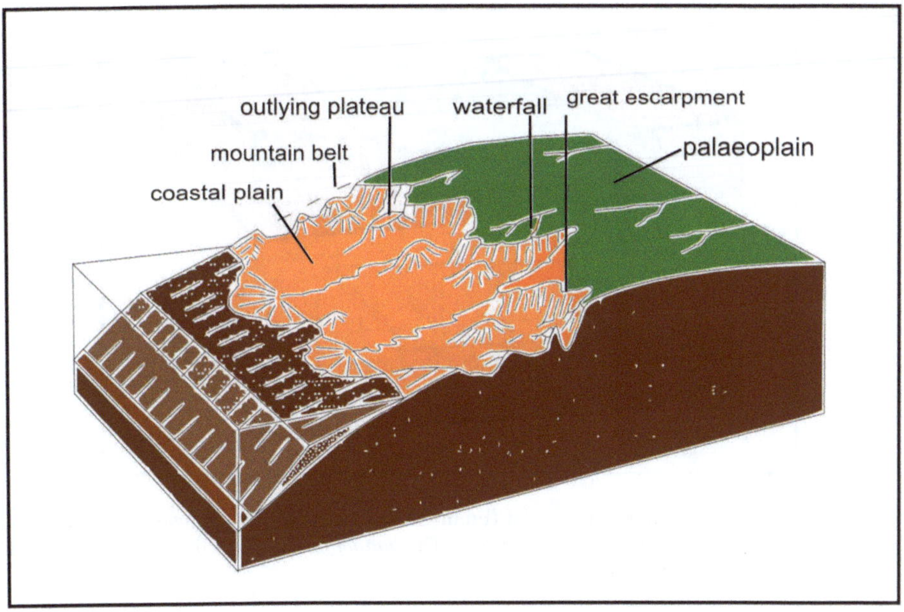

Fig. 15 In passive continental margins, a large and continuous escarpment is developed by headward erosion from the coastal plain, degrading a former paleoplain, usually an etchplain, developed during previous stages of tropical climate with deep chemical weathering. The position of the escarpment is noted by waterfalls along a very steep boundary (From Ollier and Pain 2000)

changed and a new base level is enforced due to continental uplift or sea level lowering. In Southern Africa, the uplifting process was the consequence of the passage of the continent above a hot point during the Middle to Late Cretaceous, as the rifting process made progress and the South Atlantic Ocean started to grow. Environmental changes developed and the differential response to weathering and erosion of the superficial materials. Then, an escarpment is formed as headward erosion took place from the growing coastal plain. In South Africa, it is called the "Great Escarpment"; in Brazil, the escarpment could be found in the Serra da Mantiqueira and Serra do Mar; and in India, it is represented by the Ghats Mountains. In Argentina and Uruguay, the position of the escarpment, if it existed, is still unclear and should be investigated.

Duricrusts: Ferricretes, Silcretes, Calcretes

"Duricrust" is a general term to designate hard layers found in soils and superficial deposits, basically of pedogenetic origin and related to weathering, solution and precipitation processes. Duricrusts are excellent diagnostic materials of past

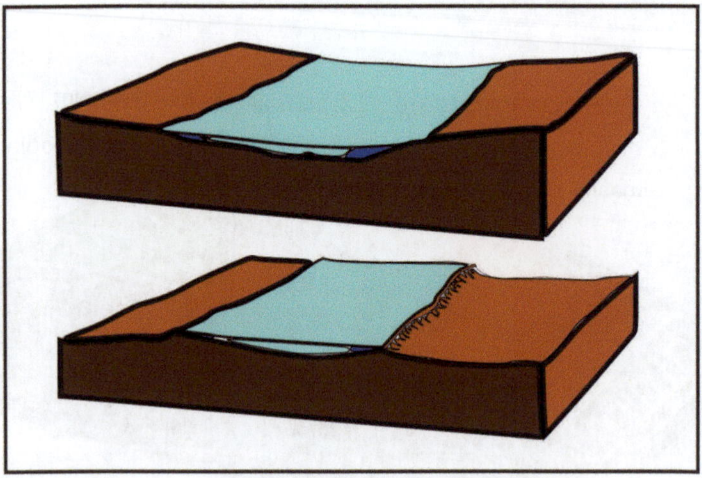

Fig. 16 Duricrusts and inversion of relief (From Ollier 1991a). In the *top figure*, ferricretes are precipitated on lower slopes and valley bottoms. The *bottom figure* shows the inversion of relief, producing a ferricrete-capped mesa or tableland

weathering conditions, and due to their varying composition, they may be found in many different bedrock conditions. Those duricrusts where iron oxide and hydroxide are dominant are called a "ferricrete". The term "laterite" includes ferricretes, but it is also used to describe soils and weathering profiles (Ollier 1991a, b, 1995; Ollier and Galloway 1990). Aluminium duricrusts are also known as "alcrete" and most commonly called "bauxites".

Ferricretes may be formed by iron translocation or by removal of other components of the soil. They indicate the existence of a weathering profile, where iron is removed and then redeposited, usually by groundwater circulation or capillary action under very warm/wet climates. Ferricretes are also common on fluvial gravels and alluvial plains. They may be originally deposited in the valley bottoms, but then exposed as capping materials in tablelands due to inversion of relief (Fig. 16). The process may be repeated through time, indicating the existence of several planation surfaces or progressively younger age with lower elevations. Since ferricretes are very resistant to weathering, they are very useful to reconstruct episodes of long-term landscape evolution, as in South Australia, where terrestrial landscapes have been exposed since the Permian glaciations and the ferricretes are Early Mesozoic in age (Ollier 1991a).

Silcretes are very hard, whitish rocks that are the result of silicification in pre-existing quartz-rich sediments. The silica is removed by highly acidic, deep weathering processes and deposited as infilling of the pores and voids of the original sediments when conditions change. Therefore, they are good indicators of long-term seasonality or cyclic changing climate. The silica content is usually very high, with a few other residual components such as titanium or zircon. Silcretes are frequently associated to kaolinized granites or basalts of different ages. Identified silcretes in South Africa and England were formed in the Paleocene, and Paleogene silcretes are common elsewhere (Ollier 1991a).

Calcretes are calcareous duricrusts, which form in many different environments and more rapidly than ferricretes or silcretes. They are related to pedogenetic processes under quite varying moisture and temperature conditions, but they usually imply seasonally dry or semiarid climates.

Bauxites are the end product of intense, deep weathering, under very wet tropical climates, where all soil components, including silica, but with the exception of the most stable alumina rich clays, have been leached away (Ollier 1991a). They are very good indicators of past climates and have been found in many different environments and ages.

Gondwana and Other Ancient Paleolandscapes in the Southern Hemisphere and Other Parts of the World

Du Toit (1954, p. 573) stated that "from the Jurassic onwards South and Central Africa underwent various cycles of prolonged planation, the most widespread one being that of the late Tertiary". Du Toit indicated that each of these planation episodes was followed by uplift, some depression and perhaps, some warping. He described erosion surfaces at 2,500 m above sea level in Rhodesia, 2,200 m a.s.l. in Southwest Africa and 1,500 m a.s.l. in the Zwartberg. He also described the "high level gravels" in southern South Africa, as remnants of former paleosurfaces, which were considered as equivalent to the "Conglomerado Rojo" of the Sierras Australes of Buenos Aires Province, Argentina, by Zárate et al. (1995, 1998) and Zárate and Rabassa (2005).

Lester C. King (1950) described global cycles of planation and extended their identification since the Jurassic (Table 1). King (1950) provided evidence for

Table 1 Planation cycles: sequence of global planation surfaces and paleolandscapes in Southern Africa (Slightly modified from L.C. King 1950)

Planation cycles	Old name	New name	Recognition
I	Gondwana	The "Gondwana" planation	Of Jurassic age, only rarely preserved
II	Post-Gondwana	The "Cretacic" planation	Early Mid-Cretaceous age
III	African	The "Moorland" planation	Current from Mid-Cretaceous to Mid-Cenozoic. Planed upland, treeless, poor soil development
IV		The rolling land surface	Mostly of Miocene age
V	Post-African	The widespread landscape	The most widespread global cycle. Pliocene in age
VI	Congo	The youngest cycle	Quaternary age, deep valleys and gorges

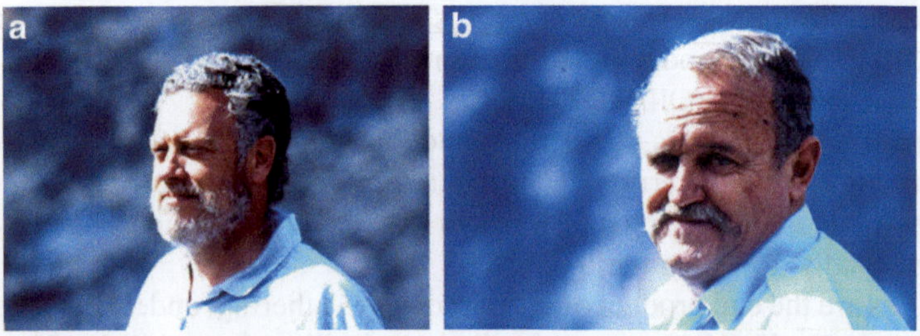

Fig. 17 (**a**) Timothy C. Partridge, who was professor of Geomorphology, University of Witwatersrand, Johannesburg, South Africa, a distinguished scholar in different fields of Gondwana Landscapes, deceased in 2009. (**b**) Rodney R. Maud, Emeritus Professor, University of Natal, Durban, who, together with Tim Partridge, opened for the present author the wonderful world of Gondwana Landscapes and made him a true "believer"

the existence of a worldwide, Jurassic age erosion surface which he called the "Gondwana surface". He also indicated that after the final rifting of Gondwana during the Cretaceous and later time, it was followed by the development of a series of younger erosion surfaces, but they would have formed independently in each continent. The second planation cycle is the post-Gondwana, of Early to Middle Cretaceous age. The most important of these surfaces is the African surface, formed through a long period from the Late Cretaceous to the Middle Tertiary. Cycles 4 and 5 are Late Tertiary and Cycle 6 is of Quaternary Age.

Tim Partridge (Fig. 17a) and Rodney Maud (Fig. 17b) (Partridge 1998; Partridge and Maud 1987, 1989, 2000) represented the South African geomorphological school that was started by Lester C. King in the 1950s. They completed the regional mapping of the larger paleosurfaces (Partridge and Maud 1989), extending the observations to other areas of Southern Africa. They distinguished the different units and their geomorphological, tectonic and economic significance due to their relationships with diamonds, bauxites, gold placers, other minerals residual concentrations, etc. They established also the importance of different types of duricrusts, such as ferricretes and silcretes, to identify different ancient surfaces and correlate them. In their papers, they established that "the inland planation surface was formed no later than the end of the Cretaceous" (Partridge and Maud 1989) and that the landscape was developed both above and below the Great Escarpment (Partridge and Maud 1987). The inland plains were developed under very wet, tropical climates during the Early Cretaceous or even before. Moreover, there are remnants of areas above the Cretaceous plains, which may have been developed during the Late Jurassic, as King (1967) had envisaged.

The Australian model of geomorphological evolution was consolidated by the work of C. R. Twidale and C. Ollier, among many other geologists and geomorphologists. A well-known example is the Gawler Ranges. This is a massif of ancient volcanic rocks located in South Australia. The area is extremely stable, and basically, the development of the present landscape began with the melting and disintegration of the Permian ice sheets (Twidale 2007a). During the Early Jurassic, the area was undergoing very intense, deep weathering in tropical climate, which generated a huge planation surface, named as the Beck Surface, which originally had a thick regolith/saprolite cover, showing differential weathering following structural controlling features. Later, uneven tectonic uplifting of the area in the Early Cretaceous forced the partial denudation of the range and the removal of most of the weathered debris, probably due to river rejuvenation after the uplifting. In the Early Tertiary, most likely during the Paleocene, climate change allowed the formation of the Nott Surface, with the almost complete removal of the Jurassic-Cretaceous regolith and development of silcretes in plains and hollows. Remnants of this ancient regolith are preserved only in a few sites at inner locations (Twidale 2007a).

In the former Soviet Union, Gorelov et al. (1970) classified the Russian erosion surfaces in two main groups, the "Ancient denudational surfaces", considered as peneplains or pediplains of Mesozoic or even pre-Mesozoic age, mostly pre-break-up to the Pangea supercontinent, and the "geomorphological surfaces", chiefly developed in the Tertiary. Likewise, Geramisov (1970) proposed three megacycles in the geomorphological development of the Earth during Mesozoic and Cenozoic times. The earliest megacycle is a Jurassic-Cretaceous basal planation surface surmounted by inselbergs, which is still present on a global scale.

Melhorn and Edgar (1975) presented a correlation of the main surfaces in the World, including North America, some of which dated from the Early Mesozoic, but their ideas were not taken in great consideration by their American colleagues, who seem to feel much more comfortable with Thornbury's (1954) classical ideas restricted to very young landscapes. Melhorn and Edgar (1975) recognized the possibility of time-synchronous, worldwide erosion surfaces, some of them as old as the Late Triassic (though in this case they are mostly covered surfaces) and the Jurassic (Table 2). For the Appalachian region, they identified periods in which appropriate conditions for net denudation and landscape planation, such as the >135–110 Ma interval (Late Jurassic-Cretaceous, which they called "Fall Zone Time"), 85–55 Ma (Late Cretaceous-Paleocene, "Schooley Time"), 45–20 Ma (Eocene-Miocene, "Harrisburg Time") and possibly between 12 and 2 Ma (Pliocene, "Somerville Time") (Table 3). Note the identification of six surfaces for the Brazil-Uruguay region, starting in the Late Triassic, with the Pre-Botucatú surface, a buried erosion surface. They agreed with King's ideas (King 1956a, b) of a Gondwana surface (Jurassic), a post-Gondwana surface (unnamed in South America, Late Cretaceous), the African surface (=the "Sul-Americana" surface; Paleocene-Eocene) and finally, two Late Cenozoic surfaces.

In Central United States, the clear delimitation of the Cretaceous Mississippi Engulfment allows to identify remnants of ancient surfaces above this ancient littoral zone and predating such transgression that developed in the non-glaciated, highly stable, cratonic areas of Southern Illinois and Arkansas, which had not been covered by the sea since perhaps the latest Paleozoic (Rabassa 2006).

In Venezuela, Schubert et al. (1986) and Schubert and Huber (1990) have described the "tepuis" of the Guiana Massif as features developed since Jurassic times (Fig. 18). This diagram suggests the possibility of having remnants of even

Table 2 World correlation of the planation surfaces and erosion landscapes (Simplified and slightly modified from Melhorn and Edgar 1975), in South Africa, West Africa, Brazil and Uruguay, Australia, India, Mongolia and China

Period	Age at the base (Ma)	South Africa	West Africa	Angola	Brazil/Uruguay	Australia	India	Mongolia	China
Pleistocene	2	Congo			Paraguaçú	Wudinna	XXX	Pankiang	Pangchaio
Pliocene	7	Coastal Plain		XXX	Velhas	Koongawa	Jamda	Gobi	
Miocene	26		Ho-Keta	Namib		Meckering	Noamundi		Tangshan
Oligocene	38					UN/Nonning	Kiriburu	Kanghai/Mongolian	
Eocene	54	African	Ashanti		Sul-Americana	Australian pediplain	Indian	Kanghai/Mongolian	Pei-Tei
Paleocene	65	Post-Gondwana			XXX	Simmens/Nott	Post-Gondwana		
Cretaceous	135	Post-Gondwana	Voltaian	Benguela		Simmens/Nott	Post-Gondwana	Post-Laurasian	Post-Laurasian
Jurassic	200	Gondwana		Planalto	Gondwana	Gondwana Mount Dale	Gondwana/Nilgiri	Laurasian	Laurasian
Triassic	250	Sub-Stormberg	Agu Moun-tain (?)		Sub-Botucatú	Lincoln			

Note that Jurassic Gondwana Landscapes have been identified in most areas in the Southern Hemisphere, but also in Asia, where Gondwana equivalent surfaces are named as "Laurasian". "XXX" refers to surfaces known but not named yet at the time the original paper by Melhorn and Edgar (1975) was published

Table 3 Correlation chart of North American erosion surfaces (Simplified and slightly modified from Melhorn and Edgar 1975)

	Age at the base (Ma)	Appalachian	Interior Low Plateaus	Interior Highlands	Central Low Plateaus	Great Plains	Rocky Mountains	Great Basin	Sierra/cascade
Pleistocene	2	Valley cycle	Deep stage	Valley cycle	Deep stage	Terraces	Canyon cycle	Pediment cycle	Kern River canyon cycle
Pliocene	7	Somerville	Parker	Post-Osage Strath	Havana Strath	Flaxville		Antler	Mountain valley (?)
Miocene	26	Harrisburg	Lexington/ Highland Rim	Osage-Strath	Central Illinois		Rocky Mountains/ subsummit		Chagoopa/broad valley
Oligocene	38			Hot Springs/ Ozark	Lancaster/ Calhoun				
Eocene	54					Prairie/Cypress Hills	Flat top/summit	"Broken Hills" (?)	Subsummit/ boreal
Paleocene	65	Schooley		Ouachita/ Springfield					Sierra/summit (?)
Cretaceous	135	Fall Zone		Boston Mtn./ Summit (?)	Dodgeville (?)	XXX			
Jurassic	200								

"XXX" refers to surfaces known but not named yet at the time the original paper by Melhorn and Edgar (1975) was published

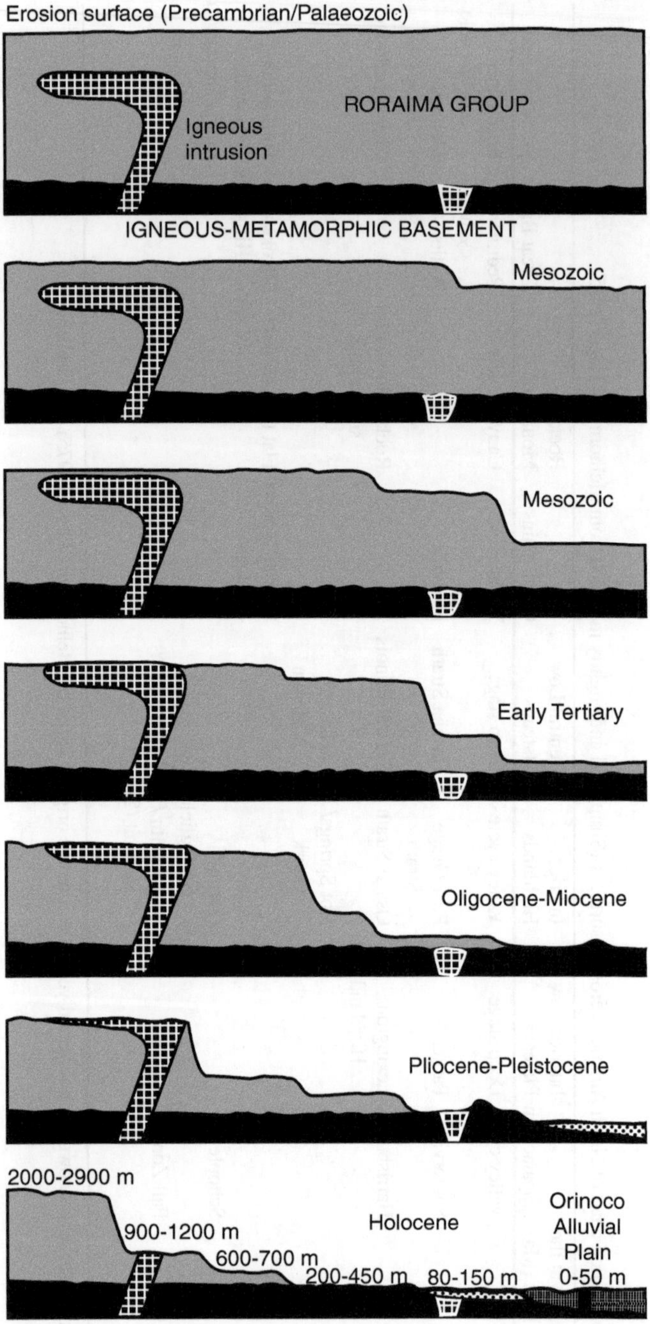

Fig. 18 Geomorphological evolution of the "tepuis" of the Gran Sabana of Venezuela, in the Guyana Massif. Development of planation surfaces since the Mesozoic (From Schubert and Huber 1990 (in Ollier and Pain 2000)). This is an example of Walther Penck's concept of "piedmont treppen". The graph implies that even remnants of Late Paleozoic surfaces may have been preserved in the uppermost portions of the massif, which have never been covered since then

Table 4 Erosion surfaces identified in the Guyana Massif, from Schubert et al. 1986 (Slightly modified from Clapperton 1993)

Elevation (m)	Name	Age
2,900–2,000	Auyán Tepui (Venezuela)	Mesozoic (?)
1,200–900	Kanuku (Guyana), Gondwana (Brazil), Kamarata-Pakaraima (Venezuela)	Mesozoic (?)
700–600	Kopinang (Guyana) Sul-Americana (Brazil), Imataca-Nuria (Venezuela)	Eocene (?)
450–200	Kaieteur/Kuyuwine-Oronoque (Guyana), Early Velhas (Brazil), Caroni-Aro (Venezuela)	Oligocene-Miocene
150–80	Rapununi (Guyana), Late Velhas (Brazil), Llanos (Venezuela)	Pliocene-Pleistocene
50–0	Mazaruni (Guyana), Paraguaçú (Brazil), Orinoco Plain (Venezuela)	Holocene

older (Paleozoic?) surfaces above the Jurassic Gondwana surface. "Tepuis" are spectacular table mountains, whose summit plateaus commonly lie above 2,500 m a.s.l. (Clapperton 1993). The tepuis and their karst-like features appear to be the result of deep chemical weathering during at least 70 Ma (Briceño et al. 1990). The landscape of the Guyana Shield is characterized by a series of planation surface remnants that are displayed in a step-like manner. George (1989; see magnificent illustrations in National Geographic Magazine, May 1989) presented a lively reconstruction of the evolution of the tepuis landscape, accepting a Jurassic age (180 Ma or older) for the summit surface. Schubert et al. (1986) identified six main levels shown in Table 4. In this table, although the ages have been disputed and considered only as tentative, the three older surfaces may be of Mesozoic-Paleogene age, thus forming part of the Gondwana Landscapes.

King (1956a, b) described Brazil planation surfaces and other features, such as inselbergs and bornhardts, which are considered to be formed under prolonged evolution under a seasonally dry to subhumid tropical to subtropical climate. The Brazilian landscape has probably developed continuously since Mesozoic times. João José Bigarella (Fig. 19) and Aziz Ab'Sáber have been the leaders of ancient landscape studies in Brazil in the second half of the twentieth century (see, e.g. Ab'Sáber 1969; Bigarella et al. 1994; Bigarella and Ab'Sáber 1964; and the papers cited there). They described planation surfaces which are essentially coincident with King's viewpoints, considering them as giant pediplains. They were named as Pd1 and Pd2, basically corresponding to the Gondwana and African surfaces. Much more recently, Rossetti (2004) has described five paleoweathering surfaces (laterites and bauxites) in northeastern Amazonia, Brazil, of which the oldest one is considered to be Campanian (Late Cretaceous), whereas the second one is of Paleogene age, corresponding to the Sul-Americana surface of southeastern Brazil, as named by King (1956a, b).

Panario (1988; for a thorough discussion of Gondwana Landscapes in Uruguay, see Panario et al. 2014) described the landscape of the Uruguayan Sierras region, mostly in the Departamento Minas, eastern Uruguay. In an overall very flat country,

Fig. 19 João José Bigarella and the summit planation surfaces in the Paraná Plateau, which he studied and described extensively

the Sierras are the areas with higher relief and potential energy. Some of the rocky ranges have very flat upper surfaces, probably reflecting very old planation processes which are very active in the Cretaceous, whereas other ranges have younger planation surfaces which are of lower elevation and Tertiary age. This set of bedrock hills and planes shows a general SE-NW orientation would have acted as a mountain front that carved pediplains and "glacis" in the ancient shield and which provided most of the sedimentary materials that are infilling the neighbouring accumulation basins. Panario (1988, p. 11) indicated the occurrence of inselbergs and other erosion features in the higher planation surfaces. Some of these ranges show inner tectonic basins where a hilly landscape was developed. The age of this tectonic subsidence is clearly postdating the formation of the higher planation surfaces. At lower elevations, several surfaces formed by pediplanation processes have been identified, which are capped by mineral reddish soils that are interpreted as formed by seasonally wetter, warmer climates. Panario and Gutiérrez (1999) concluded that the extensive planation surfaces of eastern and northern Uruguay are related to down-weathering processes during the Paleogene and particularly, the Eocene.

In Argentina, Gondwana Landscapes are recognized in all cratonic areas (Carignano et al. 1999, p. 249). Landforms of this nature have been observed in (a) the Sierras Pampeanas of Córdoba, San Luis, La Rioja, San Juan and Catamarca; (b) the Central Buenos Aires Positive area, including the Sierras Septentrionales (Tandilia), the Sierras Australes (Ventania) and the Pampa Interserrana (Demoulin et al. 2005); (c) the Sierra Pintada Block in Mendoza; (d) the Sierras de Lihuel Calel in La Pampa; (e) the Northern Patagonian Massif; (f) the Deseado Massif; and (g) the Malvinas-Falkland Islands. The nature and characteristics of the Gondwana

Landscapes in the mentioned areas are described in another paper (Rabassa et al. 2010; see also 2014).

Finally, in the Malvinas-Falkland Islands, a continental fragment which drifted away from the southernmost portion of Africa, Clapperton (1993, p. 543) described smoothly rolling uplands, at an average height of 500–600 m a.s.l., with highest summits around 700 m a.s.l., closely adjusted to underlying structure and lithology, which reflect prolonged evolution by subaerial denudation, as expected in a former portion of Gondwana. These topographic levels have been interpreted as remains of planation surfaces, but their age is still unknown, although they are clearly Triassic or younger.

Discussion and Conclusions

The nature and characteristics of Gondwana Landscapes are clearly related to the principles of the long-term landscape evolution. These paleolandscapes were developed and preserved along the passive margins of the Gondwana continents, such as Africa, South America, Australia and India. The geomorphology of passive margins assumes that these landscapes were formed during very extensive periods of tectonic and climate stability, under what it has been defined in this chapter as "hyper-tropical climates", during at least between the Late Jurassic and the Late Cretaceous, perhaps up to the Santonian (Early Senonian) and then until the Early Eocene. These extreme climates with no analogues in present times were characterized by very high greenhouse content and very high temperatures that forced unheard evaporation rates from the huge, single ocean, leading to extremely high precipitation on the continents. These conditions provided abundant moisture under very high temperatures in continental areas which, along very long stable periods, were responsible for deep chemical weathering over enormous areas that did not occurred anywhere again after the Eocene. The hyper-stable conditions were achieved because the areas where these landscapes were developed corresponded to ancient continents, with very thick crusts and deep roots in the upper mantle. When these continents started to drift due to the rifting processes in the Middle to Late Cretaceous and the South Atlantic Ocean was born, these roots scratched the mantle and generated extensive volcanic eruptions, such as the kimberlitic intrusions dykes and lavas in Southern Africa and Brazil that brought mantle diamonds up to the surface or close to it.

The deep chemical weathering was the main agent in the formation of these Mesozoic paleolandscapes, with weathering fronts reaching to depths of perhaps up to 1,000 m. When climate changed in the latest Cretaceous and then, again, later in the Paleogene, the huge thickness of weathered debris was removed by continuous denudation. The weathered materials, mostly montmorillonite-beidellite-hydromica and kaolinite, were transported by superficial runoff towards the ocean basins, most of which were opened by the rifting process in the Cretaceous, where they were deposited during most of the Tertiary. Where the denudation was complete or almost complete, the ancient weathering front became exposed, and typical landforms and deposits related to its roots are found in the most noteworthy paleolandscapes. Corestones, duricrusts of many different types (ferricretes, silcretes, calcretes), inselbergs, bornhardts, tors and domes are the most relevant landforms present in these paleolandscapes. These landforms are found as landscape elements forming part of planation surfaces, of which the most important are the etchplains, generated by deep chemical weathering and, later, by prolonged denudation. Other planation surfaces, such as pediplains, are found as well. However, in most of the studied cases, it is not possible to apply the concept of "peneplain", in the sense of Davis (1899), because the geomorphological model assumed by this author considered that these landforms were formed by lateral, sideways fluvial erosion as the dominant process, and the paleolandforms described in these regions are instead the result of deep chemical weathering (etchplains).

The observation and description of paleolandscapes formed by the aforementioned processes in all southern hemisphere continents, and even in certain areas of the northern hemisphere, allows suggesting that the landscape of cratonic regions should be reconsidered. More and renewed attention should be given for the interpretation of the genesis of extensive landforms that were formed a very long time ago, under climatic conditions nonexisting today on Earth, in very stable regions, and which were never covered again by marine transgressions, remaining steadily exposed at the atmosphere perhaps during the last 80–100 million years.

These paleolandscapes are very important because they are dominant in cratonic areas all around the world. They covered extensive areas; have very specific hydrological, hydrogeological and pedological characteristics; and, in many areas, are bearing very valuable mineral resources, such as placers of diamonds, gold and other residual minerals and thick kaolinite and bauxite deposits.

Therefore, it is very important to review the geomorphology of the cratonic areas of different parts of the world, and particularly of Argentina, with a "Gondwana vision" that replaces the presently dominant "Andean vision".

Acknowledgements This chapter is dedicated to the memory of Professor Timothy C. Partridge (University of Witwatersrand, Johannesburg, South Africa (see Fig. 17a)) who died unexpectedly in South Africa, a few days after our Paleosurface Symposium was held in September 2009, during the IV Argentine Congress of Geomorphology and Quaternary Studies, in La Plata. In several letters of the last 2 years previous to the cited meeting, Tim had encouraged me to organize such a symposium and to submit this chapter. He had also enthusiastically accepted to be the one of the main reviewers of the chapter forming the present volume, for which his experience, knowledge and support would have been extremely inspiring. His early death has deprived the geomorphology of the Ancient Surfaces of the southern hemisphere of one of its most distinguished scholars. We will certainly miss his much creative work and friendly collaboration.

To Professor João José Bigarella (Paraná State University at Curitiba, Brazil (see Fig. 19)) who firstly introduced myself in 1973 to the ideas about these topics, during a graduate course at the Universidad del Sur, Bahía Blanca, Argentina. However, regrettably, I did not believe then in his revolutionary ideas, since I could have started working on these topics many years before. Professor Bigarella also showed me his fieldwork areas along the Paraná Plateau in 1976.

To Professor Donald R. Coates, and to the memory of Professor Marie Morisawa, for allowing me to develop my Postdoctoral Fulbright Fellowship in the Department of Geology,

State University of New York at Binghamton, in 1975–1976. During such stay, they gave me wide information about the basic concepts of geomorphology and, particularly, they opened my mind to Walther Penck's ideas. Besides, they invited me to participate in the 1975 Binghamton Symposium on geomorphology devoted to "Theories on Landscape Development". The papers presented in such event were gathered in a classic volume (Melhorn and Flemal 1975).

To Professor Rodney Maud (University of Natal, Durban, South Africa [Fig. 17b]) and to the memory of Timothy C. Partridge, for providing me with all concepts about Gondwana Paleolandscapes during my academic stays in South Africa in 1991 and 1995 and thus making a true "believer" of myself. Also, I would like to thank them for their participation in fieldwork in the Pampas and Northern Patagonia in 1995. Rodney Maud also took part to a fieldwork trip to Córdoba and La Rioja provinces in 1999, where he helped us in fully understanding the regional scenarios of ancient landscapes.

To the National Geographic Society (USA), with many thanks for their generous grant for the study of the Argentine Paleolandscapes, between 1993 and 1998.

To Professor Cliff Ollier (University of Western Australia, Perth) for his strong support for our work in Argentina during his visits to Argentina in 1999, 2004 and 2012.

To Professor C. R. Twidale (University of Adelaide, Australia), for his trustful interest in publishing his "Bornhardt" paper in the Argentine Geological Association Review (RAGA), under my editorial responsibility.

To Professors Daniel Panario and Ofelia Gutiérrez (Facultad de Ciencias, Universidad de la República, Uruguay) for generously sharing with us their work in the Minas Department, Eastern Uruguay.

To Professor Alain Demoulin (Liège University, Belgium) for his participation in fieldwork in the Argentine Pampas in 1998 and his valuable support to our work. Professor Demoulin's visit to Argentina was supported by the Belgium Scientific Council and the Comisión de Investigaciones Científicas de la Provincia de Buenos Aires, La Plata.

To Professor Juan Ramón Vidal Romaní (A Coruña University, Spain) for showing me his work on the granite landscapes of Galicia, Spain, and to share with us his very instructive observations in the paleolandscapes of NW Argentina.

To my Argentine colleagues and friends, Professors Marcelo Zárate, Claudio Carignano and Marcela Cioccale for their participation in many fieldwork excursions to different areas of Argentina and sharing with me their ideas, knowledge and experience. Particularly, I would like to thank Claudio and Marcela for introducing me to some of their study areas in Córdoba and La Rioja provinces and explaining to me the nature and characteristics of spectacular, unique features in the Sierras Pampeanas.

Finally, I would like to deeply thank Mr. Ramiro López from CADIC, who did most of the master work to prepare the figures and the redrawing of some of those illustrations taken from older papers.

References

Ab'Sáber AN (1969) Participação das superfícies aplainadas nas paisagens do Rio Grande do Sul, vol 11. Universidade de São Paulo, Instituto de Geografia, São Paulo, pp 1–17

Bigarella JJ, Ab'Sáber AN (1964) Paläeogeographischer und Paläeoklimatische Aspekte des Känozoikums in Südbrasilien. Zeitschrift für Geomorphologie 8(3):286–312

Bigarella JJ, Becker RD, Friedenreich Dos Santos G (1994) Estrutura e Origem das paisagens tropicais e subtropicais. Editora da Universidade Federal de Santa Catarina (UFSC), Florianópolis, 2 vols

Blakey R (2010) Paleogeography. Northern Arizona University, Department of Geological Sciences. www.nau.edu; http://jan.ucc.nau.edu/~rcb7/globaltext2.html

Briceño H, Schubert C, Paolini J (1990) Table-mountain geology and surficial geochemistry: Chimanta Massif, Venezuelan Guyana Shield. J S Am Earth Sci 3(4):179–194
Carignano C, Cioccale M, Rabassa J (1999) Landscape antiquity of the Central Eastern Sierras Pampeanas (Argentina): Geomorphological evolution since Gondwanic times. Zeitschrift für Geomorphologie, NF, Supplement Band 118:245–268. Berlin-Stuttgart
Clapperton CM (1993) Quaternary geology and geomorphology of South America. Elsevier, Amsterdam, 779 pp
Davis WM (1899) The geographical cycle. Geogr J 14:481–504
Demoulin A, Zárate M, Rabassa J (2005) Long-term landscape development: a perspective from the southern Buenos Aires ranges of east central Argentina. J S Am Earth Sci 19:193–204
Du Toit AL (1937) Our wandering continents. A hypothesis of continental drifting. Oliver and Boyd, London
Du Toit AL (1954) The geology of South Africa. Oliver and Boyd, Edinburgh, 611 pp
Du Toit AL, Reed FRC (1927) A geological comparison of South America with South Africa. Carnegie Institution, Washington, DC
Fairbridge RW (ed) (1968) The encyclopedia of geomorphology. Reinhold Book Corporation, New York
Frakes LA (1979) Climates throughout geologic time. Elsevier, Amsterdam, 310 pp
Frakes LA (1986) Mesozoic-Cenozoic climatic history and causes of the glaciations. In: Hsü KJ (ed) Mesozoic and Cenozoic oceans, Publication No. 0131, of the International Lithosphere Program. American Geophysical Union and Geological Society of America, Washington, DC, pp 33–48
George U (1989) Venezuela's islands in time. Natl Geogr Mag 175(5):526–561. May 1989 National Geographic Society, Washington, DC
Geramisov IP (1970) Three major cycles in the history of a geomorphological stage of the development of the Earth. Geomorphology 1(1):12–17
Gilbert GK (1877) Report on the geology of the Henry Mountains. In: U.S. Geographical and Geological Survey of the Rocky Mountains region. Government Printing Office, Washington, DC
Gorelov SK, Drenev NV, Meschcheryakov YA, Tikanov NA, Fridland VM (1970) Planation surfaces of the USSR. Geomorphology 1(1):18–29
Hesselbo SP, Gröckel DR, Jenkyns HC, Bjerrum CJ, Farrimond P, Morgans Bell HS, Green OR (2000) Massive dissociation of gas hydrate during a Jurassic oceanic anoxic event. Nature 406:392–396
IPCC (2007) Fourth Assessment Report: Climate Change 2007. The physical science basis. 6.3. Pre-Quaternary climates. www.ipcc.ch/publications_and_data/ar4/wg1
King LC (1949) A theory of bornhardts. Geogr J 112:83–87
King LC (1950) The study of the world's plainlands: a new approach to geomorphology. Geol Soc Lond Q J 106(1):101–131
King LC (1953) Canons of landscape evolution. Geol Soc Am Bull 64:721–752
King LC (1956a) A geomorphological comparison between Brazil and South Africa. Q J Geol Soc Lond 112:445–474
King LC (1956b) A geomorfologia do Brasil oriental. Revista Brasileira de Geografia 18:186–263
King LC (1962) The morphology of the earth, 1st edn. Oliver and Boyd, Edinburgh, 699 pp
King LC (1963) South African scenery. 3rd edn. rev. Oliver & Boyd, Edinburgh, 308 pp
King LC (1967) The morphology of the earth, 2nd edn. Oliver and Boyd, Edinburgh
Leopold L, Wolman MG, Miller JP (1964) Fluvial processes in geomorphology. W.H. Freeman and Co., San Francisco, 522 pp
Lidmar-Bergsson K (1988) Exhumed cretaceous landforms in south Sweden. Zeitschrift für Geomorphologie, Supplement Band 72:21–40. Berlin-Stuttgart
Linton DL (1955) The problem of tors. Geogr J 121:470
Lisitzin AP (1972) Sedimentation in the world ocean: with emphasis on the nature, distribution and behavior of marine suspensions, Special publication 17. Society of Economic Paleontologists and Mineralogists, Tulsa, p, 218 pp

Melhorn WN, Edgar DE (1975) The case for episodic, continental scale erosion surfaces: a tentative geodynamic model. In: Melhorn WN, Flemal RC (eds) Theories of landform development, Publications in geomorphology. State University of New York at Binghamton/Allen & Unwin, Binghamton/New York/London, 306 pp

Melhorn WN, Flemal RC (eds) (1975) Theories of landform development, Publications in geomorphology. State University of New York at Binghamton/Allen & Unwin, Binghamton/New York/London, 306 pp

Mountain ED (1968) Geology of Southern Africa. Books of Africa, Cape Town, 249 pp

Ollier C (1960) The inselbergs of Uganda. Zeitschrift fur Geomorphologie, NF 4:43–52

Ollier C (1984) Weathering, 2nd edn. Longman, Essex

Ollier C (1990) Weathering and landforms. Macmillan, London

Ollier C (1991a) Ancient landscapes. Belhaven Press, London/New York, 233 pp

Ollier C (1991b) Laterite profiles, ferricrete and landscape evolution. Zeitschrift für Geomorphologie, NF 35(2):165–173

Ollier C (1995) New concepts of laterite formation. Geol Soc India Memoir 32:309–323

Ollier C (2004) Mountain building and climate: mechanism and timing. Geogr Fis Dinam Quat 27:139–149

Ollier C, Galloway RW (1990) The laterite profile, ferricrete and unconformity. Catena 17:97–109

Ollier C, Harrop J (1959) The soils of eastern province of Uganda. Memoirs of the Research Division, Soil series No 2. Department of Agriculture, Government of Uganda

Ollier C, Pain C (2000) The origin of mountains. Routledge, London/New York, 345 pp

Ollier C, Lawrence CJ, Beckett PHT, Webster R (1969) Terrain classification and data storage, land use systems of Uganda, M.E.X.E. report no 959. University of Oxford, Christchurch

Panario D (1988) Geomorfología del Uruguay. Avances de Investigación, Facultad de Humanidades y Ciencias, Universidad de la República, Montevideo, 32 pp

Panario D, Gutiérrez O (1999) The continental Uruguayan Cenozoic: an overview. Quatern Int 62:75–84

Panario D, Gutiérrez O, Sánchez Bettucci L, Peel E, Oyhantçabal P, Rabassa J (2014) Ancient landscapes of Uruguay. In: Rabassa J, Ollier C (eds) Gondwana landscapes in southern South America. Springer, Dordrecht, pp 161–199

Partridge T (1998) Of diamonds, dinosaurs and diastrophism: 150 million years of landscape evolution in southern Africa. S Afr J Geol 101:167–184

Partridge T, Maud R (1987) Geomorphic evolution of Southern Africa since the Mesozoic. S Afr J Geol 90:179–208

Partridge T, Maud R (1989) The end-Cretaceous event: new evidence from the southern hemisphere. S Afr J Sci 85:428–430

Partridge T, Maud R (2000) Macro-scale geomorphic evolution of Southern Africa. In: Partridge T, Maud R (eds) The Cenozoic of Southern Africa. Oxford University Press, Oxford/New York, 406 pp

Penck W (1924) Die Morphologische Analyse, vol 2, Geographisches Abh., ser. 2. J. Engelhron, Stuttgart, 283 pp

Penck W (1953) Morphological analysis of landforms. A contribution to physical geology. The Macmillan Co., London, 429 pp

Rabassa J (2006) Some ideas and notes on the ancient surfaces of Midwest U.S.A. Illinois State Geological Survey, University of Illinois at Urbana-Champaign, unpublished internal report, 10 pp

Rabassa J (2010) Gondwana paleolandscapes: long-term landscape evolution, genesis, distribution, and age. Geociências 29(4):541–570. São Paulo, UNESP, Brazil

Rabassa J, Carignano C, Cioccale M (2010) Gondwanic Paleosurfaces in Argentina: an introduction. Geociências 29(4):439–466, São Paulo, UNESP, Brazil

Rabassa J, Carignano C, Cioccale M (2014) A general overview of Gondwana landscapes in Argentina. In: Rabassa J, Ollier C (eds) Gondwana landscapes in southern South America. Springer, Dordrecht, pp 201–245

Rosetti DF (2004) Paleosurfaces from Northeastern Amazonia as a key for reconstructing paleolandscapes and understanding weathering products. Sediment Geol 169:151–174
Schubert C, Huber O (1990) The Grand Sabana. Langoven, Caracas
Schubert C, Briceño H, Fritz P (1986) Paleo-environmental aspects of the Caroni-Paragua river basin (southeastern Venezuela). Interciencia 11:278–289
Small R (1978) The study of landforms. A textbook of geomorphology, 2nd edn. Cambridge University Press, Cambridge, 502 pp
Strakhov NM (1967) Principles of lithogenesis. Oliver and Boyd, Edinburgh
Summerfield MA, Thomas MF (1987) Long-term landform development: editorial introduction. In: Gardiner V (ed) International geomorphology 1986, Part. II. Wiley, Chichester, pp 927–933
Thomas M (1965) Some aspects of the geomorphology of domes and tors in Nigeria. Zeitschrift für Geomorphologie, NF 5:37–52
Thomas M (1968) Bornhardt. In: Fairbridge RW (ed) Encyclopedia of geomorphology. Reinhold, New York
Thornbury WC (1954) Principles of geomorphology. Wiley, New York, 618 pp
Twidale CR (1968) Inselbergs. In: Fairbridge RW (ed) Encyclopedia of geomorphology. Reinhold, New York
Twidale CR (1982) Granite landforms. Elsevier, Amsterdam, 371 pp
Twidale CR (2007a) Ancient Australian landscapes. Rosenberg Publishing Co., Dural, 144 pp
Twidale CR (2007b) Bornhardts and associated fracture patterns. Revista Asociación Geológica Argentina 62(1):139–153
Uriarte Cantolla A (2003) Historia del clima de la Tierra. Servicio Central de Publicaciones del Gobierno Vasco, Vitoria-Gasteiz, 306 pp
Vidal Romaní JR, Twidale CR (1998) Formas y paisajes graníticos. Universidade da Coruña, Servicio de Publicacións, Monografías No 55. A Coruña, Spain, 411 pp
von Engeln OD (1948) Geomorphology. The Macmillan Company, New York, 655 pp
Wayland EJ (1933) Peneplains and some other erosional platforms. Annual Report and Bulletin, Protectorate of Uganda, Geological Survey Department, Notes, 1, 74, pp 376–377
Wegener A (1924) The origin of continents and oceans. Methuen, London, 212 pp
Zárate M, Rabassa J (2005) Geomorfología de la Provincia de Buenos Aires. In: XVI Congreso Geológico Argentino, Relatorio de la Geología y Recursos Minerales de la Provincia de Buenos Aires, La Plata, pp 119–138
Zárate M, Rabassa J, Partridge TC, Maud R (1995) La Brecha Cerro Colorado, es miocena? IV Jornadas Geológicas Bonaerenses, Actas, 1, Junín, pp 159–168
Zárate M, Rabassa J, Maud R, Partridge TC (1998) La silicificación de la Brecha Cerro Colorado: clasificación, génesis e implicancias ambientales. V Jornadas Geológicas Bonaerenses, Actas, Mar del Plata, pp 165–173

Some Principles in the Study of Planation Surfaces

Cliff Ollier

Abstract Despite the efforts of some workers to deny their very existence, planation surfaces are real and worthy of study. Drainage patterns provide insights into their age and deformation. They provide valuable information on the evolution of passive continental margins. Their frequent thick cover of weathering products tells us of past hydrology and climate. Several different mechanisms can cause planation, such as pedimentation and relief inversion and so lead to valuable process studies. Successive planation surfaces may indicate tectonic or climatic histories. Above all, they show that landscape formation and regolith history are on the same timescale as global tectonics and biological evolution.

Keywords Gondwana • Planation surfaces • Africa • Australia • India

Classical Geomorphology

This term is used, sometimes pejoratively, to refer to the work of geomorphologists such as Davis (1899), Cotton (1918) and King (1953). Although they differed substantially in detail, these workers saw geomorphology as a history of landscape, based on a model of erosion leading to planation, followed by new erosion phases after tectonic uplift. They argued about the processes involved in forming plains, the nature of slope retreat and the age of landscape. Some modern workers seem to be at pains to separate themselves from these ideas. 'Davis-bashing' was a theme amongst some geographers such as Chorley (1965) and Summerfield (2000), and in matters relating to the Gondwana continents, King in particular was attacked, especially by fission-track workers, for his landscape models. Brown et al. (2000,

C. Ollier (✉)
School of Earth and Environment, University of Western Australia, Perth, Australia
e-mail: cliff.ollier@uwa.edu.au

J. Rabassa and C. Ollier (eds.), *Gondwana Landscapes in southern South America*, Springer Earth System Sciences, DOI 10.1007/978-94-007-7702-6_3,

p. 274), for instance, wrote, 'Our assessment is that the approach advocated by King and adopted by other researchers is founded on unverified assumptions and therefore does not provide a viable basis for assessing the relationships between tectonics and macroscale landscape development'. Another example is provided by Fleming et al. (1999), who did cosmogenic dating in the Drakensberg region, and claimed that summit lowering is sufficient to prevent the long-term intact survival of erosion cycle surfaces formed in the Mesozoic that were previously inferred for this region by King and many other field observers. Sometimes it seems these writers are not really familiar with King's views and achievements. In fact King described orogenic axes (he invented the redundant word 'cymatogeny'), he presented a cross section of Africa similar to those of later workers, he described the evolution of the Natal Monocline in detail and he discussed the complications of isostatic adjustment (King 1955).

Nevertheless, landscapes do consist of plains, separated at times by steeper country. Anyone working on Great Escarpments must admit that the escarpment is steep, compared with the surface inland from it, which is usually a plain of some sort with local names such as Tableland, Plateau, Highveld, etc. The origin of plains and escarpments is still a valid topic for research today as in the 'classical' past.

Rivers and Drainage Patterns

On some continental margins, the major rivers were in existence before the break-up of Gondwana and can still be traced in the modern landscape, as in Australia and South Africa. It is often found that minor drainage is structurally controlled but major drainage is not and so was probably initiated on a former erosion surface or sedimentary cover. Drainage may be reversed (indicated by barbed drainage) or modified by river capture (indicated by elbows of capture and wind gaps) or evolve in more complex ways (Ollier and Pain 2000, Ch. 11). It is necessary to distinguish antecedent from superimposed rivers. Rivers can never bore their way through an uplifted tectonic ridge as suggested by Burke and Gunnell (2008). New planation surfaces start when a river grades to a new base level, and planation surfaces extend up river by erosion, increasing in area as older surfaces are eroded away. The study of the evolution of rivers is a vital component of the study of planation surfaces.

Uplift or Downwarp of Continental Margins

When new continental margins are formed, they might remain level, be downwarped or be uplifted. The downwarp model is championed by Ollier and Pain (1997). Once a new margin is formed, sediment starts to be deposited offshore. In the downwarp model the unconformity beneath the new sediment is equated with the palaeoplain on the continent, and the oldest sediment gives a minimum age of the break-up. This seems to work in many margins including the Appalachians, Southern Africa and eastern Australia.

A further feature of the downwarp model is that remnants of the downwarped surface can still be seen as triangular facets of sloping planation surface (Fig. 1). South of Bega on the New South Wales (NSW) coast, there is a superb coastal 'facet' of old downwarped surface with deep and intense weathering suggesting Mesozoic age (Kubiniok 1988). The downwarp is also shown by generalised contours in relation to drainage, as in the previous example.

Downwarp may also be shown by distinctive strata or unconformities. In southern Brazil the base of 120 Ma old basalts is at 300 m inland on the Great Escarpment and is warped down to the present coast.

Regolith and Saprolite

Deep weathering is an important feature of old erosion surfaces, often described very badly by people who do not know the subject or its vocabulary. Here, I shall try to summarise the main concepts and technical terms.

Weathered rock often remains in place with preservation of many rock structures indicating no change of volume. Weathered rock in place is called saprolite. It is commonly more than 100 m thick, but we are usually seeing only remnants of former, deeper profiles. To accumulate such a thickness of saprolite, weathering must have been occurring faster than weathering products were removed, which usually indicates low relief in the landscape. Characteristics of the regolith do not necessarily reflect climate because at depths of tens of metres, the temperature is controlled by geothermal heat more than by surface temperature, and groundwater can be present at depth even in arid regions, as in deserts today.

The upper part of weathered profiles is usually disturbed by soil-forming processes. The term 'regolith' describes all the unconsolidated surficial materials, including saprolite, disturbed layers, soil and even surficial sediments: it is the material 'between fresh rock and fresh air'. The base of weathering is irregular and penetrates along joints and other fissures, isolating corestones of less-weathered rock. The junction between weathered and fresh rock is called the weathering front and is often very sharp.

The original depth of weathering can be very great, sometimes over a kilometre. This is a constant surprise to geomorphologists, explorers and engineers, and Ollier (2010) wrote of landslides in deeply weathered terrain: 'perhaps the greatest hazard in deep weathering is the failure to recognize it'.

When erosion eventually sets in, the saprolite starts to be removed. Large blocks of unweathered rock (corestones) tend to be left behind and make distinctive landscapes, especially on granite. It is rare for erosion to remove all the saprolite and some is preserved in pockets. Some workers think there was one major period

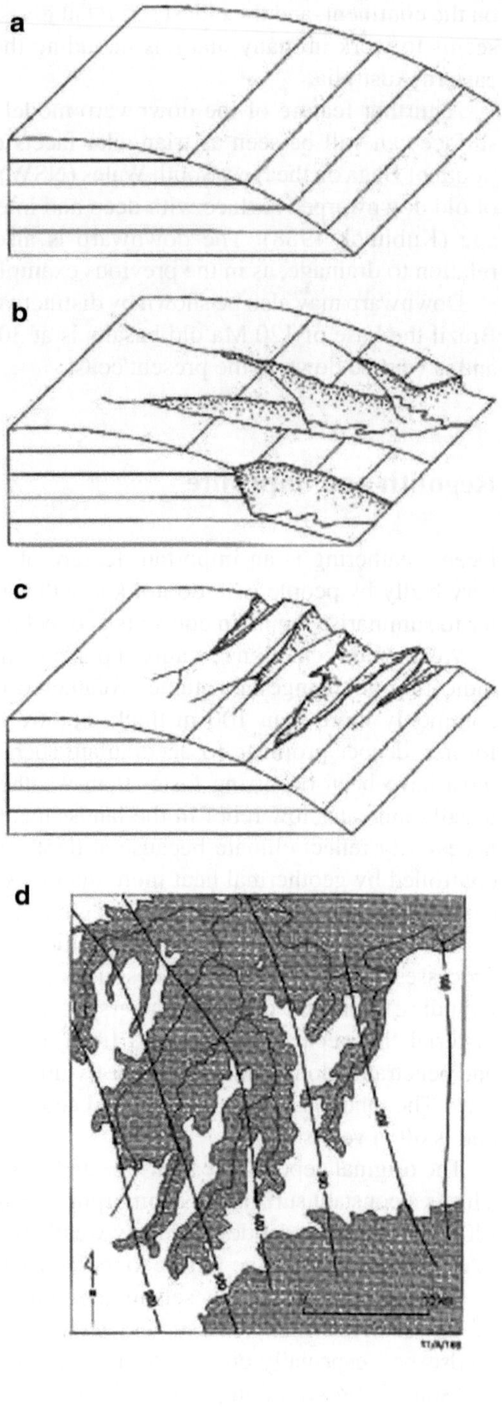

Fig. 1 Generalised contours and the evolution of the Lower Shoalhaven region, SE Australia. (**a**) Contours on a downwarped surface. (**b**) A downwarped surface dissected by valleys. Contours on the uneroded interfluves permit reconstruction of the old surface. This is the pattern expected if rivers were initiated on a simple downwarped surface. (**c**) A representation of the downwarped surface in the Lower Shoalhaven area, with valleys almost parallel to the contours of the old surface. This pattern could occur only if the valleys were incised before the downwarp, that is, they are antecedent. (**d**) A map of the gorges and generalised contours in the Lower Shoalhaven area. Note the similarity between this and **c**

of deep weathering in the Mesozoic, followed by a more or less continuous phase of stripping: others believe there were alternating periods of weathering and stripping. Alternating weathering and stripping has been likened to the process of etching, and some think that the process leads to the formation of ‘etchplains’.

Deep Weathering and Water Tables

There is much misunderstanding of this topic in the literature, so a simple review is presented here. For simplicity consider a nonporous rock such as granite. At the very base of the weathering profile is the weathering front, over fresh rock. Fresh rock does not hold groundwater, so this is the groundwater base. Above this is weathered rock and sediment, which holds some groundwater. The top of the groundwater is the water table. It is quite surprising how many authors confuse groundwater base and the water table (some examples are presented by Ollier and Pain (1996)), but they are totally distinct.

The water table separates a saturated zone from an unsaturated or aerated zone. In general, there will be reduction below the water table and oxidation above it. The upper zone is usually red or brown, the lower part grey, green or white. A zone of water table fluctuation may be mottled.

In the higher parts of many profiles, there is a soil at the surface with its own special soil-water features. In general, soil is a barrier to groundwater recharge, but will not be considered further here.

In areas of deep weathering or deep sediments, there is no reason why the water table should coincide with any soil horizons. The boundary between saprolite and disturbed zone likewise does not correspond with the water table. The soil profile nomenclature of A, B and C horizons should not be applied to weathering profiles.

Evidence for very deep weathering and chemical migration comes from many economic ore deposits. Basically, there are oxide ores above, and sulphide ores below, the junction marking a former water table that separated an oxidised zone above from a reduced zone below. Both oxidised and sulphide zones can be several hundred metres thick (Ollier and Pain 1996; Ollier 2010).

Hydrologists are well aware that lateral movement of groundwater is much more significant than vertical movement. In contrast chemists and geologists not trained in surface processes tend to deal with variation down profiles, assuming or implying that vertical movements are dominant as in a test tube. The groundwater carries dissolved minerals and is significant in making several landscape features. In regional landscapes as a whole, chemical partition is reflected in the pedological or geochemical zones: pedalfers with no carbonate in the soil profile are the soil types in well-watered areas; pedocals with carbonate in the profile occur in drier areas; soils with gypsum occur in still more arid regions and chlorides occur in the soils of the most arid areas. At a much smaller scale, solutions are carried downslope from higher ground. It is of particular significance that iron is transported in solution to valley edges where it may be precipitated as ferricrete – commonly called ‘laterite’.

Inversion of Relief

Inversion of relief is an important process in landscape evolution (Pain and Ollier 1995). Just two aspects will be treated here: the inversions associated with volcanoes and with ferricrete. Lava flows sometimes travel for many tens of kilometres, burying old alluvium. The water that originally flowed along the valley bottom is displaced to the edge of the flow, where it cuts down lateral streams and brings about inversion of relief (Fig. 2). This landscape system might seem unlikely to those unfamiliar with volcanoes and lava flows, but it has been verified over hundreds of kilometres where the alluvium contained gold (deposits known as deep leads).

Hard, iron-rich materials in sediments and weathering profiles are called ferricrete. Another word often used is 'laterite', but the definition of this is so confused that the term is better avoided. Iron is a mobile element if in solution as ferrous iron, and in combination with some organic compounds (chelation), but once it is oxidised, it is usually fixed. Many hypotheses of ferricrete formation are based on moving iron in solution to sites where it is irreversibly precipitated.

Fig. 2 Inversion of relief. *Top*. The pre-volcanic landscape with alluvium in the valley floor. *Middle*. Lava flows down the valley and covers the alluvium. *Bottom*. Streams form on both sides of the lava flow (twin lateral stream) and eventually erode new valleys leaving the former valley bottom as a ridge

Some hypotheses of ferricrete formation are based on its association with underlying weathered materials that are depleted in iron. Since the lower material is iron-poor and the upper material iron-rich, it seems reasonable to assume that the iron has somehow moved upwards in solution, aided by groundwater fluctuation, capillary action or other methods.

Another series of hypotheses suggest that lateral movement of iron is important and that it is mostly precipitated on lower slopes and in valley bottoms. There is much evidence that the ferricretes were formed in valley floors. The distribution of ferricrete in plan may resemble a drainage pattern, and there may be associated alluvial gravels, although they now cap interfluves.

Detailed observation of 'laterite' profiles often shows an unconformity between the ferricrete and the underlying saprolite, and they were not formed at the same time (Ollier and Galloway 1990). A vertical sequence of samples on such 'laterite' profiles can be quite misleading and does not show that the profile was created by upward or downward movement of iron. Many ferricretes have been dated by palaeomagnetism, yielding ages that span the Tertiary and extend into the Cretaceous (Schmidt and Ollier 1988). Nodular and concretionary ferricrete is usually unsuitable, but underlying red clays often give good results.

Climates of the Past

Climate change is on the same timescale as the fragmentation of Gondwana. The fragments would experience climate change along with changes in location and heights and the evolution of sea currents and atmospheric circulation. The disposition of the continents also affects global climate in general. Perhaps the most important change was the separation of Australia from Antarctica, which allowed the formation of the Circumpolar Current. The consequent isolation of Antarctica promoted low temperatures and enabled the build-up of an ice cap starting over 30 Ma. Glaciation in the northern hemisphere only started about 3 Ma. In palaeoclimatic studies, even the association of temperature with latitude cannot be assumed.

In the Eocene, the climate on Earth was warm and wet with little regional variation (Walker and Sloan 1992). The poles were warmer than they are today whilst the tropics were no warmer than they are today. This presents a puzzle. An enhanced greenhouse effect could warm the poles by 10 or 15 °C, but it would also warm the tropics by about as much. The problem is the marked reduction in the Eocene of the difference between tropical and polar temperatures. The Mesozoic was a similar globally warm period with a few minor exceptions. Sometime during the Cenozoic the equable climate was replaced by a more heterogeneous climate, with colder and drier areas, and in the past 3 or 4 million years, the Earth grew so cold that ice ages occurred.

The Planation of Dipping Strata and Significance of Bevelled Cuestas

In areas of folded or tilted strata, the hard rocks form ridges and the soft rocks form valleys. If the ridges tend to reach the same elevation over a wide area, it might suggest that the fold had been planated before differential erosion brought out the hills and valleys. On a simply dissected fold, a wider range of elevations would be expected. However, there is no agreement on how much variation in elevation might be attained by either process, so simple summit accordance is never more than suggestive. Some have used arguments based on the down cutting of regularly spaced rivers to deny the existence of former planation surfaces, and examples will be given later. The evidence for former planation surfaces must be considered in three dimensions. For reconstruction of old surfaces in folded sedimentary rocks, if the strike sections of ridges show accordant summits, then there is good reason to think there was a planation surface not much higher, even if bevelled cuestas are absent.

But very positive evidence that some ridges are indeed remains of a planation surface comes from bevelled cuestas. If a dipping bed of hard rock is simply eroded, it develops a dip slope and a scarp slope (Fig. 3a), with a rather sharp ridge (a cuesta) where these two slopes intersect. If the crest of this ridge has a flat top (Fig. 3b), that is to say it is bevelled, there is no doubt it is the remnant of a former erosion surface, for the rock structure would never develop a flat top unless a lateral erosion process was working at a particular base level. If there are a lot of bevelled cuestas, all at about the same elevation, then a certain confidence may be placed in the former existence of a broad planation surface. The bevelled cuestas follow the outcrops of the hard strata in plan (Fig. 4).

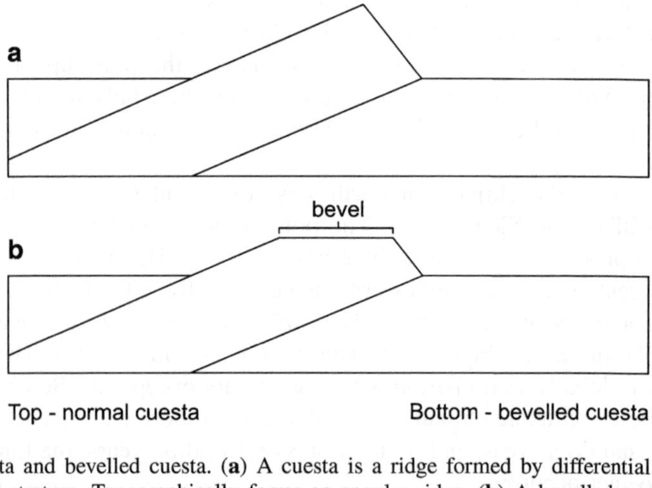

Fig. 3 Cuesta and bevelled cuesta. (**a**) A cuesta is a ridge formed by differential erosion of a dipping hard stratum. Topographically forms an angular ridge. (**b**) A bevelled cuesta has a flat bevel on the top that is a sure sign that an upper planation surface at about base level existed before differential erosion created the cuestas

Fig. 4 The Carr Boyd Ranges, Western Australia, seen from the air, showing bevelled cuestas that can be traced around the nose of the plunging anticline

The Carr Boyd bevelled cuestas (NW Australia) shown in Fig. 4 are clear indications of an old planation surface, for there is no mechanism to erode flat surfaces across the hardest rocks *after* uplift and incision of valleys. So the folding of the rocks is older than the erosion surface (and incidentally the folding did not make the mountains, which is a common misconception). Uplift of a planation surface to form a plateau made the area one of high relief, and subsequent erosion made the Carr Boyd Ranges. This simple story is repeated in a great many of the world's plateaus and ranges. The only unusual thing about the Kimberley – Carr Boyd region – is the great age of the planation surface. The ancient Kimberley Plateau had already formed and been partly dissected by the time of a Proterozoic Glaciation and has been a land area ever since (Ollier et al. 1988).

Some argue that as rivers cut down the interfluves are also lowered, and numerous 'phantom' planation surfaces may be revealed by accordant summits. This idea is shown in Fig. 5. A classic example comes from the northern hemisphere, from the Appalachians. It was put forward by Shaler (1899) and championed by Summerfield (2000, p. 8) in what appears to be another attempt to discredit the existence of planation surfaces. He wrote, 'He [Shaler] observed how the rivers of Kentucky exhibited highly regular spacing ... He went on to argue that regularly spaced streams could well explain "the origin of the coincidences in mountain crests which is so generally held to indicate the existence of ancient base levels of erosion which have been lifted to a height above the level of the sea and then dissected by rivers". He argued instead for "the emergence of an integrated stream network

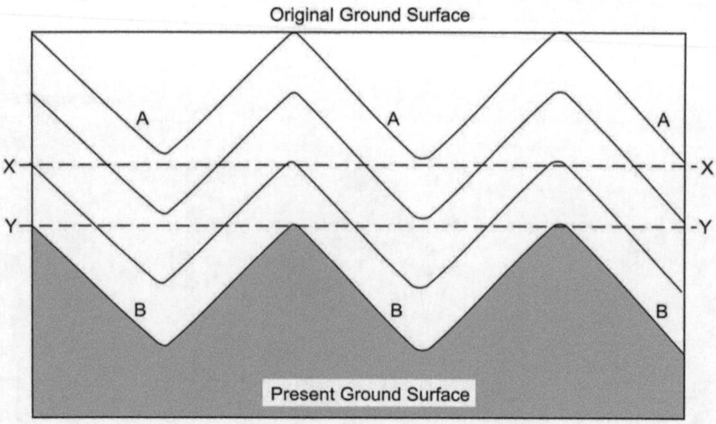

Fig. 5 Interpretation of accordant ridges. A dissected planation surface may be detected at stage A by the accordant summit levels. Without other evidence this is a risky method because at later stages of erosion, such as stage B, the approximate summit accordance may suggest the existence of a planation surface, as may stages X and Y, but such 'phantom planation surfaces' never existed in reality. In three dimensions the problem is not as difficult as the simple sections suggest, and with the assistance of correlative sediments and regolith, real planation surfaces can often be detected even after considerable erosion

pattern, whose regularly spaced components would show a tendency to be separated by divides with accordant summit levels"'. Summerfield concludes 'The clear implication of this simple geometric argument, of valley spacing and valley-side slope together determining valley depth and therefore drainage divide elevation, was that the broadly accordant summit levels in areas such as the Appalachians and the Alps could have been formed through normal processes of fluvial erosion and in the absence of an uplifted peneplain acting as a reference surface'.

Yet, the ridges of the Appalachians are beautiful-bevelled cuestas, and they vary in height by only tens of metres over many tens of kilometres along strike – the clearest planation surface you could find. I must stress that the theoretical argument of Shaler and Summerfield is utterly false, and the field evidence of bevelled cuestas, or bevelled rocks in general, is positive proof of erosion to a base level.

The Kimberley planation surface (NW Australia) is similar to the Appalachians in principle, but the bevels are on very gently dipping rocks so the bevels preserve a much larger proportion of the original surface. It is generally difficult to demonstrate planation on horizontal strata, but in areas of gently dipping rocks, the bevelling may make it clear.

This is the place to mention another approach, namely, to investigate planation surfaces by numerical modelling, such as that used by Römer (2008) in Brazil. Numerical models cannot detect bevelled cuestas, which are necessary (along with lithology, regolith and other important factors) to decipher the landscape. Numerical modelling may assist field observations, but it can never replace them.

We should also point out that planation surfaces also cut across dipping strata in mountainous areas. This is especially evident in the Andes, where the high, flat puna or altiplano surfaces are so evident. A detailed account of the planation surfaces of Ecuador and their significance is provided by Coltorti and Ollier (1999), but the significance has been recognised for a long time, as by Bowman (1916). Many more details on the Andes are provided in Chapter 6 of Ollier and Pain (2000), but this book stresses that many mountains, all over the world, are more or less dissected plateaus, uplifted planation surfaces.

Fission Track Evidence

Workers on fission-track data often come up with hypotheses at odds with the geomorphology and geology. For basic details, Gleadow and Brown (2000) have provided an excellent account of the principles, techniques and variations within the method. They mention a few of the problems but are basically optimistic. Fielding (2000) gives a better account of the problems and writes:

> Mineral cooling ages can then be combined with assumptions (or measurement) of the geothermal gradient at those dates to estimate at what depth below the surface the samples were located at those times ... Unfortunately, high rates of denudation, high local relief, or underthrusting of colder rocks can strongly affect or even temporarily reverse the normal geothermal gradient, and these effects can greatly complicate data interpretation.

An alternative hypothesis to coastal downwarp comes from researchers working on fission-track data. In eastern Australia, for instance, Moore et al. (1986) suggest Late Cretaceous uplift (not downwarp) to the east, so the coastal lowlands must be made by later erosion, removing several kilometres of rock. Bishop and Goldrick (2000) summarise the major issues of current debate. Did escarpment retreat involve dissection of an uplifted plateau or a downwarped rift shoulder? They wrote (p. 234), 'The 100 Ma old lavas and Mesozoic weathering profiles on the NSW coast are inconsistent with deep denudation having occurred along and across the whole of the coastal strip below the escarpment ...'.

On the coast, Brown (2000) reports basalts 27–30 Ma old and Oligocene to Early Miocene sediments. This is precisely where fission-track work and Apatite (UTh)/He age constraints (Persano et al. 2002) indicate kilometres of erosion at the coast, which should surely strip these materials. Similarly in north Queensland, one interpretation of fission-track data is erosion of 0.8–3.0 km of overlying rock (Marshallsea et al. 2000), but the geology of the region 'suggests that an explanation in terms of deeper burial may be untenable'. Western India is another place where massive uplift is proposed (Gunnell et al. 2003), and another example is from western South Africa (Brown et al. 2000). Similarly Japsen et al. (2012), on the basis of fission tracks, claim that the elevated passive margins are not rift shoulders but expressions of episodic, post-rift burial and exhumation.

The big problem is that to get such uplift requires a huge fault just offshore, and no such faults have been found, although they should be very obvious.

Final Remarks: Precautions in the Study of Ancient Landscapes

Taking into consideration the previously cited aspects and circumstances, it should be taken into account that in the study of planation surfaces, the following points should be borne in mind:

1. Landscape and regolith history is on the same timescale as global tectonics and biological evolution.
2. The landscape is not in equilibrium with present-day climate, hydrology or biota, but includes many historical relics.
3. Inorganic chemistry alone will not provide answers to geochemical problems. The biological component is essential.
4. In geochemical studies, it is vital to remember lateral movement of solutions and solid materials and avoid the simple vertical approach.
5. The present is not the key to the past. Landscapes and processes in the past were very different from those of today. An understanding of the past helps to understand present conditions.
6. The past must be determined from consistent geological, geomorphic and biological evidence and not derived from theory.
7. Many aspects of past climate and hydrology are so different from those of today that they are beyond modelling.

References

Bishop P, Goldrick G (2000) Geomorphological evolution of the East Australian continental margin. In: Summerfield MA (ed) Geomorphology and global tectonics. Wiley, Chichester

Bowman I (1916) The Andes of southern Peru. Henry Holt and Co., London

Brown MC (2000) Cenozoic tectonics and landform evolution of the coast and adjacent highlands of southeast New South Wales. Aust J Earth Sci 47:245–257

Brown RW, Gallagher K, Gleadow AJW, Summerfield MA (2000) Morphotectonic evolution of the South Atlantic margins of Africa and South America. In: Summerfield MA (ed) Geomorphology and global tectonics. Wiley, Chichester, pp 255–281

Burke K, Gunnell Y (2008) The African erosion surface. Geological Society of America, Memoir 201, Boulder, Colorado, 66 pp

Chorley RJ (1965) A re-evaluation of the geomorphic system of W.M. Davis. In: Chorley RJ, Huggett P (eds) Frontiers in geographical teaching. Methuen, London, pp 21–38

Coltorti M, Ollier CD (1999) The significance of high planation surface in the Andes of Ecuador. In: Smith BJ, Whalley WB, Warke PA (eds) Uplift, erosion and stability: perspectives on long-term landscape development. Geological Society, London, Special Publications, 162. The Geological Society of London, pp 239–253

Cotton CA (1918) Landscape as developed by the processes of normal erosion. Cambridge University Press, Cambridge

Davis WM (1899) The geographical cycle. Geogr J 14:481–504

Fielding EJ (2000) Morphotectonic evolution of the Himalayas and Tibetan Plateau. In: Summerfield MA (ed) Geomorphology and global tectonics. Wiley, Chichester, pp 201–222

Fleming A, Summerfield MA, Stone JO, Fifield LK, Cresswell RG (1999) Denudation rates for the southern Drakensberg escarpment, SE Africa, derived from in-situ produced cosmogenic ^{16}Cl: initial results. J Geol Soc Lond 156:209–212

Gleadow AJW, Brown RW (2000) Fission-track thermochronology and the long-term denudational response to tectonics. In: Summerfield MA (ed) Geomorphology and global tectonics. Wiley, Chichester, pp 57–75

Gunnell Y, Gallagher K, Carter A, Widdowson A, Hurford AJ (2003) Denudation history of the continental margin of western peninsular India since the early Mesozoic –reconciling apatite-fission track data with geomorphology. Earth Planet Sci Lett 215:187–201

Japsen P, Chalmers JA, Green PF, Bonow JM (2012) Elevated, passive continental margins: not rift shoulders, but expressions of episodic, post-rift burial and exhumation. Glob Planet Change 90–91:73–86

King LC (1953) Canons of landscape evolution. Geol Soc Am Bull 64:721–752

King LC (1955) Pediplanation and isostasy: an example from South Africa. J Geol Soc Lond 111:353–354

Kubiniok J (1988) Kristalinvergrusung an Bespielen aus Sudaustralien und Deutchen Mittelbergen. Kölner Geographische Arbaiten 48, University of Cologne

Marshallsea SJ, Green PF, Webb J (2000) Thermal history of the Hodgkinson Province and Laura Basin, Queensland: multiple cooling episodes identified from apatite fission track analysis and vitrinite reflectance data. Aust J Earth Sci 47:779–797

Moore ME, Gleadow AJW, Lovering JF (1986) Thermal evolution of rifted continental margins: new evidence from fission tracks in basement apatites from southeastern Australia. Earth Planet Sci Lett 78:255–270

Ollier CD (2010) Very deep weathering and related landslides. In: Calcaterra D, Parise M (eds) Weathering as a predisposing factor to slope movements, Geological Society engineering geology special publication no 23. Geological Society, London, pp 5–14

Ollier CD, Galloway RW (1990) The laterite profile, ferricrete and unconformity. Catena 17:97–109

Ollier CD, Pain CF (1996) Regolith, soils and landforms. Wiley, Chichester, 316 pp

Ollier CD, Pain CF (1997) Equating the basal unconformity with the palaeoplain: a model for passive margins. Geomorphology 19:1–15

Ollier CD, Pain CF (2000) The origin of mountains. Routledge, London

Ollier CD, Gaunt GFM, Jurkowski I (1988) The Kimberley Plateau, Western Australia: a Precambrian erosion surface. Zeitschrift fur Geomorphologie, NF 32:239–246

Pain CF, Ollier CD (1995) Inversion of relief – a component of landscape evolution. Geomorphology 12:151–165

Persano C, Stuart FM, Bishop P, Barfod DN (2002) Apatite (U-Th)/He age constraints on the development of the Great Escarpment on the southeastern Australian passive margin. Earth Planet Sci Lett 200:79–90

Römer W (2008) Accordant summit heights, summit levels and the origin of the "upper denudation level" in the Serra do Mar (SE-Brazil, São Paulo): a study of hillslope forms and processes. Geomorphology 100(3–4):312–327

Schmidt PW, Ollier CD (1988) Palaeomagnetic dating of Late Cretaceous to Early Tertiary weathering in New England, N.S.W., Australia. Earth Sci Rev 25:363–371

Shaler NS (1899) Spacing of rivers with reference to hypotheses of base levelling. Bull Geol Soc Am 10:262–276

Summerfield MA (ed) (2000) Geomorphology and global tectonics. Wiley, Chichester

Walker JGC, Sloan LC (1992) Something is wrong with climate theory. Geotimes 37(6):16–18

Planation Surfaces of the Gondwana Continents: Synthesis and Problems

Cliff Ollier

Abstract The geomorphology of Gondwana was different from that of the fragments created when it broke up. These fragments, now continents and large islands, inherited some features from the original Gondwana landscape, including major drainage patterns and planation surfaces. Other features including Great Escarpments could only form after the break-up, and some erosion surfaces are also formed after break-up and graded to new base levels. The continents of Gondwana derivation have geomorphic histories that have many features in common but also some distinct features peculiar to individual fragments.

In this chapter I describe the geomorphology of the Gondwana fragments and give an outline of the various ways the evidence has been interpreted, including the major controversies. All the fragments have a succession of planation surfaces but seldom the neat step-like succession of simplistic descriptions. Their age and correlation is difficult to determine, as is their relationship to tectonics and changes of climate. Lester King was perhaps premature in his attempts to correlate planation surfaces all over the world, but his aim is still a valuable lead. Since his day there has been a profusion of more detailed local studies, and the next task is to integrate them into a new Gondwana-wide synthesis.

Keywords Planation surfaces • Regolith • Relief inversion • Climate • Bevelled cuesta

Introduction

Today most geologists believe that the present continents were once united into a supercontinent, Pangaea, which broke into two supercontinents, Laurasia and Gondwana, and then Gondwana broke into several pieces which are the subject of

---

C. Ollier (✉)
School of Earth and Environment, University of Western Australia, Perth, Australia
e-mail: cliff.ollier@uwa.edu.au

J. Rabassa and C. Ollier (eds.), *Gondwana Landscapes in southern South America*,
Springer Earth System Sciences, DOI 10.1007/978-94-007-7702-6_4,

this chapter. The continents were separated initially by rifts, which grew wider by seafloor spreading. It is very obvious on many old continents that the landscape is very flat. In many places the rocks beneath the landscape are structurally complicated, and the flat surface must have been made by erosion. Such erosion surfaces have been given many names by geomorphologists, some local, some based on presumed manner of formation, and some based on presumed age. We believe that some landforms are inherited from the old supercontinent – Gondwana – but landscapes have evolved since the supercontinent broke up. Our task, beyond the simple description of the facts about the surfaces and their evolution, is to work out what part of the landscape is inherited from Gondwana and what has happened since the various fragments were isolated.

In plate tectonics continental margins are divided into passive margins, where break-up of a supercontinent produced new continental margins that drifted apart, and active margins where crust is thought to be lost as one plate is subducted under another. This chapter is concerned almost entirely with passive margins: only South America has an active margin, located on the western side. The passive margins can be further divided into those that are high, usually with coastal uplift and a Great Escarpment, and those that are low, with neither mountains nor escarpments. After the break-up of Gondwana the fragments – Africa, South America, India, etc. – acquired new continental edges and new base levels for erosion, so erosion set in. It is possible that younger planation surfaces were formed to new base levels, and many authors have described successions of planation surfaces, with higher ones being the oldest and the lower ones the youngest.

It is important to realise that in this chapter, as in the whole book, we are talking of the erosion surfaces of the Gondwana continents, not of the 'Gondwana surface' sensu stricto of King (1967), which is only a small part of the story.

The Geomorphology on a Supercontinent

The geomorphology on a supercontinent would have been very different from that of today's smaller and dispersed continents (Ollier 1992). Average distance to the sea would be greater, which in theory might cause more continental type climates. At average gradients inland from the coast, the centre of a supercontinent would be higher than the centre of a small continent, which again has a theoretical impact on climate. Major rivers would be longer, and so their average gradients might be lower and their erosive power less than that of their shorter successors. With reduced erosion, there would be reduced isostatic feedback to create high mountains. There is a strong possibility that temporary storage of sediment on land would be greater. Lower erosion rates would lead to greater accumulation of weathering products, and with increased continental sedimentation, there might be greater development and preservation of regolith than occurs on today's smaller continents. Changes of latitude of continents might be expected to cause changes in climate, and the distribution of the continents after break-up and drift would also have great impact on ocean currents and ultimately on climate. Since weathering profiles and geomorphic evidence such as river patterns are available from times before the break-up of Pangaea and for intermediate times up to the present, some of these speculations can be tested. We need to know about weathering in the past and climates of the past to see how they are related.

In brief we expect the supercontinents to be notably flat before their break-up and planation would be very extensive. At the time of this extensive plain, the Jurassic and Cretaceous, most of the world had a warm climate conducive to deep weathering. But as younger surfaces are formed, the higher ones are still getting rain, runoff and erosion. Is it possible that any fragments of the original Gondwana surface are preserved? We might reasonably expect them to be somewhat modified, even to the extent of being a composite surface. In contrast, Burke and Gunnell (2008, p. 29) refer to 'a time of tectonism when Pangea was a single continent long before the African Surface-related erosion began'. It seems they believe no traces of the original Gondwana surface survive.

Lester King (1967) made the greatest early contribution to the study of planation surfaces, and he also complicated the subject by changing the names he gave to the surfaces, their supposed ages and even their mapped distribution. People have used local names for local surfaces, quite rightly in my opinion. Local names are fine for local use, and the difficult task of correlation can be attempted later if it is of interest. Burke and Gunnell (2008) use a different approach and lump together all the surfaces and fragments of surfaces in Africa as the African surface, which they then regard as 'composite'.

In the following sections I describe planation surfaces in different places, but the geometric flatness is not very helpful by itself. Other features are described where they throw light on the age or origin of the surfaces, including drainage patterns and their history, overlying and underlying rocks, weathering and regolith, tectonic features like rift valleys or continental margins and Great Escarpments.

Two aspects that are important are treated minimally here, offshore correlated sediments and apatite fission-track results. They are complex and simply take too much space to discuss adequately.

We now know that geomorphology is on the same timescale as global tectonics and biological evolution, and we want to see the evolution of planation surfaces as part of the evolution of the Earth.

Africa

Much of southern Africa is a plateau, bounded by marginal swells and a Great Escarpment. The plateau is generally over 1,200 m in elevation and over 3,000 m in Lesotho. The Great Escarpment makes a huge arc all around southern Africa from Angola, through Namibia, around South Africa and to the Limpopo River. It is highest in the Drakensberg (3,299 m), where the rocks are generally horizontal, and the upper part consists of 1,500 m of Triassic basalt. The marginal swells bound the central depression of the Kalahari Basin.

Partridge (1998) reviewed the morphotectonics of southern Africa and concluded that much of Africa possessed high elevation prior to the rifting that made new continental boundaries. The suggestion that surface was already high when new continental margins appeared seems reasonable. If the Western Rift Valley of Uganda were to widen, the plateaus of Uganda and Congo would become new continental margins at elevations of about 1,000 m. In contrast, Burke and Gunnell (2008) suggest that the continent was low at the time of break-up.

Much later uplift complicates the story. Partridge wrote, 'The evidence for large-scale Neogene uplift is now beyond question … The largest movements post-date the Miocene and have contributed both to the anomalous elevation of the eastern hinterland and to the strong east–west climatic gradient across southern Africa'. This late and dramatic uplift is indicated by a wide variety of evidence. Long profiles of rivers are convex up; Early Pliocene marine deposits have been uplifted 400 m; remnants of the African surface have been warped. These and other data show total uplifts of 700–900 m along the axis of warping within the last 5 Ma (Partridge 1998).

The African continent is dominated by a basin-and-swell structure (Holmes 1965) as shown in Fig. 1. Several questions arise, including how the swells form and when they form. Are the swells on the continental margin of the same age as those inland? Their relation to the planation surfaces is important in solving these questions. Many of the swells are followed by rift valleys, and the rift system can be traced to the Red Sea where the formation of simple graben gives way to seafloor spreading.

Did the coastal swell at the continental margin, which formed when continental margins first appeared, persist to the present day although eroded by the Great Escarpment? Several authors have ascribed flexuring along the Great Escarpment axis to rift flank uplift associated with the disruption of Gondwanaland (e.g. Partridge and Maud 1987). It is very clear that the amount of slope retreat on the young rift valleys is very slight (Fig. 2) compared to the massive erosion associated with retreat of the Great Escarpments. De Wit (1999) described gravels on both the high plateau surface and on the coastal plain that are of probable Late Cretaceous age, confirming a Cretaceous age for the blocking out of the major features of southern Africa.

Moore (1999) presented a new figure of axes of epeirogenic uplift (his Fig. 8) and recognised three axes: the Escarpment axis, the Etosha–Griqualand–Transvaal axes and the Ovamboland–Kalahari–Zimbabwe axes. These roughly concentric lines of crustal flexure are broadly parallel to the edge of the subcontinent. This requires a far more elaborate tectonic explanation than simple rise to the rift as Gondwanaland broke up. Moore agrees with others that the initial uplift along the Escarpment axis was associated with the Late Jurassic–Early Cretaceous rifting that initiated the disruption of Gondwana and believes the other axes are younger. For details of suggested ages and the reasoning behind them, see Moore (1999).

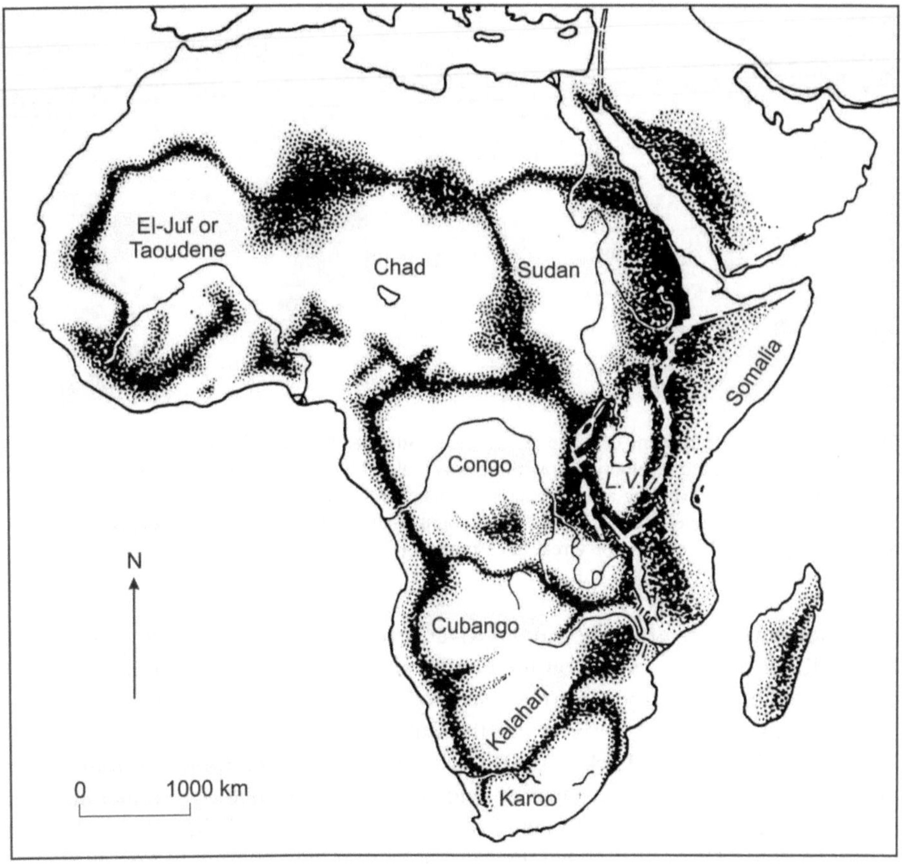

Fig. 1 The basins and swells of Africa and associated rift valleys (After Holmes 1965)

Many features of present drainage systems have been inherited from long ago, and the rivers predate the young epeirogenic movement and therefore the land surfaces that they drain. Drainage analysis can therefore indicate more general features of landscape evolution.

The major west-flowing drainage system of southern Africa is the Orange River and its main tributary the Vaal River, which rise in the mountains of the east and flow to the Atlantic Ocean, described in detail by De Wit (1999). According to him the Orange River system had established itself on a well-planated surface, more or less as it is today, by the Late Cretaceous. The lower Orange River is an antecedent river that flowed to the Atlantic, close to its present course, for at least 60 Ma.

Burke and Gunnell (2008) talk of rivers 'breaching' the continental flank uplift. Rivers cannot bore their way through a barrier: a river that crosses a barrier has to be antecedent; that is, the river was there before the uplift and maintained its course across it. (In some places a river may be superimposed, but this is not plausible for

Fig. 2 The fault escarpment of Lake Albert near Butiaba, Uganda. Most of the skyline is the African surface, and the inselbergs show there was higher land in the past, but in this area the hills do not have flat tops revealing a more ancient surface. Note the valleys have not cut down to lake level, suggesting relatively recent uplift. The amount of dissection of this rift valley margin is very trivial compared to the erosion on Great Escarpments (Photo: C. Ollier)

the major African rivers.) The Congo River provides another example. 'Where the Congo River meets the swell on the coastal side it forms Stanley Pool, whence it escapes to the Atlantic by way of a series of cataracts' (Holmes 1965).

The major east-flowing rivers are the Zambezi, Save and Limpopo, described in detail by Moore and Larkin (2001), who trace the geomorphic evolution since the disruption of Gondwana, using evidence of geomorphology, mineralogy of old alluvium, continental sediments, offshore sediments and the relationships of freshwater fish. They suggest that the Limpopo provided a conduit linking the major inland drainage to the Indian Ocean immediately following disruption of Gondwana. Moore et al. (2009) detail the evolution of drainage in Zimbabwe and present some remarkable conclusions. 'The modern drainage system to the north of the central Zimbabwe watershed is thus largely controlled by a surface that has existed since pre-Karroo times'. (The Karroo Supergroup ranges in age from Late Carboniferous to Early Jurassic.) 'Headwaters of the Zambezi tributaries were originally located well to the south of the modern divide'. 'This drainage system persisted until the Late Triassic, when rifting, linked to the early disruption of Gondwana, initiated the formation of the modern Save and Zambezi river systems. The central Zimbabwe watershed represents a Late Palaeogene (~43 to 33 Ma) asymmetric epeirogenic flexure ... which beheaded the headwaters of the early Zambezi tributaries'.

These authors suggest that much of Zimbabwe reflects two major cycles of erosion, the African and post-African. The first commenced with the disruption of Gondwana, and the second was initiated by Late Palaeogene uplift along the line of the modern central watershed. The rejuvenation of drainage led to removal of the carapace of deeply weathered saprolite that developed under the humid Middle Cretaceous climate of the Africa cycle.

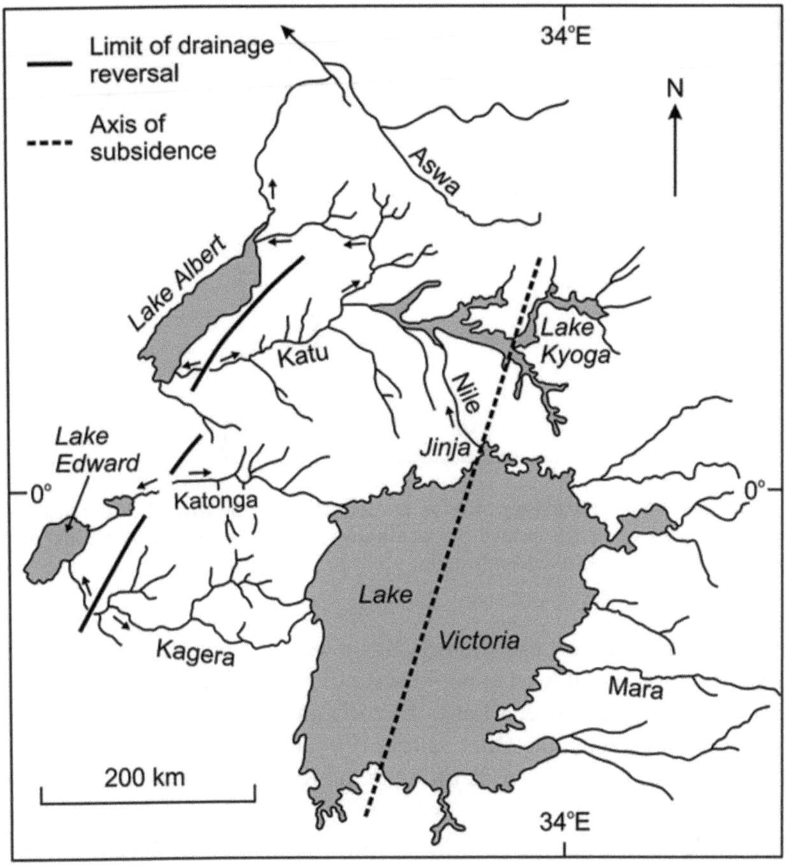

Fig. 3 Reversal of drainage in the Lake Victoria–Lake Kyoga system. The original drainage was from east to west, with the Mara continuous with the Kagera. The drainage lines of the Katonga and Kagera are continuous between Lake Victoria and Lake Edward, as is the Kafu from Lake Kyoga to Lake Albert. Much of the drainage of these rivers is now reversed as shown by the *arrows*. The middle course of the Mara–Kagera River was drowned when downwarping created the Lake Victoria Basin. Lake Victoria overflowed at the lowest point on it watershed at Jinja to form the stretch of the Nile between Lakes Victoria and Kyoga. Lake Kyoga was formed by similar back tilting of the Kafu River. It flowed up a northern tributary and overflowed into the Albert Rift Valley, forming the stretch of the Nile between Kyoga and Albert. From Lake Albert the Nile flows north along a continuation of the rift until it joins the Aswa which follows a major Precambrian mylonite band. Note the barbed drainage pattern of the reversed rivers

The headwaters of the Nile provide valuable insights into the planation surface history of Central Africa (Fig. 3). The original drainage was from east to west, with the Mara continuous with the Kagera. The drainage lines of the Katonga and Kagera are continuous between Lake Victoria and Lake Edward, as is the Kafu from Lake Kyoga to Lake Albert. Much of the drainage of these rivers is now reversed as shown by the arrows. The middle course of the Mara–Kagera River was drowned when

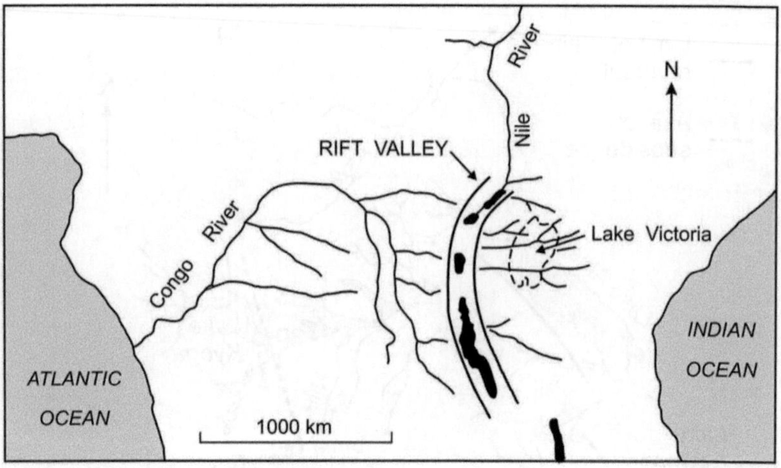

Fig. 4 The Congo–Nile relationship. Before the formation of the rift valley drainage of much of Uganda and parts of Kenya formed the headwaters of the Congo, but rifting and associated complications diverted these rivers to the Nile

downwarping created the Lake Victoria Basin which is remarkably shallow for its size. Lake Victoria overflowed at the lowest point on its watershed at Jinja to form the stretch of the Nile between Lakes Victoria and Kyoga – the only fast-flowing river in southern Uganda, the other valleys being choked by papyrus swamp. Lake Kyoga was formed by similar back tilting of the Kafu River. It flowed up a northern tributary and overflowed into the Albert Rift Valley, forming the stretch of the Nile between Kyoga and Lake Albert. From Lake Albert the Nile flows north along a continuation of the rift until it joins the Aswa which follows a major Precambrian mylonite band. Note the barbed drainage pattern of the reversed rivers. Clearly the old African surface was originally crossed by rivers with a simple dendritic drainage pattern, which was disrupted by uplift along the rift axis and downwarp along the Lake Victoria axis.

Figure 4 shows the scene at continental scale. The present headwaters of the Nile used to flow to the Congo. The palaeo-Congo flowed over a ground surface that was already in existence before Gondwana fragmented. The Nile was already flowing through Egypt by 5 million years ago, because at the time when the Mediterranean dried out, it cut its channel down. Drilling for the High Aswan Dam revealed a gorge 29 m deep with almost vertical sides, now filled with sediment.

Evidence from marine sediments is available for both east and west coasts. East coast offshore sediments range from Jurassic to Quaternary (Dingle et al. 1983). Data are summarised by Moore and Larkin (2001), with interpretation of geomorphic significance. From this, they trace changes in palaeodrainage from Jurassic to Pleistocene. Opening of the Atlantic Ocean was also associated with the early development of major depocentres on the western continental margin (Dingle et al. 1983). Sediment was supplied to these basins by the Orange River

(Dingle and Hendy 1984) and the Trans-Karoo (De Wit 1999), which both drained the interior of southern Africa. Further work on drainage evolution in Africa is provided by Dollar (1998) and Goudie (2005).

Because there is such a huge literature on planation surfaces of Africa, it is impossible to provide a concise survey of all the data on this topic, and this account is limited to some illustrative examples. Although there were many earlier studies of peneplains, it was King (1967) who kick-started planation surface studies in the modern way and provided a basis for all future work. It must be remembered that when King started work the concept of continental drift was widely reviled, and the idea of the break-up of Gondwanaland was something new.

One generalisation is that the Cretaceous African surface is generally recognised by widespread deep weathering, with many ferricretes and silcretes in places, indicating a long period of stability after planation. The stripping of the thick regolith is another important aspect of landscape evolution.

Some illustrative examples of detail come from Uganda. Ollier (1981) provided a map that showed three surfaces (Fig. 5). The highest surface is in the south, with distinct flat-topped hills rising above the African surface. In the King system this is the oldest surface and would be the Gondwana surface. Most of Uganda is eroded below this, but still with a thick weathered mantle, and is the African surface. Another map of Uganda surfaces is provided by De Swardt and Trendall (1969). An even earlier map is provided by Temple (1967) which is notable for showing the many remnants of what he calls the upland surface in greater detail than other maps which tend to draw a line around the whole group. Taylor and Howard (1998) prefer a model with several phases of deep weathering and trace the landscape evolution back to the Permian.

There is proof of later general surface lowering of the African surface. Mount Elgon in the east of the country is a huge Miocene volcano about 100 km across. In the area around Mbale township, the volcanic cover has been eroded off, revealing the sub-volcanic surface (which can be distinguished as the Mbale surface) which is slightly higher than the African surface further west.

Another example of the composite nature of the surfaces comes from northern Uganda. Here the so-called Acholi surface is largely cut across fresh rock, but to the south the 'African' planation surface retains a cover of weathered rock. Inselbergs of fresh rock rise through the weathered regolith of the 'African' surface, in a zone near the border of the two 'surfaces', and occasional pendants of deep saprolite, many kilometres across, are present on the 'Acholi' surface (Fig. 6).

This is interpreted as a rather severe complication to the erosion surface story (Ollier 1993). The original Gondwana surface extended over the whole area. It was deeply weathered to a very irregular depth. Later a younger planation surface, the African surface, cut into this and is distinguished by its thick cover of saprolite. But still further north the regolith is almost completely stripped. There is no step or escarpment between the African and Acholi surfaces, so they could be regarded as the same surface, in one place cut across saprolite and in another across fresh rock.

In fact because the original basal surface of weathering had high relief, there is a zone near the border of the two 'surfaces' where inselbergs of fresh rock rise

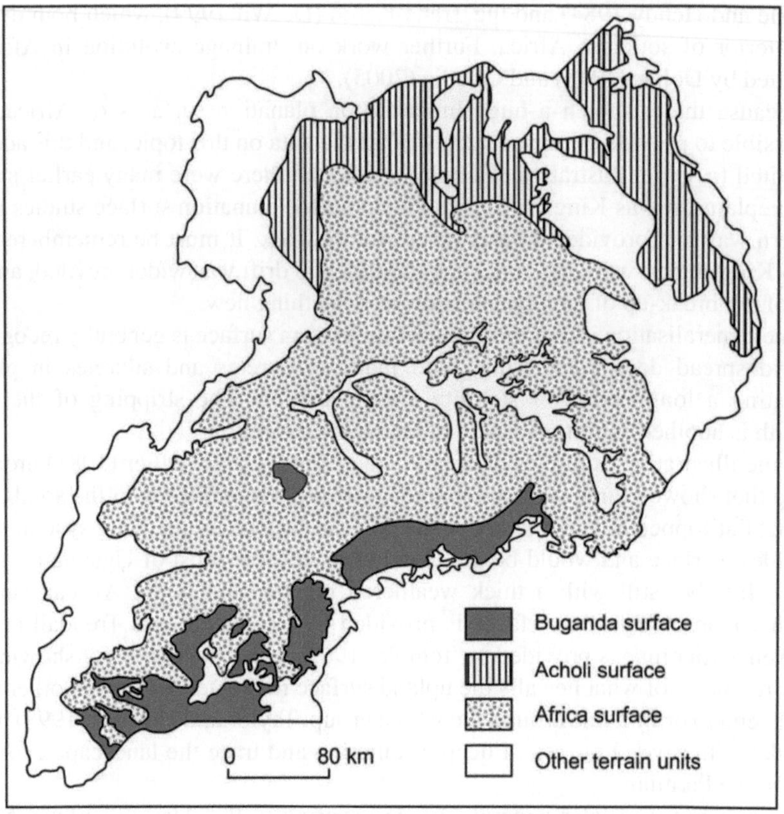

Fig. 5 Erosion surfaces of Uganda. The Buganda surface survives in places as flat remnants and has a deep weathering profile. The Africa surface is partially stripped of regolith. The Acholi surface is stripped of old regolith, and modern soils are forming on essentially fresh rock rather than saprolite

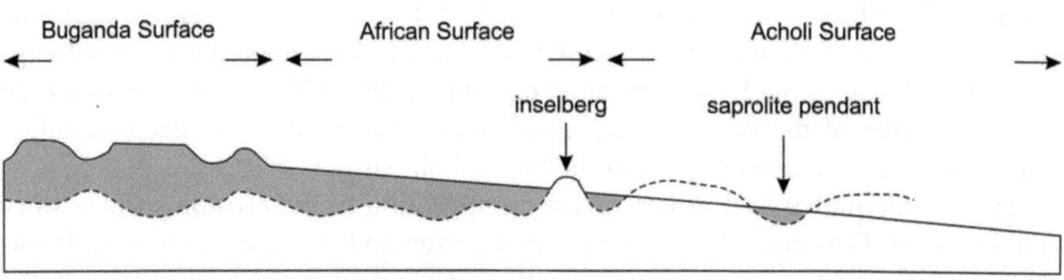

Fig. 6 Diagrammatic section of northern Uganda. Very deep weathering underlies the Buganda surface. It is partially stripped from the Africa surface and almost entirely stripped from the Acholi surface. Because the basal surface of weathering is very irregular, some inselbergs project through the saprolite on the African surface, and some pendants of thick saprolite project into the Acholi surface

through the weathered regolith of the 'African' surface, and occasional pendants of deep saprolite, many kilometres across, are present on the 'Acholi' surface. The distribution of deep weathering was only discovered by soil mapping (Ollier 1959) and is not evident in the topography. From a geomorphic origin viewpoint, the African and Acholi surfaces may be regarded as one surface, but from any agricultural or land-use point of view, they are vastly different. The irregular deep weathering was later been confirmed by groundwater drilling (e.g. Taylor and Howard 1998).

The geomorphology in the vicinity of rift valleys can offer special insights into the regional picture. There is usually a tectonic swell parallel to the rift which disrupted the older drainage. Figure 7 shows the situation on the flank of Lake Albert. On the rift side the deep saprolite has been stripped, but on the other side, it is preserved. The reversed rivers are full of clay and papyrus swamps. This surface has therefore been in existence since the drainage system was initiated, and this was long before the rift valley tectonics disrupted the drainage. Since the oldest sediments in the rift are Oligocene, the bounding surface must be at least that old.

In Ethiopia the situation is complicated. Lester King suggested there was African surface preserved here, but Coltorti et al. (2007) found that over quite a large area the surfaces were stripped surfaces, not a flight of successive erosion surfaces. Yet in other areas there are undoubted remnants of planation surfaces, and indeed on areas of Precambrian shield, it is impossible to make structural surfaces. Even in Saudi Arabia planation surfaces on granitic and metamorphic rocks are very extensive, and there are reports of very deep weathering profiles beneath the desert sands.

Alternative views about the planation surfaces of Africa are presented by Burke and Gunnell (2008). Much of their speculation is about global tectonics, not geomorphology. They talk of 'plate pinning', caused by eruption of the Karroo and Afar plumes, which is said to have arrested continental motion – as if Africa is pinned to the solid Earth. Since South America and India are getting further from Africa, they must be the ones that move as the oceans grow wider, while Africa is fixed. They also claim that the well-known basin-and-swell surface of Africa is caused by shallow convection currents related to the alleged pinning. They suggest that Africa differs from other continents in the 'absence of slab-pull force' and the 'limited role' for subduction!

They say that during a tectonic quiescent period from 130 to 30 Ma, 'denudation resulted in the formation of a composite, low lying surface of continental extent that is here called the African Surface'.

They claim that the 30-Ma Afar plume tectonism coincided with formation of the basin-and-swell structure of Africa. 'When the Afar plume erupted c 31 Ma, this Oligocene land surface, defined here as the African Surface, started to be flexed upward on newly forming swells and to be buried in sedimentary basins both in the continental interior and at the continental margins'.

In other words they deny the existence of pre-Oligocene land surfaces in Africa, they deny that any of the coastal swells and Great Escarpments relate to the break-up origin of the African margin but were formed later. 'Great Escarpments ... have formed on some swell flanks since the swells began to rise during the past 30 m.y.'.

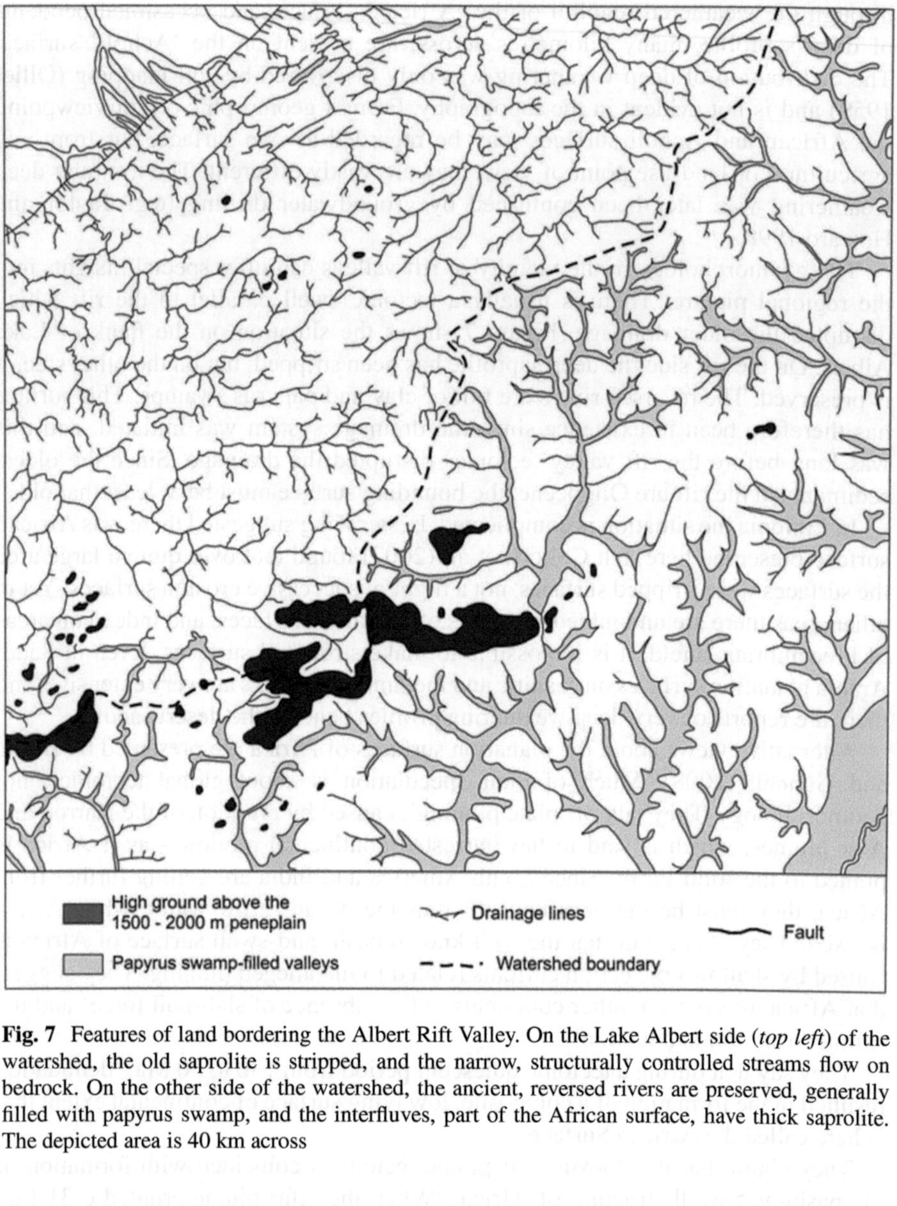

Fig. 7 Features of land bordering the Albert Rift Valley. On the Lake Albert side (*top left*) of the watershed, the old saprolite is stripped, and the narrow, structurally controlled streams flow on bedrock. On the other side of the watershed, the ancient, reversed rivers are preserved, generally filled with papyrus swamp, and the interfluves, part of the African surface, have thick saprolite. The depicted area is 40 km across

'The youthful Great Escarpments have developed in places where the original rift flank uplifts formed at the time of continental breakup. Their appearance is therefore deceptive'. They seem to imply that all observers of the Great Escarpment except themselves have been deceived.

Brown et al. (2000) provide an overview of fission-track data from southern Africa. They suggest 3–4 km of erosion along the Atlantic margin, generally increasing towards the coast, between 118 Ma and the present. This amount of erosion would be expected to remove the entire Mesozoic deep weathering mantle, which is still widely preserved. Similarly Japsen et al. (2012), on the basis of fission tracks, claim that the elevated passive margins are not rift shoulders but expressions of episodic, post-rift burial and exhumation.

Australia

Australia is mostly very flat and consists largely of erosion surfaces, even in the east where maps of Australia commonly show 'the Great Dividing Range' running parallel to the eastern coast. The Great Divide is real enough, separating coastal from inland drainage, but most of the 'ranges' are in fact parts of the Great Escarpment (Fig. 8). For most of its length, the divide crosses a plateau of low relief, and numerous lakes are located on a watershed that is equivalent to the swells parallel to rift valleys. There are mountainous and hilly areas, but even the highest point, Mt Kosciuszko, reaches only 2,282 m. The staircase of planation surfaces has not been generally applied in Australia. In this section I shall concentrate on individual surfaces, with an emphasis on how they relate to drainage evolution, epeirogenic uplift and methods of determining their age. The Eastern Highlands surface of low relief was called a palaeoplain by Hills (1975) who also termed it the Trias–Jura surface. He was the first Australian to stress the antiquity of the landscape.

Fig. 8 The abrupt junction of the plateau and the Great Escarpment at Bakers Falls, New South Wales (Photo: C. Ollier)

Most of the palaeoplain is very deeply weathered, and regolith has been dated by palaeomagnetism to at least 60 Ma (Schmidt and Ollier 1988; Idnurm 1985). Many local studies have been conducted on these surfaces and their relations to epeirogenic tectonic axes of movement, volcanoes, drainage evolution and several other aspects of geomorphology.

The geomorphology of eastern Australia is different north and south of Brisbane, though downwarp associated with the opening of the Pacific is common to both regions. To the north, in Queensland, the palaeoplain slopes inland from the Great Divide to the Great Artesian Basin which is filled with Mesozoic strata. Along the coastal zone there are many N–S faults and a Great Escarpment (Ollier 1992; Ollier and Stevens 1989). There is evidence of many reversed rivers which presumably came from land in the Pacific before the opening of the Coral Sea, perhaps 60 million years ago. The planation surface must date back to before that time, and so is a Gondwana surface.

A good example of evolutionary geomorphology is provided by southeast Australia (New South Wales and Victoria). Morphotectonic evolution of the area is a response to unique, non-cyclical events (Ollier and Pain 1994; Ollier 1995). Today three major basins, the Great Artesian Basin (Eromanga Basin), the Murray Basin and the Gippsland–Otway Basin, are separated by the Canobolas and Victoria Divides, which are intersected by the Great Divide. The divides are major watersheds. They evolved in several stages from an initial Triassic palaeoplain draining towards the Great Artesian Basin. There is evidence of former big rivers coming from the south, before Australia separated from Antarctica. Creation of the Victorian Divide cut off drainage from the south, and this divide has evolved further and is dissected into a series of plateaus known as 'High Plains', and the actual drainage divide is a ridge. Subsidence of the Murray Basin cut off drainage that formerly crossed the Canobolas Divide, and the Murray Basin is younger than the Great Artesian Basin. Downwarping of the Pacific side of eastern Australia formed the Great Divide, reversing many major rivers. The Great Escarpment was formed on the ocean side and worked backwards. The original divide was the culmination of a warp of the palaeoplain and, in places, is still in its original position. Elsewhere it has been shifted by headward erosion of rivers, river capture, volcanic activity and faulting.

Many examples of reversed drainage have been recorded, so rivers came from 'Pacifica' before the opening of the Tasman Sea about 80 Ma, and so the palaeoplain is a Gondwana surface. There are places on the Great Divide with closed depressions called 'lagoons'. To the west the rivers flow in their old direction; to the east they have been reversed, and the lagoons formed where the old riverbed had no gradient.

The sequence from low-relief peneplains to a sharp ridge is the reverse of peneplanation. With no further tectonic complications, the present topography would presumably end up as a new low-level plain. However, the first palaeoplain is Triassic in age, and the 'erosion cycle' is unlikely to end given continuing tectonic changes to interrupt the erosive processes (Huggett 2011).

The Eastern Highlands passive margin is exceptional in having abundant basalt volcanism ranging in age from the Cretaceous to 5,000 years ago, giving exceptional opportunity to date the geomorphic evolution of the landscape. There are many old lava flows, large central volcanoes that mark a hotspot trace and (in the Quaternary) areal volcanism with scores of small scoria cones and maars in northern Queensland and western Victoria. Volcanism roughly follows the main divides.

Eocene to Pliocene lava flows on the palaeoplain are preserved in inverted relief, but their very preservation shows that vertical lowering of the palaeoplain by erosion is in the order of hundreds of metres (Ollier 1978). In the mountainous areas lava flows help to elucidate the age and amount of downcutting since the lavas were erupted (e.g. Vandenberg 2010). He wrote that Palaeogene basalts originated on plateaus at about 1,000 m and flowed down valleys to an elevation below 200 m, so the plateau was uplifted and dissected by Palaeogene times. 'The Paleogene landscape contained all major geomorphic elements visible today: high plateaus bordered by steep escarpments that were breached by deep valleys . . . '. 'The uplift must therefore be considerably older, and is tied to the onset of major sedimentation into the Southern Rift System at 93 Ma (Cenomanian)'. The age of the planation surface must be older still, and is undoubtedly a Gondwana surface. The Southern Rift System is a precursor to seafloor spreading: the separation of Australia and Antarctica did not start until about 64 Ma.

An interesting story comes from western Victoria, where dated lava flows give the age of underlying gravels, much investigated for 'deep lead' gold mining. Modern stream sediments have a great range of rock types in the pebbles. As we go back in time, we find less variety and the oldest ones (Eocene) have essentially just quartz (and gold!). Even the quartz is different from that in modern streams with varieties that the miners called 'hailstone' gravels and 'jelly bean gravels' – highly rounded and etched. It seems that the Eocene streams were getting debris from only thoroughly weathered rock and that as the landscape evolved more and more bedrock was exposed. In other words in the Mesozoic there was extremely deep weathering, but through the Tertiary we have a story of stripping of the regolith with exposure of ever more bedrock.

Volcanicity in eastern Australia occurred over a long period, and not just at the time of creation of continental margins.

The Great Escarpment (Ollier 1982) is a sea-facing and largely continuous escarpment in Queensland and New South Wales. It is clearly erosional and very indented in plan where rivers cut back into the plateau. The age of the escarpment is clearly younger than the age of the plateau and presumably started to form after a new continental margin provided a new base level, about 80 million years ago. One group of workers thinks the palaeoplain was downwarped to the coast (Ollier and Pain 1997). The fission-track workers have a model of uplift towards the coast ending in a huge fault, but the necessary fault has never been found.

The Great Escarpment cuts across the 19-Ma Ebor Volcano, New South Wales, so here the escarpment is post 19 Ma. Near Innisfail in Queensland, a 3-Ma lava flow went over the Great Escarpment, so here the escarpment was in existence at that time and has not retreated very far since. Spry et al. (1999) mapped valley-filling Oligocene sediments and basalts in the Moruya area (southern New South Wales) demonstrating that the escarpment was already well developed by the Oligocene.

Brown (2000) discusses the tectonic and landscape evolution of the area and concludes there was gentle downwarp to the coast of about one degree.

Much of the drainage in Australia predates the formation of the continental margins. Before the modern eastern margin was formed, the rivers flowed from land to the east (now the Pacific) to the Great Artesian Basin (Ollier and Pain 1994). Other rivers flowed from the south (now the Southern Ocean or Bass Strait) and across what is now the east–west line of Victorian Highlands. Reversed rivers include the Clarence River (Haworth and Ollier 1992), the Hunter River (Galloway 1967) and the lower Shoalhaven River (Ollier 1978), and many other examples have been claimed, including the Daintree and Tully Rivers in Queensland and the Genoa and Cann Rivers in Victoria.

Bishop and Goldrick (2000) disagree with these reversals and wrote: '... the valley-filling lavas throughout southeast Australia indicate persistence of drainage directions from the latest Mesozoic or Early Cenozoic'. Even if they are right, it still suggests a great age for the drainage and the plains they flow across.

Drainage evolution in South Australia and neighbouring regions is traced back to the Mesozoic by Alley et al. (1999), with much detail on the stratigraphy of valley fill and major changes in catchments.

Another area that demonstrates the great age of the Australian erosion surface is a region of beach ridges on the southern rim of the continent (Kotsonis 1999; Bowler et al. 2006). In South Australia, part of the present coast consists of a beach ridge over 125 km long, behind which there is a lagoon called the Coorong. Inland from the lagoon is another beach ridge, clearly older than the active one, and further inland is another and then another and so on. These ridges can be traced inland for hundreds of kilometres into the State of Victoria (Fig. 9). The seaward ones have been shown to be Quaternary and the inland ones Tertiary, the dates being based on fossil content and palaeomagnetism. The oldest ones of this sequence are Miocene. Similar beach ridges are also present north of the Nullarbor Plain, extending from South Australia into Western Australia, and are of Eocene age, ~37 million years old (Benbow 1990).

Erosion seems to have stopped dead since then, and an Eocene beach ridge has survived at the ground surface since Eocene time. But what is the age of the flat surfaced inland? An enormous period of erosion required making it so flat – it evidently goes back to the Mesozoic at least and is truly a Gondwana surface.

Western Australia is mostly a vast plain that includes several planation surfaces, and here I shall consider just a few examples.

A prominent feature of the Pilbara region in NW Australia is the Hamersley surface. Twidale (1994) thought it may be of Early Mesozoic age (Triassic or Jurassic) and shows it on his map of Lower Mesozoic erosion surfaces. In the west of the area, Lower Cretaceous (Aptian) sediments were deposited on an older erosion surface as a result of a marine incursion. The incursion shows the area was already generally flat, but some prominent hills were probably islands in the Cretaceous sea. The sub-Cretaceous surface is being reexposed in some areas at the present time.

But indurated valley deposits, the Robe River Pisolites, are higher than the Hamersley surface because of inversion of relief. The surface must have been

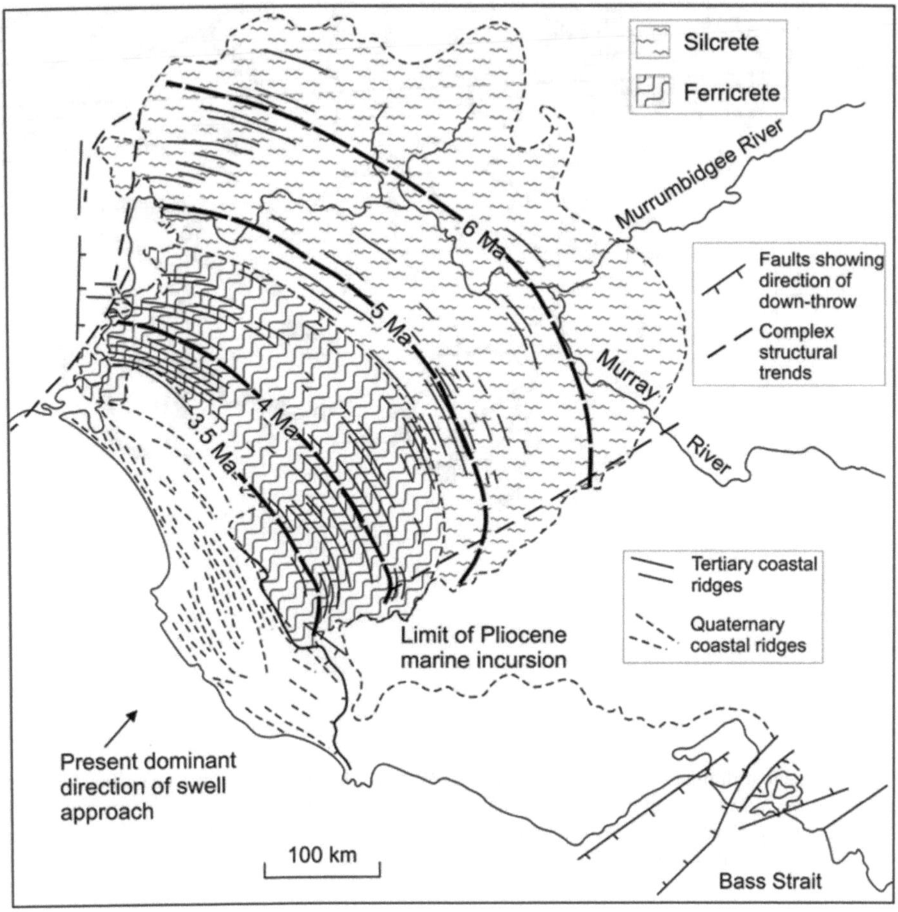

Fig. 9 Beach ridges of southern Australia. See text for explanation

regionally lowered since the valley deposits were laid down. This is Late Eocene to Early Oligocene time, so the present Hamersley surface must be about that age or younger. Further south weathering reaches at least 65 m, and Schmidt and Williams (2002) have dated the surface by palaeomagnetism and suggest an age of about 40 Ma (Late Eocene).

The land surface that existed when the Robe River Pisolites was deposited was probably already flat apart from the widely spaced valleys – the proto-Hamersley surface. Its flatness is supported by the lack of rock clasts in the old alluvium, which was fine grained. But weathering was probably intense, allowing the release of iron from the iron-rich bedrock to reach the alluvium where it was precipitated. The channel iron deposits are described in detail by Morris and Ramanaidu (2007).

To summarise, a broad, undulating to flat planation surface, the proto-Hamersley surface, covered the region at the end of the Cretaceous. In the Early Cretaceous

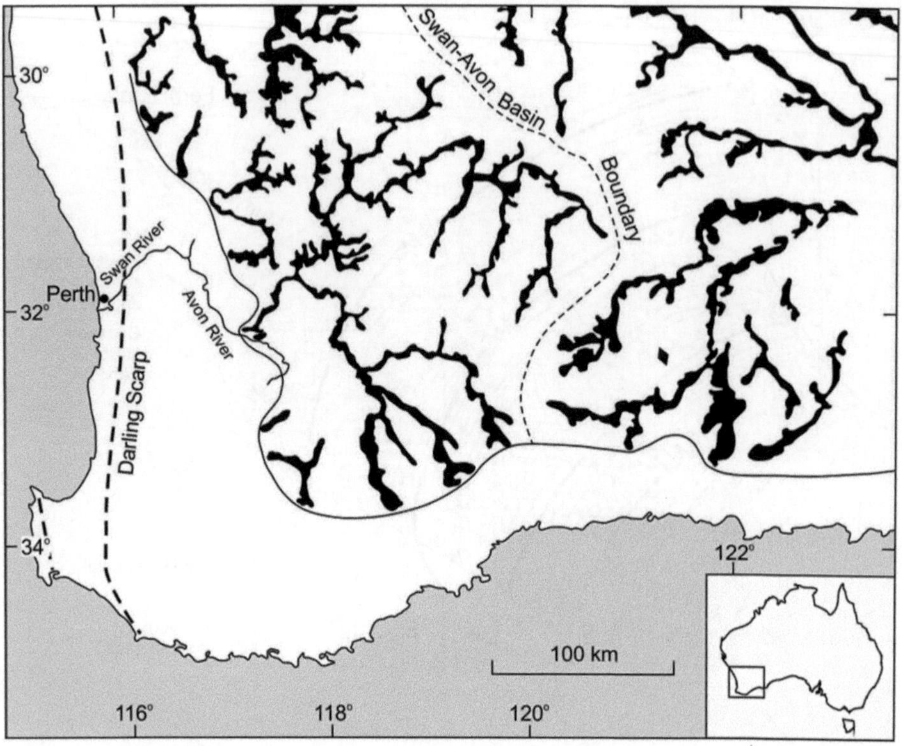

Fig. 10 Ancient drainage lines on the plateau of Western Australia. Note the beautiful dendritic pattern to the *left* indicating that the palaeodrainage came from the south, that is, from Antarctica before continental separation. Also note the rivers are about 10 km wide at source. Rivers do not start this way, and it shows that the source was far to the south

a marine incursion extended over the western part of the plateau, and then the sea receded. A drainage pattern made up of branching rivers incised into the surface, and the valleys filled with fine-grained alluvium (possibly Late Eocene), which was later indurated to ferricrete. Inversion of relief and general surface lowering created the present Hamersley Plateau.

Southwestern Western Australia consists of a vast planation surface, crossed by ‘rivers’ which might be better termed ‘palaeorivers’ as they seldom flow. Chains of salt lakes mark their course. In the southwest of Western Australia, there are two passive margins meeting at roughly 90°. The planation surface is underlain by ancient rocks and was already flat and deeply weathered by the Cretaceous. In both continental margins it can be shown the drainage pattern existed before the formation of new continental boundaries, and therefore the planation surfaces they flow across are pre-break-up and so Gondwanan.

The simplest example is shown in Fig. 10. The river pattern is dendritic and shows flow from south to north. But these rivers are over 10 km wide, and rivers do

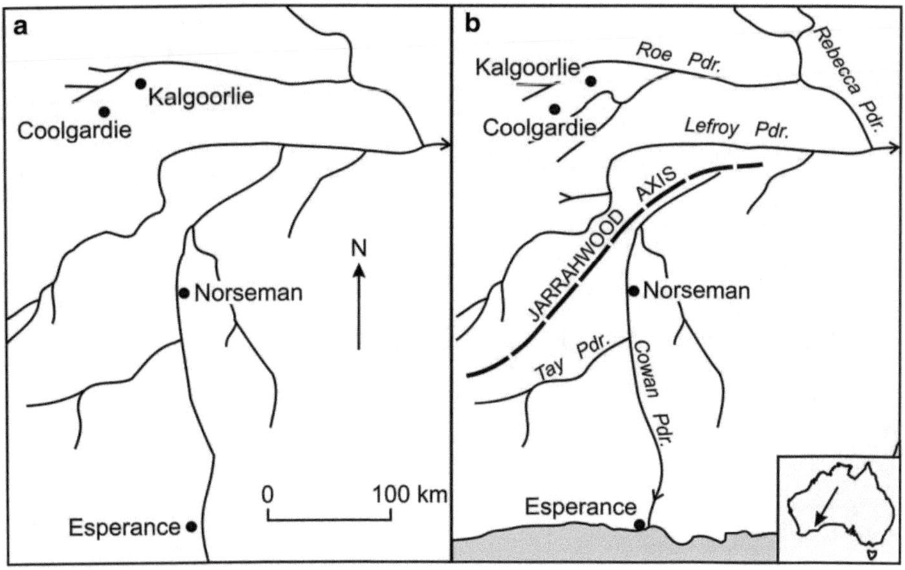

Fig. 11 Evolution of the Cowan drainage after Clarke (1994). (**a**) Palaeodrainage. (**b**) Formation of the Jarrahwood axis of warping, with downwarp to the south and reversal of drainage. Note the barbed drainage

not start with such width. The headwaters must be further south, but there is nothing further south until Antarctica. So the rivers are older than the break-up, which here started about 56 million years ago.

Clarke (1994) described a more elaborate example the region of the Cowan River (Fig. 11). With the separation from Antarctica, the southern coastal region was downwarped, creating a new watershed called the Jarrahwood axis 300–400 m high, 60–120 km inland and stretching 650 km. On the seaward side of this axis, the rivers were reversed.

Sometimes deep open pits (gold mines) reveal a cross section of such a valley, like the one at Norseman shown in Fig. 12 (Ollier et al. 1988a). The lowest sediment in the valley is terrestrial and contains leaves, fruits and pollen indicating an Eocene age of ~40 million years. Marine sediments containing sponge spicules and sharks' teeth, also ~40 million years old, overlie these. Evidently, a marine incursion caused by a rise in sea level extended from the south for hundreds of kilometres up the valleys, apparently without overflowing on to the surrounding plain. We not only know the age of the river sediments but can conclude that by 40 million years ago, not only were the valleys in existence, but they were incised in a very flat plain and had very gentle gradients. Obviously the age of the plain is much greater than Eocene. This river section is hundreds of kilometres from the coast, so the landscape was already very flat and the river gradient very gentle at the time of marine inundation, so that an arm of the sea could penetrate such a great distance from the sea.

Fig. 12 A cross section of a valley in a gold mine at Norseman. For location see Fig. 11. The lower valley fill (*white* on the photo) is terrestrial with plant remains including Eocene pollen. The *yellow* overlying layer is an Eocene marine sediment consisting mainly of sponge spicules. The *red* saprolite beneath the palaeovalley indicates intense oxidation. Such oxidation would have destroyed the organic matter in the valley fill and is therefore pre-Eocene (Photo: C. Ollier)

But below the ancient valley, the rocks are deeply weathered and highly oxidised. Such intense oxidation would have destroyed the organic material in the old valley, so the deep weathering and oxidation occurred before the valley was incised. It is pre-Eocene in age and probably Mesozoic.

We carry the story further in to the Kalgoorlie region (Fig. 13). Here the ancient, Eocene valleys are the low part of the landscape. The laterite-capped hills that rise above the valleys are obviously older. But even these are thought to be in inverted relief (Pain and Ollier 1995), so we have a landscape where the youngest part is Eocene and the older parts must be at least Mesozoic. They certainly date to before the break-up of this part of Gondwana and are a somewhat complicated Gondwana surface.

The western margin of Western Australia is dominated by the Darling Fault which separates the West Australian Plateau, cut across Archaean rocks, from a basin on the west containing over 10 km of Silurian to Cretaceous sediments. East of the Darling Fault is the Darling Uplift, a marginal swell along a N–S axis, 1,000 km long, 60–80 km wide and uplifted 150–200 m. Uplift created a depression on its eastern side, where middle Eocene fluvial sediments accumulated, thus dating the uplift (Beard 1999). Some rivers were defeated by the uplift, but the major rivers persisted as antecedent rivers across the uplifted block. Once again it is clear that the palaeoplain is much older than Eocene and a Gondwana surface.

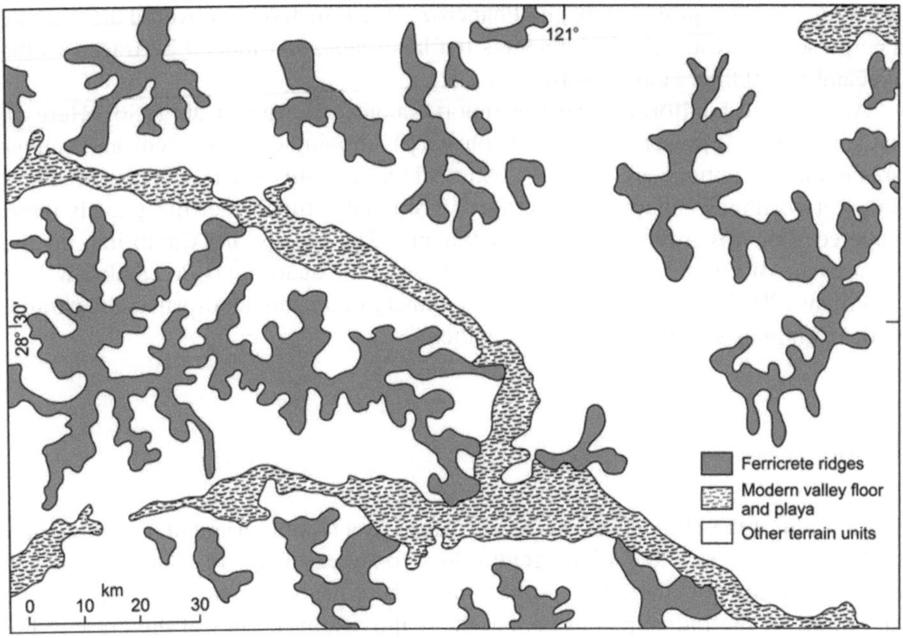

Fig. 13 Map showing ferricreted ridges near Kalgoorlie, Western Australia. The shape is like that of a dendritic drainage pattern suggesting the ferricrete was formed in ancient valleys and came to be on hilltops after inversion of relief. The modern valleys date back to at least the Eocene, so the ferricrete is even older

Twidale has written about the ancient landforms of Australia (Twidale 1994) and in particular about the role of stripping in the formation of planation surface in this chapter. He wrote, 'Cretaceous, and especially Early Cretaceous, marine transgressions played an important part in the evolution of the Australian landscape. They covered at least 40 % of the present Australian Craton and about 45 % of the entire present continent'. '[in places] the Gondwana landscapes were preserved more or less intact, as in the western Yilgarn Craton, the Mt Lofty Ranges and possibly the higher ridges of the Macdonnell Ranges'. 'The Gondwana landscape was already complex with stratigraphic conditions suitable for the exhumation of Archaean, Proterozoic, and Cambrian forms, for instance, already in place; re-exposure later took place so that very old elements form integral parts of the present landscape'.

Even older erosion surfaces are known in Western Australia. The Kimberley Plateau is an erosion surface cut across gently dipping sandstones. On softer rocks, valleys were incised below the main plateau level. Glaciation occurred after the main plateau had been created, leaving telltale glacial striations on the sandstones and patches of glacial tillite in the valleys. But this glaciation is the Precambrian Sturtian glaciation (~700 Ma). The tillite was once covered by thin Precambrian dolomites, but it is probable that no Phanerozoic sediments ever covered the plateau. The present plateau, which dominates the landscape, can indeed be traced to the Precambrian (Ollier et al. 1988b).

Another old landform is the Davenport Ranges (Stewart et al. 1986). Here an ancient surface is preserved as bevels on steeply dipping quartzite beds and valleys are cut into the softer rocks between. Alluvial terraces are preserved in these valleys and were originally mapped as Tertiary. It was later found that the gravels interfingered with fossiliferous marine sediments. The fossils are Cambrian, so the terraces are of Cambrian age and the planation of the quartzite is Precambrian.

Pillans (2007) provides a nationwide summary of ancient landforms in Australia, with an emphasis on methods of dating them.

India

India south of the Himalayas can be envisaged as a sloping triangular wedge, with highlands in the west and a long gentle slope to the east (Fig. 14). The triangle has two passive margins that meet in the south. There has been major ($\sim$1 km) uplift and tilting of the Indian peninsula, but the major drainage lines of the present day predate that uplift. The eastern side separated from Antarctica about 130 Ma; the separation of India and Madagascar was about 100–95 Ma.

The Western Highlands are known as the Western Ghats (and also as Sahyadri Range). It comprises a Great Escarpment that runs parallel to the western coast of India (Ollier and Powar 1985; Gunnell and Radhakrishna 2001; Kale 2010), backed by a gently sloping plateau. The escarpment extends for over 1,500 km and is seldom more than 60 km from the coast. The edge of the escarpment is almost entirely coincident with the divide between eastern- and western-flowing rivers, so the vast eastern slope is essentially a tilted surface. Much of it very deeply weathered and laterite-capped plateaux prevail.

In the north the escarpment is cut across the Deccan Basalts, which were erupted about the K-T boundary (Late Cretaceous), and in the south across metamorphic and igneous rocks of the Precambrian Shield, with no significant change in landform on these very different bedrocks. The length, continuity and similarity of form indicate a single, post-Cretaceous process of uplift and scarp recession. The time of extrusion of the Deccan Traps (Cretaceous–Tertiary boundary) is about the same as the formation of a new eastern continental margin, and the opening of the Arabian Sea came slightly later.

The basalt–basement contact can be observed in many places, and 'extensive horizontal Deccan flows directly overlie extensive low-relief planation surfaces cut on various older rocks (Archaean through Mesozoic) with different internal structures' (Sheth 2007). The sub-Basalt surface was deeply weathered and lateritised (Venktakrishnan 1984).

An illustrative example in detail is the Satpura Dome in the vicinity of the town of Pachmarhi in central India (Venktakrishnan 1984, 1987; Choubey 1971;

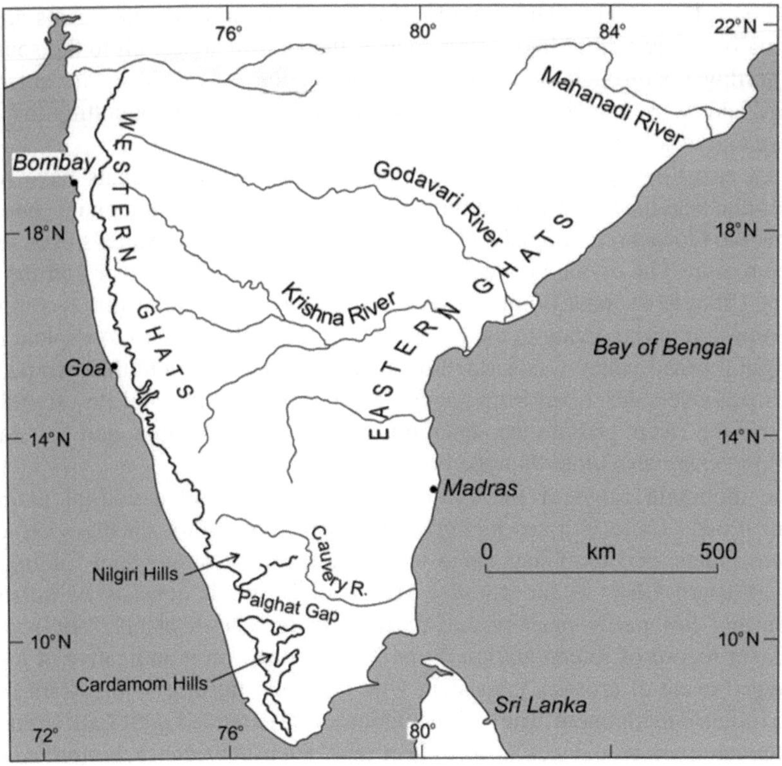

Fig. 14 India, showing the Great Escarpment closer to the west coast, and the major drainage flowing east

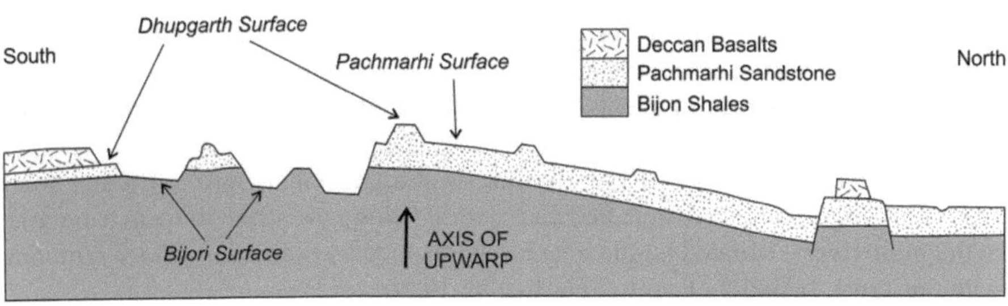

Fig. 15 Diagrammatic section of the Satpura Dome. For explanation see text

Sheth 2007). Here the Deccan Traps overlie a thick layer of Pachmarhi sandstone (Early Triassic), and the contact has been called the Dhupgarh surface (Fig. 15). This pre-Deccan surface can be traced to several outliers on the sandstone and has been equated to the African surface. A lower planation surface, the Pachmarhi surface, is cut into the sandstone, and both surfaces are warped by the formation of the Satpura Dome. The main outcrop of the Pachmarhi sandstone forms an east–west major escarpment. Major rivers such as the Denwa arise well to the south and flow northwards through the sandstone in spectacular gorges. They are antecedent streams and 'were in existence before the Satpura uplift and kept cutting the domal upwarp that was forming in their path'.

Most east-flowing streams near the crest of the Western Ghats have narrow valleys that become broad and flat only when they enter the eastern plains, but some, such as the Godavari and Krishna (Kistna) Rivers, have broad valleys even in their source region. The divide between the east-flowing Indrayani River and the west-flowing Ulhas River, near Bombay (Mumbai), is a broad flat valley that is continuous from one side of the divide to the other. This broad valley is hard to explain unless an original broad valley was formed by a single major river and then warped, like the warped rivers described from Australia and Africa (Ollier and Powar 1985). In that case the rivers predate the separation of India from Arabia, and the surfaces they flow on are also older than the break-up.

The mountain ridge of the Eastern Ghats lies behind a coastal plain 50–70 km broad. There is a sea-facing escarpment, though not on the scale of the Western Ghats, and the Ghats seem to be the result of monoclinal folding. East of the Eastern Ghats is the parallel Mahanadi Basin. It consists of horsts and graben and lies partly onshore and partly offshore. Nash et al. (1999) say the consistent record of sedimentation in the Mahanadi Basin is indicative of a single prolonged cycle of erosion, consistent with the rapid uplift of the eastern margin of the Indian subcontinent. The sedimentation cycle extended until Late Cretaceous (Cenomanian) times and was an episode of rapid basin subsidence, hinterland uplift, erosion and sedimentation. 'Much thinner Paleocene and Eocene deltaic deposits in the adjacent Mahanadi offshore basin reflect the end of continental edge rifting and the demise of the Mesozoic spreading centre'. No Palaeocene–Eocene deposits occur in the Mahanadi Onshore Basin, so it is likely that by the Middle Cenozoic a vast pediplain covered a large area, with residual massifs 'where small remnants of the original pre-rifting or Gondwana surface were (and still are) preserved as plateau cappings. This sequence of events is identical to that described for southern Africa by Partridge and Maud (1987)'.

The antecedent Godavari River divides the Eastern Ghats into north and south sections. Other rivers were defeated and diverted along the strike to form tributaries to the main rivers. Isolated highland areas are capped by planation surface remnants with duricrust, including bauxite (Nash et al. 1999).

India has a plethora of planation surfaces, and many people have studied them. Perhaps the best summary is by Gunnell (1998) who also provides his own version and interpretation.

He wrote, 'Several tiers of flat-lying denudation topographic surfaces, which are indifferent to [structure] and highlighted by a characteristic weathering mantle are identified and the conditions of their genesis discussed'. He summarises the work of 11 previous workers in his Table 1, and it is notable that several of these date surfaces back to the Jurassic.

He starts his own list of surfaces with 'an ancient primarrumpf' his S_0. In Baulig's Dictionary of Geomorphology, primarrumpf is defined as an ancient peneplain, so it really means no more than the most ancient surface of which we have record. The primarrumpf remnants are tentatively identified in two areas north and south of the Palghat Gap, the Nilgiri Hills, and the Vandaru and Anaimudi flats of the Palni Hills and also the Horton Plains of Sri Lanka. These are preserved in only small areas of 2–10 km^2 and are remarkably all at about 2,400 m. Charnockite domes rise above them indicating even higher land sometime in the past. Their summits (Doddabetta, 2,640 m; Anaimudi, 2,700 m) are the highest summits in India south of the Himalayas.

The S_1 surface. Remains of this Late Cretaceous/Palaeocene surface are found below S_0 in the Nilgiri Hills (Ooty surface), in the Palni Hills (Kodaikanal) and in Sri Lanka. These are plateau surfaces at about 2,200 m but with a rolling, down-like topography. The regolith is very significant on the Nilgiri and Palni regions because here there is bauxite, not found on any lower and younger surfaces. The landscape formed under the hot and humid climate that prevailed from Late Jurassic to the Eocene at least. Continent-wide planation was well advanced when the Deccan Traps erupted and sealed the Late Cretaceous land surface. The S_1 episode was terminated in the Early Palaeocene.

The S_2 surface is a Middle Tertiary surface. Gunnell suggests this might be termed the Indian surface, equivalent to the better-preserved African surface of King.

S_3 is the Mysore Plateau surface. In places there is a denudation escarpment between the S_2 and the S_3 surfaces. Gunnell says the S_2 and S_3 surfaces can be distinguished by their soils: the former have Ferralsols, and the latter, stripped of lateritic remains, is dominated by Chromic Luvisols.

The Tamil Nadu Miocene–Pliocene plain, S_4, 'has reached a degree of rectitude and perfection which has astonished several generations of tropical geomorphologists'. Gunnell also notes, 'knowledge of denudation rates across this huge, composite pediment rock surface is extremely limited'.

Gunnell suggests that the small size and poor preservation of the Indian surface (S_2), in comparison with the African and South American surfaces of King (1967), is a consequence of the small size of the Indian subcontinent, for landscape change propagates faster into the interior.

Laterite is extensively distributed on the old shield areas, on the Deccan traps and on the lowland strip below the Western Ghats. Evidently it formed over most of the time since the Late Cretaceous if not before. Ollier and Rajaguru (1989) found that in the type area the 'laterite' was the mottled zone of a deep weathering profile, and the term 'laterite' should be used with caution as different authors have used it with many different meanings. Some writers have tried to apply the classic 'laterite profile' of Walther, which in my experience is very rare. Others have applied the inversion of relief model (e.g. Pain and Ollier 1995).

Widdowson (1997) suggested that some laterites on the plateau of the Deccan Traps mark the original constructional surface, but his map of laterite shows a dendritic pattern that suggests a former river bed and appears to be a perfect example of inversion of relief (Ollier and Sheth 2008). If Widdowson is right we have a preserved remnant of the top of the basalt pile, albeit heavily weathered: if Ollier and Sheth are right, there has been great erosion of the basalt pile and even the highest part is a former river bed.

Gunnell et al. (2003) applied apatite fission-track data to Western India. Their results depend very much on assumptions about track length, but other fission-track workers tell me that arbitrary choice of track length is not good practice.

Antarctica

Eastern Antarctica is a fragment of Gondwana, but we cannot hope to find much comparative material because it has been under an ice sheet for about 40 million years, yet plateaus (palaeoplains) have been observed all around the continent at the margin of the ice sheet. Kerr et al. (2000) and Näslund (2001) review earlier work.

In Queen Maud Land (Dronning Maud Land to Scandinavians), a flat, high-elevation plateau has been eroded into isolated remnants bounded by escarpments. The surface existed in the Early Permian so would rate as a Gondwana surface. Näslund (2001) wrote that the morphology of Dronning Maud Land supports the idea that the escarpment formed by scarp retreat, following the rifting and separation of east Antarctica and southern Africa. Näslund's interpretation of landscape evolution may be summarised as follows:

1. Southern Africa and Queen Maud Land separated in the late Jurassic.
2. Mesozoic rifting was associated with uplift and formation of flood basalts, and an escarpment formed at the new passive continental margin.
3. Prior to the Oligocene, mountain glaciers eroded valleys on the palaeoplain and escarpment.
4. Formation of alpine landforms in the mountain ranges ceased when cold-based Cenozoic ice sheets formed in the Miocene, inhibiting glacial erosion and preserving existing landforms.

The Prince Charles Mountains consist of large flat-topped massifs with accordant summits that are the remnants of a preglacial surface of low relief. An Eocene lava flow on the surface suggests that erosion of the surface must be older than Eocene (Tingey 1985), so it could be a Gondwana surface.

The Transantarctic Mountains of eastern Antarctica run along the edge of a plateau that rises gradually from the interior, forming a dramatic, major escarpment (Kerr et al. 2000). A major debate concerns whether the Transantarctic Mountains experienced uplift of about 1 km since the Early Pleistocene (Behrendt and Cooper 1991) or whether the mountains have remained at their present level since the Miocene (favoured by Kerr et al. 2000). We have no direct evidence on the age of the planation surface, but it must predate the onset of glaciation at about 40 Ma.

Madagascar

At the simplest, Madagascar can be regarded as a huge tilt block, with a steep face in the east and a gentle slope to the west. The steep eastern side (the Angavo escarpment), has a step-like arrangement of erosion surfaces. The ages of these surfaces, as determined mainly from their relations to fossiliferous sediments, range from Quaternary to the later Tertiary, mid-Tertiary, Late Cretaceous and Jurassic (Dixey 1960).

The western side of the island is a sloping plain, divided into different physiographic regions that are sometimes rather hilly. Soil erosion is sometimes extreme in Madagascar creating huge gullies called lavaka (Fig. 16), generally supposed to be caused by past deforestation. These reveal exposures of very thick saprolite (Fig. 17), indicating the great age of the original planation surface.

It is as if a Gondwana fragment were isolated and tilted to the west, after which the western slope is eroded to slightly lower levels, while the eastern slope is a huge escarpment that reveals a flight of surfaces related to bursts of relative uplift.

According to King the main plateau was equivalent to the African surface, but since Dixey traces the history back to the Jurassic, there was certainly a Gondwana surface precursor, though it is not clear if any fragments remain.

Madagascar can be matched with Africa, from which it separated about 165 Ma and finished drifting about 121 Ma (Rabinowitz et al. 1983). The separation of Madagascar from India happened later, about 90 Ma (Torsvik et al. 2000).

Fig. 16 Huge erosion gullies (lavaka) cut in deep saprolite, near Antananarivo, Madagascar (Photo: C. Ollier)

Fig. 17 The side of a lavaka in Madagascar. The entire profile is in saprolite created by isovolumetric weathering of metamorphic rock, with preservation of structural details (Photo: C. Ollier)

The Angavo escarpment in Madagascar can be matched with the Western Ghats of India, as described by Gunnell and Harbor (2008). There is one major gap in the Western Ghats escarpment, the Palghat Gap, which matches the Ranotsara Gap in Madagascar. If these actual landforms are inherited from pre-drift landscapes, then they clearly relate to Gondwana landscape.

Sri Lanka

Most of the island's surface consists of plains between 30 and 200 m above sea level, but the heart of the country, the south-central part of Sri Lanka, is the Central Highlands. This area runs north–south for approximately 65 km and includes Sri Lanka's highest mountains, reaching 2,524 m. Some writers claim there are plateaus, including Wadia (1943), who reported three peneplains and believed that the uplands were raised by faulting though the faults are seldom exposed. My own brief visit failed to find any planation surfaces at all, but I would be delighted to be corrected.

The Horton Plains has been described as the highest plateau at 2,100–2,300 m, and across it runs the divide between Sri Lanka's two mightiest rivers, the Kelani and the Mahaweli. Some irregular isolated mountains rise from it, like Kirigalpotta (2,395 m). At one place the edge of the plateau is known as World's End, where there is a drop of about a kilometre, but the length of the cliff is just a few kilometres, and it is a singular steep feature, not an escarpment which by definition has to be long compared to its height. World's End is about 80 km from the coast.

I recently visited the area and can assure geomorphologists that there are no Horton Plains. The area could more reasonably be named the Horton Hills or even Horton Mountains. The land is rugged and there are not even small flat areas that could be regarded as remnants of planation surfaces. On the roads in the area there are hairpin bends where buses have to do a three-point turn – not a good indicator of flat country. Even the gentler grassy slopes are steep enough to develop terracettes.

The Central Highland region is a remnant of Gondwana, but it has retained no remnants of a planation surface. I would not even presume to detect a gipfelflur though some photos I have seen from Adam's Peak might encourage the idea. Whether the lower surface at 30–200 m can be related to the African surface as claimed by King (1967) or was developed entirely after the separation of Sri Lanka from India remains to be determined. Deep weathering is extensive on both lowlands and highlands, and much of the topography relates to differential erosion of the two. In the highlands I saw no trace of 'laterite', meaning indurated saprolite that can be used for building, though it is present in the humid south-west of the island (Herath and Pathirana 1983). If it were present elsewhere, it would surely have been used in construction of palaces and temples, as at Angkor Wat in Cambodia, but in Sri Lanka they are built of granite and gneiss.

New Caledonia

New Caledonia is a fragment of Gondwana swept far away from Australia by seafloor spreading (Avias 1953). There are plateau remnants with deep weathering profiles tens of metres thick generally described as laterites. The regolith has been studied extensively in part because it bears economic nickel deposits. The nickeliferous areas are on ultramafic rocks, mainly peridotites, and the ore is in the saprolite. The mineral ores can only be found at 'the peneplanation surface of Cycle 1' where it is intersected by younger erosion surfaces and the nickel ore is concentrated (Avias and Gonord 1975). We have no firm date on the old peneplains, but perhaps 'Cycle 1' relates to the Gondwana surface, though Evans (1993) thought the ore was formed in the Miocene.

Some geologists claim that for a long time, all of New Caledonia must have been below the ocean surface, but inference from the modern flora indicates that at least some land must have remained exposed, serving as a refugium. Many attributes of New Caledonia's flora, such as its high generic and familial diversity, and the presence of numerous primitive groups would be very difficult to explain otherwise. A substantial component of today's flora is thus thought to comprise the descendants of pre-Eocene Australasian groups that were able to survive on New Caledonia as it separated and drifted. Nevertheless, it would seem that the old Gondwana landscape was perilously close to sea level in part of its history.

South America

It is not my task to describe the South American evidence, which is covered in the rest of this book. But I need a few asides to show the relevance of the comparisons I am trying to make with other Gondwana fragments.

We should expect some matching of geomorphic history with other Gondwana fragments, especially Africa where the match between southern Brazil and Namibia seems incontrovertible (see, for instance, Pankhurst et al. 2008). We might also remember that the Pacific Ocean opened later than the Atlantic, and it may not be too far-fetched to match South America with New Caledonia and so to Australia (McCarthy 2003).

It is always useful to have a dated sedimentary cover to help put planation surface in a geological timescale, like the Hamersley surface in Western Australia. Aguilera et al. (2010) describe a surface near Chubut, Argentina, where Late Cretaceous sandstones show that the surface existed in the Late Mesozoic at least and is therefore a Gondwana surface. Elsewhere they write, 'This surface was generated in the Jurassic ... ' referring to a particular part and again showing a complex history like that of many other Gondwana fragments. The Cretaceous rocks have been removed in parts to form a stripped surface. Aragon et al. (2010) say the North Patagonian Plateau surface 'is covered by a thin sedimentary sequence of upper Cretaceous and lower Tertiary non-deformed continental marine sediments'. Suvires (2010) describes a flight of five ancient surface in the Western Pampean Ranges, Argentina, and records that Beltramone thought some surfaces belonged to '... an ancient pre-Triassic peneplain, which was buried under Cretaceous deposits and thereafter exhumed and dissected ...'.

Another similarity to Australia is the Miocene basalts that are flat on the plateau of the North Patagonian Massif but spill down over the rim of the plateau, like the basalts flowing over the Great Escarpment in Queensland.

In southern South America the only significant area of passive margin mountains is in southern Brazil, where the Brazilian Plateau is bounded by Great Escarpments with two names. These are very much like the Great Escarpments in other Gondwana continents. The Serra do Mar Escarpment, which extends 800 km, separates a coastal strip from the interior plateau cut across varied rocks, and all major streams drain westwards on the gentle inland slope to the Parana River. South of this is the Serra Geral Escarpment of the Planalto Plateau, underlain by the volcanics of Cretaceous age. The escarpment reaches 1,600 m and may be up to 370 km inland. Here there is undoubtedly downwarping to the coast: the base of the basalt is several hundred metres above sea level on the Great Escarpment but is at sea level at the coast. The Planalto surface was regarded by King as part of the Sul-Americana surface (= African surface).

In northern South America there are many erosion surfaces. Zonneveld (1985) provides an early summary, and Peulvast and Claudino Sales (2004) and Peulvast et al. (2006, 2009) have produced several more recent accounts. The Gran Sabana on the Guayana Shield in Venezuela is mainly developed on gently folded Precambrian quartzites. Two planation surfaces are developed, one over 2,000 m and the other about 1,000 m. Briceño and Schubert (1990) suggest that the Guayana Shield was in the tropics since the break-up of Pangaea and that both the planation surfaces are of Mesozoic age.

The Andes consist largely of great planation surfaces known as the Altiplano. The age of the surface is definitely post-Cretaceous in some parts, but the plateaus were uplifted in the Neogene: the mountain uplift is very much younger than many planation surfaces at lower levels. It is also worth noting that there is much evidence for tension in the Andes, notably the inter-Andean depression (graben), so any alleged influence of the Active Margin of South America is probably slight in the lower surfaces.

Gallagher et al. (1994) analysed fission-track data of southeast Brazil. There is a regional trend of young fission-track ages close to the coast becoming older with distance inland. Gallagher et al. concluded that 3 km has been eroded from the coastal plain and 1 km inland. Maps of fission-track age, palaeotemperature and denudation are presented in Brown et al. (2000).

Summary and Conclusions

Gondwana planation surfaces are alive and well. In the 1960s it was fashionable, at least amongst English speakers, to deny the very existence of planation surfaces (e.g. Kennedy 1964). Much of the argument was about the details of how 'peneplains' were created, but as Ollier (1981) pointed out, 'Most people who are not blind or stupid can tell when they are in an area of relatively flat country: they can recognize a plain when they see one'. Whether you call it a peneplain (which means 'nearly a plain') or some other term does not really matter. Many ancient plains have suffered a wide range of processes in their time, so we now usually call them planation surfaces or palaeoplains as they are undoubtedly old. It is virtually impossible to detect the work of a single process, such as pedimentation, on a palaeoplain.

It has also become very clear that the continents do not consist of a series of clearly defined surfaces, separated by obvious erosional escarpments, like a beautiful staircase. The planation surfaces themselves have slopes, boundaries are often indistinct, and surfaces that are clearly separate in one area may merge together in another. This situation is to be expected because of the complexities of Earth history and geomorphic processes. Uplift is not in neat jerks, erosion is not by a single process at standard rates, climates change gradually or quickly and vary through time and space, and a planation surface is usually eroded across complicated bedrock. Lester King's bold synthesis was incredibly stimulating, but premature.

Nevertheless, we are starting to see relationships between the planation surfaces of different continents. There is no doubt that Gondwana surfaces have been preserved, though modified to varying degree. In some continents we can confidently trace the geomorphic history back to the Jurassic, Triassic or even Permian or older. We also see some common landforms on most Gondwana continents. Great Escarpments are found on almost all Gondwana fragments, but not on every margin, so there are different processes and histories in different places. Inselbergs are important in some landscapes but not others. Deep weathering and stripping are geomorphic phenomena that have to be fitted into geological histories just as significantly as stratigraphy and tectonics.

The South American contribution to the study of Gondwana geomorphology is of great significance. In the past it has not been well integrated with the other regions, partly because of the language problem, but that phase is now over. The chapters in this volume show how South America compares and contrasts with other continents and presents a wealth of data and interpretation that cannot be ignored in the future.

Acknowledgements I thank Colin Pain for assistance with the illustrations and decades of discussion on the topics of this chapter. I also thank my wife Janetta for her assistance on much of my fieldwork and for help in editing this chapter.

References

Aguilera EY, Aragón E, Carretero S (2010) The paleosurface on the Paso del Sapo volcanic rocks, Chubut, Argentina. Geociências 29:479–486. São Paulo, UNESP, Brazil

Alley NF, Clarke JDA, Macphail M, Truswell EM (1999) Sedimentary infillings and development of major Tertiary palaeodrainage systems of south-central Australia. In: Thiry M, Simon-Coincon R (eds) Palaeoweathering, palaeosurfaces and related continental deposits, vol 27, Special publication of the International Association of Sedimentologists. Blackwell Science, Oxford, pp 337–366

Aragón E, Aguilera EY, Cavarozzi C, Ribot A (2010) The North Patagonian Altiplano and the Somón Curá basaltic plateau. Geociências 29:527–532. São Paulo, UNESP, Brazil

Avias J (1953) Contribution a l'étude stratigraphique et paléontologique des formations antécrétacées en Nouvelle-Calédonie. Science de la Terre no 1 and 2, Nancy, 276 p

Avias J, Gonord H (1975) New Caledonia. In: Fairbridge RW (ed) The encyclopedia of world regional geology Part I: Western Hemisphere. Dowden, Hutchinson & Ross/Wiley, Stroudsburg/New York, pp 365–371

Beard JS (1999) Evolution of the river systems of the south-west drainage division, Western Australia. J R Soc West Aust 82:147–164

Behrendt JC, Cooper A (1991) Evidence of rapid Cenozoic uplift of the shoulder escarpment of the Cenozoic West Antarctic rift system and a speculation on possible climatic forcing. Geology 19:315–319

Benbow MC (1990) Tertiary coastal dunes of the Eucla Basin, Australia. Geomorphology 3:9–29

Bishop P, Goldrick G (2000) Geomorphological evolution of the East Australian continental margin. In: Summerfield MA (ed) Geomorphology and global tectonics. Wiley, Chichester

Bowler JM, Kotsonis A, Lawrence CR (2006) Environmental evolution of the Mallee region, Western Murray Basin. Proc R Soc Vic 118(2):161–210

Briceño HO, Schubert C (1990) Geomorphology of the Gran Sabana, Guayana Shield, southeastern Venezuela. Geomorphology 3:125–141

Brown MC (2000) Cenozoic tectonics and landform evolution of the coast and adjacent highlands of southeast New South Wales. Aust J Earth Sci 47:245–257

Brown RW, Gallagher K, Gleadow AJW, Summerfield MA (2000) Morphotectonic evolution of the South Atlantic margins of Africa and South America. In: Summerfield MA (ed) Geomorphology and global tectonics. Wiley, Chichester, pp 255–281

Burke K, Gunnell Y (2008) The African erosion surface. Geological Society of America, Memoir 201, Boulder, Colorado, pp 1–66

Choubey VD (1971) Pre-Deccan Trap topography in central India and crustal warping in relation to Narmada rift structure and volcanicity. Bull Volcanol 35:660–685

Clarke JDA (1994) Evolution of the Lefroy and Cowan palaeodrainages, Western Australia. Aust J Earth Sci 41:55–68

Coltorti M, Dramis F, Ollier CD (2007) Planation surfaces in Northern Ethiopia. Geomorphology 89:287–296

De Swardt AMJ, Trendall AF (1969) The physiographic development of Uganda. Overseas Geol Miner Resour 10:241–288

De Wit MCJ (1999) Post-Gondwana drainage and the development of diamond placers in western South Africa. Econ Geol 94:721–740

Dingle RV, Hendy QB (1984) Mesozoic and Tertiary sediment supply to the western Cape Basin and paleodrainage systems in south-western Africa. Mar Geol 56:13–26

Dingle RV, Siesser WG, Newton AR (1983) Mesozoic and Tertiary history of southern Africa. Balkema, Rotterdam

Dixey F (1960) The geology and geomorphology of Madagascar, and a comparison with eastern Africa. Q J Geol Soc 116:255–268

Dollar ESJ (1998) Palaeofluvial geomorphology in southern Africa: a review. Prog Phys Geogr 22:325–349

Evans AM (1993) Ore geology and industrial minerals: an introduction. Wiley, New York

Gallagher K, Hawkesworth C, Mantovani M (1994) The denudation history of the onshore continental margin of SE Brazil inferred from fission track data. J Geophys Res 99:18117–18145

Galloway RW (1967) Pre-basalt, Sub-post-basalt and Post-basalt surfaces of the Hunter Valley, New South Wales. In: Jennings JN, Mabbutt JA (eds) Landform studies from Australia and New Guinea. ANU Press, Canberra

Goudie AS (2005) The drainage of Africa since the Cretaceous. Geomorphology 67:437–456

Gunnell Y (1998) The interaction between geological structure and global tectonics in multistoreyed landscape development: a denudation chronology of the South Indian shield. Basin Res 10:281–310

Gunnell Y, Harbor D (2008) Structural underprint and tectonic overprint in the Angavo (Madagascar) and Western Ghats (India) – implications for understanding scarp evolution at passive margins. J Geol Soc India 71:763–779

Gunnell Y, Radhakrishna BP (eds) (2001) Sahyadri: the great escarpment of the Indian subcontinent, Geological Society of India Memoir 47(1–2). Geological Society of India, Bangalore, 1054 p

Gunnell Y, Gallagher K, Carter A, Widdowson A, Hurford AJ (2003) Denudation history of the continental margin of western peninsular India since the early Mesozoic – reconciling apatite-fission track data with geomorphology. Earth Planet Sci Newsl 215:187–201

Haworth RJ, Ollier CD (1992) Continental rifting and drainage reversal: the Clarence Valley of eastern Australia. Earth Surf Process Landforms 17:387–397

Herath JW, Pathirana HCNC (1983) Genesis and constitution of Sri Lankan Laterites. J Natl Sci Counc Sri Lanka 11:277–292

Hills ES (1975) The physiography of Victoria. Whitcombe and Tombs, Melbourne

Holmes A (1965) Principles of physical geology. Nelson, London

Huggett RJ (2011) Fundamentals of geomorphology, 3rd edn. Routledge, London

Idnurm M (1985) Late Mesozoic and Cenozoic palaeomagnetism of Australia – I. A redefined apparent polar wander path. Geophys J R Astron Soc 83:399–418

Japsen P, Chalmers JA, Green PF, Bonow JM (2012) Elevated, passive continental margins: not rift shoulders, but expressions of episodic, post-rift burial and exhumation. Global Planet Change 90–91:73–86

Kale VS (2010) The Western Ghat: the Great Escarpment of India. In: Migón P (ed) Geomorphological landscapes of the world. Springer, New York, pp 257–264

Kennedy BA (1964) Inventing the Earth: ideas on landscape development since 1740. Blackwell, Oxford

Kerr A, Sugden DE, Summerfield MA (2000) Linking tectonics and landscape development in a passive margin setting: the Transantarctic Mountains. In: Summerfield MA (ed) Geomorphology and global tectonics. Wiley, Chichester

King LC (1967) The morphology of the earth. Oliver and Boyd, Edinburgh

Kotsonis A (1999) Tertiary shorelines of the western Murray Basin: weathering, sedimentology and exploration potential. In: Murray Basin mineral sands conference, extended abstracts, 21–23 April, Mildura, VIC, Australia, pp 57–63

McCarthy D (2003) The trans-Pacific zipper effect – disjunct sister taxa and matching geological outlines that link the Pacific regions. Biogeography 30:1545–1561

Moore AE (1999) A reappraisal of epeirogenic flexure axes in southern Africa. S Afr J Geol 102:363–376

Moore AE, Larkin PA (2001) Drainage evolution in south-central Africa since the break-up of Gondwana. S Afr J Geol 104:47–68

Moore AE, Cotterill FDP, Broderick T, Plowes D (2009) Landscape evolution in Zimbabwe from the Permian to present, with implications for kimberlite prospecting. S Afr J Geol 112:65–88

Morris RC, Ramanaidu ER (2007) Genesis of the channel iron deposits (CID) of the Pilbara region of Western Australia. Aust J Earth Sci 54:733–756

Nash C, Chakravartula P, Crowe W (1999) Landforms of Northern Orissa, India from analysis of satellite imagery and aircraft digital terrain data. Gondwana Research Memoir No. 5. Eastern Ghats Granulites, Osaka Field Science Publishers, pp 1–13

Näslund JO (2001) Landscape development in western and central Dronning Maud Land, East Antarctica. Antarct Sci 13:302–311

Ollier CD (1959) The soils of the Northern Province, Uganda (excluding Karamoja District), Uganda Protectorate Department of Agriculture Memoirs of the Research Division Series 1: Soils, 3, p 47

Ollier CD (1978) Tectonics and geomorphology of the eastern highlands. In: Davies JL, Williams MAJ (eds) Landform evolution in Australasia. Australian National University Press, Canberra, pp 5–47

Ollier CD (1981) Tectonics and landforms. Longman, Harlow, p 324

Ollier CD (1982) The Great Escarpment of eastern Australia: tectonic and geomorphic significance. J Geol Soc Aust 29:13–I23

Ollier CD (1992) Global change and long-term geomorphology. Terra Nova 4:312–319

Ollier CD (1993) Age of soils and landforms of Uganda. Isr J Earth Sci 41:1–5

Ollier CD (1995) Tectonics and landscape evolution in southeast Australia. Geomorphology 12:37–44

Ollier CD, Pain CF (1994) Landscape evolution and tectonics in southeastern Australia. AGSO J Aust Geol Geophys 15:335–345

Ollier CD, Pain CF (1997) Equating the basal unconformity with the palaeoplain: a model for passive margins. Geomorphology 19:1–15

Ollier CD, Powar KB (1985) The Western Ghats and the morphotectonics of Peninsular India. Zeitschrift fur Geomorphologie, Supplementband, N.F. 54:57–69

Ollier CD, Rajaguru SN (1989) Laterite of Kerala (India). Geografia Fisica e Dinamica Quaternaria 12:27–33

Ollier CD, Sheth HC (2008) The High Deccan duricrusts of India and their significance for the 'laterite' issue. J Earth Syst Sci 117:537–551

Ollier CD, Stevens NC (1989) The Great Escarpment in Queensland. In: Le Maitre RW (ed) Pathways in geology – essays in honour of Edwin Sherbon Hills. Blackwell, Melbourne, pp 140–152

Ollier CD, Chan RA, Craig MA, Gibson DL (1988a) Aspects of landscape history and regolith in the Kalgoorlie region, Western Australia. BMR J Aust Geol Geophys 10:309–321

Ollier CD, Gaunt GFM, Jurkowski I (1988b) The Kimberley Plateau, Western Australia: a Precambrian erosion surface. Zeitschrift für Geomorphologie 32:239–246

Pain CF, Ollier CD (1995) Inversion of relief –a component of landscape evolution. Geomorphology 12:151–165

Pankhurst RA, Trouw RAJ, de Brito Nemes BB, de Wit MJ (2008) West Gondwana: pre-Cenozoic correlations across the South Atlantic Region. Geol Soc Lond Spec Publ 294:1–8

Partridge TC (1998) Of diamonds, dinosaurs and diastrophism: 150 million years of landscape evolution in southern Africa. S Afr J Geol 101:167–184

Partridge TC, Maud RR (1987) Geomorphic evolution of southern Africa since the Mesozoic. S Afr J Geol 90:179–208

Peulvast JP, Claudino Sales V (2004) Stepped surfaces and palaeolandforms in the northern Brazilian "Nordeste": constraints on models of morphotectonic evolution. Geomorphology 62:89–122

Peulvast JP, Claudino Sales V, Bezerra H, Betard F (2006) Landforms and neotectonics on a passive margin: the equatorial side of northeastern Brazil. Geodin Acta 19:51–71

Peulvast JP, Betard F, Lageat Y (2009) Long-term landscape evolution and denudation rates in basement and platform areas: the contribution of the morphostratigraphic approach. Géomorphol Relief Processus Environ 2:95–108

Pillans B (2007) Pre-Quaternary landscape inheritance in Australia. J Quat Sci 22:439–447

Rabinowitz PD, Coffin MF, Falvey D (1983) The separation of Madagascar and Africa. Science 220:67–69
Schmidt PW, Ollier CD (1988) Palaeomagnetic dating of Late Cretaceous to Early Tertiary weathering in New England, NSW, Australia. Earth Sci Rev 25:363–371
Schmidt PW, Williams GE (2002) Palaeomagnetic dating of the Hamersley Surface and deep weathering in the Pilbara-Northern Yilgarn region, W.A. In: Preiss WV (ed) Geoscience 2002: expanding horizons. 16th Australian Geological Convention, Adelaide, 1–5 July 2002. Geological Society of Australia Abstracts no 67, Sydney, p 431
Sheth HC (2007) Plume-related regional prevolcanic uplift in the Deccan Traps: absence of evidence, evidence of absence. Geol Soc Am Spec Pap 430:785–813
Spry MJ, Gibson DL, Eggleton RA (1999) Tertiary evolution of the coastal lowlands and the Clyde River palaeovalley in south-east New South Wales. Aust J Earth Sci 46:173–180
Stewart AJ, Blake DH, Ollier CD (1986) Cambrian river terraces and ridgetops in central Australia: oldest persisting landforms? Science 233:758–761
Suvires GM (2010) Paleosurfaces and relief evolution in cratonic areas of the western Pampean Ranges (Province of San Juan, Argentina). Geociências 29:501–509. São Paulo, UNESP
Taylor RG, Howard KWF (1998) Post-Palaeozoic evolution of weathered landsurfaces in Uganda by tectonically controlled deep weathering and stripping. Geomorphology 25:173–192
Temple P (1967) Geomorphology Map. Atlas of Uganda. Department of Lands and Survey, Uganda, pp 8–9
Tingey R (1985) Uplift in Antarctica. Zeit. f. Geomorph., Supplementband 85:85–99
Torsvik TH, Tucker RD, Ashwal LD, Carter LM, Jamtveit B, Vidyadharan KT, Venkataramana P (2000) Late Cretaceous India-Madagascar fit and timing of break-up related magmatism. Terra Nova 12:220–225
Twidale CR (1994) Gondwana (Late Jurassic and Cretaceous) palaeosurfaces of the Australian Craton. Palaeogeogr Palaeoclim Palaeoecol 112(1–2):157–186
Vandenberg AHM (2010) Paleogene basalts prove early uplift of Victoria's eastern uplands. Aust J Earth Sci 57:291–315
Venktakrishnan R (1984) Parallel scarp retreat and drainage evolution, Pachmarhi area, Madhya Pradesh, central India. J Geol Soc India 25:401–403
Venktakrishnan R (1987) Correlation of cave levels and planation surfaces in the Pachmarhi area, Madhya Pradesh. A case for base level control. J Geol Soc India 29:240–249
Wadia DN (1943) The three superposed peneplains of Ceylon: their physiography and geological structure. Ceylon Department of Mineralogy, Professional paper 1:25–38
Widdowson M (1997) Tertiary palaeosurfaces of the SW Deccan, western India: implications for passive margin uplift. In: Widdowson M (ed) Palaeosurfaces: recognition, reconstruction, and palaeoenvironmental interpretation, Geological Society special publication 120. Geological Society of London, London, pp 221–248
Zonneveld JIS (1985) Geomorphological notes on the continental border in the Guyanas. Z Geomorph, Supplementband 54:71–84

A Brief Review of the Development of Gondwana Landscape Studies in Africa, the Centrepiece of the Former Gondwana

Rodney R. Maud

Abstract This chapter presents a summary of the development of landscape studies in Africa since their inception in the 1930s. The role of Lester C. King in the progress and advancements of geomorphology studies in Africa, both in terms of processes and landforms, particularly the ancient landscape complexes and the evolution of geomorphological thinking, is herein highlighted. Later scientific developments in these fields are recognised as well.

Keywords Gondwana • Southern Africa • Lester King • Great Escarpment • Planation surfaces

Introduction

> 'Ex Africa semper aliquid novi' – Pliny, the Elder in 'Historia Naturalis', Pliny having been a notable Roman natural history and geography writer.

The Pliny quotation above continues to be applicable (usually in an adverse context) to the present day! What Pliny was unaware of, however, is that Africa, to the present day, carries some very old features in the form of some of its elevated, mainly high plain landscapes of erosional origin which have remained substantially unchanged, subaerially, since their initiation in the Cretaceous, about 100 Ma ago. That such landscapes could exist thus, effectively largely unchanged, for such a very long time and that they still exist today have been incomprehensible to most Europe-based geomorphologists, although in recent times, acceptance of this fact has become a little more widespread.

R.R. Maud (✉)
School of Geological Sciences, University of KwaZulu-Natal, Durban, South Africa
e-mail: Reception@dmpconsulting.co.za; cillagm@iafrica.com

J. Rabassa and C. Ollier (eds.), *Gondwana Landscapes in southern South America*, Springer Earth System Sciences, DOI 10.1007/978-94-007-7702-6_5,

Fig. 1 Lester King (1907–1989)

This is in no small measure due to the works of Lester King in the field of landscape and related studies, this writer having been fortunate enough to have been one of his students (although described by him on occasion as one of the most 'misguided' thereof) during his main period of productive effort and publication in the 1950s and 1960s, as a result of which he acquired, later, the reputation of being one of the most influential (if at times controversial) geomorphologists of the twentieth century.

What follows below is a resume of the development of landscape studies in Africa since their inception in the 1930s, the contribution of Lester King thereto necessarily being concentrated on (Fig. 1).

Early Landscape Studies in Africa (1930–1942)

Unsurprisingly, because of their excellent development in places there, the high plains of Central and East Africa obviously drew the attention to themselves of the early workers in the landscape development field.

Probably the earliest worker in the field in Africa was Wayland (1931) who recognised three planation surfaces of different ages in the vicinity of the rift valley in Uganda, these being one of post-Karoo to Late Cretaceous age, one of Early Miocene age (dated from fossiliferous sediments around Lake Victoria), and one of Late Pliocene age. It is remarkable that these deductions have been largely confirmed to be the case in a wider context elsewhere in Africa by most of the later workers. The foundations for landscape development studies in Africa were thus well and truly laid.

Wayland was followed by Beetz (1933) who identified a sub-escarpment Namib surface in southern Angola which is warped down beneath Late Cretaceous marine sediments along the coast. He extrapolated this surface to South-West Africa (Namibia) and to South Africa and noted that it was there associated with deeply penetrating leaching and silicification – an early very perceptive and important observation in the light of subsequent work by others in this part of Africa (Partridge and Maud 1987).

Veatch (1935) described an extensive land surface of Cretaceous age, extending over much of Central Africa, which remained undisturbed until the Middle Miocene on the evidence of a continuous, undisturbed, Early Cretaceous to Middle Miocene coastal sedimentary sequences preserved in Angola. He also identified an end-Tertiary surface produced through the erosion of some 3,000 m of earlier sediments following the Middle Miocene up-warping of the coastal margin (shades of what was later to be found to be the case in South Africa).

Jessen (1936) recognised five main surfaces in Angola, which he believed to be produced by intermittent uplift of the region since the Middle Triassic. The main plateau surface he regarded as being of Jurassic to Early Cretaceous age, while higher remnants were considered to date to the Late Triassic and Early Jurassic. Below the plateau surface was a plain of Early to Middle Cretaceous age, which was followed at a lower elevation by a surface of Albian to Early Eocene age (the Namib surface of Beetz). Along the coast, a surface of Mio-Pliocene age was preserved.

Willis (1936) postulated a high-level planation surface based on a remnant feature in Tanzania which he regarded as being of Jurassic age. The main plateau of Tanzania was considered to be produced by an erosion cycle which culminated in the Miocene. He also described a Late Tertiary surface more than 100 m below the main plateau surface.

Dixey (1938) (later Sir Frank Dixey, a geologist in the British Colonial Service, in Central Africa, who was later to become the main disputant with Lester King in the matter of African erosion surfaces and landscape development) entered the scene. On evidence chiefly from Central Africa, but also from southern Africa, he proposed the existence of a late Jurassic peneplain which truncated Karoo boundary faults in the vicinity of the Nyika-Vipya plateau in Malawi. Valleys incised into the Jurassic surface, as well as the down-faulted Nyasa-Shiré and Luangwa-Zambezi troughs, were filled with sediments of Early Cretaceous age. The main plateau surface was formed by a cycle lasting from the Early Cretaceous to the Middle Miocene, but most of the planation occurred at the beginning of this interval and was interrupted by warping and volcanism in the Middle Cretaceous, which resulted in a new planation of the surface during the Early Tertiary. A lower end-Tertiary surface, in the form of broad valleys, was developed chiefly on softer lithology in response to uplift along both the east and the west coasts at the end of the Pliocene and the beginning of the Pleistocene.

Dixey (1942) added to his 1938 list of erosion surfaces an end-Cretaceous-Early Tertiary surface and an incised Early Pleistocene surface (typified by gorge cutting below the Victoria Falls). The earlier of these accommodated the high remnants above the main plateau surface, which had previously been ascribed to end-Tertiary up-warping. Hence, a distinction was drawn between an Early Cretaceous 'trough' surface and an end-Cretaceous upland surface. The latter was recognised as occurring around the margins of Lesotho, in South Africa, at 2,250–2,300 m elevation. Attention was also drawn to the stripping of large areas of Karoo sediments during the main cycle which ended in the Middle Miocene, resulting in the exposure of a pre-Karoo surface. Previous continuity between the main plateau surface and its coastal equivalent was proposed and was ascribed to up-warping along an axis coinciding generally with the 'Great Escarpment'. Tectonic influences were considered to be limited to Middle Cretaceous faulting, with little subsequent disturbance until the main period of uplift at the end of the Tertiary.

Thus, by the beginning of the 1940s, the basic groundwork regarding the development of the African erosional landscapes had been laid by diverse authors in a short period of only some 10 years. Having had its origin in Central and East Africa, with time its focus moved south in large measure to Southern Africa.

In effect all later work in this regard comprised refinement to varying degree of this groundwork, with the addition to, or the subtraction from, the pre-existing lists of landscape erosional and depositional cycles, and the refinement of their ages.

The stage was now set for the bursting on to this scene of one, Lester King who was to leave an indelible mark on both the African, and the general worldwide, geomorphological scene.

Lester C. King (1907–1989; Publishing Activity: Major African and General Erosion, 1942–1983)

Lester Charles King was born in London in 1907, but in 1909 his family migrated with him to New Zealand, where he obtained his B.Sc. and M.Sc. degrees in due course. Initially his main interest was in palaeontology, but he also had a peripheral interest in geomorphology while a lecturer at the then Victoria University College, Wellington. During this time he published a number of papers in these, his subjects of interest.

In 1935, he migrated to South Africa taking up an appointment as lecturer in geology and geography at the then Natal University College in Pietermaritzburg. In 1946, he was appointed professor in geology; in 1948, he transferred to Durban to a new Department of Geology and Mineralogy, the institution becoming the University of Natal the following year. He was to remain in this position until his retirement therefrom in 1973.

He published one of his first papers in South Africa on the 'Monoclinal Coast of Natal' in 1941. In 1942, he published his first geomorphological book, 'South African Scenery' (King 1942), which later ran to three editions, two revised, but this first edition being very probably the best. In all, King published personally (mainly), and in collaboration, of the order of some 50 major papers and books in his field of geomorphology, the last, in 1983, being entitled 'Wandering Continents and Spreading Sea Floors on an Expanding Earth'. He also produced numerous 'minor' local publications.

King based his interpretations of landforms and landscapes very firmly in the concept of erosion by parallel scarp retreat, for although he acknowledged the

Fig. 2 Dynamic geomorphology – scarp retreat in action! Note landslide on lower near slope and rockfall above it in right middle distance, Drakensberg escarpment with crest at 3,000 m on left skyline, near Bergville, KwaZulu-Natal

possibility of slope decline in certain (lithological) circumstances, he used that mechanism little, if at all, in understanding landscape. Rather, pediments and inselbergs, for instance, as well as entire landscapes, he construed in terms of scarp recession. He considered pediplanation (scarp retreat and pedimentation) to be active in varying degree in all regions where running water is responsible for shaping the land surface. King's studies in denudation chronology are consistent with his commitment to scarp retreat, implicit in which is the possibility of the survival of very old, flat or near flat land surfaces (Fig. 2).

Perhaps the best known single paper to the wider audience of geomorphologists interested in landscape evolution and chronology is King's 'Canons of Landscape Evolution' published in 1953 in the Bulletin of the Geological Society of America,

Fig. 3 The summit of the Lesotho Highlands ±3,000 m originally considered by King to be of Jurassic age. Actually of structural origin on flat-lying basalt. Near Underberg, KwaZulu-Natal

a paper in which he developed the concept of backwearing of slopes and spelled out his geomorphological credo in 50 laws or canons. His 'magnum opus', a weighty tome, *The Morphology of the Earth* (1962) consolidated his tectonic and geomorphological interpretations on a global scale and to every continent.

Some of the more important other publications of King include the following: King (1950, 1957, 1966 and 1976). See also King (1950, 1951, 1953, 1956 and 1963).

Regarding landscape development in Southern Africa, King's views changed significantly with time. Thus, in his 1942 book, King concluded that the landscape of Southern Africa could be resolved into three major cycles and surfaces. These were the Gondwana surface of Cretaceous and Late Miocene age; the African surface of Early to Middle Tertiary age, frequently carrying laterite or silcrete duricrusts; and the multiphase Late Cenozoic surface. In his early 1944 paper, he considered that the crests of the Lesotho Highlands represent a Jurassic erosion surface that has survived since prior to the break-up of Gondwana in Late Jurassic to Early Cretaceous time and that the disparity of elevations of erosion surfaces below the Great Escarpment and on the interior plateau (Highveld) was due to the respective differences in distances to base level in each case (Fig. 3).

In 1949, King argued that the coexistence of land surfaces of different ages is possible only if landscape evolution proceeds through backwearing under a process of pediplanation. For the first time, names were proposed for the various erosion cycles as expressed by the land surfaces: Miocene, Gondwanaland; end-Tertiary, African; Pleistocene, Victoria Falls; and present, present. In 1951, he more or less reiterated the same situation adding a Gondwanaland surface which also predated the break-up of Gondwana, it occurring at different levels above and below the Great Escarpment. The African surface originated with the fragmentation of Gondwana, the subsequent Victoria Falls and Congo cycles being manifested in river incision. Fair and King (1954) renamed the 'marginal' Gondwanaland surface the 'post-Gondwana', it having been initiated in the Middle Cretaceous based on the evidence of coastal marine sequences. Subsequent cycles were the 'African' (Middle Tertiary), 'Victoria Falls' (end-Tertiary), 'Congo' (Pleistocene) and the 'Latest' (recent).

King and King (1959) (the second King being Lester King's daughter, Linley) postulated that the African surface was carried across the Great Escarpment by end-Tertiary uplift that in the continental marginal areas having been continuous with the higher Highveld surface of the interior. The summits of the Lesotho Highlands were now considered to represent the Gondwana cycle of Jurassic age. The post-Gondwana was initiated by the fragmentation of Gondwana in the Early Cretaceous. Epeirogenic rejuvenation of the marginal monocline, with some faulting, in the Middle Cretaceous gave rise to the African cycle. Uplift in the Early Miocene produced renewed incision and planation in the post-African cycle, these cycles being correlated with unconformities in coastal marine sediments of Cretaceous and younger age. Rejuvenation and enlargement of the marginal monocline during the Pliocene produced local planation in a Late Tertiary phase II cycle. Finally major up-warping at the end of the Pliocene initiated major gorge incision in the sub-escarpment zone.

In a subsequent paper (1976), and in his final synthesis, King (1983) proposed a new (somewhat bizarre) nomenclature for the various surfaces which he considered as being of global applicability, he having correlated the African landscape development with what he had observed on visits to eastern Australia (1950) and Brazil (1956). This was his last erosional chronology scheme in which he identified six major surfaces and cycles: the 'Gondwana' planation of Jurassic age, the 'Kretacic' (Early-Middle Cretaceous), the 'Moorland' (Late Cretaceous to Middle Cenozoic), the 'Rolling' (Miocene), the 'Widespread' (Pliocene) and the 'Youngest' (Quaternary, essentially modern). Needless to say this last scheme of King's has not received the same degree of support as did his original one. It is not surprising that King's many of views frequently generated controversy in certain more conventional quarters. (The reader of all the above is to be readily forgiven, if at the end of it, he finds himself more than a little confused.)

Lester King himself was an excellent lecturer, his lectures being clear, informative, well organised and laced with amusing anecdotes. He was very versatile, his expertise extending over the whole range of the geological sciences. In later life, however, as is often the case, sadly, he became more than a little authoritative. This writer earned the title of 'his most misguided student' by questioning (Maud 1961) his interpretation of his beloved Natal 'hard rock' Monocline (1941, 1972) and showing that in fact the region in question is of tensional faulted origin (break-up of Gondwana age), it being part of what is now termed a rifted passive continental margin. Lester King died in 1989 after a long illness, and his ashes were scattered at the foot of the erosional Drakensberg Mountains which form part of the Great Escarpment in what is now KwaZulu-Natal; a more fitting place for Lester King's ashes to rest in, below a mighty erosional scarp, could not be found.

Later Landscape Studies in Africa

Given the prominence of work of Lester King, mainly in Southern Africa, it must not be supposed the similar landscape and related studies were not being undertaken elsewhere in Africa. Thus, in West Africa, Pugh (1954) worked in northern and eastern Nigeria (Jos State there is known as the 'Plateau State') as did Thomas (1994), while in East Africa, Uganda and Kenya notable workers included Ollier (1959) and Macfarlane (1976) (Figs. 4 and 5).

No doubt many other workers were active in these parts of Africa whose publications are unknown to this author. This author though has visited those parts of Africa and observed the close similarity of the landscape features prevailing there to those extant in the southern portion of the continent.

Fig. 4 Residual of laterite duricrusted African surface with underlying deep weathering. Kano, Northern Nigeria

Fig. 5 Islands of laterite duricrust of the African surface downwarped as part of Lake Victoria basin, near Entebbe, Uganda

The Works of Partridge and Maud (1987, 1989, 2000), Partridge (1998) and Maud (2012)

Partridge and Maud modified and revised the work of King in Southern Africa in the light of more recently available relevant information and their own extensive observations throughout this part of the continent.

Thus, the idea of King that landscape on the top of Drakensberg had survived since prior to the disruption of Gondwana was rejected on the evidence of the morphology of diamondiferous kimberlite pipes of Middle Cretaceous age which penetrated to at least this level (3,000 m), the indication therefrom being that at least 300 m of basalt has been removed by erosion from above this surface.

On the same basis and in the light of xenoliths in such pipes of country rock and of relevance in alluvial diamond exploration in Central Southern Africa, it has been established that at least 2,500 m of Karoo basalt and sediments which formerly covered this area has been removed by eastward scarp retreat since the Cretaceous (Hanson et al. 2009; in Maud 2012). Since the Late Cretaceous the landscape over much of the central portion of the subcontinent has suffered negligible erosion as is shown by the survival on the present landscape surface of silicified wood of Late Cretaceous age in fluvial gravels (diamondiferous) at Mahura Muthla west of Kimberley (Partridge 1998).

Fig. 6 Outcrop of duricrust laterite on the African surface on right. African surface also on distant flat skyline. Post-Africa I surface at lower level in the middle distance with younger river incision. Total relief approaching 1,800 m. Valley of 1000 Hills northwest of Durban

Fig. 7 Near flat skyline of African surface etchplain with remnants of deep weathering (kaolinisation) in road cutting. Eastern Transvaal, near Ermelo

Fig. 8 Silcrete duricrust overlying deep weathered (kaolinised) rock. About 50 m above sea level, sea visible in the distance. Near Hermanus, Western Cape (east of Cape Town)

Fig. 9 Massive silcrete duricrust on intermontane valley-ride African surface shoulder, as can be seen in the middle-far distance. Near Oudtshoorn, Western Cape

Fig. 10 African surface with duricrust silcrete capping and deep weathering (kaolinisation). Note: large granite core-stones in weathered material which will become tors with further backwearing erosion. Lower surface on left, post-African I. Near Platbakkies (Springbok), Northern Cape Province

Fig. 11 Remnant of marine Eocene, shelly limestone on African surface at 400 m elevation, 25 km from present coastline. Note the rise of the planed skyline from right to left, in an inland direction, of the African surface. Near East London, Eastern Cape

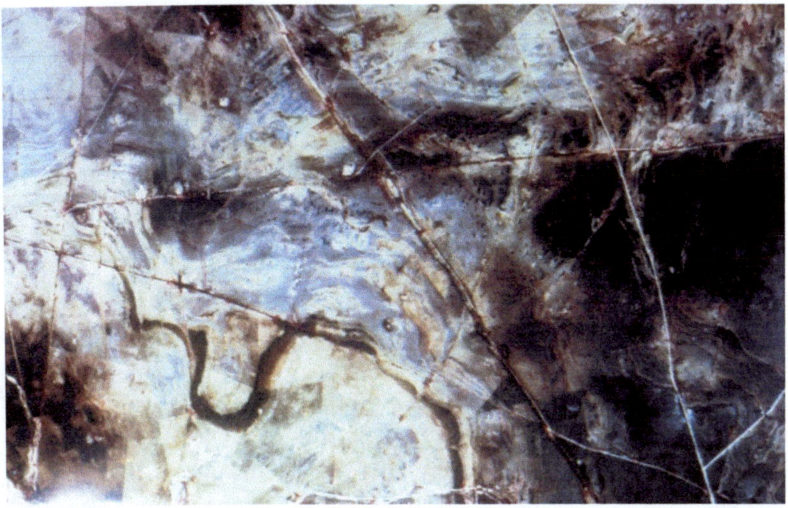

Fig. 12 Landsat view of former river meander of Cretaceous age on the etched African surface. This has survived subaerially for some 80 Ma, which is incomprehensible to most European geomorphologists. Many such palaeo-river channels, in their calcreted fluvial gravels, are diamondiferous. Mahura Muthla near Vryburg west of Kimberley, Northern Cape Province

The findings of Partridge and Maud, as given in the publications indicated above, may be briefly summarised as follows:

1. Africa in the centre of Gondwana stood relatively high in relation to surrounding areas.
2. Break-up of Gondwana in the Late Jurassic to Middle Cretaceous by rift faulting, after associated increase in elevation of the rift shoulders, initiated the African cycle of erosion, which is multiphase, the resulting surface of advanced planation being characterised by duricrust laterite in the north and silcrete in the south, together with deep weathering (kaolinisation) of underlying rocks. Some minor lowering of the surface due to removal by erosion of weathered material (etching) has taken place in some areas. Corresponding sedimentation has taken place off continental margins.
3. The African cycle was terminated by moderate epeirogenic uplift of varying amount along axes parallel to and in the hinterland of the coast. The east coast hinterland experienced greater uplift than the southern and west coasts, leading to topographic asymmetry of the subcontinent.
4. The African cycle was succeeded by the post-African I cycle of erosion which did not achieve the advanced planation of the African cycle.
5. Termination of the post-African I cycle of erosion by major epeirogenic uplift in the Early Pliocene along some axes in the coastal hinterland previously occurred. Total amount of combined uplifts on the east coast was about 1,150 m, on southern coast about 400 m and on the west coast about 250 m.

Fig. 13 Silicified fossil wood in the fluvial gravels at Mahura Muthla of Late Cretaceous age, on the etched African surface. Near Vryburg, west of Kimberley, Northern Cape Province

6. The post-African I cycle was terminated by the post-African II erosion cycle of major valley incision, especially in the southeastern coastal hinterland.
7. The post-African II erosion cycle was terminated in the Late Pliocene and Pleistocene by river valley erosion consequent on climatic and glacio-eustatic changes of ocean and local base levels, which situation continues to the present day (Figs. 6, 7, 8, 9, 10, 11, 12, 13 and 14).

As far as possible, Partridge and Maud attempted to recognise the succession of erosion cycles first described by King and, where appropriate (as in the name of the African cycle), continue his nomenclature. However, his more exotic nomenclature was replaced, where applicable, by one of a more easily understandable chronological type.

Acknowledgements This writer wishes to acknowledge with thanks some of the information contained in a manuscript version of a paper given to him by C.R. (Rowl) Twidale, Australia, some years ago now, entitled 'Lester King – Last of the Explorer Scientists'.

He also acknowledges with deep gratitude all the work and companionship of his long-time friend and colleague Tim Partridge (1942–2009), who was also a very observant scientist and who researched and published in quite a few scientific disciplines other than geomorphology (archaeology, soil science, palaeoanthropology, palaeoclimates, engineering geology and geology).

Fig. 14 Outcrop of pipe of volcanic alnoite (melilite basalt) 63 Ma. Formerly capped by silcrete duricrust on Africa flat surface visible in the background. Near Swellendam, Eastern Western Cape Province

He was indeed a prolific publisher, like King, but sadly he died of a sudden heart attack, all too soon, while still in his productive prime.

This writer also wishes to thank Mrs. Sandy du Toit and Mrs. Sally Padayachee for their valuable assistance in the production of this review.

References[1]

Beetz PFW (1933) The geology of south-west Angola between Cunene and Lunda axis. Trans Geol Soc S Afr 36:137–176

Dixey F (1938) Some observations on the physiographic development of central and southern Africa. Trans Geol Soc S Afr 41:113–172

Dixey F (1942) Erosion cycles in central and southern Africa. Trans Geol Soc S Afr 45:151–181

Fair TJD, King LC (1954) Erosion land-surfaces in the eastern marginal areas of South Africa. Trans Geol Soc S Afr 62:19–26

Hanson EK, Moore JM, Bordy JS, Marsh JS, Hayworth G, Robey JVA (2009) Cretaceous erosion in central South Africa: evidence from upper crustal xenoliths in kimberlite diatremes. S Afr J Geol 112:125–140

Jessen O (1936) Reisen Und Forschungen in Angola. Reimer, Berlin, 397 pp

King LC (1941) The monoclinal coast of Natal, South Africa. J Geomorphol 3:144–153

[1]It must be appreciated that the references cited here are only a small proportion of those covering the subject of this review.

King LC (1942) South African scenery. Oliver & Boyd, Edinburgh, 308 pp
King LC (1950) The study of the world's plainlands: a new approach in geomorphology. Q J Geol Soc 106:101–131
King LC (1951) South African scenery, 2nd rev edn. Oliver & Boyd, Edinburgh, 379 pp
King LC (1953) Canons of landscape evolution. Geol Soc Am Bull 64:721–752
King LC (1956) Rift valleys of Brazil. Trans Geol Soc S Afr 59:199–208
King LC (1957) The uniformitarian nature of hillslopes. Trans Geol Soc Edinb 17:81–102
King LC (1962) The morphology of the Earth. Oliver & Boyd, Edinburgh, 699 pp
King LC (1963) South African scenery, 3rd rev edn. Oliver & Boyd, Edinburgh, 308 pp
King LC (1966) The origin of bornhardts. Zeitschrift für Geomorphologie 10:97–98
King LC (1972) The Natal monocline: explaining the origin and scenery of Natal, South Africa. Geology Department, University of Natal, Durban, vii, 113 p
King LC (1976) Planation remnants on high lands. Z Geomorphol 20:133–148
King LC (1983) Wandering continents and spreading sea floors on an expanding earth. Wiley, Chichester, 232 pp
King LC, King LA (1959) A re-appraisal of the Natal monocline. S Afr Geogr J 41:15–30
Macfarlane MJ (1976) Laterite and landscape. Academic, London
Maud RR (1961) A preliminary review of the structure of coastal Natal. Trans Geol Soc S Afr 64:247–256
Maud RR (2012) Macroscale geomorphic evolution. In: Meadows M, Holmes P (eds) Southern African geomorphology; recent trends and new directions. Sun Press, Bloemfontein, pp 7–19
Ollier CD (1959) A two cycle theory of tropical pedology. J Soil Sci 10:137–148
Partridge TC (1998) Of diamonds, dinosaurs and diastrophism: 150 million years of landscape evolution of Southern Africa since the Mesozoic. S Afr J Geol 90:179–208
Partridge TC, Maud RR (1987) Geomorphic evolution of Southern Africa since the Mesozoic. S Afr J Geol 90:179–208
Partridge TC, Maud RR (1989) The end-Cretaceous event; new evidence from the southern hemisphere. S Afr J Sci 85:428–430
Partridge TC, Maud RR (eds) (2000) The Cenozoic of Southern Africa, Oxford monograph no. 40. Oxford University Press, New York, 406 pp
Pugh JC (1954) High-level surfaces in the eastern highlands of Nigeria. S Afr Geogr J 36:31–42
Thomas MF (1994) Geomorphology in the tropics. A study of weathering and denudation in low latitudes. Wiley, Chichester
Veatch AC (1935) Evolution of the Congo Basin. Geol Soc Am Mem 3, 183 pp
Wayland EJ (1931) Summary of progress for the years 1919 to 1929. Geol Surv, Uganda
Willis B (1936) Studies in comparative seismology – East African Plateau and rift valleys. Carnegie Institution of Washington Publication, Washington, DC, 470 pp

Modeling the Atmospheric Circulation and Climatic Conditions over Southern South America During the Late History of the Gondwana Supercontinent

Rosa Hilda Compagnucci

Abstract The different processes responsible for climate and atmospheric circulation forcing and their relevance on the general circulation of the Southern South America together with the conditions over Patagonia, for the period of the Gondwana supercontinent, are identified in this chapter. During the history of this supercontinent, the main paleoclimate forcings were as follows: (1) the continental drift that affected latitude, elevation, and topography; (2) changes in the amount of greenhouse gases in the Earth's atmosphere; and (3) volcanic activity. The paleoatmospheric circulation is analyzed in special sections according to age, Early Triassic to Early Jurassic, Middle to Late Jurassic, and Cretaceous, accordingly with the key changes in the ocean–land distribution and locations of the continents. Different paleoclimatic modeling scenarios through the periods are reviewed and compared with proxy data. From both sources of information, it arises that the opening of the Hispanic Corridor and the formation of the Atlantic Ocean were the chief factors that produced the strong climatic changes registered from the Triassic to the Cretaceous and the remarkable difference with current climate conditions. Other important factors were the variations in the volume of greenhouse gases, especially CO_2, which is related to volcanic activity and changes in the heat transport through the oceans. The observed results suggest that strong monsoon conditions dominated this period of the Gondwana supercontinent. However, there are large differences with respect to the impact of the various climatic forcings between model simulations of circulation general conditions in the Cretaceous. An extensive list of references provides detailed and updated information on the topics covered in this chapter.

Keywords Paleoclimate modeling • Palaeoatmospheric circulation • Forcing of climatic change • Gondwana • Greenhouse gases

---

R.H. Compagnucci (✉)
DCAO/Facultad de Ciencias Exactas y Naturales, Universidad de Buenos Aires—CONICET, Guiraldes 2160, Ciudad Universitaria, C1428EGA, Buenos Aires, Argentina
e-mail: rhc@at.fcen.uba.ar

J. Rabassa and C. Ollier (eds.), *Gondwana Landscapes in southern South America*, Springer Earth System Sciences, DOI 10.1007/978-94-007-7702-6_6,

Introduction

Different processes are responsible for climate and atmospheric circulation forcing, and their relevance depends on the specific period analyzed and the frequency of climate change considered. Viewing the Earth's climate as a global system, Frakes (1999) described the evolution of climate throughout the past 600 Ma. He highlighted the complex interactions between the carbon cycle, continental distribution, tectonics, sea-level variation, ocean circulation, and temperature change as well as other processes. Valdes (2000) provided an overview of climatic forcing mechanisms and explored their possible role in Phanerozoic climate variations.

Continental drift and orography are important for low-frequency processes, those involving changes over hundreds of millions of years. Glacial eras characterized times when continental masses were located at the pole due to the capacity of land to support and retain ice sheets. During such periods polar ice caps can maintain themselves and grow through the snow and ice accumulation during repeated annual cycles. Warm climate with Earth free of permanent ice sheets characterized the times when the oceans dominated the polar and subpolar regions of the world and continental masses were located in tropical and subtropical latitudes (Crowley and North 1999).

Forcing through atmospheric greenhouse gases is another important factor of climatic change. Two hundred and fifty million years ago, global CO_2 concentrations were approximately 2,000 ppmv (i.e., 3–8 times higher than today; Royer 2006), producing pronounced warming and enhancing seasonal monsoon circulation. Both forcing processes were present especially during the Triassic and Jurassic periods. These times were characterized by a Gondwana supercontinent without significant landmasses located at the southern polar region, together with exceptionally high CO_2 concentrations. Furthermore, sulfur dioxide (SO_2) is the most voluminous chemically active gas emitted by volcanoes, and the major volcanic eruptions have usually formed sulfuric acid aerosols in the lower stratosphere that cooled the Earth's surface. According to Ward (2009), major volcanic activity in Silicic Volcanic Provinces (SVP) has typically preceded an increase in glaciation and a subsequent decrease in sea level throughout the last 600 Ma.

In this chapter, the influence of the different forcing processes in the Southern South America climate and atmospheric circulation during the Gondwana supercontinent are described.

The paleoatmospheric circulation taken from climatic modeling scenarios through the ages is reviewed and compared with proxy data. Detailed and updated reference information on the relevant topics analyzed in this chapter is provided as well.

Note that the ages indicated for each geological period cited in this chapter correspond to those presented at the US Geological Survey Geologic Names Committee (2010), as Divisions of Geologic Time—major chronostratigraphic and geochronological units.

Forcing Description

Paleoclimatic differences from the Triassic to the Cretaceous periods are due principally to changes in the land–ocean distribution. The Pangaea supercontinent was the largest landmass in the Earth's history (Fig. 1). According to plate tectonics and continental drift theory, the Pangaea supercontinent began breaking up about 225–200 Ma ago and rifted into two major landmasses during the Jurassic, Laurasia to the north and Gondwana to the south, roughly separated by the equator.

During the Middle to Late Jurassic, new oceanic gateways were formed with, in particular, the opening of the proto-Atlantic Ocean (the so-called Hispanic Corridor) (Fig. 1), connecting the Pacific Ocean to the western Tethys Ocean (Ziegler 1988; Scotese 2001, 2012; http://www.scotese.com/earth.htm). The reorganization of the Tethys–Atlantic oceanography was triggered by the opening and deepening of the Hispanic Corridor and produced paleoclimatic changes (Rais et al. 2007).

This single continent stretched in a north–south direction across every part of the zonal atmospheric circulation, thereby producing an extraordinary effect on global paleoclimate (e.g., Dubiel et al. 1991; Sellwood et al. 2000; Valdes 1993; Sellwood and Valdes 2006).

Although the continents have not always been in their present positions, the most extreme land of Southern South America, Patagonia, has always been within latitudes actually affected by the westerlies belt area. Iglesias Llanos et al. (2006) suggested that Patagonia shifted between the Earliest and Late Jurassic from latitudes $\sim$50° S to around 30° S affecting its climate (for more details see Volkheimer et al. 2008).

Furthermore, the warm Mesozoic Era (230–65 Ma ago) was likely associated with high levels of CO_2 (Fig. 2) that became stabilized around 2,000 ppm. The Early Jurassic to Late Cretaceous times at 184–66.5 Ma were characterized by very high CO_2 levels (6,000 ppm), though with cool pulses lasting less than 3 Ma (Frakes 1999). Thereafter, CO_2 levels oscillated between very high ($\sim$2,000 ppm) and low (500 ppm) values (Royer 2006).

A major expansion of Antarctic glaciations starting around 35–40 Ma was likely a response, in part, to declining atmospheric CO_2 levels from their previous peak in the Cretaceous ($\sim$100 Ma) (DeConto and Pollard 2003). Furthermore, atmospheric CH_4 concentration, land surface albedo, and ocean heat transport may all have played major, but not mutually exclusive, roles (Valdes 2000). Another important factor was the opening of the Drake Passage, between 49 and 17 Ma (Scher and Martin 2006).

Another important influence on climatic change is volcanic activity. It has been suggested that increased volcanic and seafloor-spreading activities during the Jurassic period released large amounts of carbon dioxide—a greenhouse gas—and led to higher global temperatures. In Fig. 3 the data of the major volcanic activity in Silicic Volcanic Provinces (SVP) is presented, together with the average $\delta^{18}O$ (thick black line) (per million year = 1.45‰—[per mil]) which represent times of

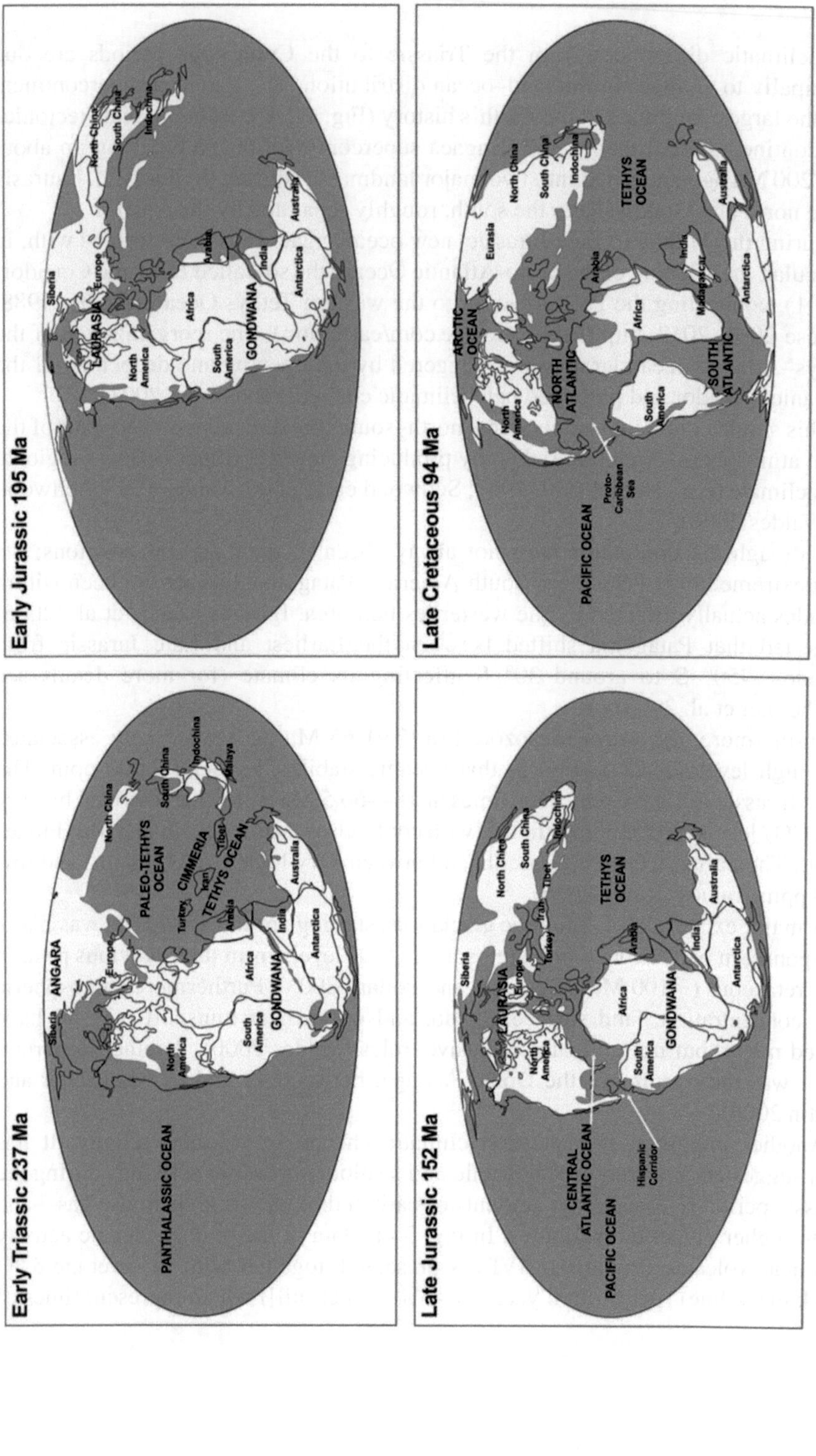

Fig. 1 World map from the Early Triassic (*upper left panel*) to the Late Cretaceous (*lower right panel*) (Adapted from Scotese (2012; http://www.scotese.com/earth.htm))

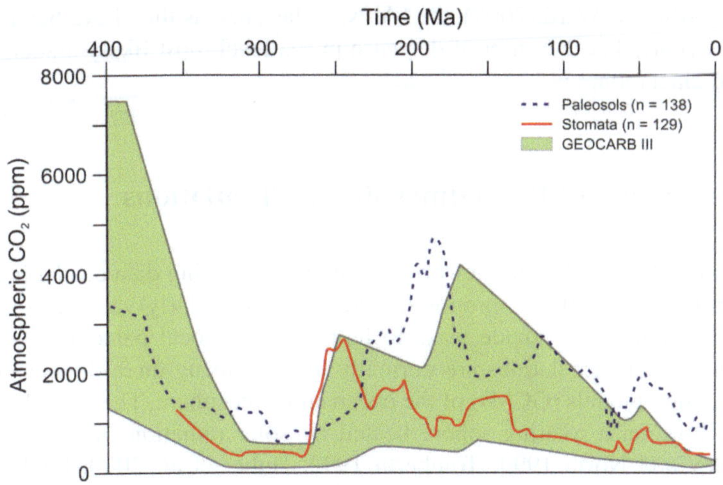

Fig. 2 Atmospheric CO_2 and continental glaciation 400 Ma to present. Plotted CO_2 records represent five-point running averages from each of the four major proxies (Adapted from Royer 2006). Also plotted are the plausible ranges of CO_2 from the geochemical carbon cycle model GEOCARB III (Adapted from Berner and Kothavala 2001). All data have been adjusted to the Gradstein et al. (2004) timescale

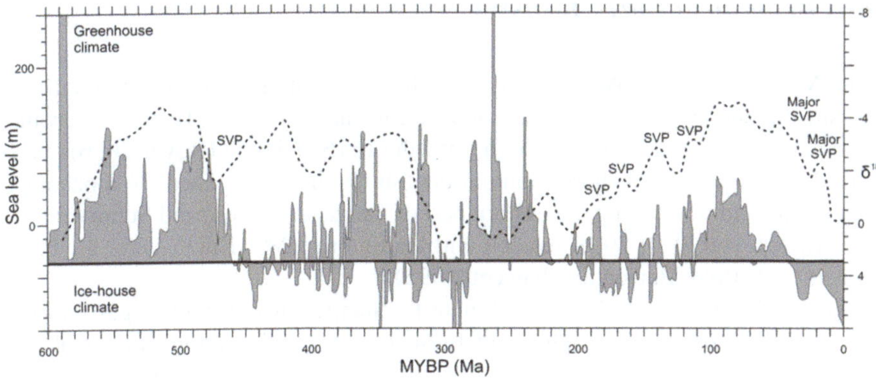

Fig. 3 The times of major volcanic activity in Silicic Volcanic Provinces (SVP). The *shaded gray* areas show the $\delta^{18}O$ proxy for tropical sea-surface temperature (From Veizer et al. 1999). Values below the *black line* show times of glaciation ("icehouse world"), whereas over such line, the values show times of little of no glaciation ("greenhouse world"). The *dotted black line* shows sea level (Haq et al. 1987; Ross and Ross 1987, 1988; Adapted from Ward 2009)

extensive glaciation or icehouse climate (below the solid back line) or warmer times of limited glaciation or greenhouse climate (above the solid back line) and sea-level oscillations (dotted line). The Gondwana supercontinent began breaking up in the Late Jurassic; the opening of the Atlantic Ocean is coincident with the start of major

SVP. According to Ward (2009), the SVPs, as far back as they have been mapped, typically occurred at the onset of decrease in sea level most likely associated with increased glaciation.

Climate Models of Paleoatmospheric Circulations

Climate models provide a framework within which existing data can be interpreted and appropriate hypotheses proposed. They are particularly valuable for regions with limited data and provide ways to help interpret local paleoclimate records. Paleoatmospheric circulations are normally inferred using three-dimensional general circulation models (GCMs) of the ocean and atmosphere. These models predict the fluid flow on a rotating sphere heated by solar radiation (more information in Crowley and North 1999; Trenberth 1992; Huber et al. 2000; Poulsen 2008, among others). All the different GCM simulations for the Gondwana supercontinent (Moore et al. 1992a, b; Chandler et al. 1992; Valdes and Sellwood 1992; Kutzbach and Gallimore 1989) showed strong seasonal alteration of summer monsoon lows and winter monsoon highs.

Early Triassic to Early Jurassic

In the Middle Permian (Wordian; at ~265 Ma ago), the Pangaean supercontinent was surrounded by a Panthalassa super-ocean, and the CO_2 level was around 1,000+ ppm and remained quite high until the Early Triassic (Royer 2006). Most of Early Jurassic atmospheric simulations show the dominance of monsoonal circulation conditions along the eastern part of Pangaea (Scotese and Summerhayes 1986; Crowley et al. 1989; Kutzbach and Gallimore 1989; Kutzbach et al. 1990; Fawcett et al. 1994; Barron and Fawcett 1995).

The sensibility of the Wordian climate to changes in greenhouse gas concentrations, high-latitude geography, and Earth orbital configurations was analyzed by Winguth et al. (2002) using a coupled atmosphere–ocean model. The $1\times CO_2$ concentration (present level) would more likely be associated with a glacial climate, whereas high CO_2 ($8\times CO_2$) concentrations simulate very warm conditions, even in high latitudes.

The simulated climate with $4\times CO_2$ present level (Fig. 4) usually agrees well with sediments and phytogeographical patterns and has a significantly warmer polar climate than the present day, owing to increased flux of heat from ocean to atmosphere in high latitudes. The simulated ocean conditions is characterized by the following: a strong westward equatorial current which is blocked by islands at the eastern Tethys Sea, a warm poleward-directed currents along the Tethys coast and east coast of Gondwana, a cold equatorward current along the east coast of Angara, large meridional overturning circulation cells with deep water formation at

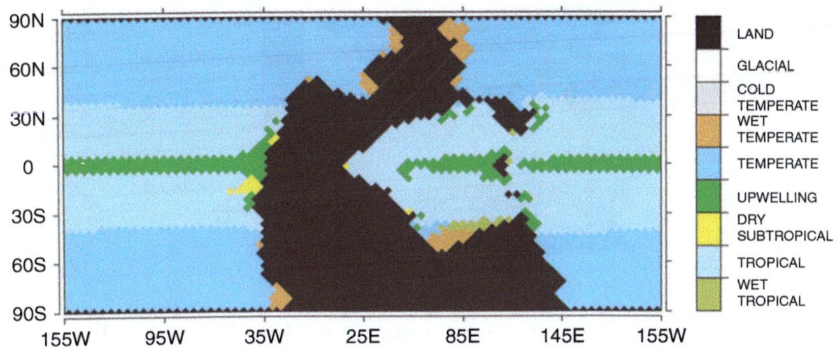

Climate Zone	Water Mass Essentials	Geological Recognitions
Glacial	Surface permanent frozen; SST<-1.8°C	Marine till
Cold temperate	Winter ice flows; SST -1.8 – 0°C	Dropstones, rhythmites
Wet temperate	Brackish surface water; SST 0 – 20°C, salinity <32	Temperate peats, organic rich shales
Temperate	Mixed water column; SST 0 – 20°C, salinity 32–37	Clastics
Cool subtropical	Upwelling currents; vertical velocity w $>1\text{x}10^{-6}$ m s^{-1}	Organic-rich shales, phosphorites, cherts
Dry subtropical	High evaporation; salinity > 37	Gypsum, halite, sabkha facies
Tropical	Deep light penetration; SST > 20°C, salinity 32–37	Carbonates, oolites, coralgal reefs
Wet tropical	High precipitation or continental runoff; SST > 20°C, salinity <32	Tropical peats, muddy sediments

Fig. 4 Water mass classification according to Ziegler et al. (1998) derived from the climate simulation with 4× CO_2 concentrations (Winguth et al. 2002)

both poles, and a warm deep ocean with low current speed in the eastern equatorial Panthalassa. The simulations suggest that a moderate climate over South Gondwana could be reproduced by an atmospheric CO_2 concentration of at least four times present levels and/or the existence of a south polar seaway that further increases the heat flux to the atmosphere from the ocean.

The climate of the Late Permian and Triassic times (~250–200 Ma) has been described in terms of strong continental climate conditions and monsoons. The paleoclimate conditions were simulated by Kutzbach and Gallimore (1989) using idealized Pangaean continent with changes in CO_2 and solar luminosity. In spite of

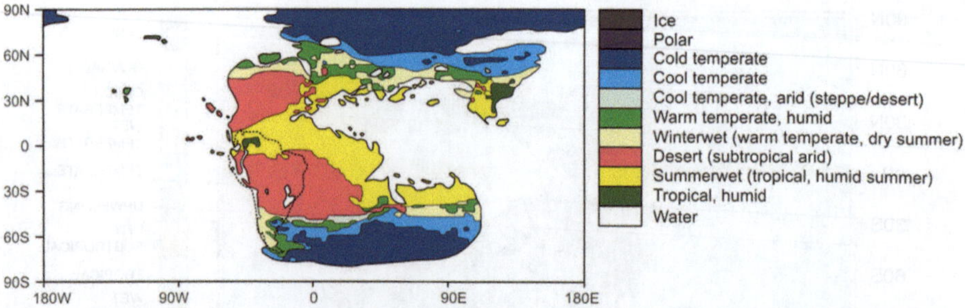

Fig. 5 Following Walter (1985), the biome zones for the Late Triassic based on the model predicted temperatures and precipitation are shown. At the *right*, the key to Walter's (1985) biome zones (Adapted from Sellwood and Valdes 2006)

the various limitations of the simulations carried out by these authors, the resulting scenarios show the mechanism whereby the large Pangaean continent produces the conditions needed for massive summer and winter monsoonal circulations and cross-equatorial flows. The presence of the large continent also helps to produce the high summer temperatures and evaporation rates and the long overland trajectories that deplete the available atmospheric moisture. Acting together, these factors generate the large arid region in the continental interior and in the equatorial zone. The presence of the large continent also helps to produce the very cold winters of middle and high latitudes and the very hot summers of the tropics.

Sellwood and Valdes (2006) simulated the climate responses to the Mesozoic changes in geography. For the Triassic period, the Central Pacific (Panthalassa) is modeled as having a super, and semipermanent, El Niño system.

Figure 5 shows the biome zones according to the precipitation, temperature, and evaporation model simulated. Southwestern Gondwana is winter–wet. The eastern parts of Gondwana are moister than the western parts, being generally wetter during the summer months (particularly December, i.e., a modeled summer–wet biome/climate). The balance between evaporation and precipitation reflects these seasonal changes. Southern Gondwana is in balance, or has an excess of evaporation, from November through February (summer dryness). But from March to October, the southern parts of the continent in particular have an excess of precipitation (winter–wet). This embraces the months of winter darkness for the southern polar area.

In the Early Jurassic three-dimensional model simulation of Chandler et al. (1992), the atmospheric composition, solar radiation, and orbital parameters were set at present-day values. The resulting major features include warm surface-air temperatures, extreme continental aridity in the low and middle latitudes of western Pangaea, and monsoons which dominate along the midlatitude coast of Tethys and Panthalassa and also affect conditions deep into the continental interiors in some regions. They also experimented with variations in topography, sea-surface temperature, vegetation, and CO_2 content and discovered that none of the scenarios yielded continental interiors in which winter temperatures remained above freezing

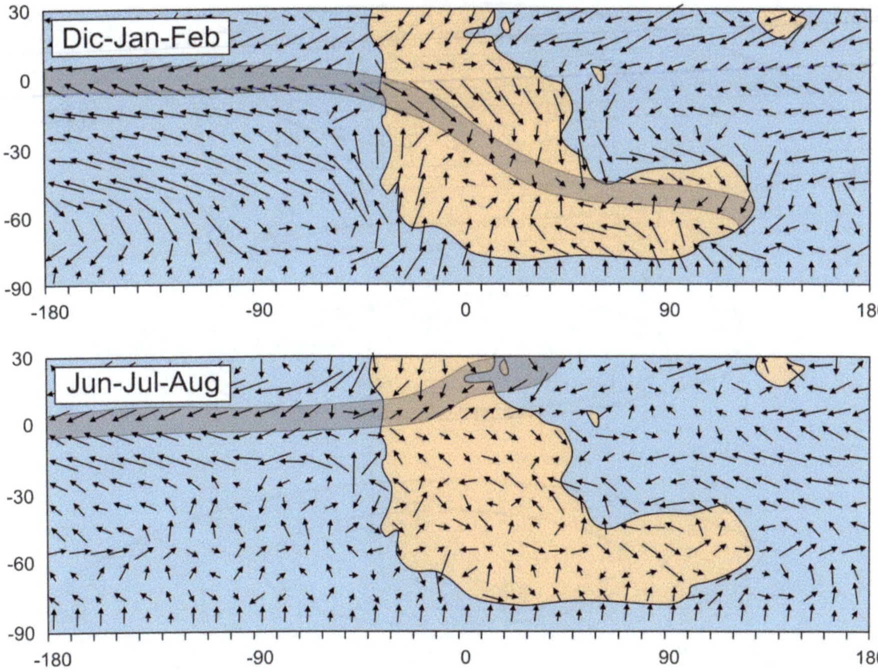

Fig. 6 Simulated global distribution of Early Jurassic wind vectors (**a**) surface wind; hatched line indicates approximate position of ITCZ for the South Hemisphere summer (Dec–Jan–Feb) and South Hemisphere winter (Jun–Jul–Aug) (Adapted from Chandler et al. 1992)

everywhere. The most important outcome of the simulations is that the warm climate of the Early Jurassic could be maintained without additional forcing from CO_2. The results show that increased ocean heat transport may have been the primary force generating warmer climates during the past 180 Ma. The mega-monsoons are found to be associated with localized pressure cells whose positions are controlled by topography and coastal geography. The simulated sea-level pressure (not shown) and surface winds (Fig. 6) shown in the Southern Hemisphere, as in the Northern Hemisphere, pressure centers localized above high topography regions, the winter high-pressure cells being shifted poleward from the summer lows. During winter the winds are generally light and below freezing surface-air temperature extends across most of the southern Pangaea, although seasonally averaged coastal temperatures remain above freezing at all latitudes. During summer, monsoon circulations are responsible for increases in Southern Hemisphere precipitation. Cyclonic circulation about the subtropical low in the southwestern Pangaea causes surface winds over the western Tethys Ocean to flow southward, where they invade upon Pangaea in the region of the Indian subcontinent. Winds from the polar oceans, despite their cooler source, also cause prominent summer precipitation maximum over southeastern Pangaea, primarily as a result of the uplift of air over

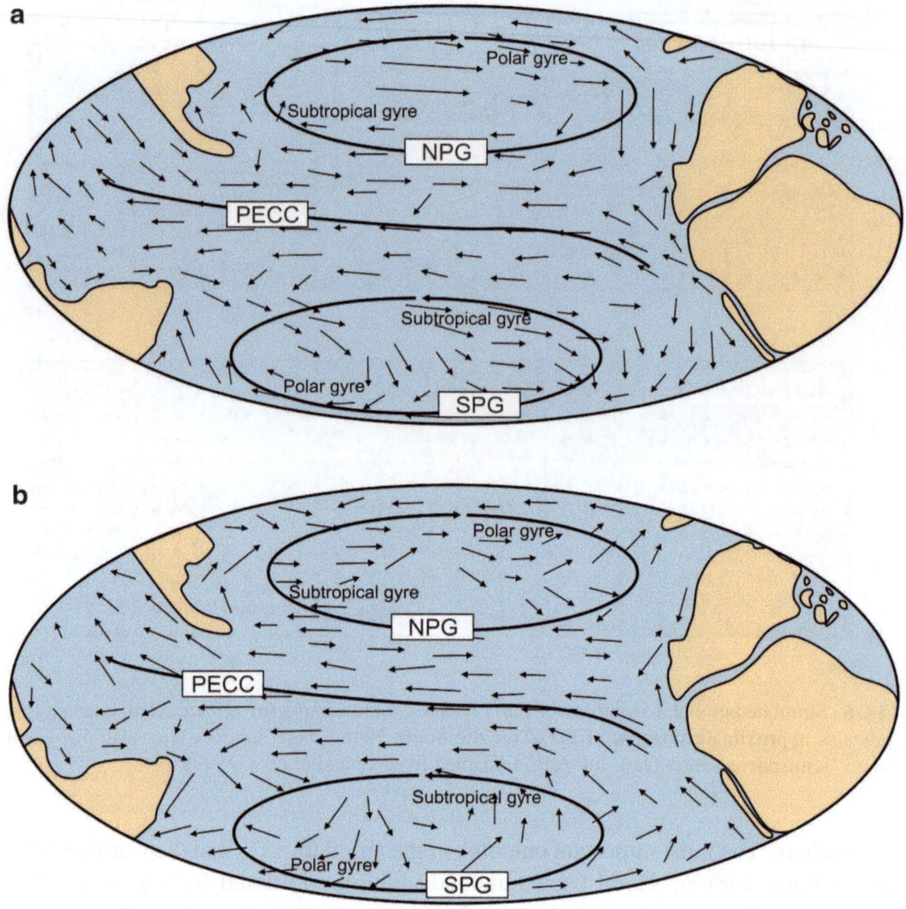

Fig. 7 Schematic representation of the surface oceanic circulation in the Panthalassa Ocean for the Northern Hemisphere (**a**) winter monsoon and (**b**) summer monsoon: Panthalassa Equatorial Counter Current (*PECC*); North Panthalassa Gyre (*NPG*) and South Panthalassa Gyre (*SPG*) or South Central Gyre (Adapted from Arias 2008)

coastal mountains. This monsoon circulation causes great seasonal variations in the Southern Hemisphere precipitation fields.

The surface oceanic circulation during the Early Jurassic Panthalassa oceans modeled by Arias (2008) is shown in Fig. 7. The proposed conceptual paleoceanographic model was based on fundamental physical–oceanographic principles, a global paleogeographic reconstruction, the atmospheric wind paleo-patterns, and paleobiogeographic data. The resulting Panthalassic oceanic circulation pattern presents approximately hemispherical symmetric structure of two large subtropical gyres that rotates clockwise in the Northern Hemisphere and anticlockwise in the Southern Hemisphere. During the Northern Hemisphere summer, the surface Tethyan Ocean circulation is dominated by monsoonal westerly directed equatorial surface currents that reached its western corner and drove them to the north, along the northern side of the Tethys Ocean, and in opposite direction during the winter.

Middle and Late Jurassic

In the Middle to Late Jurassic the Hispanic Corridor (Fig. 1), connecting the Pacific Ocean to the western Tethys Ocean, affected the paleo-oceanic currents permitting a globe-circling, mainly west-flowing current system in low latitudes (Winterer 1991; Rais et al. 2007) and consequently the atmospheric circulation mainly in the tropical regions.

Moore et al. (1992a, b) used a GCM to obtain two Kimmeridgian/Tithonian ($\sim$154.7–145.6 Ma) paleoclimate seasonal simulations, with geologically inferred paleotopography: one simulation used a CO_2 concentration of 280 ppm (preindustrial level) and the other used 1,120 ppm. Increasing the CO_2 content fourfold, it warms virtually the entire planet.

The simulations show greatest warming over the higher-latitude oceans and a smaller one over the equatorial and subtropical regions (Fig. 8). During the Southern Hemisphere summer (Fig. 8c), latitudes north of 30° S were characterized by the influence of the Intertropical Convergence Zone (ITCZ), while southwestern South America was dominated by the movement of the Southern Panthalassa ocean semipermanent anticyclone, which shifted to the south, affecting the southwest margins of Gondwana, that is, Patagonia. The monsoon lows were centered poleward of the western Tethys Sea, near 35° S. This location is just east of the region of summer maximum temperature. The wind pattern over Patagonia for June to August was opposite to that during summer. During winter the Panthalassa subtropical high was centered near 25° S. South of it, a subpolar low, centered near 60° S, dominated the south Panthalassa Ocean. These high- and low-pressure systems along with the continental winter monsoon high, which was located poleward of the western Tethys Sea and was centered near 40° S, were responsible for the winds from the northwest (Fig. 8d), forcing advection of wet and warm air over Patagonia. Furthermore, in the middle troposphere the axes of the Southern Hemisphere midlatitude storm track are located over 60° S. These factors are related to high positive net precipitation where precipitation exceeds evaporation. The model-simulated precipitation rate in the region is $\geq$5 mm/day (Valdes 1993) in concordance with the coal localities that are proxies for wet environments (Scotese 2012).

Moore et al. (1992a) proposed that the Jurassic warmth suggested by climate proxies may be explained by elevated atmospheric CO_2. In contrast, Chandler et al. (1992), for the Early Jurassic simulation, suggested that the resulted warm SSTs in energy balance without high atmospheric CO_2 imply that a warm Jurassic climate could have been the product of enhanced poleward heat transport through the oceans.

In other words, two feedback mechanisms are presumed to be primarily responsible for the very warm climate over Patagonia during the Jurassic: the elimination of sea and land ice that resulted from the warm polar sea-surface temperatures (SSTs) and the equatorward shift of Antarctica resulting in a decrease in surface albedo.

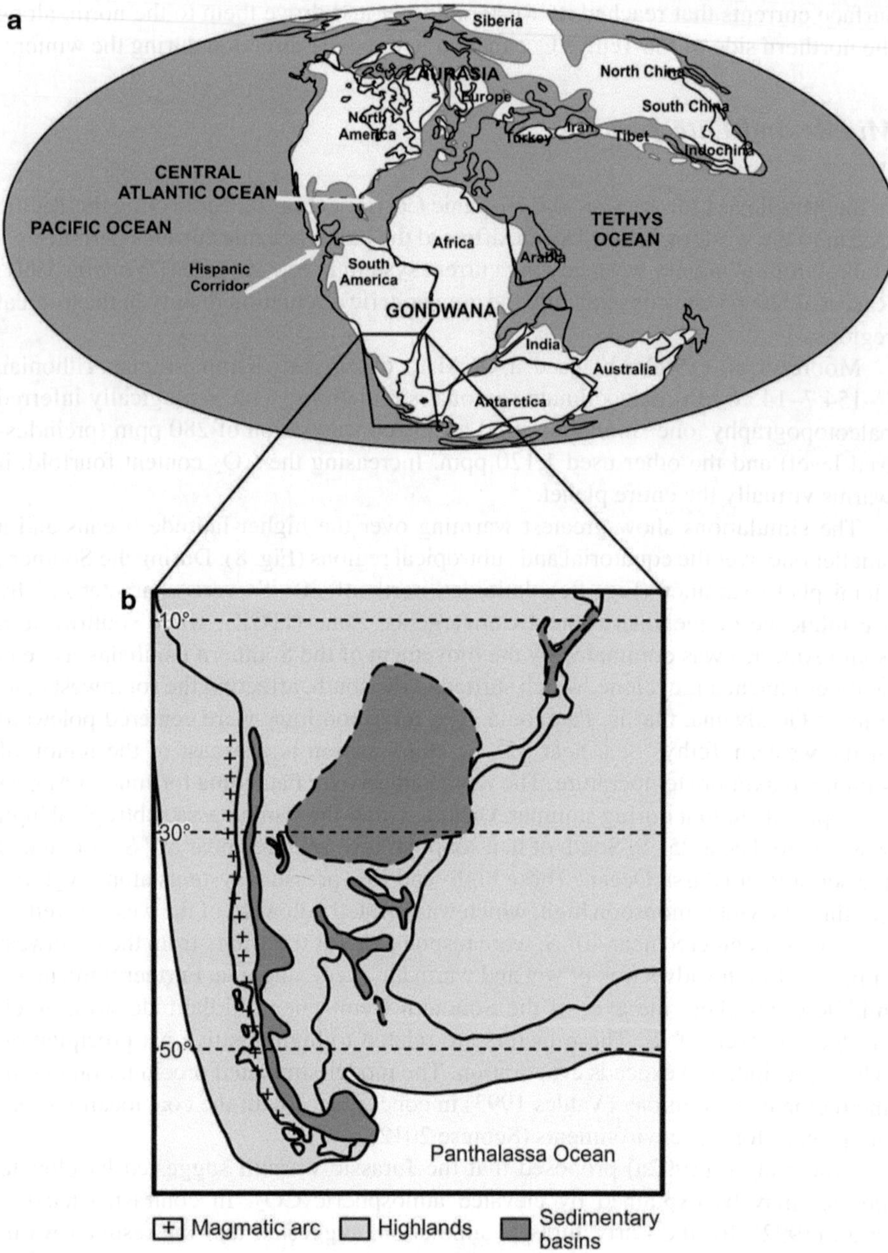

Fig. 8 Latest Jurassic–Earliest Cretaceous: (**a**) Paleolatitude and paleogeography (Adapted from Scotese 2012); (**b**) Paleogeographic map of the southwestern Gondwana continent (Adapted from Scherer and Goldberg 2007) and maps of simulated Kimmeridgian/Tithonian surface wind vectors using 1,120 ppm CO_2 for (**c**) December, January, and February; (**d**) June, July, and August (Adapted from Moore et al. 1992a)

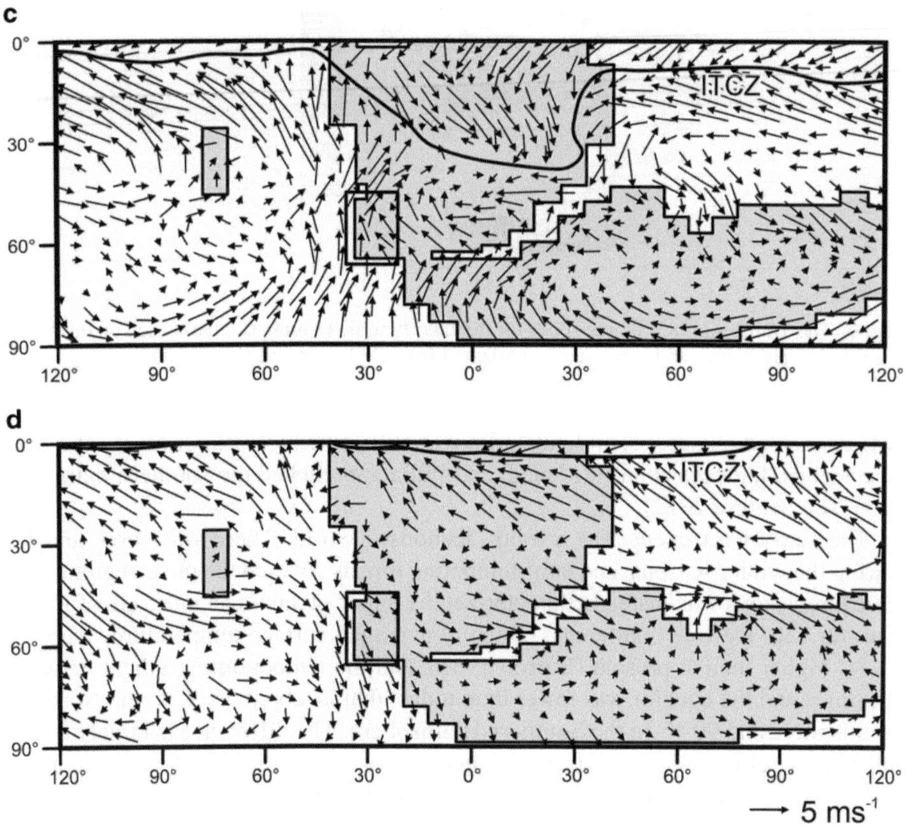

Fig. 8 (continued)

The modeled Kimmeridgian climate of Sellwood and Valdes (2006), between about 10° and 30° S, shows evaporation exceeding precipitation and a return to excess precipitation in the mid-to-high southern latitudes (Fig. 9). From fossil evidence, Jurassic plant productivity and maximum diversity were concentrated at mid-latitudes, reflecting a migration of the zone of peak productivity from low to higher latitudes during greenhouse times (reviewed and modeled in Rees et al. 2000).

Cretaceous

During the Late Jurassic to Early Cretaceous times, the separation between Gondwana and Laurasia was already well in progress, and the South Atlantic Ocean had already started to develop between those continents. India separated from Madagascar and raced northward on a collision course with Eurasia. North America

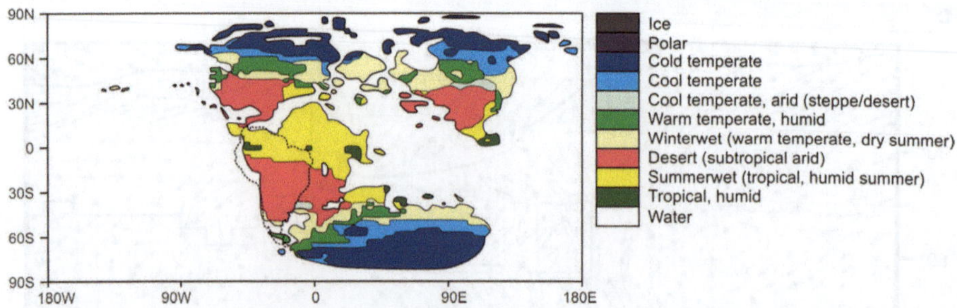

Fig. 9 The Walter's (1985) biome zones for the Late Jurassic (Kimmeridgian) based on the model predicted temperatures and precipitation (Adapted from Sellwood and Valdes 2006)

was connected to Europe, and Australia was still joined to Antarctica. By the Late Cretaceous, the oceans had widened, and India approached the southern margin of Asia (see Scotese 2012).

The Early Cretaceous was a mild "icehouse" world. There was snow and ice during the winter seasons, and cool-temperate forests covered the polar regions. The Late Cretaceous was instead characterized by super greenhouse intervals of global warmth with ice-free continents. Globally averaged surface temperatures were 6–14 °C higher than at present (Barron 1983), and the temperature gradient between the poles and the equator was lower than today, that is, about 50 °C in the Northern Hemisphere and 90 °C in the Southern Hemisphere. The difference is largely due to adiabatic cooling reflecting the elevation of Antarctica. Frakes (1999) summarized the data on estimates of Cretaceous sea-surface and terrestrial temperatures.

There are four different assumptions concerning Cretaceous temperatures and meridional gradients: (1) tropical sea-surface temperatures were the same as today, but polar temperatures were warmer (5–8 °C) except when ice was present (0–5 ° C); (2) the tropics were significantly cooler and midlatitudes warmer than today; (3) tropical sea-surface temperatures were 32–34 °C, with polar regions 10–18 °C; and (4) tropical sea-surface temperatures were about 42 °C and polar temperatures >18 °C. Furthermore, different hypothesis can explain the drastic warming and equable high latitudes during super-greenhouse intervals of the Cretaceous and Early Cenozoic. On the basis of coupled ocean–atmosphere model simulations of the Middle Cretaceous, Poulsen et al. (2003) hypothesized that the formation of an Atlantic Gateway could have contributed to the Cretaceous thermal maximum. Kump and Pollard (2008), by simulating the Middle Cretaceous using 4× CO_2 from preindustrial atmospheric level, failed to produce results to explain the extreme high-latitude warmth implied by temperature proxy data. However, simulations with the combined increases in cloud droplet radii, which mainly affect cloud optical depth, and precipitation efficiency, resulted in a reduction in global cloud cover from 64 to 55 % with optically thinner clouds which reduced planetary albedo from 0.30 to 0.24. The ensuing warming was dramatic, both in the tropics and in high latitudes, where warming was augmented by surface albedo. In other words, warming was produced by albedo reduction due to diminishing cloud cover. Then, surface albedo feedback augmented warming in the tropics and high latitudes and produced almost the vanishing of snow and sea-ice cover, thus forcing less albedo and more intense warming. Otto-Bliesner et al. (2002) altered the models by the inclusion of high-latitude forest thus changing the paleogeography. These low-albedo forests warmed the high-latitude continents, which then transferred more heat to the high-latitude oceans, impeding sea-ice formation and warming coastal regions.

The scenarios for the Turonian (~93.5–89.3 Ma) paleogeography were obtained by Floegel (2001) using GCM simulations with different orbital configurations and 1,882 ppm CO_2 (5× AD 2,000 CO_2, 7× preindustrial CO_2) concentrations (Fig. 10). The simulated atmospheric circulation resulted in a much more complex circulation than that observed today. In this model, the tropical easterlies remain constant and strong throughout the year. However, at higher latitudes the circulation varies with the seasons. Due to the absence of polar highs during the winter, strong westerly wind belts develop between 50° S and the high-pressure zones at 30° S. Another important difference in Floegel's scenarios lies in the strong trade winds, which developed during each hemispheric winter.

In Floegel's model, in the Southern Hemisphere summer (December to February), the polar region is under the influence of a low atmospheric pressure system, and therefore the westerly winds were weak and variable and may have even had a reverse direction. Also, the subtropical to polar frontal systems would not exist (Fig. 10). This would result in disruption of the mid- and high-latitude wind systems. In the winter (June to August), the southern polar region is under the influence of a high-pressure system and the westerlies are well developed (Hay et al. 2005; Hay 2008).

The Maastrichtian (65.5–70.6 Ma) paleowind scenario in Fig. 11 was obtained by Bush (1997) using an atmospheric–oceanic GCM, dynamically and thermodynamically coupled, with four times the present-day value of atmospheric CO_2, as indicative of Cretaceous levels. Even with these relatively new results, it is necessary to be cautious with the adopted paleogeographic reconstruction for the Maastrichtian (Ziegler et al. 1982), which was used in the model, if the Drake Passage were considered as it was already open.

Some GCM simulations of the Cretaceous super-greenhouse also considered as a boundary condition that the Drake Passage was already open (Sewall et al. 2007; Poulsen et al. 2007; Kump and Pollard 2008; Zhou et al. 2008; among others), while others considered that the Drake Passage was still closed (e.g., Haywood et al. 2004). For instance, Bush and Philander (1997) had estimated a later age for the opening of the Drake Passage, ranging from 49 to 17 Ma ago; see also more details and additional references in Cavallotto et al. (2011).

The earliest connection between the Pacific and Atlantic oceans at the Drake Passage is still controversial but very important, because the gateway opening probably had a profound effect on the global circulation and climate (see Cavallotto et al. 2011). The Drake Passage influence on climate was studied by Sijp and England (2004) by means of three main model simulation experiments, all of them set up identically with the exception of bathymetric data. They either kept (a) the

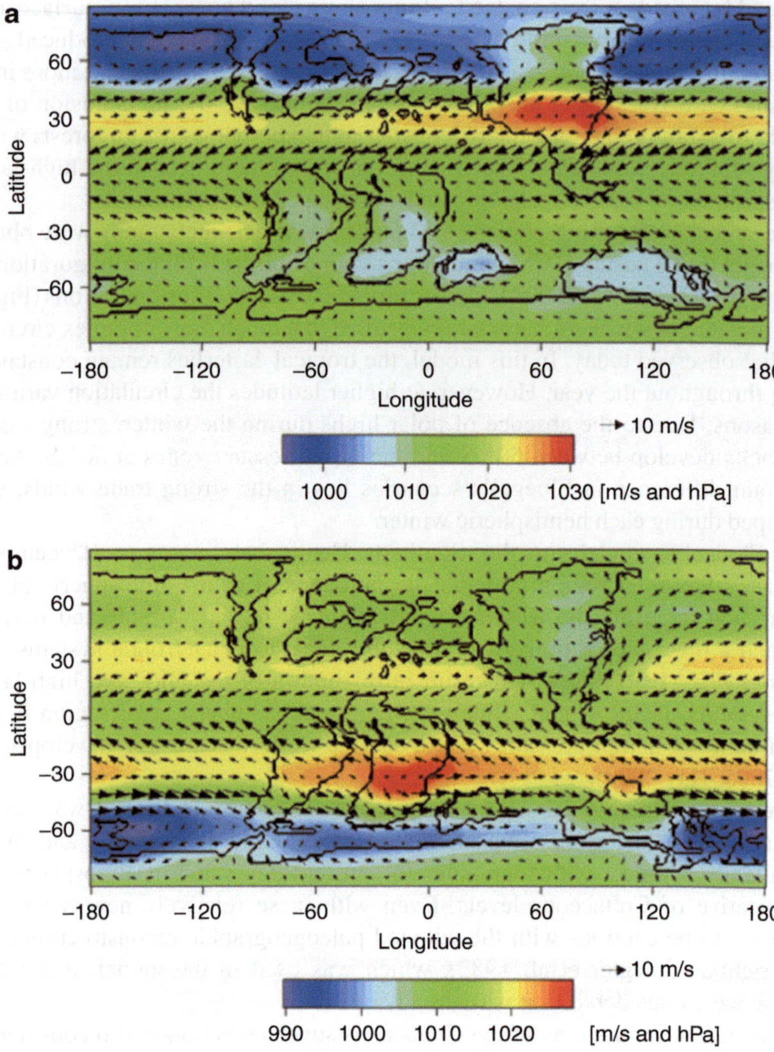

Fig. 10 Turonian (93.5 Ma) wind speed and pressure at sea level for (**a**) Dec, Jan, Feb and (**b**) Jun, Jul, Aug [m/s and hPa] (From Floegel 2001; © Sascha Floegel, IFM-GEOMAR, Kiel, Germany)

Drake Passage closed by a land bridge between the Antarctic Peninsula and South America, (b) the Drake Passage open to a maximum depth of 690 m, or (c) the Drake Passage open at its present-day depth, which was modeled as an uninterrupted through flow at a depth of ca. 2,316 m.

The modeled climate with the Drake Passage closed is characterized by warmer Southern Hemisphere surface-air temperature and little Antarctic ice. An increase in Antarctic sea ice and a significant cooling of the Southern Hemisphere would have

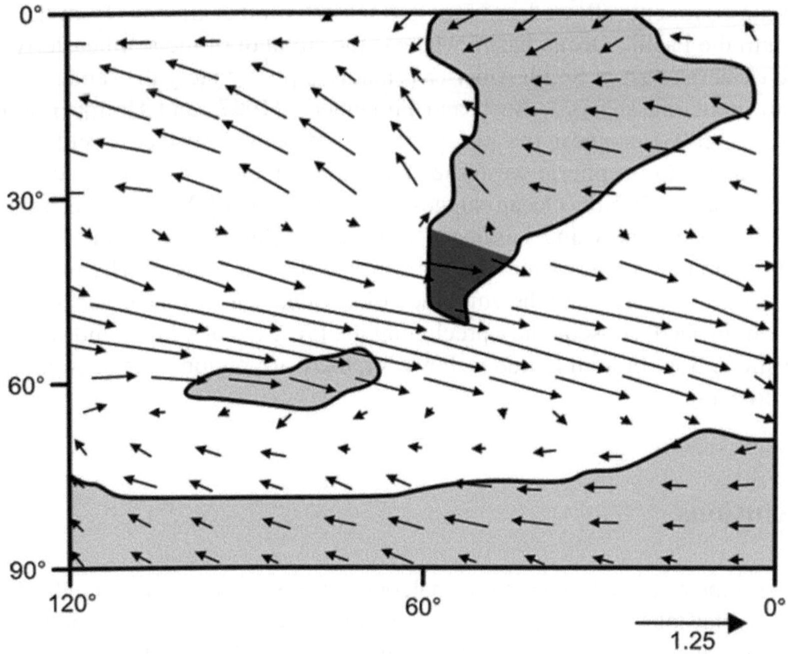

Fig. 11 Annual mean wind stress vectors (in dynes per square centimeter) for the Late Cretaceous. *Arrows* are scaled as displayed in the *bottom right* (Adapted from Bush 1997)

taken place when the Drake Passage was open to a much shallower depth of 690 m. As the Drake Passage opened, the climate would have become mostly similar in the Southern Hemisphere to the conditions of the Drake Passage if open at 690 m.

Bush's (1997) simulated pattern of annual mean wind stress (Fig. 11) resulted in a pattern similar to that of the present day with only some changes as a consequence of the modified continental–ocean distribution. The annual mean fields show the tropical–subtropical region circulation characterized by the ITCZ and the easterly winds occupying the latitudinal band from 30° S to 30° N. The Southern Hemisphere westerlies occupy the latitudinal band from 40° to 70° S and the midlatitude tropospheric jets dominate the atmospheric circulation over Patagonia (Fig. 11). The simulation shows no poleward shift of the midlatitude westerlies in response to an ice-free planet.

In the Bush and Philander's (1997) simulation, using a coupled atmosphere–ocean GCM with quadrupled CO_2, the annual mean Hadley circulation indicates a general reduction in strength of the Southern Hemisphere cells in the Cretaceous, with the south equatorial Hadley cell weakening by 20 %. Subsidence in the high southern latitudes decreased dramatically in the Cretaceous in response to the elimination of the Antarctic ice sheet, which in the present climate induces strong subsidence poleward of 75° S. As a consequence, the strengths of the middle- and high-latitude cells decrease. In addition, the separation of the North and South

American continents allowed stronger northeasterly trades flowing from the Tethys basin into the Pacific Ocean basin, whereas the strength of the southeasterly winds off the western coast of South America remains approximately the same.

Both modeling analysis, Bush and Philander's (1997) and Hotinski and Toggweiller's (2003), suggest the possibility of intensified surface circulation. The stronger upper tropospheric westerlies and stronger lower tropospheric easterlies over the tropical Pacific Ocean suggest a "permanent El Niño" state, as Davies (2006) found by analyzing laminated sediments. The model results showed the global precipitation approximately 10 % higher than at present. The mean annual temperatures increased and the amplitude of seasonal cycle in near-surface temperatures diminished, consequently precluding the presence of year-round snow or ice in the simulation. In high latitudes, however, there are regions that seasonally drop below freezing.

Conclusions

A major transformation in the global paleogeography occurred between the Triassic and the Cretaceous, which involved opening of the Hispanic Corridor and formation of the Atlantic Ocean. These are the main factors that produced the climatic changes registered in this period. Other important factors were the variations in the greenhouse gases, especially in CO_2 which is related to the volcanic activity and the heat transport through the oceans. Strong monsoon conditions dominated during the Mesozoic in the Gondwana supercontinent.

The major climate features of the Early Jurassic, as simulated by means of GCM, comprised warm surface-air temperatures, extreme continental conditions in the low and midlatitudes, and monsoonal environments, which basically dominated along the midlatitude coasts of the Tethys and the Panthalassa oceans. The sea-level pressures over the continent show high-pressure cells during winter and opposite low pressure in summer.

Over the oceans, a dipole of subpolar low-pressure cells and subtropical high-pressure zone is present. Despite the similarity between the hemispheric patterns, a substantial degree of asymmetry is caused by land–sea distribution and topographic differences. The strong differences between winters and summers generated a large annual cycle. During the Southern Hemisphere winter, faint insolation and strong heat loss caused a decrease of the temperature over the interior of the southern Pangaea, and consequently, the sinking cooled air could create a high-pressure belt over the Gondwana continent. At the same time during the Northern Hemisphere summer, the warm temperatures can also generate an extensive low-pressure zone over central Laurasia, northeastern Pangaea, (between 10° N and 30° N) that deepens above the higher elevations in the eastern part of the continent. In the Southern Hemisphere winter, the situation was exactly the inverse.

In the Cretaceous greenhouse world, the paleogeography was dramatically different from that of the Gondwana continent, which was by now breaking up and when the Atlantic Ocean was already well developed. Results from a coupled ocean–atmosphere model indicate that the opening of the Atlantic Equatorial Gateway could have caused a large-scale reorganization of the tropical climate and regional warming. This mechanism was considered as a cause of the Cretaceous Thermal Maximum. Both the oceanic and atmospheric circulation was affected. The westerly winds developed only seasonally. In the absence of persistent westerlies, the subtropical ocean gyres would have weakened leading to an ocean circulation dominated by eddies. As a result, the pycnocline would have been more diffuse than today, contributing to enhanced thermohaline circulation and global oceanic heat transport (Hay 2008).

A caution note must be highlighted about the very large differences in the response of the general circulation models to Cretaceous boundary conditions. For example, the GENESIS model (Bice et al. 2006) must have had much higher pCO_2 than previously thought ($\sim$12 $\times$ present-day levels) to match the warmest tropical paleotemperature. Otherwise, high atmospheric pCO2 levels are not required to achieve a very warm Cretaceous climate. With moderate pCO_2 levels ($\sim$3 $\times$ present-day levels), the Hadley coupled ocean–atmosphere model predicts a hot Cretaceous world dominated by latent heat (Markwick and Valdes 2002; Haywood et al. 2004). The NCAR model with moderate pCO_2 levels ($\sim$3 $\times$ present-day levels) predicts deep-ocean temperatures of 9–11 °C and surface temperatures that are 3–4 and 6–14 °C warmer than modern at low and high latitudes, respectively. Ocean heat transports in the model are diminished in the Northern Hemisphere and enhanced in the Southern Hemisphere (Otto-Bliesner et al. 2002). A similar spectrum of climate models used for the Cretaceous is being used to predict future climates, each of them with different sensitivity to increased greenhouse gas concentration.

In summary, the climate of the Gondwana supercontinent was quite varied and significantly different from present-day conditions in most of it. Very high mean annual temperatures and increased precipitation characterized the climate of the second half of the Mesozoic, which have had strong influence in the landscape and ecosystem evolution.

References

Arias C (2008) Palaeoceanography and biogeography in the Early Jurassic Panthalassa and Tethys Oceans. Gondwana Res 14:306–315

Barron EJ (1983) A warm, equable Cretaceous: the nature of the problem. Earth Sci Rev 19:305–338

Barron EJ, Fawcett PJ (1995) The climate of Pangaea: a review of climate model simulations of the Permian. In: Scholle PA, Peryt TM, Ulmer- Scholle DS (eds) The Permian of Northern Pangea, vol 1. Springer, Berlin, pp 37–52

Berner RA, Kothavala Z (2001) GEOCARB III: a revised model of atmospheric CO_2 over phanerozoic time. Am J Sci 301:182–204

Bice KL, Birgel D, Meyers PA, Dahl KA, Hinrichs K, Norris RD (2006) A multiple proxy and model study of Cretaceous upper ocean temperatures and atmospheric CO_2 concentrations. Paleoceanography 21, PA2002. doi:10.1029/2005PA001203

Bush ABG (1997) Numerical simulation of the Cretaceous Tethys circumglobal current. Science 275:807–810
Bush ABG, Philander SGH (1997) The late Cretaceous: simulation with a coupled atmosphere–ocean GCM. Paleoceanography 21:475–516
Cavallotto JL, Violante RA, Hernández-Molina FJ (2011) Geological aspects and evolution of the Patagonian continental margin. Biol J Linn Soc 103:346–362
Chandler M, Rind D, Ruedy R (1992) Pangean climate during the Early Jurassic: GCM simulations and the sedimentary record of paleoclimate. Bull Geol Soc Am 104:543–559
Crowley TJ, North GR (1999) Paleoclimatology. Oxford University Press, New York, 360 pp
Crowley TJ, Hyde WT, Short DA (1989) Seasonal cycle variations on the supercontinent of Pangea. Geology 17:457–460
Davies A (2006) High resolution palaeoceanography and palaeoclimatology from mid and high latitude Late Cretaceous laminated sediments. Unpublished doctoral dissertation, Faculty of Engineering Science and Mathematics, School of Ocean and Earth Science, University of Southampton, Southampton, 274 pp
DeConto RM, Pollard D (2003) Rapid Cenozoic glaciation of Antarctica induced by declining atmospheric CO_2. Nature 421:245–249
Dubiel RF, Parrish JT, Parrish JM, Good SC (1991) The Pangaean megamonsoon—evidence from the Upper Triassic Chinle Formation, Colorado Plateau. Palaios 6:347–370
Fawcett PJ, Barron EJ, Robison VD, Katz BJ (1994) The climatic evolution of India and Australia from the Late Permian to Mid-Jurassic: a comparison of climate model results with the geologic record. Geol Soc Am Spec Pap 288:139–158
Floegel S (2001) On the influence of precessional Milankovitch cycles on the Late Cretaceous climate system: comparison of GCM-results, geochemical, and sedimentary proxies for the Western Interior Seaway of North America. Universitätsbibliothek der Christian-Albrechts-Universität Kiel, Kiel
Frakes LA (1999) Estimating the global thermal state from Cretaceous sea surface and continental temperature data. Spec Pap-Geol Soc Am: 49–58
Gradstein F, Ogg J, Smith A, Bleeker W (2004) A new Geologic Time Scale, with special reference to Precambrian and Neogene. Episodes 27:83–100
Haq BU, Hardenbol J, Vail PR (1987) Chronology of fluctuating sea levels since the Triassic (250 million years ago to present). Science 235:1156–1167
Hay WW (2008) Evolving ideas about the Cretaceous climate and ocean circulation. Cretac Res 29(5–6):725–753
Hay WW, Flögel S, Söding E (2005) Is the initiation of glaciation of the Cretaceous Ocean-Climate System on Antarctica related to a change in the structure of the ocean? Glob Planet Change (Geol Soc Am Spec) 45:23–33
Haywood AM, Valdes PJ, Markwick PJ (2004) Cretaceous (Wealden) climates: a modeling perspective. Cretac Res 25:303–311
Hotinski RM, Toggweiller JR (2003) Impact of a Tethyan circumglobal passage on ocean heat transport and "equable" climates. Paleoceanography 18(1):1007. doi:10.1029/2001PA000730
Huber BT, MacLeod KG, Wing SL (eds) (2000) Warm climates in earth history. Cambridge University Press, Cambridge, p 462
Iglesias Llanos MP, Riccardi AC, Singer SE (2006) Palaeomagnetic study of Lower Jurassic marine strata from the Neuquén Basin, Argentina: a new Jurassic apparent polar wander path for South America. Earth Planet Sci Lett 252:379–397
Kump LR, Pollard D (2008) Amplification of Cretaceous warmth by biological cloud feedbacks. Science 320:195
Kutzbach JE, Gallimore RG (1989) Pangaean climates: megamonsoons of the megacontinent. J Geophys Res 94(D3):3341–3357. doi:10.1029/JD094iD03p03341
Kutzbach JE, Guetter PJ, Washington WM (1990) Simulated circulation of an idealized ocean for Pangaean time. Paleoceanography 5(3):299–317

Markwick PJ, Valdes PJ (2002) A quantitative evaluation and application of the results of a Maastrichtian (Late Cretaceous) coupled ocean–atmosphere experiment using the HadCM3 AOGCM, Cretaceous Climate and Oceans Dynamics Workshop, 14–17 July 2002, The Nature Place, Florissant, CO, USA

Moore GT, Hayashida DN, Ross CA, Jacobson SR (1992a) Palaeoclimate of the Kimmeridgian/Tithonian (Late Jurassic) world. I. Results using a general circulation model. Palaeogeogr Palaeoclimatol Palaeoecol 93:113–150

Moore GT, Sloan LC, Hayashida DN, Umrigar NP (1992b) Paleoclimate of the Kimmeridge/Tithonian (Late Jurassic) world. II. Sensitivity tests comparing three different paleotopographic settings. Palaeogeogr Palaeoclimatol Palaeoecol 95:229–252

Otto-Bliesner BL, Brady EC, Shields C (2002) Late Cretaceous ocean: coupled simulations with the National Center for Atmospheric Research Climate System Model. J Geophys Res 107. doi:10.1029/2001JD000821

Poulsen CJ (2008) Paleoclimate modeling, pre-quaternary. In: Gornitz V (ed) Encyclopedia of paleoclimatology and ancient environments. Kluwer Academic, Dordrecht, pp 700–709

Poulsen CJ, Gendaszek AS, Jacob R (2003) Did the rifting of the Atlantic Ocean cause the Cretaceous thermal maximum? Geology 31:115–118

Poulsen CJ, Pollard D, White TS (2007) General circulation model simulation of the $\delta^{18}O$ content of continental precipitation in the middle Cretaceous: a model-proxy comparison. Geology 35:199–202

Rais P, Louis-Schmid B, Bernasconi SM, Weissert H (2007) Palaeoceanographic and palaeoclimatic reorganization around the Middle-Late Jurassic transition. Palaeogeogr Palaeoclimatol Palaeoecol 251:527–546

Rees PM, Zeigler AM, Valdes PJ (2000) Jurassic phytogeography and climates: new data and model comparisons. In: Huber BT, MacLeod KG, Wing ST (eds) Warm climates in earth history. Cambridge University Press, Cambridge, pp 297–318

Ross CA, Ross JRP (1987) Late Paleozoic sea levels and depositional sequences. In: Ross CA, Haman D (eds) Timing and depositional history of eustatic sequences: constraints on seismic stratigraphy, Special Publication 24. Cushman Foundation for Foraminiferal Research, Washington, DC, pp 137–149

Ross CA, Ross JRP (1988) Late Paleozoic transgressive regressive deposition. In: Wilgus CK, Hastings BS, Kendall CGSC, Posamentier HW, Ross CA, Van Wagoner JC (eds) Sea level change: an integrated approach, Special Publication vol 42. Society of Economic Paleontologists and Mineralogists, Tulsa, pp 227–247

Royer DL (2006) CO_2-forced climate thresholds during the Phanerozoic. Geochim Cosmochim Acta 70:5665–5675

Scher HD, Martin EE (2006) Timing and climatic consequences of the opening of Drake Passage. Science 312:428–430

Scherer CMS, Goldberg K (2007) Palaeowind patterns during the latest Jurassic-earliest Cretaceous in Gondwana: evidence from aeolian cross-strata of the Botucatu Formation, Brazil. Palaeogeogr Palaeoclimatol Palaeoecol 250(1–4):89–100

Scotese CR (2001) Atlas of earth history. PALEOMAP Project, Arlington

Scotese CR (2012) PALEOMAP, Earth history and climate history. [WWW document]. URL http://www.scotese.com/. Accessed Mar 2012

Scotese CR, Summerhayes CP (1986) A computer model of paleoclimate to predict upwelling in the Mesozoic and Cenozoic. Geobyte 1:28–42

Sellwood BW, Valdes PJ (2006) Mesozoic climates: general circulation models and the rock record. Sediment Geol 190:269–287

Sellwood BW, Valdes PJ, Price GD (2000) Geological evaluation of GCM simulations of Late Jurassic palaeoclimate. Palaeogeogr Palaeoclimatol Palaeoecol 156:147–160

Sewall JO, van de Wal RSW, van der Zwan K, van Ooosterhout C, Dijkstra HA, Scotese CR (2007) Climate model boundary conditions for four Cretaceous time slices. Clim Past 3:647–657

Sijp WP, England MH (2004) Effect of the Drake Passage throughflow on global climate. J Phys Oceanogr 34:1254–1266

Trenberth KE (1992) Climate system modeling. Cambridge University Press, New York, 788 p
U.S. Geological Survey Geologic Names Committee (2010) Divisions of geologic time—major chronostratigraphic and geochronologic units: U.S. Geological Survey Fact Sheet 2010–3059, 2 p
Valdes PJ (1993) Atmospheric general circulation models of the Jurassic. Philos Trans R Soc B 341(1297):317–326
Valdes PJ (2000) Warm climate forcing mechanisms. In: Huber BT, MacLeod KG, Wing SL (eds) Warm climates in earth history. Cambridge University Press, Cambridge, pp 3–20
Valdes PJ, Sellwood BW (1992) A palaeoclimate model for the Kimmeridgian. Palaeogeogr Palaeoclimatol Palaeoecol 95:47–72
Veizer J, Ala D, Azmy K, Bruckschen P, Buhl D, Bruhn F, Carden GAF, Diener A, Ebneth S, Godderis Y, Jasper T, Korte C, Pawellek F, Podlaha O, Strauss H (1999) $^{87}Sr/^{86}Sr$, $\delta^{13}C$ and $\delta^{18}O$ evolution of Phanerozoic seawater. Chem Geol 161:59–88
Volkheimer W, Rauhut OWM, Quattrocchio ME, Martínez MA (2008) Jurassic Paleoclimates in Argentina, a review. Rev Asoc Geol Argent 63(4):549–556
Walter H (1985) Vegetation of the earth. Springer, Berlin, 318 pp
Ward PL (2009) Sulfur dioxide initiates global climate change in four ways. Thin Solid Films 517:3188–3203
Winguth AME, Heinze C, Kutzbach JE, Maier-Reimer E, Mikolajewicz U, Rowley D, Rees A, Ziegler AM (2002) Simulated ocean circulation of the Middle Permian. Paleoceanography 17(5):1057
Winterer EL (1991) The Tethyan Pacific during Late Jurassic and Cretaceous times. Palaeogeogr Palaeoclimatol Palaeoecol 87:253–265
Zhou J, Poulsen CJ, Pollard D, White TS (2008) Simulation of modern and middle Cretaceous marine $\delta^{18}O$ with an ocean–atmosphere general circulation model. Paleoceanography 23, PA3223. doi:10.1029/2008PA001596
Ziegler PA (1988) Evolution of the Arctic–North–Atlantic and the western Tethys. American Association of Petroleum Geologists, Tulsa
Ziegler AM, Scotese CR, Barrett SF (1982) In: Brosche F, Sundermann J (eds) Tidal friction and earth's rotation II. Springer, Berlin
Ziegler AM, Gibbs MT, Hulver ML (1998) A mini-atlas of oceanic water masses in the Permian period. Proc R Soc Aust 110(1/2):323–343

Gondwana Paleosurfaces in the State of Rio Grande do Sul, Southern Brazil

Francisco Sergio Bernardes Ladeira

Abstract Several papers in Brazil have identified the importance of the planation surfaces in the evolution of the landscape, particularly those referring to the Gondwana Surface and the "Sul-Americana" Surface. These works were especially common between the 1950s and 1970s. Starting with the 1980s, such contributions became much scarcer. More recently, several, more appropriate dating techniques stimulated again the publication of papers related with these themes. This study has the aim of reviewing the knowledge about the paleosurfaces of the state of Rio Grande do Sul. The investigation discusses the summit surfaces named by Ab'Sáber as the Vacaria Surface and the Caçapava do Sul Surface. The Vacaria Surface is carved into Mesozoic effusive rocks, whereas the Caçapava do Sul Surface has a much more complex evolution, basically eroded on Precambrian rocks. Contrarily to other areas of Brazil, the surface carved in the oldest rocks is located at intermediate elevations. The Vacaria Surface evolved during the Late Cretaceous and the Paleogene, whereas the Caçapava do Sul Surface has an even older development, typically corresponding in this area to the Gondwana Surface, with some particular characteristics which are unique in regional terms, as remnants of Triassic sedimentary rocks on top and evidence of lateritic profiles, which today occur only in the lowest parts of these sections, but which had influence on the development of some landforms until today.

Keywords Gondwana • Brazil • Rio Grande do Sul • Passive margin geomorphology • Planation surfaces

---

F.S.B. Ladeira (✉)
Geography Department, Geosciences Institute, University of Campinas – UNICAMP, João Pandiá Calógenas Street, 51 – Barão Geraldo, Zip Code 13083-870 Campinas, São Paulo, Brazil
e-mail: fsbladeira@ige.unicamp.br

J. Rabassa and C. Ollier (eds.), *Gondwana Landscapes in southern South America*, Springer Earth System Sciences, DOI 10.1007/978-94-007-7702-6_7,

Introduction

Planation surfaces are topographic surfaces that cut across rock types of different nature and age (Godard et al. 2001). The Gondwana paleolandscapes are related to long-term landscape evolution of planation surfaces that were developed over many millions of years, under very warm and wet climatic conditions and with tectonic stability as typical passive margins (Rabassa 2010, 2014; Ollier 2014a, b). These surfaces in South America were developed at least until the Early Cretaceous, when the rifting processes started the destruction of Gondwanaland and the Southern Atlantic Ocean was formed.

Several works have identified these surfaces in Brazil, as part of the relief, particularly the Gondwana Surface and the "Sul-Americana" (i.e., South American) Surface (Moraes Rêgo 1932; De Martonne 1943; Freitas 1951; Ab'Sáber 1955a, b; King 1956; Barbosa 1965; Dorr 1969; Braun 1970; Bigarella 2007), among many other papers by these authors and other scientists). A synthesis of several proposals which covers the entire Brazilian territory is presented in Table 1, showing a certain analytical complexity when a diversity of interpretations according to the different authors is observed.

Particularly in Southern Brazil, Brasil (1986) indicated that it is a region which presents a polycyclic relief characteristic that brings as a consequence the difficulties of identification, correlation, and genetic determination of several areas, what also adds to the polygenetic character of the larger planation surfaces, the result of superposition of morphoclimatic systems. Concerning this situation, Ab'Sáber (1969) noted that even when these surfaces are widely represented and with well-defined geographic distribution and topographic position in the state of Rio Grande do Sul, the southernmost of Brazilian states (Fig. 1), the high degree of preservation of these landforms occurring in the states of Goiás, Mato Grosso, or Bahia is not observed here.

In Rio Grande do Sul, Ab'Sáber (1969) indicated that the paleosurface scenario is much more complex, since higher surfaces occur in the area of the huge Paraná Basin, which may be older than the highest summits of the Uruguay-Rio Grande do Sul shield.

In the state of Rio Grande do Sul, the areas with higher altimetry correspond to two, very characteristic summit surfaces. On one side, the highest surface is found with altimetry normally higher than 1,000 m above sea level (m a.s.l.), on rocks of the Serra Geral Formation (rhyolite and basaltic flows with ages between 110 and 160 Ma), located at the northeastern sector of the state (a surface that extends into the State of Santa Catarina) and another summit surface, with lower altimetry (around 450 m a.s.l.), carved onto crystalline basement rocks, with a much longer and more complex evolutionary history than the first one, found in the central-southeastern portion of the state of Rio Grande do Sul (a surface that continues into Uruguay).

The central objective of this chapter is to present a revision of the knowledge of these paleosurfaces, with the aim of organizing the information about the summit surfaces of the state of Rio Grande do Sul, and to show data belonging to field studies in the area. In this chapter, the two summit surfaces indicated by Ab'Sáber (1969) will be discussed: (a) the Vacaria Surface, which presents portions with elevations of 1,000 (m a.s.l.) on the volcanic flows of the Serra Geral Formation (which would correspond to the surfaces of the top of the "Planalto" – that is, high plain – dos Campos Gerais, as stated by Dantas et al. 2010), and (b) the Caçapava do Sul Surface, with altimetry between 450 and 460 m a.s.l. and which cuts across the rocks of the Sul-Rio-Grandense Shield (Ab'Sáber 1969) (being the equivalent to the uppermost surface of the Planalto Sul-Rio-Grandense, according to Dantas et al. 2010). Brasil (1986) has worked as if these two surfaces were correlated, whereas in this chapter, they will be treated separately because they are associated with different materials and are located at very different topographic levels. Nevertheless, it should be noted that the question of their correlation is not considered here.

Summit Paleosurfaces of Rio Grande do Sul

In altitude terms, the state of Rio Grande do Sul presents its highest summits in the northeastern portion of the state (Fig. 2), at the border with the State of Santa Catarina, commonly with elevations higher than 1,000 m a.s.l., reaching up to 1,800 m a.s.l. in some places of Santa Catarina itself. In this sector, the altitudes diminish towards the west and southwest. The Planalto Sul-Rio-Grandense extends further southwards with lower altimetry, slightly higher than 400 m a.s.l. In a clear way, these high plains are separated by the valley of Rio Jacuí, where lower elevations are observed, together with the tributaries of the Rio Uruguai that divide the state territory in a west–east direction.

In climatic terms, the state of Rio Grande do Sul has a humid subtropical climatic régime, with cold winters in the southern part and along the higher surfaces of the high plains resulting in moderate chemical weathering, due to the dominating relatively low temperatures (Dantas et al. 2010). In the higher portions of the landscape, it is common to have transitory snow cover during the winter.

According to Dantas et al. (2010), four major geomorphological dominions are observed in the state of Rio Grande do Sul: (1) the coastal plains, composed of marine, eolian, and fluviolacustrine deposits of Quaternary age; (2) the Planaltos Alçados (i.e., "uplifted plains"), developed upon the volcanic and volcano-sedimentary sequences of Mesozoic age of the Paraná Basin; (3) the inter-plain depressions, which have evolved on the Paleozoic and Mesozoic sedimentary sequences of the Paraná Basin; and (4) the high plains and the Serras Baixas (i.e., "low hills"), modeled upon the crystalline basement rocks of Precambrian age of the Sul-Rio-Grandense Shield. This classification is similar to that developed by Suertegaray and Fujimoto (2004), who proposed a synthetic scheme about the evolution of the morphogenesis of Rio Grande do Sul. Basically, these authors outlined three morpho-structures that characterize the landscape of the state of Rio Grande do Sul: (1) the Rio de La Plata Craton and the Dom Feliciano Belt,

Table 1 Synthesis of the different interpretations concerning the Brazilian paleosurfaces (Ponçano and Almeida 1993; Leonardi et al. 2011.)

Period			Epoch	Moraes Rêgo (1932)	De Martonne (1943)	Freitas (1951)	King (1956)
Cenozoic	Quaternary		Holocene	Uplift		Quaternary Cycle	Paraguaçu Cycle
			Pleistocene				
	Tertiary	Neogene	Pliocene	Peneplanation	Neogenic Surface	Level A peneplain or Tertiary peneplain	Velhas Cycle
			Miocene	Uplift			
		Paleogene	Oligocene		Cristas Médias Surface		Uplift
			Eocene	Peneplanation post-Cretacic			
			Paleocene				Sul-Americana Surface
Mesozoic	Cretaceous		Late		Campos Surface	Level B peneplain or Cretacic peneplain	
			Middle				Post-Gondwana Surface
			Early				
	Jurassic		Late				Gondwana Surface
			Middle				
			Early				
	Triassic		Late				Surface aggradation under desert regime
			Middle				
			Early				
Paleozoic	Permian		Late				
			Early				
	Carboniferous		Late		Pré-Permian Surface		
			Early				
	Devonian		Late				
			Middle				
			Early				
	Silurian		Late				
			Early				
	Ordovician		Late				
			Middle				
			Early				
	Cambrian		Late				
			Middle				
			Early				

which correspond to Precambrian igneous and metamorphic rocks belonging to the Uruguayan-Sul-Rio-Grandense high plain; (2) the Paraná sedimentary basin, characterized by Paleozoic and Mesozoic sedimentary rocks,corresponding to the so-called Peripheric Depression, the Planalto Meridional and the Cuesta de Haedo. (3) the Pelotas sedimentary basin constituted by Cenozoic sedimentary rocks, corresponding to the coastal plains and lowlands. The basic geomorphological units, based upon Dantas et al. (2010), are shown in Fig. 3.

The two geomorphological units discussed here are the Planalto dos Campos Gerais, whose summits would correspond to the Vacaria Surface (according to

Table 1 (continued)

Ab'Sáber (1962) and Bigarella (2007)	Almeida (1964)	Barbosa (1965)	Braun (1970)	Bigarella (2007)	Valadão (2009)
Jundiaí Surface		Cycle XII – Paraguaçu of King	Velhas Cycle		Sul-Americana II Surface
Neogene Surface	Several surfaces along the valleys	Pediplan X – Velhas Surface of King	Sul-Americano Cycle	Pd_1	Sul-Americana I Surface
		Pediplan VIII – Sul-Americana Surface		Pd_2	Uplift
Cristas Médias Surface	Faulting				Sul-Americana Surface
	Japi Surface			Pd_3	
Altos Campos Surface					
		Pediplan VII – Superfície Culminante (Pós-Gondwana)			
		Pediplano V (?) – Superfície Gondwana Surface of King and Campos Surface of Martonne	Post-Gondwana Cycle		
			Gondwana Cycle		
	Itagua Surface				
	Itapeva Surface				

Ab'Sáber 1969), and the Planalto Sul-Rio-Grandense, where the highest points would belong to the Caçapava do Sul Surface (Ab'Sáber 1969). The Planalto dos Campos Gerais would correspond to the highest portion of the "Planalto das Araucárias" (IBGE 1995; Dantas et al. 2010). It is composed of volcanic flows pertaining to the Serra Geral Formation (Cretaceous). According to Dantas et al. (2010), its elevations vary from 600 to 1,300 m a.s.l., with a smooth inclination to the west. In these areas, freezing is common and snow is not rare in the highest points.

The Planalto Sul-Rio-Grandense is located in the south-southeast sector of the state of Rio Grande do Sul, composed of Precambrian igneous and metamorphic

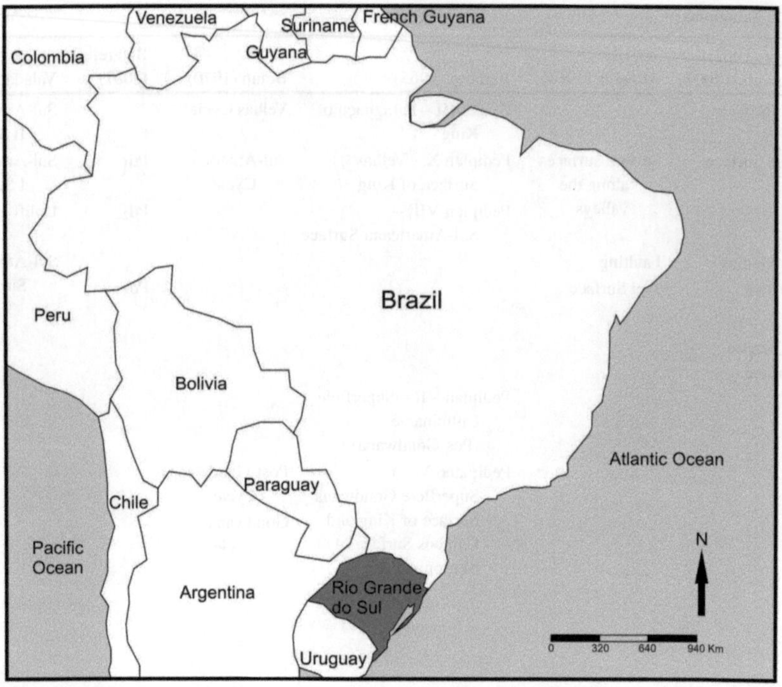

Fig. 1 Location of the study area

rocks. This area is characterized by the Rio de la Plata Craton and the Dom Feliciano Belt. This belt corresponds to an ancient collision area between two continents and these rocks correspond today to the cratons of Rio de la Plata (state of Rio Grande do Sul and Uruguay) and Kalahari (Southern Africa) (Fig. 4). The rocks corresponding to the Rio de la Plata Craton and the Dom Feliciano Belt became the source of the Paleozoic sedimentation that took place within the Intercratonic Depression of Paraná (Paraná Basin). These sediments are basically of continental nature and are essentially periclinal structures, diminishing mainly towards the west, that is, towards the Paraná and Lower Uruguay river basins (Suertegaray and Fujimoto 2004). In those times, in geomorphological terms, the area of Rio Grande do Sul comprised two units, the Planalto Sul-Rio-Grandense, which was undergoing erosion, and a large sedimentary plain corresponding to the Paraná sedimentary basin.

The present configuration of the relief of Rio Grande do Sul has its origin in the breakup of Gondwanaland, which in this area is characterized by the opening of the South Atlantic Ocean that started around 132 Ma (Suertegaray and Fujimoto 2004). Since then, the geomorphological scenario became much more complex, with the occurrence of the rhyolite and basaltic flows and the formation of faults which generated elevated and depressed regions. Between the Middle Jurassic and

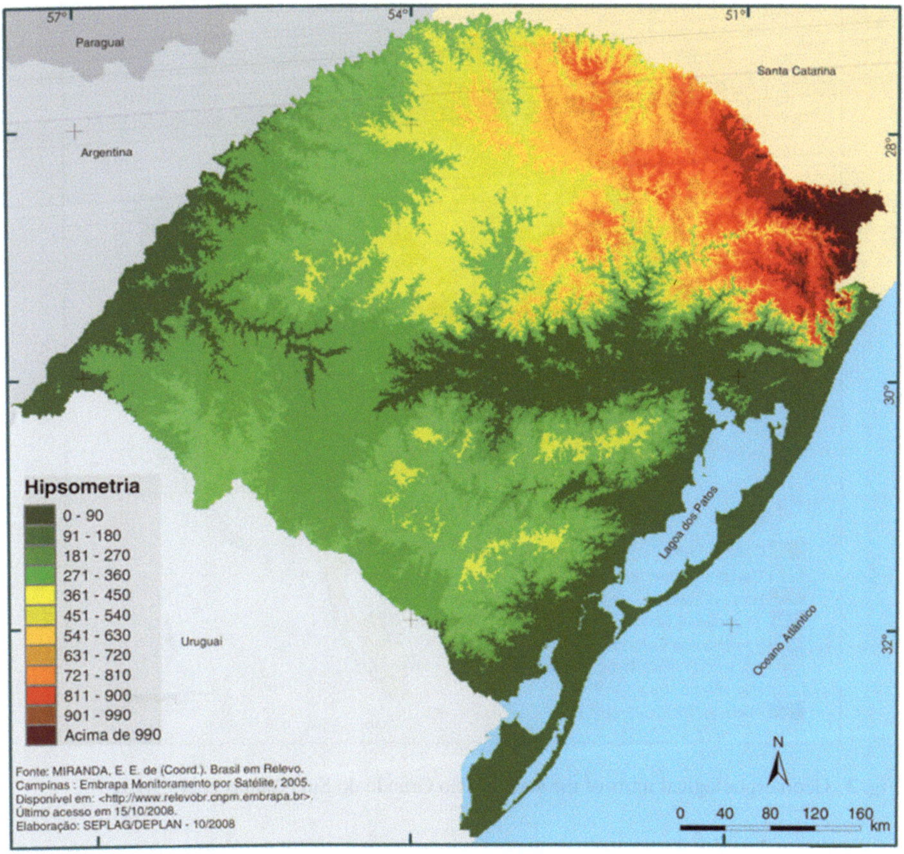

Fig. 2 Hypsometric map of the state of Rio Grande do Sul

the Middle Cretaceous, the fragmentation of Gondwanaland forced a series of fault lines and reactivation that defined the main compartments of the present regional landscape.

The opening of the South Atlantic Ocean created a new direction for the erosion of the Planalto Sul-Rio-Grandense, where a new surface was formed to the east, associated with the creation of a depressed area which the ocean would occupy later on. The reactivation of fault alignments followed by uplifting and down-warping promoted the reorganization of the drainage, which initiated the erosion processes that would generate the Periphery Depression of Rio Grande do Sul. Thus, towards the Late Cretaceous, a strong erosion phase became installed in the area, ending with a subsequent dynamics with dominance of depositional processes (Suertegaray and Fujimoto 2004). The area of the Planalto Sul-Rio-Grandense passed this phase with intense vertical erosion, forming the Rio Camaquã valley, which drains eastwards cutting the Planalto Sul-Rio-Grandense in two portions, north and south.

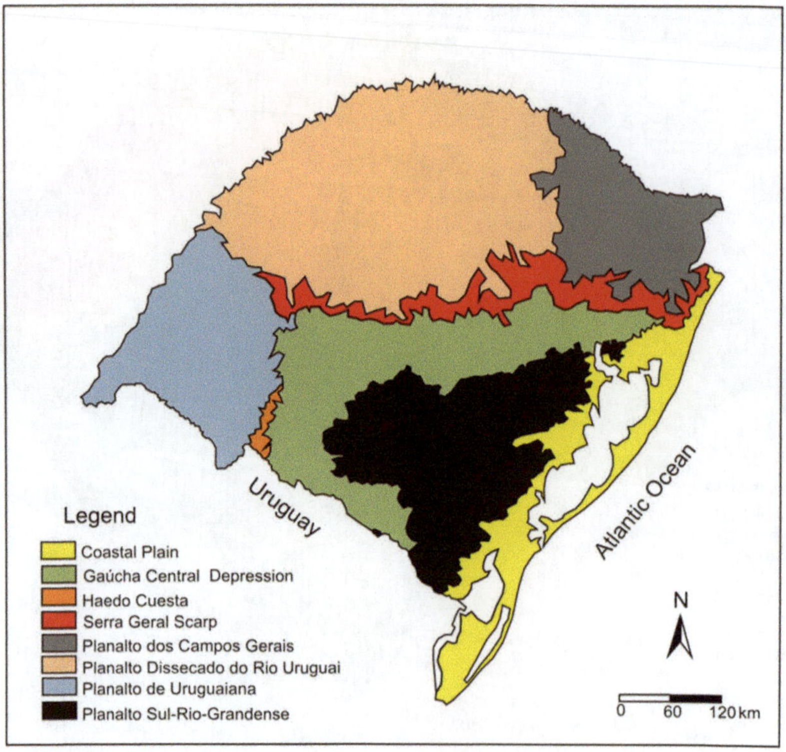

Fig. 3 Geomorphological units of the state of Rio Grande do Sul (Dantas et al. 2010)

According to Ab'Sáber (1949), towards the end of the Cretaceous, the area would have been a vast lowlands extending across planated remains of the crystalline nuclei and basaltic plains, with drier climate and endorheic drainage. In the Tertiary, the climate would have been more humid and the uplifting of the area would have promoted the establishment of exorheic drainage, forcing marginal denudation.

The Vacaria Surface

The Vacaria Surface corresponds to the highest portions in the entire state of Rio Grande do Sul (Fig. 3), with altimetry between 950 and 1,100 m a.s.l., smoothly sloping towards the west, south, and southeast, preserving the same landscape pattern until altitudes of 750–800 m a.s.l. are reached, when this high plain is abruptly interrupted by high scarps (Ab'Sáber 1969).

In the eastern portion of the area the highest elevations (close to 1,200 m a.s.l.) occur, and in the State of Santa Catarina several isolated points are even higher, as it is the case of the Morro da Igreja, with 1,822 m a.s.l. (Brasil 1986).

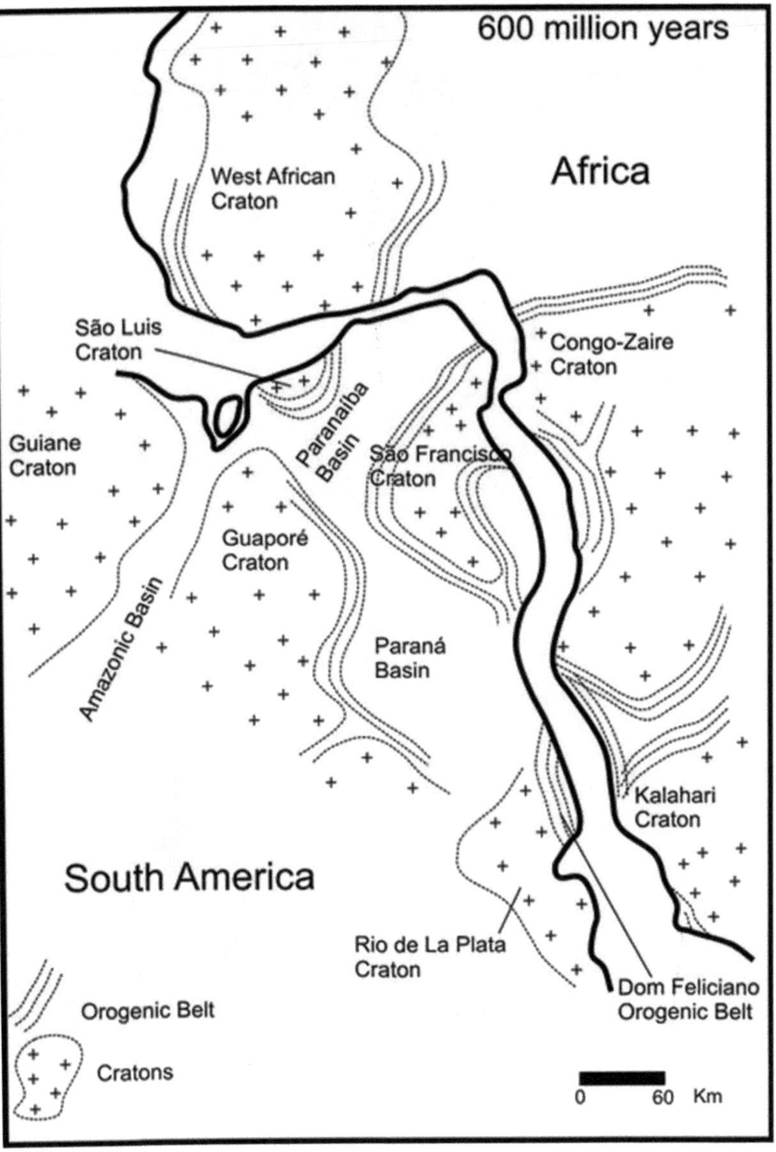

Fig. 4 Position of the Rio de la Plata and Kalahari cratons and the Dom Feliciano Belt, approximately ca. 600 Ma (Salgado-Laboriau 1994)

According to Brasil (1986), this surface is well preserved and it was formed by pediplanation processes, truncating rocks with little or no weathering. The evolution of the relief in this area indicates the existence of different phases of recurrent erosion processes, with the persistency of a well-preserved planation surface, interfingering with areas where erosion was able to elongate the valleys and create

Fig. 5 Surface elaborated on rhyodacites at the top of the Planalto dos Campos Gerais (Vacaria Surface) (**a**) and at its scarps (**b**), with the summit surface around elevations of 850 m a.s.l.

topographic thresholds with scarps of minor relief. In some areas, erosion succeeded in extensively expanding the valleys, leaving remnants of the old surface. The latter ones correspond to large areas with isolated hills, separated by broad valleys with flat bottoms. These hills usually preserve flat summits and breaking of inclination along the slopes (Brasil 1986; see Fig. 5).

This surface would correspond, in a broad sense, to the geomorphological region named as "Planalto das Araucárias" (or that of the Campos Gerais, according to Dantas et al. 2010), with extension to the State of Santa Catarina, presenting homogeneous geomorphological characteristics, with the higher parts corresponding to broad planation surfaces which sometimes end in scarps with more than 500 m of local relief (Brasil 1986).

Brasil (1986) indicated that this surface is developed on top of rocks belonging to the Serra Geral Formation. Its upper part consists of acid lava flows, such as rhyolites, rhyodacites, and felsic gabbros. In the basal portions of the Serra Geral Formation, in this area, basic rocks occur. The Vacaria Surface has been mainly developed on the acid rocks of the Serra Geral Formation, but locally it has evolved on more basic rocks where relief is flatter. According to Ab'Sáber (1969), the remnants of that surface at elevations close to 800 m a.s.l. show an inclination from north to south and from north-northeast to south-southwest, indicating the impact of a modern deformation on the ancient planation surface. For this author, this surface would correspond to an important period of exorheic pediplanation, postdating the lava flows, and most likely of Cretaceous age.

Ab'Sáber (1969) believed that this summit surface would have been affected by pediplanation at the time when the different sectors of the present Brazilian states of Rio Grande do Sul, Santa Catarina, and Paraná of the southern high plains served as watersheds of the Cretaceous deposits in the northern portion of the craton (states of São Paulo, Mato Grosso, southwest of Goiás and the Triângulo Mineiro) and of the southern sector (Uruguay and southwestern portion of Rio Grande do Sul) of the Paraná sedimentary basin.

According to Ab'Sáber (1969), this surface would have been very well preserved, because for a long time the drainage was scarcely incised. But fast epirogenetic uplifting movements in the Paleogene and the Neogene generated strong incision of the stream channels, whereas the water divides of the high plains would have been preserved. In spite of the canyon deepening, the summit areas in between them remained quite flat and untouched by stream erosion.

Suertegaray (2010) noted, with information based in more recent mapping projects (DNPM 1989), that there is a larger diversification of the rock types related to this surface, not only basalts as it was believed before. More specifically reddish porphyritic rhyodacites occur on the Vacaria Surface. This acid volcanic rock is much more resistant than basalt to weathering and erosion. Ab'Sáber (1969) noted that in this area the soils were quite thin which could be associated, on one hand, to the lower temperatures that occur in this area (Dantas et al. 2010) and also to the resistant rock. In spite of their shallow nature, the soils of the region are intensively used for agricultural purposes (Fig. 6).

In this surface, the fluvial channels are strongly incised following structural alignments (Brasil 1986) (Fig. 7).

Caçapava do Sul Surface

The area of crystalline basement in Rio Grande do Sul is the so-called Planalto Sul-Rio-Grandense, which extends for over 46,000 km^2, bounded to the north and west by the Gaúcha Periphery Depression, to the east by the Inner Coastal Plain and to the

Fig. 6 Vacaria Surface with relatively thin soils, normally Cambisols and Neosols (**a**) and a relatively flat topography being used for soya crops in the neighborhood of the city of Vacaria (**b**)

south by the Uruguayan territory where it continues. The area of higher elevations (above 400 m a.s.l.) belongs to the better preserved portion of the Caçapava do Sul Surface, with a total extent of 15,070 km^2 (Brasil 1986). Normally, the contact of this high plain takes place smoothly without large topographic irregularities, as happens with the coastal plain (Fig. 8).

The complexity of the Precambrian lithology of the area is reflected in the variety of the landforms. The area is intensely and deeply dissected, especially following the SW-NE and NW-SE directions. Similarly to this area, other less incised landscapes

Fig. 7 A deep canyon eroded in the Vacaria Surface

Fig. 8 Planalto Sul-Rio Grandense (at the background), with a substratum of granite in contact with the Coastal Plain, which locally presents altitudes lower than 10 m a.s.l.

occur, in summit positions, corresponding to pediplain surfaces (Brasil 1986). This better preserved surface located in summit positions was named the Caçapava do Sul Surface by Ab'Sáber (1969); Brasil (1986) named it as the "Planaltos Residuais" ("residual high plains") of Canguçu and Caçapava do Sul, because these higher high plains are dissected by the valley of the Rio Camaquã (Figs. 9 and 10).

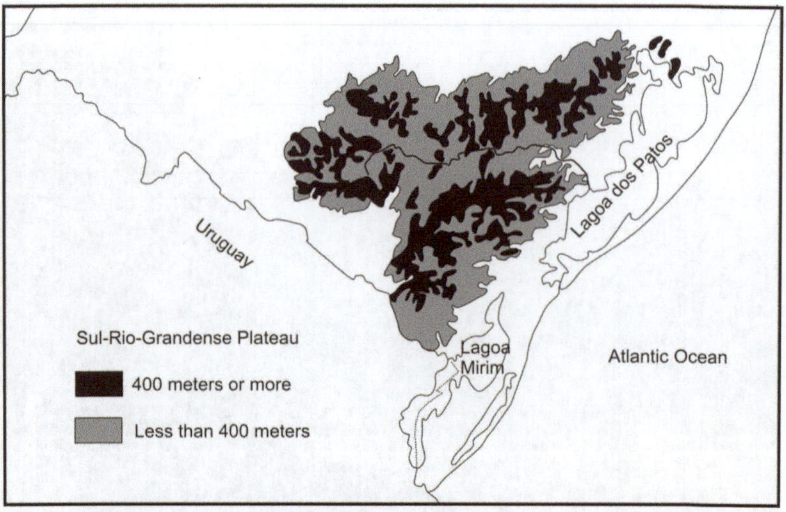

Fig. 9 The Planalto Sul-Rio-Grandense and the Rio Camaquã. The *darker colors* correspond to elevations above 400 m a.s.l. and the *lighter* ones represent the lower elevations

Fig. 10 Summit surface (*A*) near 420 m a.s.l., with the southern portion at the foreground (Canguçu) and the northern portion of the Planalto Sul-Rio Grandense (Caçapava do Sul) at the background; (*B*): the valley of the Rio Camaquã, with elevations lower than 200 m a.s.l.

Brasil (1986) noted that residual portions with elevations above 400 m a.s.l. were surrounded by another surface with altitudes lower than 400 m a.s.l., both eroded on the crystalline basement rocks. The better dissected and with lower elevation portions of the Uruguayan-Sul-Rio-Grandense Shield appear in between

Fig. 11 The Canguçu region. Top of the surface at 463 m a.s.l., with outcropping granites (**a**), associated with Neosols forming tors (**b**)

100 and 200 m a.s.l. These eroded areas of the relief are intensively dissected in the Precambrian rocks, and they have steep slopes where ridges, inselbergs, and inclined rocky plains are found (Brasil 1986).

Brasil (1986) indicated that the areas with higher elevations of the Uruguayan-Sul-Rio-Grandense (Planalto Sul-Rio-Grandense) are only slightly dissected in portions of the area, being formed by hills with flat tops that are remnants of the ancient pediplanation surface. Hermann and Rosa (1990) noted that the most densely dissected portions are those between the contour lines of 100 and 200 m a.s.l.

The less dissected areas or those with flat summits are characterized by thin superficial formations or exposed bedrock in the form of rocky outcrops and detrital pavements (Hermann and Rosa 1990) (Fig. 11).

In spite of their modest elevation (450–460 m a.s.l.), this is one of the summit surfaces most characteristic of the country, corresponding to a surface that levels the central portion of the Uruguayan-Sul-Rio-Grandense Shield, indicating a high degree of perfection in the ancient planation processes. This surface is present in both the territories of Rio Grande do Sul and Uruguay, with no residual landforms in any point of its entire extent. It cuts indiscriminately across all sorts of structures and lithological types (Ab'Sáber 1969).

Ab'Sáber (1969) indicated that, in its original extent, it would have been a vast pediplain, plunging towards the west, south, and southwest, although it also slopes

Fig. 12 (**a, b**) Tors positioned at the slopes of the uppermost surface

towards the northeast and the east. The eastern portion of the region is strongly affected by the Atlantic (passive margin) tectonics. Suertegaray (2010) stated that this inclination towards the east of the summit surface of the Uruguayan-Sul-Rio-Grandense Shield is directly related to the opening of the Atlantic Ocean (Middle Jurassic to Cretaceous), when a reactivation of the faulting zones parallel to the present coast of the state of Rio Grande do Sul took place, sloping the surface eastwards, in the same manner as block down-warping affected the Paraná Basaltic high plain.

It is worth noting that it is quite rare to observe tors in the higher portions of this surface, although these are very common in the lateral slopes of the highest portions (Fig. 12). These blocks are rather abundant in several areas, particularly in those on granites.

Fig. 13 Images of the Santa Tecla Formation (silicified sediments of the Early Tertiary) in the area of Caçapava do Sul, positioned next to the crystalline rocks of the Planalto Sul-Rio-Grandense, being thus associated with them (Photos by Constantino Brião de Oliveira).

Ab'Sáber (1969) indicated that, contrarily to what happens in the Vacaria Surface (which does not have correlative deposits in its surroundings), the Caçapava do Sul Surface shows its relationship with the deposits of the Santa Tecla Formation, which would be quite important for its geomorphological characterization. When Ab'Sáber described this association, the Santa Tecla Formation was considered to be Cretaceous in age, but it is considered today as of a probably Early Tertiary age (Kaul 1990) (Fig. 13).

Ab'Sáber (1969) believed that at the time of formation of the summit surface of the Uruguayan-Sul-Rio-Grandense Shield, in its northern portion, the surface of the rhyolite and basaltic plain was above the Caçapava do Sul Surface. The space where the Gaúcha Peripheral Depression is today would have a much larger volume of Gondwana deposits, which would have surrounded laterally or partially covered the axis of the shield. In such times, the Vacaria Surface would have evolved affected a much larger amount of geological formations and structures. With renewed Cretaceous uplift and deformation, there was also a renewal of the planation processes, which exhumed various buried areas of the shield and created vast pediplains across the summits. The planation of the Caçapava do Sul Surface would have conquered geographical areas which previously belonged to the Vacaria Surface. This author believed that the carving of this surface would correspond to the end of the Cretaceous sedimentation in Uruguay and the state of Rio Grande do Sul.

After the elaboration of this surface, Ab'Sáber (1969) noted that uplift of this area forced intense vertical erosion, in both in the central area of the massif and its surroundings, forming broad surfaces in between the high plains. Several of those inter-"planalto" surfaces, younger than the Caçapava do Sul Surface, were well preserved due to the silicification and ferrugination processes on the sandstones of the Rio Bonito Formation.

The down-wasted surface, which Ab'Sáber (1969) discussed, underwent a relatively intense process of lateritic profile formation. This profile affected in an undifferentiated manner crystalline basement rocks and also Triassic sedimentary rocks of the Santa Maria Formation (Figs. 14 and 15). The base of these lateritic profiles is preserved exclusively on the lower surfaces, below 300 m a.s.l. Most likely, this profile covered the entire high plain during the Early Tertiary but was later partly eroded.

Also in this largely down-wasted portion, little, shallow depressions are very common, which have been formed directly on the crystalline basement (Fig. 16). In these areas, the soils are very shallow, indicating that they are not related to recent karst processes. These small depressions may be related to the most depressed portions of the weathering front, at the contact between the weathering profile and the fresh bedrock (Jorge Rabassa, personal communication, 2012). At these elevations, lateritic accumulations also occur, as was identified by Ladeira and Santos (2006). These are the remains of former, extensive lateritic profiles which were formed in a distant past, and in most places, the entire lateritic profile has been eliminated. The present soils are very thin, some of them evolving on top of the remnants of Tertiary lateritic profiles.

Geomorphological Evolution: From the Permian to the Paleogene

During the Paleozoic and the Mesozoic, the South American Platform was affected by uplifting and subsidence processes which were fundamental in the sedimentation and configuration of the Paraná Basin (Almeida 1964, 1981). In this period, three

Fig. 14 (**a**) Eroded lateritic profile on gneiss, at 110 m a.s.l.; (**b**) Detailed view

well-defined, tectono-sedimentary cycles have been recorded, which marked periods of deposition or erosion (Soares et al. 1978) that were very important in the long-term sculpting of the center-south portion of Brazil. In Fig. 17, it is possible to observe a proposed evolutionary scheme for the Southern Brazil surfaces, from the Paleozoic onwards.

In the studied area, Devonian sediments are absent, being the surface elaborated only on the crystalline basement materials. By the Late Permian, deposition processes started concerning the materials corresponding to the present Planalto Sul-Rio-Grandense as source area of the sediments, generating deposits of this age

Fig. 15 (**a**) Profile of lateritic weathering affecting the Triassic sediments of the Santa Maria Formation, within the valley of the Rio Camaquã, in the inner portion of the Planalto Sul-Rio-Grandense; (**b**) Detailed view

towards the west, northwest, and southeast of the shield. Eastwards, these materials do not occur because the Atlantic Ocean was not yet open. These deposits have a well-defined, marine character (Kaul 1990).

In the Late Triassic, continental deposits of the Santa Maria Formation occurred, which in the state of Rio Grande do Sul cover the crystalline basement, a fact confirmed by the presence of terrestrial deposits of this formation, with fossilized tree trunks in the interior of the Planalto Sul-Rio-Grandense, especially in its western portion. In the Late Jurassic eolian deposits are associated with the Botucatú Formation. The source area for these sediments should be positioned farther away towards the east, where the crystalline basement was still outcropping. These areas are positioned today in Africa and in the submarine platform of the Southern Atlantic Ocean.

At the end of the Jurassic and in the Early Cretaceous, the rifting processes started that would lead towards the genesis of the Atlantic Ocean. In this process, volcanic activity took place that generated vast areas with lava flows belonging to the Serra Geral Formation. Ab'Sáber (1969) did not rule out the possibility that these flows would have covered even the crystalline basement area of the present Planalto Sul-Rio-Grandense, since these rocks were not observed during field studies. According to Brasil (1986), on the contrary, in the Late Jurassic, with the beginning of the Gondwana breakup processes, the area of the Planalto Sul-Rio-Grandense was undergoing erosion processes, whereas the rest of the areas were receiving continental sedimentation. In that moment, fissure volcanism was in-filling the Paraná Basin. Synchronically to these deposits, subsidence processes were taking place which allowed a significant piling-up of volcanic rocks. Thicker deposits occurred within the syncline, whereas they were thinner in the borders. At this time the Pelotas Basin started to form.

From this time erosion of the basement took a new direction, going in an eastward direction towards the Pelotas Basin. During the Cretaceous, the Triassic and Jurassic deposits on top of the basement were almost totally eroded, remaining only at just

Fig. 16 (**a**) Depressions in a landscape of smooth slopes; (**b**) A depression in the foreground, below elevations of 300 m a.s.l.; in the background, the higher surface of the Planalto Sul-Rio-Grandense

a few specific points. Thus, during the Cretaceous the Planalto Sul-Rio-Grandense became again an important positive area provider of sediments for the surroundings; therefore, it had been exhumed during this period.

During the Early Tertiary, the area of the Planalto Sul-Rio-Grandense was still a source area for the continental deposits of the Santa Tecla and Tupanciretã formations, and it continued to be the source area for the coastal and marine deposits of the eastern face, with the Rio Camaquã cutting the entire basement in an eastward direction.

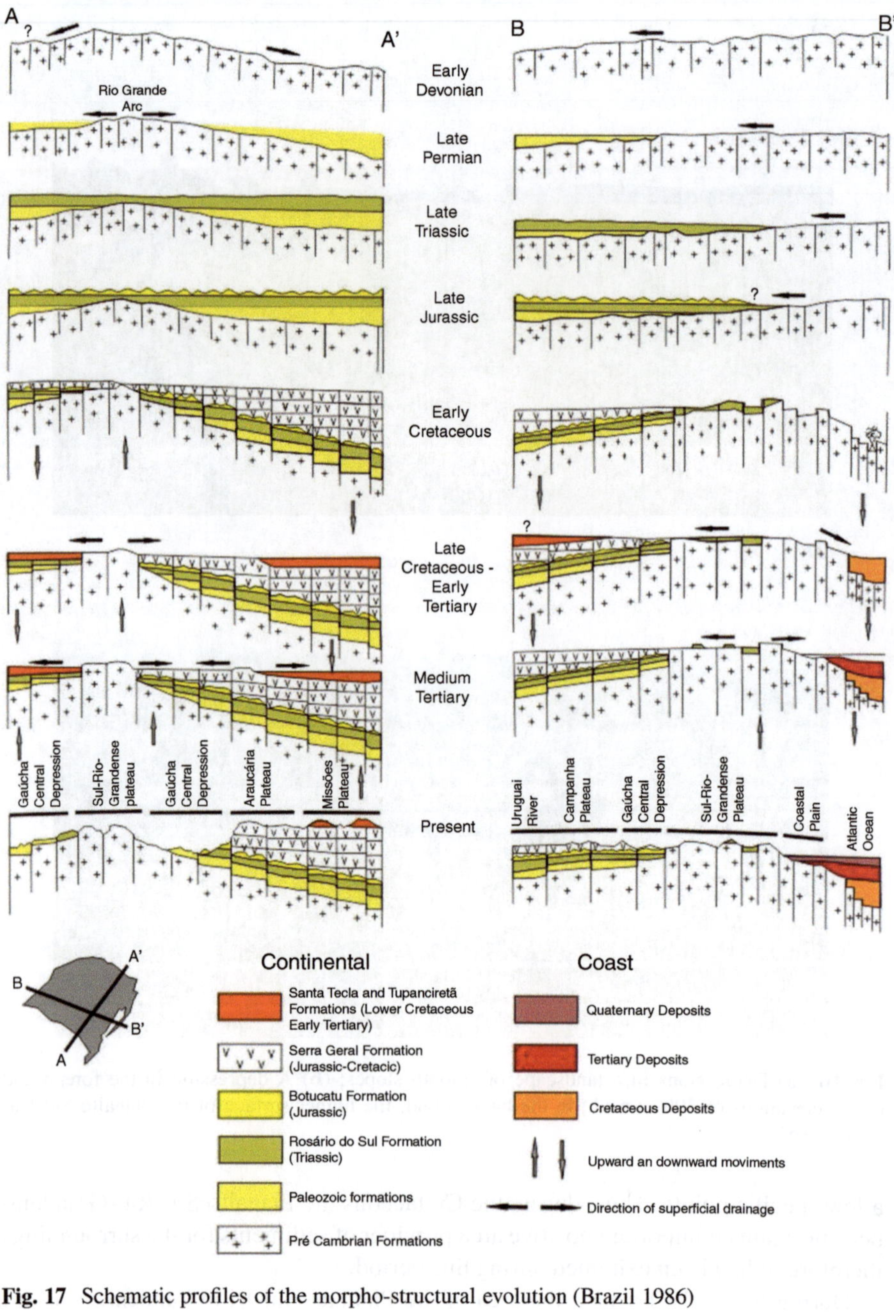

Fig. 17 Schematic profiles of the morpho-structural evolution (Brazil 1986)

By the Early Tertiary, the present geomorphological configuration of the state of Rio Grande do Sul was established, with its different units. It is likely that in the Early Tertiary the lateritic weathering profiles were developed, affecting very clearly the Planalto Sul-Rio-Grandense. Presently, all these profiles have been truncated, indicating erosion action after this period.

The tectono-magmatic events stopped during the pre-Aptian times in the Cretaceous, starting then a long period of quietness that lasted until the Late Cretaceous. In this period, the Pelotas Basin was receiving sediments of a tafrogenic sequence, and in the hinterland, a polygenetic erosion surface was developing, whose evolution continued even into the Paleogene. In the interior, both in the north and the south of the Rio Grande Arc, correlative sediments were deposited which corresponded to the Santa Tecla and Tupanciretã formations (Brasil 1986).

During the Late Cretaceous the pediplanation surface in the Planalto Sul-Rio-Grandense was affected by tectonic reactivation and alkaline magmatic processes (Brasil 1986).

According to Brasil (1986), during the Paleogene more humid and cooler conditions dominated, following the uplift of the Andean Cordillera to the west and the presence of the large oceanic mass of the Atlantic. Then, an exorheic drainage network was established on a surface without large topographic irregularities, being almost horizontal. The slow uplifting that affected the area allowed the superimposition of the drainage. This action, associated to the impact of alternating aggressive climates allowed the occurrence of down-cutting phases with moments of lateral degradation and intense erosion of the slopes, which provoked the removal of the capping of effusive rocks and the Gondwana deposits in the periphery of the Planalto Sul-Rio-Grandense, initiating an extensive peripheral denudation.

Following Brasil (1986), the endorheic conditions that evolved in the Paleogene allowed the elaboration of the Caçapava-Vacaria Surfaces. The highest topographic levels correspond to evolutionary conditions in an endorheic system, which was modified when the exorheic conditions were established.

As stated by Brasil (1986), the Vacaria and Caçapava Surfaces followed a similar evolutionary pathway, and therefore, they would be part of the same surface. The surface would have had a polygenetic evolution, according to the occurrence of successive erosion phases and the climatic variations dominating during their formation, which would have taken place from the Cretaceous until the Oligocene.

Final Remarks

The paleosurfaces of the state of Rio Grande do Sul deserve a more dedicated analysis, with mapping and research in greater detail. These surfaces are characterized by specific conditions that help in interpretation and are more complex than other areas of Brazil. The understanding of these surfaces includes situations in which two sets of lithology with totally different ages were deeply involved and closely associated.

From one side, the surface named by Ab'Sáber (1969) as the Caçapava do Sul Surface cuts through a varied lithology and served as a source area for the Paleozoic sedimentary units. At some moment of the Early Mesozoic, this surface was covered by sediments (and perhaps also by lava flows related to the opening of the South Atlantic Ocean). In the Late Cretaceous, this surface was exhumed (the present Planalto Sul-Rio-Grandense) and became the source area for the sediments of the Late Cretaceous and the Tertiary, a process that still continues today. Likewise, the exhumation was not able to fully eliminate the Triassic sediments found today in the hinterland of the Planalto, a strong indication of this ancient covering.

It would be of great importance if joint research projects between Uruguayan and Brazilian geomorphologists could be developed for the observation and correlation of the Gondwana Surfaces in the territory of these two adjacent countries.

Acknowledgements To FAPESP for funding the project Geomorphic Paleosurfaces Correlation of Summit in Brazil, Uruguay and Argentina – dating, morphology, cartography, deposits and associated alteration profiles (2011/23325-7). To CNPq for productivity scholarship (307465/2012-8).

References

Ab'Sáber AN (1949) Regiões de circundesnudação pós-cretácea no Planalto Brasileiro. Boletim Paulista de Geografia 1:2–18

Ab'Sáber AN (1955a) As altas superfícies de aplainamento do Brasil Sudeste. Boletim Geográfico 126:295–300

Ab'Sáber AN (1955b) Problemas Paleogeográficos do Brasil Sudeste. Boletim Geográfico 127:392–402

Ab'Sáber AN (1962) Revisão dos conhecimentos sobre horizonte sub-superficial de cascalhos inhumados do Brasil Oriental. Curitiba. Bol Univ Paraná Geografia Física 2:1–32

Ab'Sáber AN (1969) Participação das Superfícies Aplainadas nas Paisagens do Rio Grande do Sul. Geomorfologia 11:1–17

Almeida FFM de (1964) Fundamentos Geológicos do Relevo Paulista. Série Teses e Monografias. Instituto de Geografia, Universidade de São Paulo, São Paulo, 99 p

Almeida FFM de (1981) Síntese sobre a tectônica da Bacia do Paraná. In: Simpósio Regional de Geologia 3, Curitiba, 1981. Atas, vol 1, pp 1–20

Barbosa O (1965) Quadro Provisório de Superfícies de Erosão e Aplainamento no Brasil (inferências paleoclimáticas e econômicas). Revista Brasileira de Geografia 27(4):641–642

Bigarella JJ (2007) Estrutura e origem das paisagens tropicais e subtropicais, 2nd edn. Edic. da Universidade Federal de Santa Catarina, Florianópolis

BRASIL. Projeto RadamBrasil. 1986. Levantamento de Recursos Naturais, vol 35. Folha SH.22 Porto Alegre e parte das folhas SH.21 Uruguaiana e SI.22 Lagoa Mirim. IBGE, Rio de Janeiro, 791 p

Braun OPG (1970) Contribuição à Geomorfologia do Brasil Central. Revista Brasileira de Geografia 32(3):3–40

Dantas ME, Viana AC, Silva DRA (2010) A Origem das Paisagens. In: Viero AC, Silva DRA (eds) Geodiversidade do Estado do Rio Grande do Sul. CPRM, Porto Alegre, pp 37–50

De Martonne E (1943) Problemas Morfológicos do Brasil Tropical Atlântico. Revista Brasileira de Geografia 4:523–550

DNPM (1989) Mapa Geológico do Estado do Rio Grande do Sul e parte do Escudo Sul-Rio-Grandense. 1 mapa, Porto Alegre

Dorr JV (1969) Physiographic, Stratigraphic and Structural Development of the Quadrilatero Ferrifero, Minas Gerais, Brazil. United States Government Printing Office, Washington, DC

Freitas RO (1951) Relevos Policíclicos na Tectônica do Escudo Brasileiro. Boletim Paulista de Geografia 7:3–19

Godard A, Simon-Coinçon R, Lagasquie J-J (2001) Planation surfaces in basement terrains. In: Godard A, Lagasquie J-J, Lageat Y (eds) Basement regions. Springer, Berlin, pp 9–34

Hermann MLP, Rosa RO (1990) Relevo. In: IBGE – Geografia do Brasil, vol 2. Região Sul. IBGE, Rio de Janeiro, pp 55–84

IBGE (1995) Mapa geomorfológico do Brasil. Escala 1:5.000.000. IBGE, Rio de Janeiro

Kaul PFT (1990) Geologia. In: IBGE – Geografia do Brasil, vol 2. Região Sul. IBGE, Rio de Janeiro, pp 29–54

King LC (1956) A Geomorfologia do Brasil Oriental. Revista Brasileira de Geografia 2:147–265

Ladeira FSB, Santos M (2006) Tectonics and Cenozoic paleosols in Itaqueri's Hill (São Paulo-Brazil): implications for the long-term geomorphological evolution. Zeitschrift für Geomorphologie, Supplementband 145:37–62

Leonardi FA, Ladeira FS, Santos M (2011) Paleosurfaces and bauxite profiles in the Poços de Caldas Plateau, São Paulo/Minas Gerais, Brazil. Geociências 30(2):147–160

Moraes Rêgo LF (1932) Notas sôbre a Geomorfologia de São-Paulo e sua gênesis. Boletim Geográfico 4(37):9–17

Ollier C (2014a) Planation surfaces of the Gondwana continents: synthesis and problems. In: Rabassa J, Ollier C (eds) Gondwana landscapes in southern South America. Springer, Dordrecht, pp 61–96

Ollier C (2014b) Some principles in the study of planation surfaces. In: Rabassa J, Ollier C (eds) Gondwana landscapes in southern South America. Springer, Dordrecht, pp 47–59

Ponçano WL, Almeida FF (1993) Superfícies erosivas nos planaltos cristalinos do leste paulista e adjacências: uma revisão. Cadernos IG/UNICAMP 3(1):55–90

Rabassa J (2010) Gondwana paleolandscapes: long-term landscape evolution, genesis, distribution and age. Geociências 29(4):541–570

Rabassa J (2014) Some concepts on Gondwana Landscapes: long-term landscape evolution, genesis, distribution and age. In: Rabassa J, Ollier C (eds) Gondwana landscapes in southern South America. Springer, Dordrecht, pp 9–46

Salgado-Laboriau ML (1994) História Ecológica da Terra. Edgard Blücher, São Paulo, 307p

Soares PC, Landim PMB, Fúlfaro VJ (1978) Tectonic cycles and sedimentary sequences in the Brazilian intracratonic basins. Geol Soc Am Bull 89:181–191

Suertegaray DMA (2010) Geomorfologia do Rio Grande do Sul, o saber de Ab'Sáber. In: Modenesi-Gauttieri MC, Bartorelli A, Mantesso-Netto V, Carneiro CDR, Lisboa MBAL (eds) A Obra de Aziz Nacib Ab'Sáber. Beca, São Paulo, pp 334–343

Suertegaray DMA, Fujimoto NSVM (2004) Morfogênese do relevo do estado do Rio Grande do Sul. In: Verdum R, Basso LA, Suertegaray DMA (eds) Rio Grande do Sul: Paisagens e territórios em transformação. Universidade Federal do Rio Grande do Sul Editora, Porto Alegre, pp 11–26

Valadão RC (2009) Geodinâmica de superfícies de aplanamento, desnudação continental e tectônica ativa como condicionantes da megageomorfologia do Brasil Oriental. Rev Brasil Geomorfol 10(2):77–90

Ancient Landscapes of Uruguay

Daniel Panario, Ofelia Gutiérrez, Leda Sánchez Bettucci, Elena Peel, Pedro Oyhantçabal, and Jorge Rabassa

Abstract In this chapter, based on the available geological information, a model for the genesis and evolution of the Uruguayan landscape is proposed. A structural framework of the landscape evolution is provided and the record of such evolution in the most representative geological units is considered. A brief summary of the Uruguayan geology and its location in the regional context is performed, from Precambrian to Cenozoic times.

From the analysis of the geological record, it may be observed that the climate was very arid during part of the Jurassic and the Early Cretaceous. Together with the lava flows of the Arapey Formation, the climate became less arid as the Gondwana continents were moving apart from each other. However, the geological record suggests that semiarid climates were still prevailing. In the Middle Cretaceous, semiarid and wetter climates progressively alternated, until the Early Tertiary, when very wet and warm conditions were established, in coincidence with the "Palaeocene–Eocene Thermal Maximum (PETM)", followed by semiarid climates in the Oligocene, wetter conditions in the Miocene and semiarid again in the Pliocene, with alternating semiarid and humid conditions during the entire Quaternary.

D. Panario (✉) • O. Gutiérrez
UNCIEP, Instituto de Ecología y Ciencias Ambientales (IECA), Facultad de Ciencias, Universidad de la República, Montevideo, Uruguay
e-mail: daniel.panario@gmail.com; panari@fcien.edu.uy; oguti@fcien.edu.uy

L. Sánchez Bettucci • E. Peel • P. Oyhantçabal
Departamento de Geología, Facultad de Ciencias, Universidad de la República, Montevideo, Uruguay
e-mail: leda@fcien.edu.uy; elena@fcien.edu.uy; oyhantca@gmail.com

J. Rabassa
Laboratorio de Geomorfología y Cuaternario, CADIC-CONICET, Ushuaia, Tierra del Fuego, Argentina

Universidad Nacional de Tierra del Fuego, Ushuaia, Tierra del Fuego, Argentina
e-mail: jrabassa@gmail.com

J. Rabassa and C. Ollier (eds.), *Gondwana Landscapes in southern South America*, Springer Earth System Sciences, DOI 10.1007/978-94-007-7702-6_8,

Based on the palaeoclimatic evolution, the development of relief is discussed, considering the analysis of different morphostructural units in which the country is divided. Due to their size, shape and location (passive margin) of Uruguay, climate uniformity is assumed for each period throughout the entire territory. It is also assumed that the surfaces around elevations of 500 metres above sea level (m a.s.l.) correspond to relicts of probably pre-Cretaceous etchplains, strongly denudated, which are observed only in the surroundings of Aiguá area.

The landforms situated below the oldest surfaces, for instance, those below 320 m a.s.l. in the Eastern Hills Region (Sierras del Este), correspond to a new generation of geomorphological surfaces that may be considered of Cretaceous age, according to the information presently available. This surface may be correlated with the oldest surface developed on top of the lava flows of the Arapey Formation.

The extremely warm and wet climate of the Eocene prepared the conditions for the planation processes that covered most of the Uruguayan territory during the Oligocene, generating pediplains which were later reworked during the Late Cenozoic, up to the Quaternary, generating a landscape of smooth hills.

The morphogenetic potential of each morphostructural region determined the available energy of the resulting landscape, this being at a minimum in the Santa Lucía Basin, which continued to be under subsidence condition until the Tertiary, and almost nonexistent in the Laguna Merín Basin, where subsidence remains active until the Holocene.

Keywords Gondwana landscapes • Cenozoic landscapes • Uruguay • Paraná Basalt • Cratonic areas

Introduction

Uruguay lies on the West Atlantic Ocean coast of South America, between 30° and 35° South latitude and 53° and 58° West longitude (Fig. 1). It has a total land area of 176,215 km^2. The Uruguayan relief is quite reduced, between sea level and maximum elevation around 500 m a.s.l. (Fig. 2). Most of the territory is smoothly undulated, and it is developed within a range of 0–200 m a.s.l.

The climate of the region is temperate with an annual rainfall of 1,200 mm year^{-1} and a mean temperature of 18 °C. It is of the humid subtropical type (*Cfa* according to the classical Köppen climate classification). Seasons are properly well separated: spring is frequently humid, cool and windy; summers are warm; autumns are mild; and winters are chilly and uncomfortably damp. Bidegain and Caffera (1997) suggested the following climatic classifications: (1) mild climate, moderate and rainy (the cooler temperatures standing between −3 and 18 °C); (2) wet climate (rain is irregular, intermediate conditions between w and Köppen s types), "F type"; and (3) a temperature of the warmest month above 22 °C, "A type".

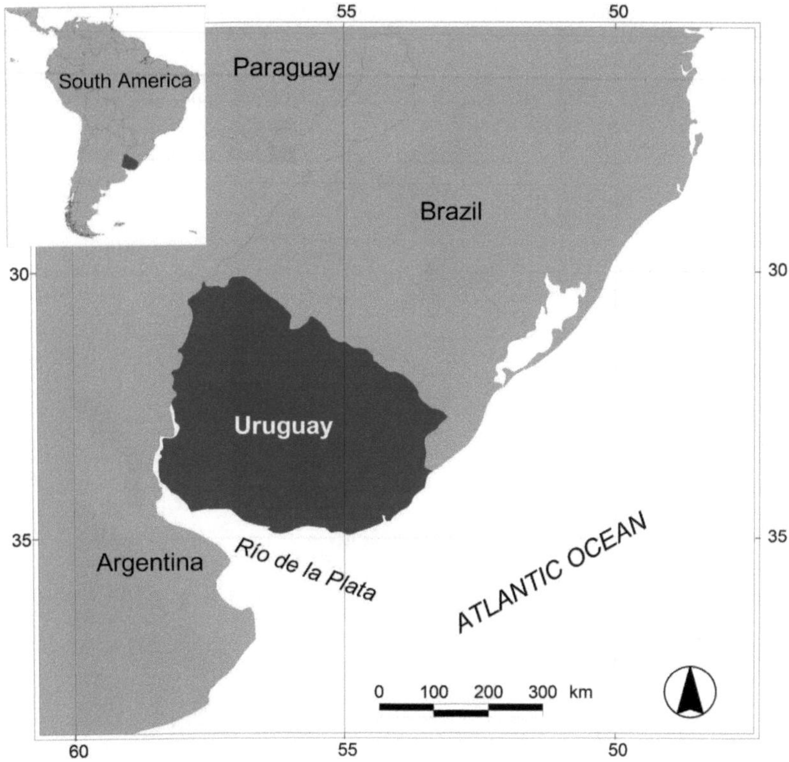

Fig. 1 Location map

Regional Geology

Precambrian Geology

Uruguay is part of the South American Platform and its geology consists of a Precambrian basement cropping out in the southern part and Palaeozoic to Mesozoic sediments and Mesozoic basaltic flows in the northern region, the latter being part of the Paraná Basin. Two main Mesozoic rift basins, related to the opening of the South Atlantic Ocean, are present in the southern portion (Santa Lucía Basin) and in the eastern portion (Laguna Merín Basin) of the country (Figs. 3, 4 and 5).

The Precambrian basement comprises nearly approximately 45 % of the country surface, and different approaches have been used within the last 30 years to define its main units. A first division was postulated by Ferrando and Fernández (1971), who considered two groups of ages defining two main domains, one of them of Palaeoproterozoic age (2.2–2.0 Ga) in the southwest and the other of Neoproterozoic age (900–550 Ma) in the East. Afterwards, Fragoso-Cesar (1980) defined the Dom Feliciano Mobile Belt (Neoproterozoic), located at the east of the Río de la Plata Craton (RPC).

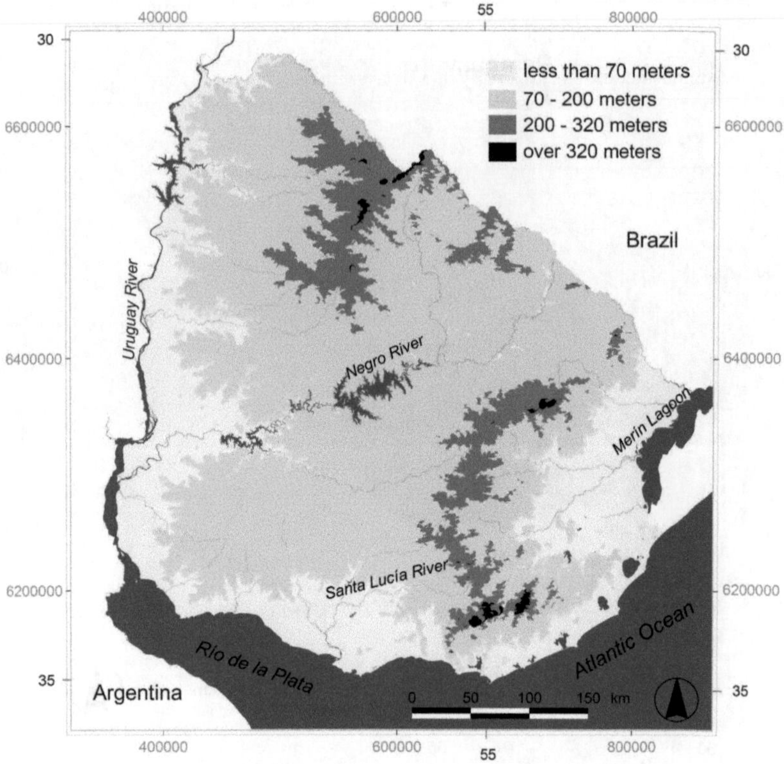

Fig. 2 Hypsographic map. Uruguay presents a landscape that occurs within a quite reduced altitudinal range, between sea level and maximum elevations around 500 m a.s.l. This hypsographic map has been prepared using 10 m contour lines in maps provided by the Servicio Geográfico Militar (SGM) of Uruguay

The Río de la Plata Craton (RPC) was originally defined by Almeida et al. (1973) including the older cratonic areas. Later, Bossi and Campal (1992) considered it as a build-up of two main terranes, the Piedra Alta Terrane (PAT) on the western side of the Sarandí del Yí Shear Zone (SYSZ) and the Nico Pérez Terrane (NPT) developed between the Sarandí del Yí and the Sierra Ballena Shear Zones (SBSZ) (see Fig. 5). Recently, Oyhantçabal et al. (2011) proposed the redefinition of the Río de la Plata Craton including only the juvenile Palaeoproterozoic rocks which were not tectonically reworked during the Neoproterozoic. According to this new definition, the Río de la Plata Craton (RPC) crops out only in the Piedra Alta Terrane of Uruguay (see Fig. 5) and in the Tandilia system in Argentina (Cingolani 2011). The Nico Pérez Terrane on the other hand includes Archaean and Palaeoproterozoic rocks, was strongly tectonically reworked during the Neoproterozoic and Brasiliano granitic intrusions are widespread; it should therefore be considered as an allochthonous basement unit, latter accreted to the Río de la Plata Craton (Oyhantçabal et al. 2011; Rapela et al. 2011).

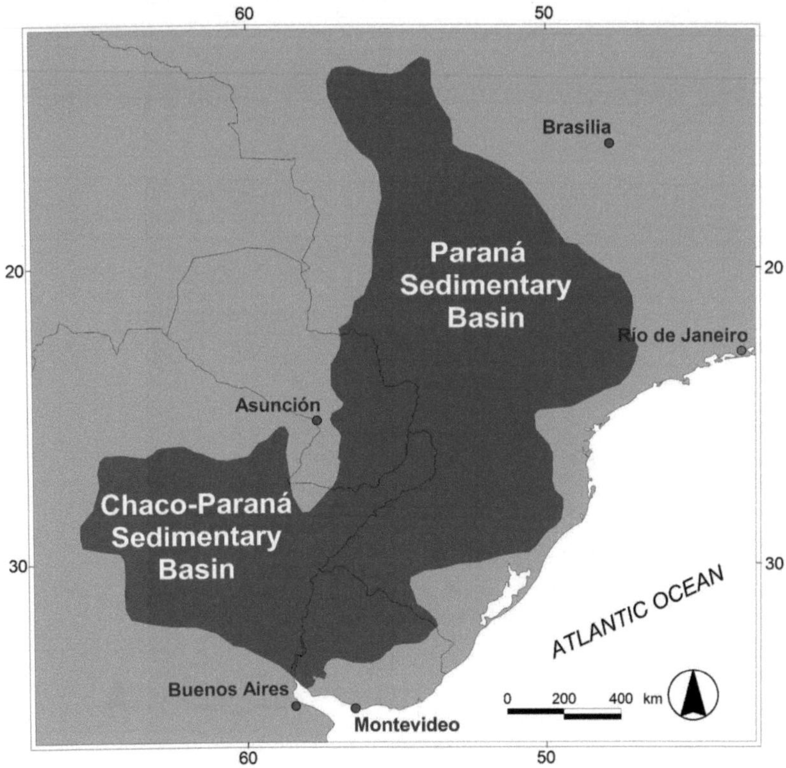

Fig. 3 Tectonic domains of Uruguay. Spatial distribution of the Paraná and the Chaco–Paraná sedimentary basins (Palaeozoic to Mesozoic) (Modified from Milani 1997)

The Dom Feliciano Belt (DFB) crops out in eastern Uruguay (see Fig. 5) and extends for more than 1,000 km along the Atlantic coast of Uruguay and southern Brazil. It was developed between ca. 750 and 550 Ma (Sánchez Bettucci et al. 2010a) and represents the Brasiliano/Pan-African orogenic cycle. It is genetically related to tectonic episodes that occurred during the convergence of the Río de la Plata, Congo and Kalahari cratons (Fig. 6) during Neoproterozoic times (Sánchez Bettucci et al. 2010a).

The basement of the Dom Feliciano Belt in the southern portion is named as the Campanero Unit (Sánchez Bettucci 1998; Sánchez Bettucci et al. 2010b) and comprises mainly orthogneisses with protolith age around 1.7 Ga (U/Pb SHRIMP in zircon; Mallmann et al. 2003). Similar ages were obtained by Sánchez Bettucci et al. (2004). In the easternmost part of the area, a pre-Brazilian Basement Inlier, the Cerro Olivo Complex (Masquelin 2002; Masquelin et al. 2012), consists of gneisses, migmatites and granulites of Neoproterozoic age.

The Dom Feliciano Belt on a regional scale is subdivided into three main tectonic units, from East to West (Basei et al. 2000): (a) Granite Belt, (b) Schist Belt and (c) Foreland Belt.

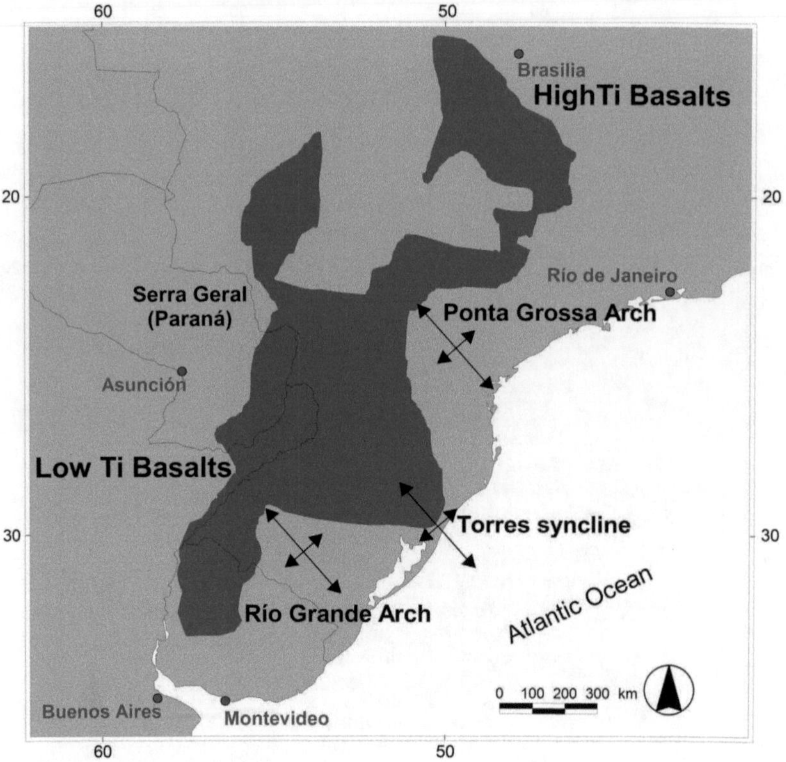

Fig. 4 Schematic map showing the geographic distribution of the Paraná igneous province displaying the distribution of high- and low-Ti areas. The main basement highs of the Basin (Ponta Grossa, Torres and Río Grande archs) are shown (Modified from Piccirillo and Melfi 1988)

The Granite Belt is represented by three large batholiths known as the Aiguá Batholith (Uruguay), Pelotas Batholith (Rio Grande do Sul State, Brazil) and Florianopolis Batholith (Santa Catarina State, Brazil). Ages between 630 and 550 Ma have been reported. These batholiths show calc-alkaline affinity.

The Schist Belt comprises pre-collisional Neoproterozoic meta-volcanic and sedimentary sequences showing metamorphism under greenschist to lower amphibolite facies. Three lithostratigraphic units are defined in this belt: the Lavalleja (Uruguay), Porongos (Rio Grande do Sul) and Brusque (Santa Catarina) groups of southern Brazil.

The Neoproterozoic Lavalleja Group is composed mainly of basic volcanics, schists, calc-schists and limestones, conforming three formations (Minas, Fuente del Puma and Zanja del Tigre; Sánchez Bettucci et al. 2001). Recently, the Zanja del Tigre Formation (Meso- to Neoproterozoic) integrated by limestones, quartzites, pelites, sandstones and minor BIF's ("Banded Iron Formation") and acid volcanic rocks, metamorphosed in greenschists to lower amphibolite facies

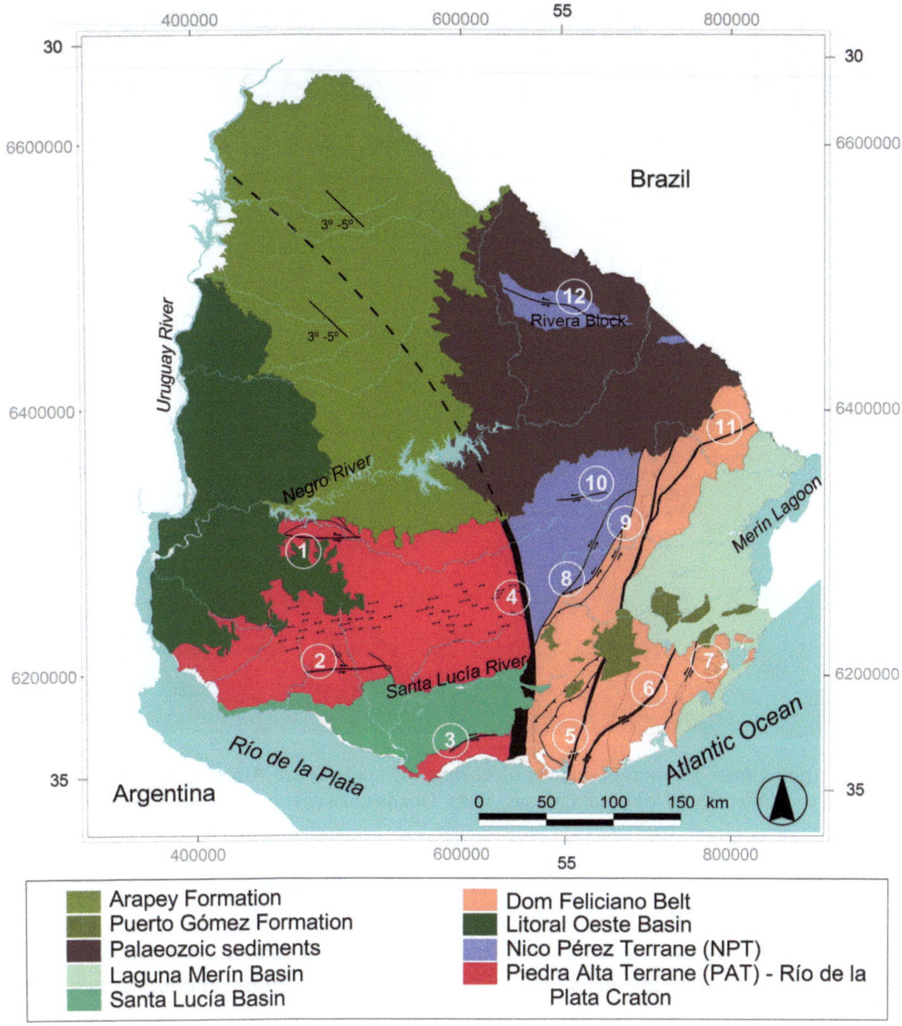

Fig. 5 Main geological units of Uruguay (Cenozoic cover is not shown): Precambrian terranes and shear zones, Palaeozoic sediments and Mesozoic basaltic flows and rift-related basins (Modified from Sánchez Bettucci et al. 2010b, after Preciozzi et al. 1985 and Bossi and Ferrando 2000). Shear zones: *1* Paso Lugo, *2* Cufré, *3* Mosquitos, *4* Sarandí del Yí, *5* Sierra Ballena, *6* Cordillera, *7* Rocha, *8* Cueva del Tigre, *9* Fraile Muerto-María Albina, *10* Tupambaé, *11* Cerro Amaro, *12* Rivera

(Sánchez Bettucci and Ramos 1999; Sánchez Bettucci et al. 2001, 2010a), is considered as a basement inlier of the Dom Feliciano Belt based on isotopic data (Oyhantçabal et al. 2009; Sánchez Bettucci et al. 2010a).

The Foreland Belt consists of several volcano-sedimentary and sedimentary successions located between the Schist Belt and the Palaeoproterozoic domains

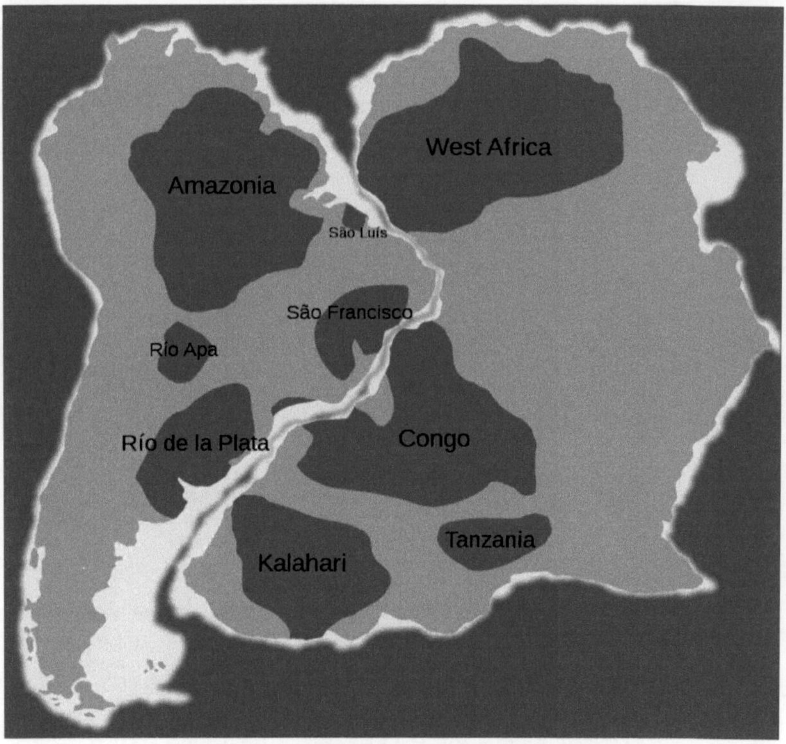

Fig. 6 Approximate location of cratons older than 1.3 Ga in South America and Africa (https://commons.wikimedia.org/wiki/File:Cratons_West_Gondwana.svg)

of the Río de la Plata Craton (Basei et al. 2000). These basins include marine to molasse Ediacaran deposits of the Arroyo del Soldado (Gaucher 2000; Gaucher et al. 2003, 2004) and Maldonado Groups (Pecoits et al. 2004, 2008; Teixeira et al. 2004). These groups are affected by very low- to low-grade metamorphism and deformation.

The Sierra de Las Animas – Aiguá area – is considered the region of Uruguay where the relicts of Gondwana age palaeosurfaces are best preserved.

Overview of the Phanerozoic Geology of Uruguay

Palaeozoic Paraná Foreland Basin

The Palaeozoic Paraná Basin is located at the central southern region of South America. It is a foreland basin with sedimentary deposition ranging in age from Neo-Ordovician to Tertiary. This basin occupies about 1.7 million km^2 in Argentina,

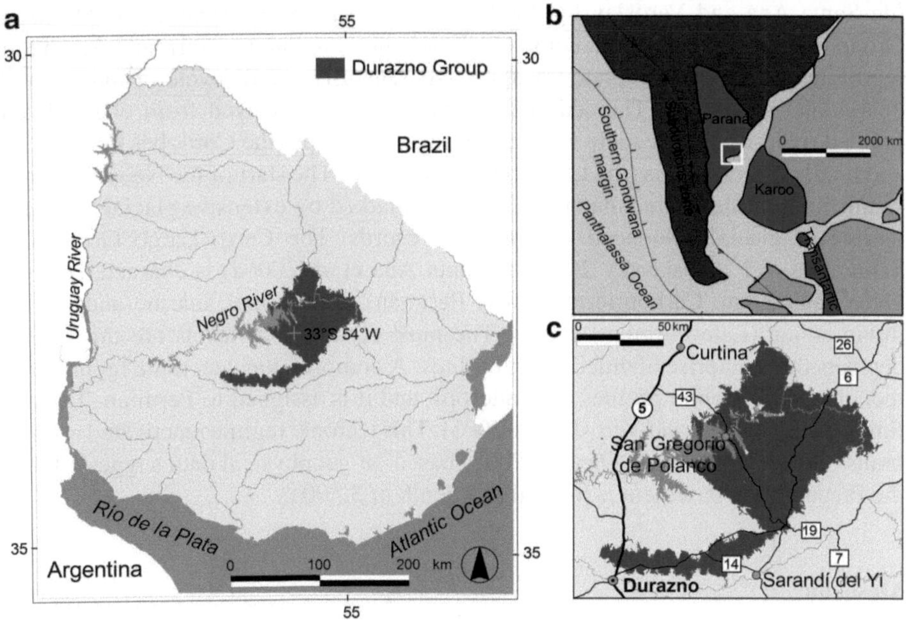

Fig. 7 (**a**) Regional distribution of Devonian sedimentary units in Uruguay (Durazno Group) which is more extensive than it had been established so far. Its surface has been inferred from their spectral response in satellite imagery (Landsat TM) and field observations (**b**) Palaeogeographic setting in the framework for the Western Gondwana (the present approximate location is indicated by a white square) (**c**) Details of the geographical location of the Durazno Group

Bolivia, Brazil, Paraguay and Uruguay. The basin has a NNE-SSW-trending elliptical form with two-thirds of its area covered by Mesozoic basaltic lavas. The stratigraphic record of this vast basin reaches 7,000 m in thickness in the central depositional centre, just under the Paraná River (Milani and Zalán 1999). Milani et al. (1998) suggested that the Paraná Basin comprises six stratigraphic mega-sequences delimited by interregional unconformities (Vail et al. 1977). The eastern border of the Paraná Basin corresponds to a crustal region deeply affected by the South Atlantic Ocean rifting (see Fig. 3). Consequently, the uplift and erosion have been responsible for the removal of large amounts of Palaeozoic sedimentary rocks. The western border of this basin is defined by the Asunción arch, a flexural bulge related to the loading of the Cenozoic Andean thrust belt nearby Argentina and Bolivia, whereas the northern and southern borders, these deposits on-lap the Precambrian basement (Milani and Zalán 1999). The arrangement of this basin has led some authors to postulate foreland basin deposits (Catuneanu 2004), together with the Karoo (South Africa), Beacon (Antarctic) and Bowen (Australia) basins.

The sedimentary record in Uruguay begins in the Lower Devonian to Lower Permian. The Devonian units constitute the Durazno Group (Veroslavsky et al. 2006) (Fig. 7) and the Carboniferous–Permian units form the Cerro Largo Group (de Santa Ana and Veroslavsky 2003; de Santa Ana et al. 2006a). The Durazno Group comprises the Cerrezuelo, Cordobés and La Paloma formations, and it represents an almost complete transgressive–regressive (T-R) cycle of marine and continental sediments. The sedimentary environments evolved from channelized braided rivers (the Cerrezuelo Formation) to clayey slope (the Cordobés Formation) and finally littoral plains (the La Paloma Formation). The start of the Neopalaeozoic sedimentation (de Santa Ana et al. 2006b) is marked by extensive glacial, glacial–marine or glacial-influenced sedimentary records. The Cerro Largo Group (de Santa Ana and Veroslavsky 2003; de Santa Ana et al. 2006a) is characterized by glaciogenic (Late Carboniferous–Early Permian), transitional, marine and finally fluvio-eolian (Late Permian) cycles. The most conspicuous levels are the glacial deposits that comprise diamictites and tillites. A compressional tectonic regime was recognized in seismic profiles and outcrops, and it is assigned to Permian–Triassic times (de Santa Ana and Veroslavsky 2003). This tectonic regime reactivated normal faults. On the other hand, Oleaga (2002) based on geophysical data suggested that the Precambrian basement is located at a depth of 3,500 m.

Mesozoic

The Atlantic Ocean Uruguayan margin, a portion of the eastern margin of the South American platform, corresponds to a passive or Atlantic-type margin. According to Turner et al. (1994), the thermal anomaly or Tristan da Cunha mantle plume was responsible for the opening of the South Atlantic Ocean and had its peak between 137 and 127 Ma. Thomaz-Filho et al. (2000) suggested that magmatic activity occurred in different stages during the break-up of South America and Africa (Cesero and Ponte 1997). The most important extensional event in Uruguay related to the break apart of Pangea took place in the mid-Triassic and is represented by Cretaceous magmatism related to continental rifting and is part of the Paraná–Etendeka magmatic province. The deformation is dominated by brittle faulting that affected all linked units and is characterized by normal faults, usually of short length and average East–West orientation dipping towards both the North and South. Also, there is a series of N 350° faults with westward to subvertical inclinations. Some brittle features are evidenced by gouge formation. The direction of preferential fault is N 75° to N 120° that generates hemi-graben-type basins filled by clastic deposits and alkaline and peralkaline magmatism.

Extensional Magmatism

The extensional magmatism was related to the continental rifting (Tristan da Cunha mantle plume) (e.g. O'Connor and Duncan 1990; Peate et al. 1990; Hawkesworth et al. 1992), and it is part of the Paraná–Etendeka magmatic province. The Paraná–Etendeka igneous province is one of the main flood volcanic provinces in the world covering an area of 1.2×10^6 km^2, with its magmatic activity peak at ca. 132 Ma

(Erlank et al. 1984; Bellieni et al. 1984; Renne et al. 1992, 1996a, b). The South American portion of this province (Paraná) contains an estimated acidic volcanic rock of 3 % of the total volume (Bellieni et al. 1984, 1986), whereas in the African portion (Etendeka), it is estimated in more than 5 % of the total volume. This difference of proportions would be related to the rift geometry asymmetry (Turner et al. 1994). The Paraná basalts were defined as aphyric tholeiitic basalts (Comin-Chiaramonti et al. 1988). Based on the criteria of separation in low TiO_2 ($\approx$1) and high TiO_2 (>3) proposed by Bellieni et al. (1984), Fodor (1987), Cox (1988), Mantovani et al. (1985) and Turner and Hawkesworth (1995), among others, the existing data in Uruguay fall in the field of low TiO_2 (*sensu* Sánchez Bettucci 1998).

Unimodal Extensional Magmatism

The unimodal extensional magmatism is named in Uruguay as the Arapey Formation (Bossi 1966; see Fig. 5), and it is outcropping in the NW region of the country. The ages obtained for this formation are ca. 132 Ma (Creer et al. 1965; Umpierre 1965, in Bossi 1966; Stewart et al. 1996; Féraud et al. 1999). The ~134 Ma corresponds with main geodynamic changes in the Earth's history where large igneous provinces (LIPs) are developed (Renne et al. 1996a, b). Contemporaneously with these flood basalts, alkaline complexes were emplaced around the margin of the Paraná Basin. The Paraná Province displays characteristics of bimodality with a strong geographical correlation. The volcanic suite includes andesitic basalts to andesites. The volcanic rocks of Arapey Formation are emplaced above aeolian sandstones (Tacuarembó Formation, Jurassic–Cretaceous). A latest tectonic event determined that these basalts were tilted between 3° and 10° to the WSW. A major tectonic lineament (Sarandí del Yí Shear Zone) controlled not only the emplacement of basalts but also the further development of the Littoral Basin.

Bimodal Extensional Magmatism

The bimodal extensional magmatism is represented by the Puerto Gómez and Arequita formations and the San Miguel and Valle Chico complexes. These units in SE Uruguay are linked to aborted rifts (failed arms) associated with the opening of the South Atlantic Ocean.

The Arequita Formation is represented by acidic volcanic rocks including lava flows and pyroclastic rocks with rhyolitic to dacitic compositions. The high Zr concentrations indicate that these rocks show peralkaline affinity (Kirstein et al. 1997, 2000). The peralkaline rhyolites suggest an important late magmatic episode in the continental rifting event (Sánchez Bettucci 1998). The Puerto Gómez Formation is constituted by olivine and alkaline basalts (hawaiite), of strongly amygdaloid aspect, suggesting shallow submarine environments. Sánchez Bettucci (1998) suggested the occurrence of flows with pillow lavas.

The Valle Chico Complex (Muzio 2000; Lustrino et al. 2005) is composed of felsic plutonic rocks (quartz monzonites to syenites, quartz syenites and granites), volcanic rocks and dykes (quartz latites to trachytes and rhyolite). Lustrino et al. (2005) suggested chemical similarities between the Valle Chico Complex and the Arequita Formation. Lustrino et al. (2003) suggested that the existence of these mildly alkaline to transitional basic rocks is clear evidence that the Puerto Gómez and Arequita formations are atypical among the Paraná–Etendeka igneous province.

Litoral Oeste Intracratonic Basin

Intracratonic sag sedimentary basins occur in the middle of stable continental or cratonic blocks and are infrequently fault bounded, although strike-slip faulting can occur within them (Middleton 1989). The Litoral Oeste Basin of Uruguay occupies an area just over ca. 50,000 km^2 continuing westwards in the "Mesopotamia" region of Argentina. The basement of the basin in the southern portion is the Piedra Alta Terrane (Palaeoproterozoic), whereas in the North and Northeast, the basement is the Arapey Formation. The evolution of this basin apparently was controlled by thermo-tectonic subsidence (Goso and Perea 2004).

This basin is filled by Cretaceous and Cenozoic deposits. The Cretaceous units are the Guichón and Mercedes formations, both representing fluvial deposits (Goso and Perea 2004). Moreover, the Cenozoic deposits are represented by the Fray Bentos, Salto and Raigón Formations. The Fray Bentos Formation (Late Oligocene) comprises aeolian silts and scarce fluvial deposits developed in dry environments.

Rift Deposits (Santa Lucía and Laguna Merín Basins)

The Santa Lucía and Laguna Merín basins (see Fig. 5) are located in the South and East of Uruguay, respectively. Both basins present an elongated E-NE shape and are considered a failed rift formed during the Gondwana break-up (Sprechmann et al. 1981). They were controlled by the *Sa*nta *L*ucía–*A*igua–*M*erín (*SaLAM*) tectonic alignment (see Rossello et al. 1999) related to the Paraná–Etendeka volcanic province (O'Connor and Duncan 1990). In the Santa Lucía rift, the Santa Rosa structural high (parallel to the basin borders) is located in the central region of the basin and divides it in two subbasins. The Cretaceous volcanic and sedimentary infilling is up to 2,500 m thick, whereas the Cenozoic sediments are only a few tens of metres thick (de Santa Ana et al. 1994). The Early Cretaceous sequence (the Migues Formation, 1,800 m thick; Jones 1956) represents the deepest levels of the basin, and it is composed of sandstones, siltstones and mudstones. The Migues Formation is overlain by siltstones and sandstones of the Oligocene Fray Bentos Formation.

The limestone sandstone deposits (the Mercedes Formation, Bossi et al. 1975, 1998) found in the Santa Lucía Basin were considered as part of the Upper

Cretaceous (Veroslavsky et al. 1997) and were formerly correlated to the "Calizas del Queguay" deposits that crop out in western Uruguay. Recent studies considered that these siltstones are the result of calcrete formation, post-depositional processes that occurred during the Tertiary (Goso and Perea 2004) or Early Pleistocene (Panario and Gutiérrez 1999). Different authors (Lambert 1940; Jones 1956; Goso 1965; Goso and Bossi 1966a, b; Gómez Rifas et al. 1981; Preciozzi et al. 1985; de Santa Ana et al. 1994; Peel et al. 1998) assigned a lacustrine origin to these deposits. Also, the Mercedes Formation records the most significant pedogenetic processes occurred in the Cenozoic times such as ferrification, silicification (silcrete formation) and calcretization.

The Laguna Merín Basin is filled primarily by volcanic rocks: basalts (the Puerto Gómez Formation), rhyolites, dacites, ignimbrites (the Arequita Formation) and to a lesser extent conglomerates and red sandstones (Veroslavsky 1999) and Quaternary loess and sands units.

Cenozoic

Towards the end of the Cretaceous, subsidence processes slowed down as the basins were filled and during the Cenozoic deposition and sedimentation were limited by uplift and erosion. The preserved sedimentary deposits are linked to successive transgressive and regressive eustatic cycles recorded at regional and global scale during the Cenozoic. Based on drilling information of the continental shelf, a detailed and fairly continuous record of marine sediments appears, corresponding to the Cretaceous–Tertiary boundary. Many successive variations in sea level were recognized during the rest of the Cenozoic (Ubilla et al. 2004).

The base of the Palaeogene is poorly represented. The scarcity of Palaeogene geological records is related to nondepositional processes that indicate climate variations at the beginning of the Palaeogene. Examples include the development of oxysol and ferricrete formation in the Eocene (Panario and Gutiérrez 1999) or in the Late Palaeocene–Eocene, and particularly on Cretaceous continental sediments (already mentioned above), the development of silcretes, fossiliferous pedogenetic calcretes, limestone and lacustrine deposits.

In the Oligocene, due to a basement reactivation linked to the Andean orogeny, alluvial and fluvial deposits, landslide processes and loess materials occurred. During the Late Miocene, there was a new marine transgression (Martínez 1989; Ubilla et al. 2004), and in the Pliocene–Pleistocene continental evolution, processes occurred, mainly developing extensive fluvial systems.

The Quaternary is characterized by the development of continental deposits on the coast of the Río de la Plata and the Atlantic Ocean. Associated with frequent oscillations of sea level, barrier islands, lake sedimentation, marsh and lagoon deposits occurred (Ubilla et al. 2004).

The Fray Bentos Formation (Bossi 1966) outcrops in western Uruguay in the Paraná Basin and to the South and East in the Santa Lucía and Laguna Merín basins. It lies unconformably on the Mercedes Formation and on the Precambrian basement.

Fig. 8 Details of the sedimentary structures of the Camacho Formation (Miocene), with sediments ranging from very fine to coarse sandstones, siltstones and mudstones with fossil marine bivalves among other groups

It is covered unconformably by the Camacho (Miocene) and Salto (Pliocene-Pleistocene) formations. The Fray Bentos Formation consists of fine sandstones, loess siltstones, mudstones, conglomerates and diamicton levels. It represents the first significant depositional episode during the Cenozoic (Goso 1965; Goso and Bossi 1966a; Veroslavsky and Martínez 1996) only preceded by the removal of oxisols and associated ferricretes and alterites off the main features as alluvial fans (Panario and Gutiérrez 1999). The thickness in outcrops is less than 15 m, but in the subsurface, it reaches 100 m (Bossi and Navarro 1991).

The Camacho Formation (Fig. 8) is composed of a succession of very fine to coarse sandstones, siltstones and mudstones (Martínez 1994; Ubilla et al. 2004). This unit outcrops along the coasts of the Colonia and San José departments, but it is also found in subsurface in San José, Maldonado and Rocha. The maximum outcropping thickness is about 15 m, whereas in the continental shelf, it reaches ca. 200 m (Gaviotín and Lobo drill holes: Stoackes et al. 1991; Ucha et al. 2004). It lies unconformably over the Precambrian basement or on the Fray Bentos Formation (Late Oligocene).

The Raigón Formation (Goso 1965) conformably overlies the Camacho Formation and it is uncomformably deposited over the Fray Bentos Formation and the Precambrian basement (Spoturno and Oyhantçabal 2004). The Raigón Formation is exposed at the coastal cliffs of the Río de la Plata with a maximum thickness of 30 m. This pile of sediments is of fluvial and transitional origin, and it is unconformably covered by the Libertad Formation, which developed in semiarid continental climatic conditions and has been assigned to the Pleistocene. This formation has been assigned to the Pliocene (Panario and Gutiérrez 1999), but, however, some authors like Perea and Martínez (2004) have considered as belonging to younger land-mammal ages (even Pliocene–Middle Pleistocene) those sediments formed following the re-transportation process of the Raigón Formation or otherwise to relate them with deposits of similar colour, grain-size characteristics and sedimentary environment of those corresponding to the genesis of such formation.

Andreis and Mazzoni (1967), following Francis and Mones (1966), named this unit as the San José Formation, dividing it into two sections: the bottom unit formed by clays, silts, sandy-silts and subordinate greenish-grey sands and the upper portion composed of medium to very coarse pink to yellow sandstones. According to Bossi and Navarro (1991), the Raigón Formation consists of green clay, medium-fine sand, coarse sands and conglomerate levels. Besides, Tófalo et al. (2006) indicated that these fluvial sediments can be divided into two sections predominantly sandy, separated by a regional discontinuity, pointing out to an episode of sedimentation reactivation.

The Salto Formation is attributed to the Late Pliocene and the Pleistocene, having also a fluvial origin. It is exposed in small outcrops near the Río Uruguay, and it was correlated with the Raigón Formation by Goso (1965) and Panario and Gutiérrez (1999). It also correlates with the Salto Chico and Ituzaingó formations in Argentina. According to Veroslavsky and Montaño (2004), it represents deposits of braided rivers distinguishing two depositional cycles. These deposits present lenticular geometry, are multi-episodic and have normal grading (Tófalo and Morrás 2009).

The Salto, Salto Chico and Ituzaingó formations are all clearly related to the Río de la Plata Basin, formed by the Paraná and Uruguay rivers, whose basins are only differentiated since their middle portions and whose sediments have continued to be deposited until today, according to Herbst (2000), which makes it difficult to establish the chronostratigraphic location of its deposits, which have been assigned both to the Pliocene and to the Pleistocene by different authors. Thus, the Salto Formation (Goso 1965; Panario and Gutiérrez 1999) and the Salto Chico Formation (Iriondo 1996) have been considered to be of Late Pliocene–Pleistocene age, as it is the case of the Ituzaingó Formation (Iriondo 1980).

The Libertad Formation (Early to Middle Pleistocene; Fig. 9) was defined by Goso (1965). This formation has a generalized distribution throughout the territory, but its greatest expression takes place in southwestern Uruguay. It has a thickness of about 20 m, lying unconformably over the Raigón Formation, several Cretaceous formations and both Palaeozoic rocks and the Precambrian Basement. It is also covered unconformably by Middle and Late Quaternary formations (Spoturno and Oyhantçabal 2004). According to Bossi and Ferrando (2000), it includes massive friable mudstones with scattered gravel and abundant calcium carbonate. According to Tófalo et al. (2006), it corresponds to loess deposits accumulated in semiarid regions of gentle slope undergoing significant pedogenetic processes.

Zárate (2003) suggested that this loess, mainly represented by a 1–2 m thick mantle, has similar composition to similar units of the Northern Pampas loess (Entre Ríos and Corrientes provinces of Argentina). Two main loess units have been identified, named Libertad I and Libertad II, of Early and Middle Pleistocene

Fig. 9 Loessic sediments may be observed in the cliff, showing a continuous process of soil formation, corresponding to the Libertad I Formation (Quaternary). The *dashed line* indicates the unconformity with the Late Pliocene Raigón Formation

age, respectively (Goso 1965). The Libertad I Formation is composed of poorly calcareous edaphized loess while the Libertad II Formation shows evidences of water reworking and pedogenetic modifications.

On the other hand, Sánchez Bettucci et al. (2007) presented preliminary magnetostratigraphic results of the Camacho, Raigón and Libertad formations (Neogene). Reverse polarity signal was found in the Camacho Formation, ascribed to the Gilbert magnetic zone. The sediments of the Raigón Formation have normal polarity interpreted as belonging to the Gauss magnetic zone. Finally, the Libertad I Formation shows reverse magnetic polarity, which is referred to the Matuyama magnetic zone. The palaeomagnetic pole obtained by these authors is located at 88.2° S lat., 189.7° W long, Dp 5° Dm 7.2° N = 39. The Libertad II Formation showed normal polarity, and it has been assigned to the Brunhes palaeomagnetic age, according to Sánchez San Martín (2010).

In Uruguay, neotectonic studies have not been performed, but some evidence of tectonic activity is known. Brazilian studies suggested that the Neotectonic period (Eocene–Oligocene) should be related to the episode at which the last major tectonic reorganization occurred. The Neotectonic period presents a possible correlation between events of the Andean orogeny (Bezerra et al. 2001, 2003; Bezerra and Vita-Finzi 2000). Hasui (1990) suggested that the maximum age of the neotectonic period in Brazil should be the Oligocene, which corresponds to the most recent extensional pulses of the South Atlantic Ocean extension. However, the depth at which Cenozoic units are located (at the west and east) suggests a steady continuous dominant subsidence since the Cenozoic mainly in the eastern part, whereas in the western region uplifting dominated. In this last region displacement direction and low-magnitude reverse faulting have been identified. In addition, the historical seismic data in Uruguay include low-intensity movements that certainly should have left their mark in the landscape.

Geomorphology of Uruguay

Landscape Modelling

The evolution of the Uruguayan landscape is the result of a variety of regional climates throughout its geological history. These climates had a strong influence upon the landscape modelling and modification of the pre-existing landforms. The sedimentary materials generated in the different periods and resulting landforms allow the inference of several palaeoenvironmental features. The time climate reconstruction based solely upon the observed landforms is only possible when those landforms have been preserved. Even though only at a relict level, those remnants are a clear expression of the dominant palaeoclimate.

These features are only possible under intense conditions or of long enough duration so as to imprint clear features of undoubted genesis which would provide a reliable interpretation.

Many landforms have certainly been eroded and erased from the surface: the oldest relict landforms are mainly represented by isolated elevations, generally thoroughly denudated. These relicts may be interpreted as either positional inselbergs, bornhardts, whereas others are considered as etchplains, which are the major landscape features.

Palaeoclimates

Palaeozoic

Some palaeoclimatic evidence may be established for this region since the Devonian. In this sense, from the Early Devonian to the Early Permian, several transgressive marine events have been identified. Continental deposits formed by braided rivers are also found, thus indicating alternating relatively arid conditions and presumably wetter climates. During the Early Permian, fluvio-aeolian deposits occurred as well, which are related to arid and semiarid conditions (Goso and Perea 2004). The wetter and warmer periods which would have taken place may be associated to the clayey facies, due to the landscape stability during the marine transgressive stages. There were also moraines and till deposits of Carboniferous–Permian age, which indicate the existence of higher relief, probably located further north.

Mesozoic

The cold and wet conditions of the Permian slowly changed to warmer and drier climates during the Late Permian and the Triassic. The climate conditions during most of the Jurassic were clearly those of a large desert, as it is shown by the sandstones of the Tacuarembó Formation, known as the Botucatú Formation in Brazil, mostly composed of rubified aeolian sands, which were then active dune fields. This formation also presents lagoonal environment facies of less extreme conditions (Bossi 1966).

The arid conditions were maintained during the Early Cretaceous, as it is proven by the existence of silicified barkhan dunes and sand sheets (inter-trap sandstones) coming from the north, interbedded with the Paraná volcanic province basalts.

Later on, the climate seemed to have evolved towards more semiarid conditions, related to the opening of the South Atlantic Ocean, exposed also by rubified fluvial sandstones (the Guichón and Migues formations). The semiarid conditions allowed the discontinuous development of incipient soils (Goso and Perea 2004) which persist until the end of the Cretaceous, but presumably under a temperate climate according to the sedimentology data pertaining to the Mercedes Formation. These circumstances suggest that the conditions needed for the genesis of planation surfaces were relatively continuous from some time in the Jurassic to the end of the Cretaceous if previous humid condition prevailed.

Cenozoic

The dominant climatic conditions during the Palaeocene are still somewhat unclear, since the geological record has not enough continuity. Deep drilling data coming from the submarine shelf will be undoubtedly very useful in this interpretation. The origin and development of the most extensive geomorphological features of Uruguay may be tracked back to Eocene (Panario and Gutiérrez 1999) or Late Palaeocene times. A widespread Cenozoic planation of the Uruguayan landscape was possible under the warm and humid Eocene climate, with deep weathering accompanied by oxysol development and ferricrete formation. Eocene ferricretes have developed over Cretaceous and Precambrian rocks in Uruguay and on basaltic rocks in the provinces of Corrientes and Misiones in Argentina. Ferricretes appear also as isolated boulders in Jurassic sandstones (the Tacuarembó Formation; Caorsi and Goñi 1958).

Oligocene erosion of the Eocene soils under generally arid and semiarid conditions resulted in the deposition of alluvial fans of plintite cobbles (Ford 1988), which pass upwards through a decimetre transition zone into the loess-dominated

Fray Bentos Formation. These erosion processes were facilitated by the intense Eocene weathering yielding extensive planation surfaces in metamorphic, igneous and sedimentary domains (Table 1).

During the Miocene, the geological record (Camacho Formation) indicates a marine transgression, whose mollusc fauna and the presumed associated continental fauna would indicate warm and wet climate conditions.

Based on palaeontological data, this unit was considered by Rodrigues et al. (2008) as deposited in subtropical marine provinces, ranging from intertidal to middle-shelf setting.

The Pliocene erosion, again under generally arid conditions, resulted in the formation of coarse braided river deposits known as the Raigón Formation (Goso 1965), alluvial fans (Malvín Formation; Antón and Prost 1974) and probably the Salto Formation related with the Uruguay River as well as other fluvial sediments in southwestern Uruguay, comparable to the Ituzaingó Formation as defined by De Alba (1953), Herbst (1971) and Herbst et al. (1976) in Argentina (see Krohling and Iriondo 1998; Brea and Zucol 2011).

The Structural Framework

The landscape evolution in Uruguay presents different characteristics basically due to the structural framework and mainly because of the size of its territory, which suggests that climatic conditions were relatively uniform for the entire surface of the country for each studied period. The main morphostructural regions are characterized by tectonic events and within each region, for the variety of rock types involved, which provide the landscape with their peculiar characteristics (Panario 1988).

The following eight main structural features present in almost the entire extent of the country clearly transitional zones, of 17–20 km in width, with the exception of the western margin of the Eastern Hills Region (Sierras del Este) and the Río Uruguay (the boundary with Argentina), which does not allow the boundary definition at the cartographic resolution of this scale. In the present graphical representation, the boundaries were determined by changes in the spectral response of the Landsat images at the chosen scale.

Landscape Characteristics of the Different Morphostructural Regions

North Eastern Sedimentary Basin

The Gondwanic Sedimentary Basin was stable in terms of sediment accumulation since times long before those that modelled the landscape during the Cenozoic, which allowed the process expression according to the resistance of the pre-existing

Table 1 Cenozoic units of Uruguay (Modified from Panario and Gutiérrez 1999, and Ubilla et al. 2004)

Era	System/period	Epoch	Tectono-sedimentary processes		Ages
Cenozoic	Quaternary	Holocene	Sea level fluctuation, local tectonic reactivations	Fluvial terraces – coastal sand dunes *Villa Soriano Formation*	11,700 BP to present
		Pleistocene	Sea level fluctuation, local tectonic reactivations	*Dolores-Sopas* *Chuy Formation* *Libertad Formation, Bellaco unit*[a]	11,700 year to 2.59 Ma
	Neogene	Pliocene	Sea level fluctuation, local tectonic reactivations	*Salto Formation – Raigón Formation*	5.33 to 2.59 Ma
		Miocene	Marine ingression. Development of Río de la Plata Basin	*Camacho Formation*	23.03 to 5.33 Ma
			Uplift, minor faulting, erosion (Miocene unconformity)		
	Palaeogene	Oligocene	Tectonic reactivation, formation of small basins	*Fray Bentos Formation*	33.9 to 23.03 Ma
		Eocene	General stability conditions	*Ferricretes: del Palacio Paleosols* (Oxisols)	55.8 to 33.9 Ma
		Palaeocene	General stability conditions	*Calizas del Queguay*[b] *Gaviotín Formation*	65.5 to 55.8 Ma

[a]It corresponds to a soil unit of the 1:1,000,000 scale soil map of Uruguay (Dirección de Suelos y Fertilizantes 1976), but still not stratigraphically formally defined

[b]Both the calcrete and silicification formation processes may be attributed to several episodes during the Cenozoic; thus, the assignment of these formations to a certain age may later on be modified

materials. The absence of later accumulation processes of certain relevance suggests that the morphogenetic potential of the region has not been modified during the Quaternary, when the main incision of the landscape took place presumably, and, therefore, it is composed of strong slopes and large hills. According to Panario (1988), a large portion of the main drainage lines are born in remnants of the basaltic "cuesta" front as described in the Sierra de Ríos, thus suggesting that the role of the uplift of the Rivera Crystalline Island (Fig. 9) in the basin modelling the relief was of a secondary significance.

Basaltic "Cuesta"

The main structural events in the region are the tilting of the Arapey basaltic flows (of Cretaceous age), which provides the region with a dominant "cuesta" structure which is facing eastwards (see Fig. 10). These flows covered sedimentary rocks of the previously mentioned basin.

The characteristic of these lava flows is a dominance of horizontal structures and the strong resistance of such fresh rocks to fluvial incision, which have favoured in this region the preservation of planar landforms, which has motivated doubts about the morphoclimatic origin of these landforms. Nevertheless, when a lower resistance to weathering is available, large ranges and hills with nonplanar upper surfaces are found. Several higher hills, such as Cerro Travieso, have lost their planar upper surface. In those regions in which the basaltic flows have a certain inclination, they occur at the surface with relatively parallel boundaries, which in general is interpreted as of erosive origin. With the exception of the alterite accumulation zones, the soils in this area are very thin (Fig. 11) which has favoured a slope retreat of the concave type, characteristic of the dominance of erosion processes under semiarid conditions (Fig. 12). Some of the accumulation surfaces, such as accumulation glacis ("glacis d'accumulation"), are slightly dissected, generating smooth hills at the divides, as in Recta de Cunha.

Litoral Oeste Sedimentary Basin

This unit is composed of thick packages of Cretaceous sandstones and Tertiary sediments with very thin Quaternary cover (see Fig. 10). This sedimentary basin is also related to the Cretaceous tectonics, possibly accordingly to the tilting of the basaltic cuesta.

As in the previous unit, this basin received only small sediment supply during the Quaternary, and, therefore, the drainage lines became more entrenched here than in the southern and southwestern tectonic basins. The frequent existence of layers of varied hardness within the accumulated sediments, usually formed by boulder pavements, was the result of scarp recession during previous epochs, of which very little evidence still remains, such as Cerro del Clavel, or small elevations of the ferricretes named as the Asencio Sandstones, or sub-horizontal calcareous duricrusts

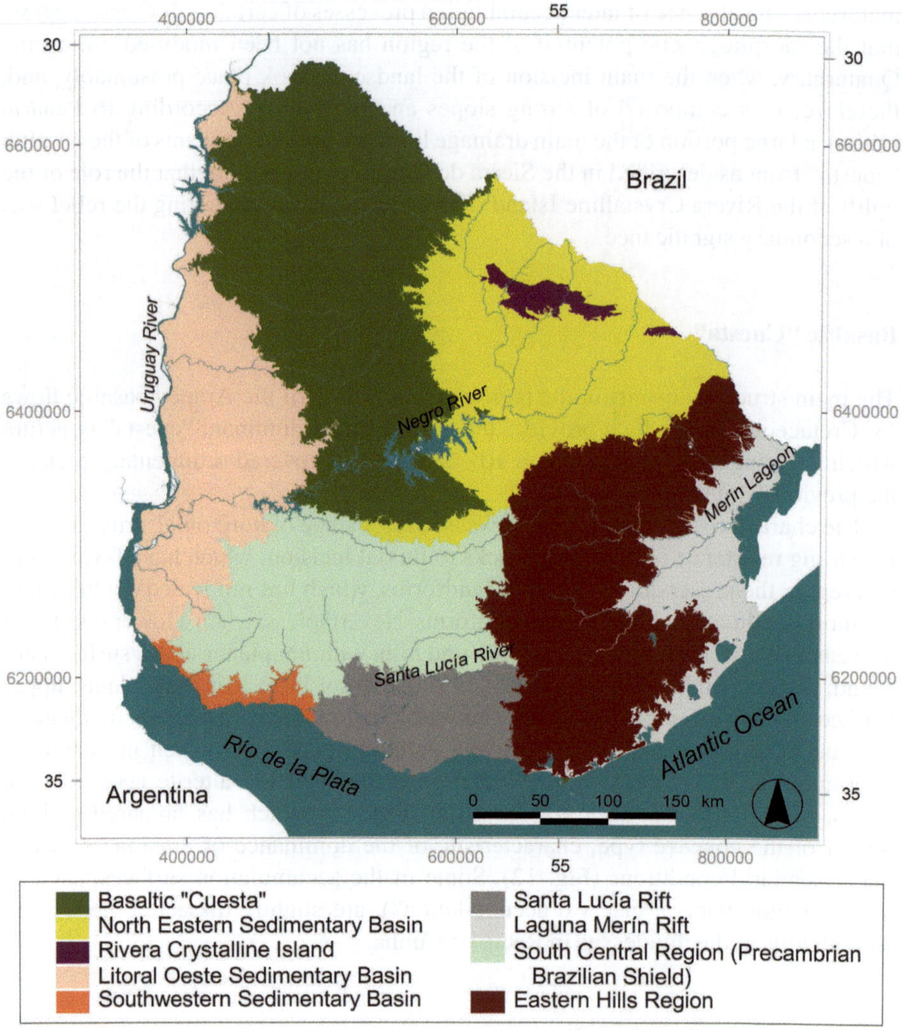

Fig. 10 Structural framework of Uruguay. The boundaries of the units have been depicted following CONEAT (1979) cartography and the topography generated from the 10 m contour lines in maps provided by the Servicio Geográfico Militar (SGM) of Uruguay, satellite imagery (Landsat TM), photointerpretation of aerial photograph (1:40,000) and field observation

with rugged borders, when preserving a surface of sufficient extension and generate hilly interbasin divides, such as those in the Camino de la Cuchilla, Department of Río Negro. When this surface is smaller, tabular hills are present, and when the scarp recession allowed the generation of a landscape at a lower level, smooth hilly valleys occur, generally without much area expression, as those existing in the Department of Río Negro (Mellizos), the Sánchez Grande and Sánchez Chico River basin, and Quebracho, at the Department of Paysandú.

Fig. 11 Very flat landscape with superficial soils in the basaltic zone of northern Uruguay, formed from an erosion glacis

Fig. 12 Scarp retreat with recessional concave profile characteristic of the basaltic zone of northern Uruguay

Southwestern Sedimentary Basin

Towards the southwest, another sedimentary basin of smaller significance is found (see Fig. 10), based on its territorial extent as well as for the thickness of its sedimentary accumulations, mainly very thick Tertiary and Quaternary deposits.

This region has acted as a sediment reception basin until recent times, late Middle Quaternary. The present dissection of the landscape does not agree with its morphogenetic potential or with the fragility of the composing materials, what suggests that it could have been affected by tectonic uplift until very recent times. This hypothesis is supported by: (i) the existence of paleo-coastlines and coastal lagoons that are clearly in-filled by sediments even at elevations above present sea level, (ii) the occurrence of marine units such as the Camacho Fm., several meters above their corresponding stratigraphic units in Argentina (the Paraná Formation) and, at different levels in Uruguay (Antón and Goso 1974), (iii) the existence of creeks that still have entrenching capabilities in unconsolidated materials, and (iv) Quaternary marine deposits that occur at higher levels than those found in the rest of the country. This uplifting process is perhaps continued irregularly eastwards, at least along a narrow coastal fringe until the Merín Rift.

Santa Lucía Rift

Southwards, the basin of Santa Lucía is found (see Fig. 10), more likely one of the two most important of the Cretaceous basins within the continental portion of the country, from the point of view of the Cretaceous, Tertiary and Quaternary sediments included in it. Subsidence and sedimentation were very active in the Santa Lucía Tectonic Basin until the Early Quaternary. This means it had no morphogenetic potential in this period and that after it, such potential was very reduced, which determined a landscape composed mainly by smooth hills of gentle slopes, with the exception of those found at the margins of the basin and the Santa Rosa Basement high (Rosello et al. 2000).

Laguna Merín Rift

Eastwards, another rift with similar age for the beginning of the event and size is located (see Fig. 10); this basin, however, presents Tertiary and Cretaceous sediments in its continental side as the oldest materials. Eastern ranges and the Laguna Merín Tectonic Basin, a system of hills and low ranges is located, which are composed of crystalline rocks with a thin Quaternary cover, whose genesis could be related to the tectonic events that formed the cited basin. Studies on the Uruguayan continental shelf in the region have shown that this rift has materials whose age also dates back to Cretaceous (Rosello et al. 2000) their geomorphological characteristics, which has allowed the interpretation that it has been active until present times with organic sediments in its most depressed areas. The capture of part of the Cebollatí River Basin during the Holocene (Bracco et al. 2012) is a clear demonstration of their recent activity, compared with the Santa Lucía rift, as well as other smaller basins located in between, such as those of Valle Fuente, Valle Aiguá, that were remodelled during the Pleistocene.

The nature of the sediments, their diagenetic evolution and the resistance of the crystalline and consolidated materials to weathering and the morphogenetic potential of each of these regions are the conditions that are responsible for their geomorphological profile.

The landscape of this region is practically flat due to its almost null morphogenetic potential. The deposition of the Pleistocene and Holocene sediments in it is largely developed under the shape of stepped sedimentary terraces, which allows the identification of at least four levels of plains separated by breaks in slope, which vary from a few centimetres to a few metres.

South Central Region (Precambrian Brazilian Shield)

The Southern Central Region is occupied by rocks belonging to the Brazilian Shield (see Fig. 10) which have kept under conditions relatively stable at least during Cretaceous times. These relatively stable conditions, as well as the characteristics of the morphoclimatic systems dominating the area since those times, have provided the landscape with a "senile" aspect, which determined that Chebataroff (1955) described it as a "crystalline peneplain", in accordance with the genetic interpretations of those times. At present it is defined as dissected and reworking plains.

The arid and semiarid periods that occurred with short interruptions during most of the Tertiary and the Quaternary must have modelled the palaeolandscape into erosional plains with a few local smooth elevations, characteristic of planation on crystalline rocks. During the early Quaternary, this area received a sedimentary cover of alterites coming from the hilly areas, these materials being still preserved on the main interfluvial divides. After the formation of this pediment, it was strongly dissected, a process favoured by deep weathering processes generated during the Eocene (Panario and Gutiérrez 1999) and earlier. This dissection produced an undulating relief, interrupted by smooth hills at the interfluvial divides at the areas with thicker Quaternary accumulation.

Eastern Hills Region

This region is composed by a complex of folded emerged structures and other uplifted features as Dom Feliciano Belt, of which the oldest one is undoubtedly the Carapé Massif which corresponds to the main water divide in the region (see Fig. 10), due to the fact that the drainage lines which have their sources in the region are cross-cutting other features, including highly deformed granites and quartzites as the Sierra de la Ballena and Sierra de las Cañas chains.

This unit represents the landscape with higher potential energy. Notwithstanding, the uppermost portion of the Sierras shows rather flat top surfaces, which correspond to very old planation (etchplains) processes developed probably during the Cretaceous or even older, with others at lower elevation which may have been formed during the Middle Tertiary. This group of elevations shows a clear SW-NE orientation and they would have acted as a mountainous region of the southernmost Brazilian Shield from which the glacis were carved, providing most of the infilling sedimentary materials of the Santa Lucía and Laguna Merín rift.

Within this area, certain areas of tectonic down-warping are found which generated smooth hilly valleys, such as Valle Fuentes and Valle Aiguá.

Palaeosurfaces

Gondwana Palaeosurfaces

The uppermost palaeosurface on the Granite Batholiths (see "Precambrian Geology") is located on granite exposures with two "treppen" in the sense of Penck (1953). The second surface is located on deeply weathered granite. These surfaces could be of the same age or, alternatively, of quite close ages, with little time difference in between their formations.

There are obvious dating problems concerning the palaeosurfaces, and the correlation with Southern Brazil has not been established yet.

The existence of a volcanic explosions in this region with an $^{40}Ar/^{39}Ar$ age of ∼130 to 128 Ma (Cernuschi Rodilosso 2011), the lack of evidence of it on the ancient surfaces, suggests that these surfaces are planation surfaces, probably etchplains, which suffered later on intensive denudation, presumably since the Oligocene until part of the Pleistocene, but for this, it is necessary to assume a denudation rate of 5–10 m per million years, only possible under extremely stable condition.

The first palaeosurface is located approximately between 320 and 500 m a.s.l., whereas the second palaeosurface is found between 280 and 320 m a.s.l.

The elevation difference between them is very small, but this would not be too rare in a tectonically very stable, as it happens in the Tandilia and Ventania ranges of the Buenos Aires province, Argentina (Demoulin et al. 2005; Rabassa et al. 2005, 2010, 2014).

The Cerro Campanero, in the Department of Lavalleja, shows a perfect example of weathering front remnants, on which corestones have been left after removal of the weathered materials. These corestones are a common feature in the granitic batholiths (e.g. Carapé region) (Fig. 13) and are part of dismantled tors, and some of them may have also reached the state of rocking stones during their evolution. Looking northwest in Fig. 14, the clear flatness of the supposed Gondwana palaeosurface is exposed forming the horizon, with very little local relief, as mentioned before.

Fig. 13 Examples of (**a**) tors and (**b**) corestones which may be observed on granitic rocks at the summits of the hills of Sierras de Carapé

In the northern part of the country, the inselbergs modelled on basalts of the Arapey Formation prove that they were developed after the eruption of these rocks (Early Cretaceous). At a lower altitude compared with these relict features, but in accordance with them, degraded surfaces assigned to arid climates have been described and named as "Charqueada" (Antón 1975). This name has been given to this surface due to their occurrence in a site in the Department of Artigas where these features are found, extending to the Eastern and Northeastern hills. It is presently considered that this surface may be subdivided in two units, separated by an entrenchment. It is herein proposed the preliminary denomination of "Charqueada I" for the highest, supposed oldest, extensive surface and "Charqueada II" for the younger (lower) unit. The scarce preserved soils in the uppermost surface are of the mineral, reddish type, which indicate very strong weathering produced under very warm climatic conditions and, at least, seasonally very humid environments. Most of these soils occur in such positions that indicate colluvial processes along associated slopes and valleys. However, it should be taken into account that these

Fig. 14 (**a**) the *dashed line* depicts the change in landscape surface. Relicts of two palaeosurfaces are found above, indicated by a *bluish line* (*the lower one*) and a *reddish line* (*the higher one*). The small relief in between them suggests that these two surfaces were essentially coeval or separated by a very short time span. (**b**) Panoramic view from the uppermost topographic surface, in which the two lower surface levels may be observed

soils are perhaps the result of superposition of several red alteration (lateritic) processes. In the second palaeosurface, which occurs at a lower level, the soils are better developed, although formed by a brownish material, same times more or less lixiviated mollisols. These palaeosurfaces are clearly exposed when the summits of the regional ranges are linked in a graph, such as the Eastern, Aiguá and Yerbal Sierras.

However, tectonic action has deformed these landscapes in a great manner, due to their antiquity. Thus, overlying sediments are not always preserved, making very difficult the correlation of the surface relicts. Younger relocation and transport of the sediments make even more questionable their identification and correlation. Precisely, the entrenchment and development of a new surface does not freeze the evolution of the older one, but it may accelerate it instead, although under varying conditions with respect to the original ones, frequently removing sediments from the upper zones to the lower landforms. The humid periods responsible for the entrenchment that separates the Charqueada I and II surfaces, and other surfaces of the region (Masoller), could have been also responsible for the aforementioned red alterite formation during de Eocene. These surfaces, when they suffered the action of alternated periods of wet and dry climates, originated most of the landscape of the Eastern Hills Region, which had been previously uplifted by tectonic processes. When the valley incision did not affect the upper surface, highland ranges were formed (Sombroek 1969). Contrarily when the valley incision affects the upper surface, typical "sierras" (steep hills) landscape is developed.

Cenozoic Palaeosurfaces

Separated from the old surfaces by an entrenchment, perhaps favoured by the Eocene alteration process, another surface of similar genesis (arid morphogenesis) occurs, which was named as the Masoller surface by Antón (1975). Erosion and accumulation glacis that formed it are found in many localities, as it may be observed in the geomorphological map by Antón (1975). According to Panario and Gutiérrez (1999), this surface may be assigned to a more intense planation process that developed during periods of semiarid climate in the Tertiary (perhaps, the Oligocene), simultaneously with the conglomerates, limestones and aeolian deposits of the Fray Bentos Formation. This process continued during the Pliocene, when fluvial deposits also of semiarid conditions were formed, such as conglomerates and sandstones of the Salto and Raigón formations.

The deposits of the Salto and Ituzaingó formations have been defined as of subtropical climate by several authors (Iriondo 1980; Jalfin 1988; Herbst 2000). However, it should be taken into consideration that the Río de la Plata Basin extends over a wide latitudinal band and it reaches much lower latitudes at its mouth. Therefore, even if the provenance of the materials may be from tropical or subtropical areas, the conditions in the depositional areas could have been very different.

The Salto and Raigón formations present a higher variability of their sedimentary materials which indicates environmental rhythmicity. During their genesis, periods with sufficient aridity developed so as to transport and deposit coarse materials and other wetter periods in which the transport and deposition of the finer sediments took place, thus favouring the formation of large glacis. The deposition of very fine (clayey) materials seems to correspond to lacustrine environments, characteristics of these climatic conditions when closed depressions are available (Raigón Formation). The fact that aeolian silts were herein incorporated suggests that there were some periods in which, even though locally, a certain plant cover developed. Towards the later portion of this period and in coincidence perhaps with the earlier major glaciations, the deposition of the Libertad I Formation took place, most likely under semiarid conditions. From a genetic point of view, the Libertad Formation was formed during several Pleistocene glacial periods, without clear internal unconformities, perhaps with the exception of the events known as Libertad I and Libertad II, which points towards a loess unit with continuous soil formation, as it has been noted by Blasi et al. (2001) under similar conditions in the Argentine Pampas.

Between the Salto and Raigón formations and the Libertad Formation there is no entrenchment which may indicate the necessary conditions for landscape dissection. The Libertad I Formation is generally composed of finer materials than the Raigón Formation. This would imply that a loss of competence of the transportation agents would have taken place, due to a loss of morphogenetic potential or climatic changes in the region; the latter interpretation would be preferred. Apparently, the deposition of the final portion of the Libertad Formation would have taken place under somewhat more humid conditions, whose more evident relicts are the clayey deposits occurring under seasonally confined, shallow waters where vegetation and/or evaporation would be responsible for their deposition or later weathering of finer sediments into montmorillonite clays. The smaller amount of illite in relation with smectites would indicate a warmer climate than during the deposition of the Libertad sediments.

The deposition of clays and fine materials requires very special conditions which are related to lakes, ponds or marshes with dense vegetation. The latter case would be the one better adapted to the conditions in this country, perhaps reconstructing ancient drainage basins. After the deposition, due to the difficulties to erode the clayey sediments when climate changed, drainage channels tended to entrench the margins of the swampy areas but not their deposits. In the long term, a process of relief inversion took place, with the clayey deposits in the uppermost areas. Considering the crystalline zone, the Risso and La Carolina units of the 1:1,000,000 scale soil map of Uruguay (Dirección de Suelos y Fertilizantes 1976) may be considered, together since they are zones with vertisols and calcium–montmorillonite-dominated soils. A palaeobasin may be reconstructed which, starting at the Eastern Ranges, would extend southwestwards until approximately the present mouth of the Uruguay River (Panario and Gutiérrez 1999). The dry period in which the Libertad I Formation deposition took place would be associated to the glacial periods at the beginning of the Pleistocene, as low sea levels would be related to glaciation and dry climates. The increase of the morphogenetic potential implied by lowering sea level is compensated in dry areas by the loss of erosion potential of the streams, due to loss of yield and detrital load. The entrenchment under these conditions would have taken place during wetter periods at the end of the glaciations, before sea level rises. The subsequent climatic alternating periods modelled the thus formed surfaces, originating most of the present smooth hills like the Cuchilla Grande. Some relict surfaces are found even in the neighbourhood of the city of Montevideo (the La Tabla Range, among others), connected to position inselbergs such as El Cerrito de la Victoria. The higher energy of the hilly landscape may be attributed to successive periods of entrenchment affecting the same drainage lines previously established, which forced frequent changes in slope inclination in the landscape. In those places where the landforms are due to a varying rock resistance, larger high plains were preserved, such as Cuaró, Recta de Cunha and Masoller. After the formation of these surfaces, marine transgressions took place, since then, alternating wetter–drier, warmer–colder climates related to glacial–interglacial periods represent the dominant conditions during the rest of the Pleistocene and the early Holocene.

Final Remarks

The existence of pre-Cenozoic palaeosurface relicts has been largely discussed from a neo-Darwinian and classic thermodynamics point of view, still perceived in modern geomorphology. Although the absolute ages of the older surfaces are difficult to establish at our present state of the art, some conclusions may be obtained:

1. For the first time, the nature, characteristics and distribution of Gondwana landscapes in Uruguay has been presented within the framework of the long-term landscape evolution of this country.
2. The different stratigraphic units found in the various morphostructural regions of Uruguay have been presented, and their relationship with the occurrence and distribution of landscapes and landforms has been discussed and analyzed.
3. Several features emerged from such analysis. The Cretaceous lava flows of the northern portion of the country show clear evidence of tilting.
4. In the topographically higher area, the existence of palaeosurface relicts with recessional scarps of the knick-point type may be observed, carved on the basaltic flows of the upper section, thus the younger ones.
5. The topographically lower area of the tilted Cretaceous lava flows is covered by fluvial deposits pertaining to a Middle Cretaceous sedimentary basin, clearly genetically separated by the scarp.
6. Part of the sediments present here is related to the denudation processes that originated the relicts. Thus, it may be clearly assumed the existence of at least extensive surfaces of Late Cretaceous age.
7. In those place were the Cretaceous lavas are overlying the northwest margins of the Dom Feliciano Belt, they are found at elevations around 200 m a.s.l., whereas the maximum elevations of this structure and its corresponding palaeosurface may reach 500 m a.s.l., which could be interpreted as an Early Cretaceous or even a pre-Cretaceous age for these surfaces, in which corestones, tors and other landforms indicating pre-existing deep alteration mantles over highly quartzose, granitic rocks are found.
8. The existence of Carboniferous–Permian glacial sediments of the mountain glaciation type suggests that very high mountain summits were already present in those times. On the other hand, the occurrence of Eocene ferricrete clasts in the matrix of Oligocene fine-grained aeolian deposits and the distribution of surfaces framed by iron mantles at elevations corresponding to the general landscape planation during an Oligocene semiarid period are also according with the extensive planation of the emerged landscape.
9. Absolute dating and/or clear correlation among the palaeosurfaces of the South American passive margin with surfaces genetically and geographically related, located in other parts of South America and Southern Africa, will be undoubtedly needed to establish a reliable genetic chronosequence.
10. The study of the provenance of Cretaceous and pre-Cretaceous sediments would also be a significant input in the future to understand the timing of

the development and denudation of these ancient landscapes. The cratonic areas of Uruguay were affected by deep chemical weathering during perhaps millions of years in the Late Mesozoic and the Early Palaeogene. An enormous cover of saprolite, perhaps many hundreds of metres thick, was removed by subaerial denudation during the Tertiary. These weathering products are mostly lying today in the surrounding ocean basins. The sedimentary sequences of these marine basins will inform us about the characteristics and thickness of the weathered materials, but understanding the ancient weathering processes and their products will enable us to interpret the provenance, nature and age of the sediments infilling those basins. Needless to say, regional studies on the geomorphology of the cratonic areas of Uruguay should be paired to the investigation of the marine basins of the South Atlantic Ocean.

Acknowledgements The authors are greatly indebted to the Universidad de la República, Montevideo, Uruguay, and CADIC-CONICET, Ushuaia, Argentina, for the access to various facilities, data and materials that were used for the preparation of this chapter.

Likewise, they are extremely grateful to the late Professor Tim Partridge (then at the University of Witwatersrand, Johannesburg, South Africa) and Professor Rodney Maud for introducing us in the study, analysis and discussion of Gondwana palaeolandscapes. Moreover, Professor Maud and Dr. Claudio Carignano (CONICET and Universidad de Córdoba, Argentina) took part of a memorable field trip to the region of Minas, eastern Uruguay, in 2000, when some of the ideas exposed in this chapter were discussed for the first time. Funding for such field trip was provided by a National Geographic Society field exploration grant to Jorge Rabassa, which is deeply acknowledged.

The authors are also thankful to the FAPESP research project presently developed under the leadership of Professor Francisco Ladeira (Universidade de Campinas, São Paulo, Brazil) for allowing the exchange of experience on the Gondwana landscapes of southern South America for the first time at the regional level.

References

Andreis RR, Mazzoni MM (1967) Sedimentología de las formaciones Kiyú y San José, Departamento de San José, R. O. del Uruguay. Revista del Museo de La Plata, Sección Geología 6(41):41–96

Antón D (1975) Evolución Geomorfológica del Norte del Uruguay. Unpublished internal report, Dirección de Suelos y Fertilizantes, Ministerio de Agricultura y Pesca, Montevideo

Antón D, Goso H (1974) Estado actual de los conocimientos sobre el Cuaternario en el Uruguay. Dirección de Suelos y Fertilizantes, Ministerio de Agricultura y Pesca, Montevideo. In: Anais do 25º Congresso Sociedade Brasileira de Geologia, Porto Alegre, vol 3

Antón D, Prost T (1974) Observaciones sobre las formaciones cuaternarias de la Sierra de las Ánimas. Unpublished report presented to the 28th. Brazilian Geological Congress

Basei MAS, Siga O Jr, Masquelin H, Harara OM, Reis Neto JM, Preciozzi Porta F (2000) The Dom Feliciano Belt and the Rio de la Plata Craton: tectonic evolution and correlation with similar provinces of southwestern Africa. In: Cordani UG, Milani EJ, Thomaz-Filho A, Campos DA (eds) Tectonic evolution of South America, 31st International Geological Congress, Rio de Janeiro, pp 311–334

Bellieni G, Comin-Chiaramonti P, Marques LS, Melfi AJ, Nardy AJR, Piccirillo EM, Roisemberg A (1984) High- and low- TiO_2 flood basalts from the Paraná plateau (Brazil): petrology and geochemical aspects bearing on their mantle origin. Neues Jahrb Mineral-Abh 150:273–306

Bellieni G, Comin-Chiaramonti P, Marques LS, Melfi AJ, Nardy AJR, Papatrechas C, Piccirillo EM, Roisemberg A, Stolfa D (1986) Petrogenetic aspects of acid and basaltic lavas from the Paraná plateau (Brazil): geological, mineralogical and petrochemical aspects. J Petrol 27(4):915–944. doi:10.1093/petrology/27.4.915

Bezerra FHR, Vita-Finzi C (2000) How active is a passive margin? Paleoseismicity in northeastern Brazil. Geology 28:591–594, 10.1130/0091-7613(2000)28<591:HAIAPM>2.0.CO;2

Bezerra FHR, Amaro VE, Vita-Finzi C, Saadi A (2001) Pliocene-Quaternary fault control of sedimentation and coastal plain morphology in NE Brazil. J South Am Earth Sci 14:61–75

Bezerra FHR, Barreto AMF, Suguio K (2003) Holocene sea-level history on the Rio Grande do Norte State coast. Brazil Mar Geol 196:73–89

Bidegain M, Caffera RM (1997) Clima del Uruguay y la región. http://www.rau.edu.uy/uruguay/geografia/Uy_c-info.htm. Accessed 10 Mar 2011

Blasi AM, Zárate MA, Kemp RM (2001) Sedimentación y pedogénesis cuaternaria en el noreste de la pampa bonaerense: la localidad Gorina como caso de estudio. Rev Asoc Argent Sedimentol 8(1):77–92

Bossi J (1966) Geología del Uruguay, vol 2. Departamento de Publicaciones de la Universidad de la República, Montevideo, 469 pp

Bossi J, Campal N (1992) Magmatismo y tectónica transcurrente durante el Paleozoico inferior del Uruguay. In: Gutiérrez J, Saavedra J, Rábano I (eds) Paleozoico Inferior de Ibero-América. Universidad de Extremadura, Alicante, pp 343–356

Bossi J, Ferrando LA (2000) Carta Geológica del Uruguay, escala 1/500.000, 2nd edn. Geoeditores, Montevideo

Bossi J, Navarro R (1991) Geología del Uruguay. Departamento de Publicaciones, Universidad de la República, Montevideo, 970 pp

Bossi J, Ferrando LA, Fernández A, Elizalde G, Morales H, Ledesma J, Carballo E, Medina E, Ford I, Montaña J (1975) Carta geológica del Uruguay (1:1.000.000): Dirección de Suelos y Fertilizantes, Ministerio de Agricultura y Pesca, Montevideo, 25 pp

Bossi J, Ferrando LA, Montaña J, Morales H, Campal N, Gancio F, Piñeiro D, Schipilov A, Sprechmann P (1998) Carta Geológica del Uruguay, escala 1/500.000. Geoeditores, Montevideo

Bracco R, del Puerto L, Inda H, Capdepont I, Panario D, García-Rodríguez F (2012) Evolución ambiental y constructores de cerritos en la región de India Muerta. Un replanteo. In: III Jornadas del Cenozoico, Montevideo, June 2012, CD-Rom, pp 14–15

Brea M, Zucol AF (2011) The Paraná-Paraguay basin: geology and paleoenvironments. In: Albert JS, Reis RE (eds) Historical biogeography of neotropical freshwater fishes. University of California Press, Berkeley, pp 69–88

Caorsi J, Goñi J (1958) Geología uruguaya. Boletín del Instituto Geológico del Uruguay 37:1–73

Catuneanu O (2004) Basement control on flexural profiles and the distribution of foreland facies: the Dwyka Group of the Karoo Basin, South Africa. Geology 32(6):517–520. doi:10.1130/G20526.1

Cernuschi Rodilosso F (2011) Geology of the Cretaceous Lascano-East intrusive complex: magmatic evolution and mineralization potential of the Merin Basin, Uruguay. Unpublished Master of Science dissertations, Oregon State University, 340 pp. http://hdl.handle.net/1957/20581. Accessed 11 Dec 2012

Cesero PD, Ponte FC (1997) Análise comparativa da paleogeologia dos litorais atlânticos brasileiro e africano. Boletim de Geociências da Petrobras, Rio de Janeiro 11(1–2):1–18

Chebataroff J (1955) Evolución del relieve del Uruguay y de Río Grande del Sur. Revista Uruguaya de Geografía 39–96. http://www.periodicas.edu.uy/Revista_Uruguaya_Geografia/pdfs/Revista_Uruguaya_Geografia_08_1955.pdf. Accessed 4 Dec 2012

Cingolani CA (2011) The Tandilia system of Argentina as a southern extension of the Río de la Plata Craton: an overview. Int J Earth Sci 100:221–242

Comin-Chiaramonti P, Bellieni G, Piccirillo EM, Melfi AJ (1988) Classification and petrography of continental stratoid volcanics and related intrusives from the Paraná basin (Brasil). In: Piccirillo EM, Melfi AJ (eds) The Mesozoic flood volcanism of the Paraná Basin: petrogenetic and geophysical aspects. Instituto Astronômico e Geofísico USP, São Paulo, pp 47–72

CONEAT (1979) Grupos de Suelos. Índices de Productividad. Comisión Nacional de Estudio Agroeconómico de la Tierra (CONEAT), Ministerio de Agricultura y Pesca, Montevideo, 167 pp

Cox KG (1988) The Karroo province. In: MacDougall JD (ed) Continental flood basalts. Kluwer Academic, Dordrecht, pp 239–271

Creer KM, Miller JA, Gilbert-Smith A (1965) Radiometric age of the Serra Geral Formation. Nature 207:282–283. doi:10.1038/207282a0

De Alba E (1953) Geología del Alto Paraná en relación con los trabajos de derrocamiento entre Ituzaingó y Posadas. Revista de la Asociación Geológica Argentina 8:121–161

de Almeida FFM, Amaral G, Cordani UG, Kawashita K (1973) The Precambrian evolution of the South American cratonic margin south of the Amazon River. In: Nairn AEM, Stehli FG (eds) The ocean basin and margins. The South Atlantic, vol 1. Plenum Press, New York, pp 411–446

de Santa Ana H, Veroslavsky G (2003) La tectosecuencia volcanosedimentaria de la Cuenca Norte de Uruguay. Edad Jurásico-Cretácico Temprano. In: Veroslavsky G, Ubilla M, Martínez S (eds) Cuencas Sedimentarias de Uruguay: Geología, Paleontología y Recursos Naturales – Mesozoico. DIRAC, Facultad de Ciencias, Universidad de la República, Montevideo, pp 51–74

de Santa Ana H, Goso C, Muzio R, Oyhantçabal P, Veroslavsky G (1994) Bacia do Santa Lucia (Uruguai): Evolução tectônica e sedimentar. Geociências 13(1):37–52, São Paulo, UNESP

de Santa Ana H, Goso C, Daners G (2006a) Cuenca Norte: estratigrafía del Carbonífero-Pérmico. In: Veroslavsky G, Ubilla M, Martínez S (eds) Cuencas Sedimentarias de Uruguay: Geología, paleontología y recursos naturales – Paleozoico. DIRAC, Facultad de Ciencias, Universidad de la República, Montevideo, pp 147–208

de Santa Ana H, Veroslavsky G, Fulfaro V, Rossello E (2006b) Cuenca Norte: Evolución tectónica y sedimentaria del Carbonífero-Pérmico. In: Veroslavsky G, Ubilla M, Martínez S (eds) Cuencas Sedimentarias de Uruguay: Geología, Paleontología y Recursos Naturales – Paleozoico. DIRAC, Facultad de Ciencias, Universidad de la República, Montevideo, pp 209–256

Demoulin A, Zárate M, Rabassa J (2005) Long-term landscape development: a perspective from the southern Buenos Aires ranges of east central Argentina. J South Am Earth Sci 19:193–204

Dirección de Suelos y Fertilizantes (1976) Carta de Reconocimiento de Suelos del Uruguay a escala 1:1.000.000, y Leyenda que la acompaña. Ministerio de Agricultura y Pesca, Montevideo

Erlank AJ, Marsh JS, Duncan AR, Miller RMcG, Hawkesworth CJ, Betton PJ, Rex DC (1984) Geochemistry and petrogenesis of the Etendeka volcanic rocks from SWA/Namibia. In: Erlank AJ (ed) Petrogenesis of the Volcanic Rocks of the Karoo Province. Geological Society of South Africa, Special publication 13, pp 195–245

Féraud G, Bertrand H, Martínez M, Ures C, Schipilov A, Bossi J (1999) ^{40}Ar/^{39}Ar age and geochemistry of the southern extension of Paraná traps in Uruguay. In: Actas II Simposio Sudamericano de Geología Isotópica, Córdoba, pp 57–59

Ferrando L, Fernández A (1971) Esquema tectónico cronoestratigráfico del Pre-Devoniano en el Uruguay. In: Anais do 25° Congresso Brasileiro de Geología, São Paulo, 1, pp 199–210

Fodor RV (1987) Low- and high-TiO_2 flood basalts of southern Brazil: origin from picritic parentage and a common mantle source. Earth Planet Sci Lett 84(4):423–430. doi:10.1016/0012-821X(87)90007-0

Ford I (1988) Conglomerados con nidos de insectos fósiles: Formación Palmitas (prov.); Terciario inferior (tentativo). In: Memorias del 6° Panel de Geología del Litoral, Montevideo, pp 47–49

Fragoso-Cesar ARS (1980) O craton do Rio de la Plata e o Cinturão Dom Feliciano no Escudo Uruguaio-Sul-Riograndense. In: Anais do 31° Congresso Brasileiro de Geologia, Camboriú, 5, pp 2879–2892

Francis JC, Mones A (1966) Artigasia magna n. g., n. sp. (Eumegamyinae) un roedor gigantesco de la época Pliocena Superior de las barrancas de San Gregorio, departamento de San José, República Oriental del Uruguay. Kraglieviana 1(3):89–100

Gaucher C (2000) Sedimentology, palaeontology and stratigraphy of the Arroyo del Soldado Group (Vendian to Cambrian, Uruguay). Beringeria 26:1–20

Gaucher C, Boggiani PC, Sprechmann P, Sial AN, Fairchild TR (2003) Integrated correlation of the Vendian to Cambrian Arroyo del Soldado and Corumbá Groups (Uruguay and Brazil): palaeogeographic, palaeoclimatic and palaeobiologic implications. Precambrian Res 120:241–278. doi:10.1016/S0301-9268(02)00140-7

Gaucher C, Chiglino L, Pecoits E (2004) Southernmost exposures of the Arroyo del Soldado Group (Vendian to Cambrian, Uruguay): palaeogeographic implications for the amalgamation of W-Gondwana. Gondwana Res 7:701–714. doi:10.1016/S1342-937X(05)71057-1

Gómez Rifas C, Heinzen W, Roth W, Spoturno J (1981) Las calizas del Uruguay. Boletín Instituto Geológico del Uruguay 38:5–37

Goso H (1965) El Cenozoico en el Uruguay. Instituto Geológico del Uruguay, M.I.E. Ed. Mimeogr. Montevideo, 36 pp

Goso H, Bossi J (1966a) Cenozoic stratigraphy of the Rio Grande do Sul coastal province. Boletim Paranaense de Geociências 33:54–55

Goso H, Bossi J (1966b) Relevamiento geológico del Departamento de San José. Instituto Geológico del Uruguay. Unpublished technical report, 40 pp

Goso C, Perea D (2004) El Cretácico post-basáltico de la Cuenca Litoral del Río Uruguay: Geología y Paleontología. In: Veroslavsky G, Ubilla M, Martínez S (eds) Cuencas sedimentarias de Uruguay: Geología, paleontología y recursos naturales -Mesozoico, 2da ed. DIRAC, Facultad de Ciencias, Universidad de la República, Montevideo, pp 143–169

Hasui Y (1990) Neotectônica e Aspectos Fundamentais da Tectônica Ressurgente no Brasil. In: Workshop sobre Neotectônica e Sedimentação Cenozóica Continental no Sudeste Brasileiro, Sociedade Brasileira de Geologia-Núcleo Minas Gerais, Belo Horizonte, 1, pp 1–31

Hawkesworth CJ, Gallagher K, Kelley S, Mantovani M, Peate DW, Regelous M, Rogers NW (1992) Paraná magmatism and the opening of the South Atlantic. In: Storey BC, Alabaster T, Pankhurst RJ (eds) Magmatism and the causes of continental break-up. Geological Society, London, Special publication 68, pp 221–240. doi:10.1144/GSL.SP.1992.068.01.14

Herbst R (1971) Esquema estratigráfico de la provincia de Corrientes, República Argentina. Revista de la Asociación Geológica Argentina 26(2):221–243

Herbst R (2000) La Formación Ituzaingó (Plioceno). Estratigrafía y distribución. In: Aceñolaza FG, Herbst R (eds) El Neógeno de Argentina, Instituto Superior de Correlación Geológica INSUGEO, San Miguel de Tucumán, 14, pp 181–190

Herbst R, Santa Cruz JN, Zabert LL (1976) Avances en el conocimiento de la estratigrafía de la Mesopotamia argentina, con especial referencia a la Provincia de Corrientes. Revista de la Asociación de Ciencias Naturales del Litoral 7:101–121

Iriondo MH (1980) El Cuaternario de Entre Ríos. Revista de la Asociación de Ciencias Naturales del Litoral 11:125–141

Iriondo MH (1996) Estratigrafía del Cuaternario de la Cuenca del Río Uruguay. Actas XIII Congreso Geológico Argentino y III Congreso de Exploración de Hidrocarburos, Buenos Aires, 4, pp 15–25

Jalfin GA (1988) Formación Ituzaingó (Plio-Pleistoceno) en Empedrado, provincia de Corrientes: un paleorío arenoso entrelazado tipo Platte. In: Actas de la II Reunión Argentina de Sedimentología, Buenos Aires, pp 130–134

Jones GH (1956) Memoria explicativa y mapa geológico de la región oriental del Departamento de Canelones. Boletín del Instituto Geológico del Uruguay 34:1–193

Kirstein LA, Hawkesworth CJ, Turner SP (1997) CFB magmatism in southern Uruguay: marginal to a mantle plume? In: J Conf Abs Continental Magmatism, VSG – Minsoc '97, Cambridge, 2, pp 1, 44

Kirstein L, Peate DW, Hawkesworth CJ, Turner SP, Harris C, Mantovani MSM (2000) Early Cretaceous basaltic and rhyolitic magmatism in southern Uruguay associated with the opening of the South Atlantic. J Petrol 41(9):1413–1438

Krohling D, Iriondo M (1998) Guía de Campo N° 1: Pampa Norte, Cuenca del Carcarañá. International Joint Field Meeting Loess in Argentina: Temperate and Tropical. Paraná, Entre Ríos, Argentina, 36 pp

Lambert R (1940) Memoria explicativa de un mapa geológico de reconocimiento del Depto. de Paysandú y los alrededores de Salto. Instituto Geológico del Uruguay, Montevideo, 27, pp 1–41

Lustrino M, Gomes CB, Melluso L, Morbidelli L, Muzio R, Ruberti E, Tassinari CCG (2003) Early Cretaceous magmatic activity in southeast Uruguay: trace element and Sr-Nd isotopic constraints. In: Proceeding of IV South American Symposium on Isotopic Geology, Salvador, 2, pp 596–597

Lustrino M, Melluso L, Brotzu P, Gomes GB, Morbidelli L, Muzio R, Ruberti E, Tassinari CCG (2005) Petrogenesis of the early Cretaceous Valle Chico igneous complex (SE Uruguay): relationships with Paraná-Etendeka magmatism. Lithos 82(3–4):407–434. doi:10.1016/j.lithos.2004.07.004

Mallmann G, Chemale F Jr, Armstrong RA, Kawashita K (2003) Sm-Nd and U-Pb SHRIMP zircon studies of the Nico Pérez Terrane, reworked Rio de la Plata Craton, Uruguay. In: IV South American symposium on Isotope Geology, Short Papers, Salvador, 1, pp 207–209

Mantovani MSM, Marques LS, De Sousa MA, Civetta L, Atalla L, Innocenti F (1985) Trace element and strontium isotope constraints on the origin and evolution of Paraná continental flood basalts of Santa Catarina state (southern Brazil). J Petrol 26(1):187–209. doi:10.1093/petrology/26.1.187

Martínez S (1989) Los depósitos de la "transgresión entrerriana" (Mioceno de Argentina, Brasil y Uruguay). Comparación de sus principales áreas fosilíferas a través de los bivalvos y los gastrópodos. Ameghiniana 25(1):23–29

Martínez S (1994) Bioestratigrafía (Invertebrados) de la Formación Camacho (Mioceno, Uruguay). Unpublished doctoral thesis (PhD), Universidad de Buenos Aires, 346 pp

Masquelin H (2002) Evolução estrutural e metamórfica do Complexo Gnáissico Cerro Olivo, Sudeste do Uruguai. Unpublished doctoral thesis (PhD), Universidade Federal do Rio Grande do Sul (UFRGS), 2 vols, t1: 1–227, t2: 1–117, 1 map

Masquelin H, D'Avila Fernandes LA, Lenz C, Porcher CC, McNaughton NJ (2012) The Cerro Olivo Complex: a pre-collisional Neoproterozoic magmatic arc in Eastern Uruguay. Int Geol Rev 54(10):1161–1183. doi:10.1080/00206814.2011.626597

Middleton MF (1989) A model for the formation of intracratonic sag basins. Geophys J Int 99(3):665–676. doi:10.1111/j.1365-246X.1989.tb02049.x

Milani EJ (1997) Evolução tectono-estratigráfica da Bacia do Paraná e seu relacionamento com a geodinâmica fanerozóica do Gondwana sul-ocidental. Unpublished doctoral thesis (PhD), Universidade Federal do Rio Grande do Sul (UFRGS), 255 pp

Milani EJ, Zalán PV (1999) An outline of the geology and petroleum systems of the Paleozoic interior basins of South America. Episodes 22(3):199–205

Milani EJ, Faccini UF, Scherer CM, Araujo LM, Cupertino JA (1998) Sequences and stratigraphic hierarchy of the Paraná Basin (Ordovician to Cretaceous), Southern Brazil. Boletim IG/USP, Série científica 29:125–173

Muzio R (2000) Evolução petrológica e geocronologia do Maciço Alcalino Valle Chico, Uruguai. Unpublished PhD dissertation, Universidad Estadual Paulista, Rio Claro, SP, 171 pp

O'Connor JM, Duncan RA (1990) Evolution of the Walvis Ridge -Rio Grande rise hot spot system: implications for African and South American plate motions over plumes. J Geophys Res 95(B11):17475–17502

Oleaga A (2002) Contribución a la hidrogeología del Acuífero Guaraní en el sector Uruguay. Un enfoque integral. Unpublished Master of Science dissertations, Universidad Nacional Autónoma de México, México DF, 119 pp

Oyhantçabal P, Siegesmund S, Wemmer K, Presnyakov S, Layer P (2009) Geochronological constraints on the evolution of the southern Dom Feliciano Belt (Uruguay). J Geol Soc 166:1075–1084. doi:10.1144/0016-76492008-122

Oyhantçabal P, Siegesmund S, Wemmer K (2011) The Río de la Plata Craton: a review of units, boundaries, ages and isotopic signature. Int J Earth Sci 100:201–220

Panario D (1988) Geomorfología del Uruguay. Departamento de Publicaciones, Facultad de Humanidades y Ciencias, Universidad de la República, Montevideo, 32 pp

Panario D, Gutiérrez O (1999) The continental Uruguayan Cenozoic: an overview. Quatern Int 62(1):75–84. doi:10.1016/S1040-6182(99)00025-7

Peate DW, Hawkesworth CJ, Mantovani MSM, Shukowsky R (1990) Mantle Plumes and flood- basalt stratigraphy in the Paraná, South America. Geology 18(12):1223–1226. doi:10.1130/0091-7613(1990)018<1223:MPAFBS>2.3.CO;2

Pecoits E, Aubet N, Oyhantçabal P, Sánchez Bettucci L (2004) Estratigrafía de sucesiones sedimentarias y volcanosedimentarias Neoproterozoicas del Uruguay. Revista Sociedad Uruguaya de Geología 11:18–27

Pecoits E, Gingras M, Aubet N, Konhauser K (2008) Ediacaran in Uruguay: palaeoclimatic and palaeobiologic implications. Sedimentology 55:689–719. doi:10.1111/j.1365-3091.2007.00918.x

Peel E, Veroslavsky G, Fúlfaro VJ (1998) Geoquímica de las Pelitas de las Formaciones Castellanos y Migues (Cretácico), Cuenca del Santa Lucía – Uruguay: Consideraciones Paleoambientales. In: Actas II Congreso Uruguayo de Geología, Punta del Este, pp 151–157

Penck W (1953) Morphological analysis of land forms: a contribution to physical geology (trans: Czech H, Cumming Boswell K). Macmillan, London, 429 pp

Perea D, Martínez S (2004) Estratigrafía del Mioceno-Pleistoceno en el Litoral Sur-Oeste de Uruguay. In: Veroslavsky G, Ubilla M, Martínez S (eds) Cuencas Sedimentarias de Uruguay: Geología, paleontología y recursos naturales – Cenozoico. DIRAC, Facultad de Ciencias, Universidad de la República, Montevideo, pp 105–124

Piccirillo EM, Melfi AJ (1988) The Mesozoic flood volcanism from the Paraná Basin (Brazil): petrogenetic and geophysical aspects. Universidade de São Paulo. Instituto Astronômico e Geofísico, São Paulo, 600 pp

Preciozzi F, Spoturno J, Heinzen W, Rossi P (1985) Memoria explicativa de la Carta Geológica del Uruguay a escala 1:500.000. Dirección Nacional de Minería y Geología, Ministerio de Industria, Energía y Minería, Montevideo, 92 pp

Rabassa J, Carignano C, Cioccale M (2010) Gondwana paleosurfaces in Argentina: an introduction. Geociências 29(4):430–466, São Paulo, UNESP

Rabassa J, Carignano C, Cioccale M (2014) A general overview of Gondwana landscapes in Argentina. In: Rabassa J, Ollier C (eds) Gondwana landscapes in southern South America. Springer, Dordrecht, pp 201–245

Rabassa J, Coronato AM, Salemme M (2005) Chronology of the Late Cenozoic Patagonian glaciations and their correlation with biostratigraphic units of the Pampean Region (Argentina). J South Am Earth Sci 20:81–103. doi:10.1016/j.jsames.2005.07.004

Rapela CW, Fanning CM, Casquet C, Pankhurst RJ, Spalletti L, Poiré D, Baldo EG (2011) The Rio de la Plata craton and the adjoining Pan-African/brasiliano terranes: their origins and incorporation into south-west Gondwana. Gondwana Res 20(4):673–690. doi:10.1016/j.gr.2011.05.001

Renne PR, Ernesto M, Pacca IG, Coe RS, Glen JM, Prévot M, Perrin M (1992) The age of Paraná flood volcanism, rifting of Gondwanaland and the Jurassic–Cretaceous boundary. Science 258:975–979

Renne PR, Deckart K, Ernesto M, Féraud G, Piccirillo EM (1996a) Age of the Ponta Grossa dike swarm (Brazil), and implications to Paraná flood volcanism. Earth Planet Sci Lett 144(1–2):199–211. doi:10.1016/0012-821X(96)00155-0

Renne PR, Glen JM, Milner SC, Duncan AR (1996b) Age of Etendeka flood volcanism and associated intrusions in southwestern Africa. Geology 24(7):659–662. doi:10.1130/0091-7613(1996)024<0659:AOEFVA>2.3.CO;2

Rodrigues SC, Simões MG, Kowalewski M, Petti MAV, Nonato EF, Martínez S, del Río CJ (2008) Biotic interaction between spionid polychaetes and bouchardiid brachiopods: paleoecological, taphonomic and evolutionary implications. Acta Palaeontologica Polonica 53(4):657–668

Rossello EA, de Santa Ana H, Veroslavsky G (1999) El lineamiento Santa Lucía–Aiguá–Merín (Uruguay): un rifting transtensivo mesozoico abortado durante la apertura Atlántica? In: Actas del V Simpósio sobre o Cretáceo do Brasil-I Simposio sobre el Cretácico de América del Sur, Serra Negra, 1, pp 443–448

Rossello EA, de Santa Ana H, Veroslavsky G (2000) El lineamiento Santa Lucía-Aiguá-Merín (Uruguay): un corredor tectónico extensivo y transcurrente dextral precursor de la apertura atlántica. Revista Brasileira de Geociências 30(4):749–756

Sánchez Bettucci L (1998) Evolución tectónica del Cinturón Dom Feliciano en la región Minas-Piriápolis, Uruguay. Unpublished PhD dissertation, Universidad de Buenos Aires, Buenos Aires, 344 pp

Sánchez Bettucci L, Ramos VA (1999) Aspectos geológicos de las rocas metavolcánicas y metasedimentarias del Grupo Lavalleja, Sudeste de Uruguay. Revista Brasileira de Geociências 29(4):557–570

Sánchez Bettucci L, Cosarinsky M, Ramos V (2001) Tectonic setting of the Late Proterozoic Lavalleja Group (Dom Feliciano Belt), Uruguay. Gondwana Res 4(3):395–407

Sánchez Bettucci L, Oyhantçabal P, Loureiro J, Ramos VA, Preciozzi F, Basei MAS (2004) Mineralizations of the Lavalleja Group (Uruguay), a probable Neoproterozoic volcano–sedimentary sequence. Gondwana Res 7(3):745–751. doi:10.1016/S1342-937X(05)71060-1

Sánchez Bettucci L, Orgeira MJ, Sánchez G, Bertoni-Machado C, Fariña RA (2007) Magnetostratigraphy of the Neogene sediments of SW Uruguay. In: American Geophysical Union, Spring Meeting 2007, abstract #GP21A-06 http://adsabs.harvard.edu//abs/2007AGUSMGP21A..06S. Accessed 12 Dec 2012

Sánchez Bettucci L, Peel E, Masquelin H (2010a) Neoproterozoic tectonic synthesis of Uruguay. Int Geol Rev 52(1):51–78. doi:10.1080/00206810903358095

Sánchez Bettucci L, Peel E, Oyhantçabal P (2010b) Precambrian geotectonic units of the Río de La Plata craton. Int Geol Rev 52(1):32–50. doi:10.1080/00206810903211104

Sánchez San Martín G (2010) Estudio magnetoestratigráfico de los sedimentos Cenozoicos del Departamento de San José (balnearios Kiyú, Arazatí y Mauricio). Unpublished graduation thesis, Universidad de la República, Montevideo, 184 pp

Sombroek WG (1969) Soil studies in the Merin Lagoon basin: Merin lagoon regional project. Treinta y Tres. CLM/PNUD/FAO, LM 131, Treinta y Tres, Uruguay, vol 1

Spoturno J, Oyhantçabal P (Coord) (2004) Mapa de Recursos Minerales del Departamento de San José a escala 1:100.000. Proyecto CONICYT 6019 – Fondo Clemente Estable. Convenio Departamento de Geología (Facultad de Ciencias – UdelaR) – Dirección Nacional de Minería y Geología (Ministerio de Industria, Energía y Minería), Montevideo

Sprechmann P, Bossi J, Da Silva J (1981) Cuencas del Jurásico y Cretácico del Uruguay. In: Volkheimer W, Musacchio EA (eds) Cuencas sedimentarias del Jurásico y Cretácico de América del Sur. Comité Sudamericano del Jurásico y Cretácico, Buenos Aires, 1, pp 239–270

Stewart K, Turner S, Kelley S, Hawkesworth C, Kirstein L, Mantovani M (1996) 3-D, ^{40}Ar–^{39}Ar geochronology in the Paraná continental flood basalt province. Earth Planet Sci Lett 143(1–4):95–109. doi:10.1016/0012-821X(96)00132-X

Stoackes FA, Campbell CV, Cass R, Ucha N (1991) Seismic stratigraphic analysis of the Punta del Este Basin, offshore Uruguay, South America. Bulletin of the American Association of Petroleum Geologists 75(2):219–240

Teixeira AL, Gaucher C, Paim PSG, Fonseca MM, Parente CV, Silva Filho WF, Almeida AR (2004) Bacias do Estágio de Transição da Plataforma Sul-Americana. In: Mantesso-Neto V, Bartorelli A, Carneiro CDR, Brito Neves BB (eds) Geologia do Continente Sul-Americano: Evolução da Obra de Fernando Flávio Marques de Almeida. Beca Produções Culturais, São Paulo, pp 487–537

Thomaz-Filho A, Mizusaki AMP, Milani EJ, de Cesero P (2000) Rifting and magmatism associated with the South America and Africa break up. Revista Brasileira de Geociências 30(1):17–19

Tófalo OR, Morrás HJM (2009) Evidencias paleoclimáticas en duricostras, paleosuelos y sedimentitas silicoclásticas del Cenozoico de Uruguay. Rev Asoc Geol Argent 65(4):674–686

Tófalo OR, Morrás H, Sánchez L, Pecoits E, Aubet N, Zech W, Moretti L (2006) Litofacies y Paleosuelos de las Fms. Raigón (Plioceno Tardío-Pleistoceno Medio) y Libertad (Pleistoceno Inferior-Medio?), Uruguay. In: Actas 3° Congreso Argentino del Cuaternario y Geomorfología, Córdoba, 1, pp 807–816

Turner S, Hawkesworth C (1995) The nature of the sub continental mantle: constraints from the major-element composition of continental flood basalts. Chem Geol 120(3–4):295–314. doi:10.1016/0009-2541(94)00143-V

Turner S, Regelous M, Kelley S, Hawkesworth C, Mantovani M (1994) Magmatism and continental break-up in the South Atlantic: high precision 40Ar-39Ar geochronology. Earth Planet Sci Lett 121(3–4):333–348. doi:10.1016/0012-821X(94)90076-0

Ubilla M, Perea D, Goso Aguilar C, Lorenzo N (2004) Late Pleistocene vertebrates from northern Uruguay: tools for biostratigraphic, climatic and environmental reconstruction. Quatern Int 114(1):129–142. doi:10.1016/S1040-6182(03)00048-X

Ucha N, de Santa Ana H, Veroslavsky G (2004) La cuenca Punta del Este: geología y potencial hidrocarburífero. In: Veroslavsky G, Ubilla M, Martínez S (eds) Cuencas sedimentarias de Uruguay: Geología, Paleontología y Recursos Naturales. Mesozoico, 2da ed. DIRAC, Facultad de Ciencias, Universidad de la República, Montevideo, pp 173–192

Vail PR, Mitchum RM, Todd RG, Widmier JM, Thompson S, Sangree JB, Bubb JN, Hatlelid WG (1977) Seismic stratigraphy and global changes on sea-level. In: Payton CE (ed) Seismic stratigraphy: application to hydrocarbon exploration. American Association of Petroleum Geologists, Tulsa, Okla, Memoir, vol 26, pp 49–212

Veroslavsky G (1999) Geologia da Bacia de Santa Lucía-Uruguai. Unpublished PhD dissertation, Universidad Estadual Paulista, Rio Claro, SP, 152 pp

Veroslavsky G, Martínez S (1996) Registros no depositacionales del Paleoceno-Eoceno del Uruguay: nuevo enfoque para viejos problemas. Revista Universidade Guarulhos Serie Geociências 1:32–41

Veroslavsky G, Montaño J (2004) Sedimentología y estratigrafía de la Formación Salto (Pleistoceno). In: Veroslavsky G, Ubilla M, Martínez S (eds) Cuencas Sedimentarias de Uruguay: Geología, paleontología y recursos naturales – Cenozoico. DIRAC, Facultad de Ciencias, Universidad de la República, Montevideo, pp 147–166

Veroslavsky G, Martínez S, de Santa Ana H (1997) Calcretas de aguas subterráneas y pedogénicas: génesis de los depósitos carbonáticos de la Cuenca de Santa Lucía, sur del Uruguay (Cretácico Superior? – Paleógeno). Revista de la Asociación Argentina de Sedimentología 4(1):25–35

Veroslavsky G, Fulfaro V, de Santa Ana H (2006) El Devónico en Uruguay: estratigrafía, correlación geológica y recursos minerales. In: Veroslavsky G, Ubilla M, Martínez S (eds) Cuencas Sedimentarias del Uruguay, Geología, paleontología y recursos naturales – Paleozoico. DIRAC, Facultad de Ciencias, Universidad de la República, Montevideo, pp 107–132

Zárate MA (2003) Loess of southern South America. Quat Sci Rev 22:18–19, 1987–2006. doi:10.1016/S0277-3791(03)00165-3

A General Overview of Gondwana Landscapes in Argentina

Jorge Rabassa, Claudio Carignano, and Marcela Cioccale

Abstract Gondwana Landscapes in Argentina were already identified by Juan Keidel and Walther Penck at the beginnings of the twentieth century, as well as by other geologists and naturalists of the different European schools that worked in this country. These studies were continued at a very good level in Brazil, thanks to the work of Lester C. King, later on intensively followed by João José Bigarella. However, these concepts gradually disappeared from the Argentine geological scene, dominated by the influence of American geomorphologists, and particularly William Thornbury, who doubted the existence of such ancient landforms, when one of the paradigms of the time was that "practically there is no landscape older than the Pleistocene." These landforms are the result of the process of both deep chemical weathering, developed in very stable tectonic and climatic environments, under hyper-tropical climates, and pediment processes in semiarid to humid environments.

The Gondwana Landscapes or their fragmented remnants have been recognized in Argentina, from north to south, in the basaltic hills of the province of Misiones; the Sierras Pampeanas of the provinces of Catamarca, La Rioja, and San Juan; the Sierras Chicas, Sierras Grandes, and Sierra Norte of Córdoba province; the Sierras

J. Rabassa (✉)
Laboratorio de Geomorfología y Cuaternario, CADIC-CONICET, Ushuaia, Tierra del Fuego, Argentina

Universidad Nacional de Tierra del Fuego, Ushuaia, Tierra del Fuego, Argentina
e-mail: jrabassa@gmail.com

C. Carignano
Centro de Investigaciones en Ciencias de la Tierra (CICTERRA), CONICET and Universidad Nacional de Córdoba, Córdoba, Argentina

Facultad de Ciencias Exactas y Naturales, Universidad Nacional de Córdoba (UNC), Córdoba, Argentina

M. Cioccale
Facultad de Ciencias Exactas y Naturales, Universidad Nacional de Córdoba (UNC), Córdoba, Argentina

J. Rabassa and C. Ollier (eds.), *Gondwana Landscapes in southern South America*, Springer Earth System Sciences, DOI 10.1007/978-94-007-7702-6_9,

de San Luis, the Sierra Pintada, or San Rafael Block of Mendoza province; the Sierras de Tandil, Sierra de la Ventana, and the Pampa Interserrana of Buenos Aires province; the Sierras de Lihuel Calel of the province of La Pampa; the Somuncurá or Northern Patagonian Massif in the provinces of Río Negro and Chubut; the Deseado Massif of Santa Cruz province; and the Malvinas-Falklands archipelago. In other regions of Argentina, these surfaces have been downwarped in tectonic basins and are covered by sedimentary and/or volcanic units of various ages. The ages for the development of the Gondwana Landscapes have been estimated in between the Middle Jurassic and the Paleogene.

The Argentine Gondwana Landscapes were uplifted, fragmented, and eroded during the Middle to Late Tertiary. They have remained as mute testimony of the past above extensive pediplains and piedmont deposits, as climates and environments became more arid and cooler, approaching the present conditions.

Keywords Gondwana landscapes • Argentina • Cratonic areas • Planation surfaces • Etchplains

Introduction

Gondwana Landscapes were originally defined by Fairbridge (1968a, p. 483) as an "ancestral landscape" composed of "series of once-planed remnants" that "record traces of older planation" episodes, during the "late Mesozoic (locally Jurassic or Cretaceous)." This has been called the "Gondwana cyclic land surface" in the continents of the southern hemisphere, occurring extensively in Australia, Southern Africa and the cratonic areas of South America, and other smaller regions (see Ollier 2014a). Remnants of these surfaces are found also in India and the Arabian Peninsula, and it is assumed they have been also preserved in Eastern Antarctica, both on the surface and also underneath the Antarctic ice sheet which covers that region with an average thickness of 3,000 m. These landscapes were generated when the former Gondwana supercontinent was still intact and similar tectonic conditions in its drifted fragments have allowed their preservation (Fig. 1). These are ancient, Mesozoic landscapes that were later never fully covered by marine sediments and most of them have been exposed at the surface since their genesis.

This chapter is a revision of a former contribution by Rabassa et al. (2010) at the light of new evidence and information. The Mesozoic paleoclimates and tectonic conditions in Gondwana allowed the formation of landscape systems in all portions of the ancient supercontinent. In South America, these surfaces have been studied in Brazil, Uruguay, Venezuela, the Guyana Massif, and Argentina. These geomorphological systems were generated when the former Gondwana supercontinent was still intact and similar tectonic conditions in its drifted fragments have allowed their preservation. The nature and genesis of these landscapes have been thoroughly discussed by Rabassa (2010, 2014) and Ollier (2014b).

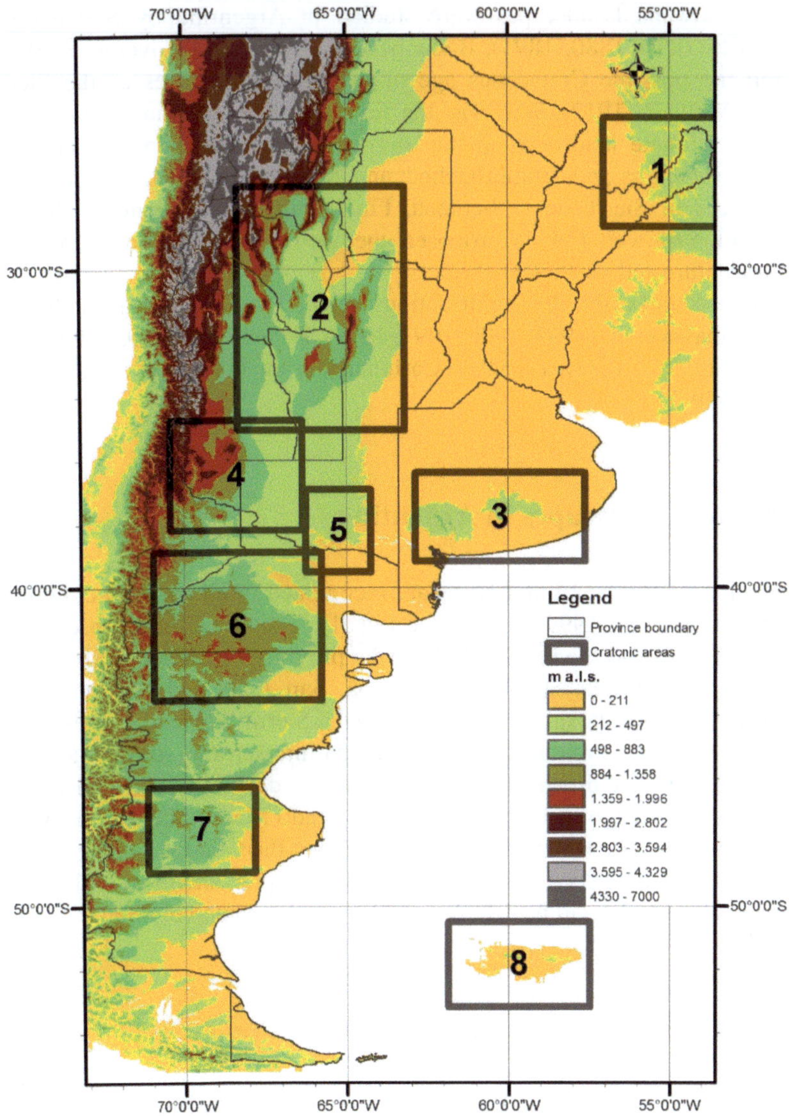

Fig. 1 Map of Argentina, showing the area in which significant extensions of Gondwana Landscapes are present. All ancient surfaces rise at least above 300–500 m a.s.l. These paleosurfaces reach their highest locations in the western Pampean Ranges. (*1*) The basaltic hills of the province of Misiones; (*2*) the Sierras Pampeanas of Córdoba, San Luis, La Rioja, San Juan, and Catamarca; (*3*) the Central Buenos Aires Positive area, including the Sierras Septentrionales (Tandilia), the Sierras Australes (Ventania), and the Pampa Interserrana; (*4*) the San Rafael or Sierra Pintada Block in Mendoza; (*5*) the Sierras de Lihuel Calel in La Pampa; (*6*) the Northern Patagonian Massif; (*7*) the Deseado Massif; and (*8*) the Malvinas-Falklands Islands

The Gondwana Landscapes were studied in Argentina by Stelzner (1885), Brackebush (1879, 1880, 1891), Bodenbender (1905), and Rovereto (1911), who analyzed the possible Cretaceous age of the summit surfaces of the Sierras de Córdoba; Walther (1912), who was the first one in mentioning the existence of subtropical climate landforms in the Sierras de Tandil; Schmieder (1921), who identified inselbergs and "rumpffläche"; and Keidel (1916, 1922), who described paleosurfaces in Sierra de la Ventana. Finally, it should be noted that Rolleri (1975) and Yrigoyen (1975) proposed the idea of the Buenos Aires Positive morphostructure and its ancient surfaces.

In modern times, the first contributions devoted to Gondwana Landscapes are those of Rabassa et al. (1995, 1996, 1997, 1998), Zárate et al. (1995, 1998), Pereyra and Ferrer (1995), Pereyra (1996), Carignano et al. (1999), and Carignano and Cioccale (1997).

Gondwana Landscapes in Argentina

In Argentina, Gondwana Landscapes are recognized in all exposed cratonic areas (see Carignano et al. 1999, p. 249; see Fig. 1). They have been observed in the following regions: (a) the basaltic hills of the province of Misiones; (b) the Sierras Pampeanas of Córdoba, San Luis, La Rioja, San Juan, and Catamarca; (c) the Central Buenos Aires Positive area, including the Sierras Septentrionales (Tandilia), the Sierras Australes (Ventania), and the Pampa Interserrana; (d) the San Rafael or Sierra Pintada Block in Mendoza; (e) the Sierras de Lihuel Calel in La Pampa; (f) the Northern Patagonian Massif; (f) the Deseado Massif; and (g) the Malvinas-Falklands Islands. The nature and characteristics of the Gondwana Paleolandscapes in these areas are described and discussed in this chapter, and most of these regions have been studied in detail by different authors in this same volume.

Basaltic Hills of the Province of Misiones

The province of Misiones, in the northeastern end of Argentina (Figs. 1 and 2), is presently under a wet tropical climate. These conditions may have persisted throughout the Tertiary, maintaining this region under environmental conditions quite similar to those that existed during the Late Mesozoic. For a careful discussion, see the chapter by Kröhling et al. (2014). The entire region was covered by a huge basaltic plateau, the Paraná Plateau. These volcanic eruptions produced more than 1.5 million km^3 in less than 1 million years, starting at 133 ± 1 Ma, closely after the Jurassic-Cretaceous boundary (Renne et al. 1992) and immediately before the rifting of Gondwana and the opening of the South Atlantic. These volcanic

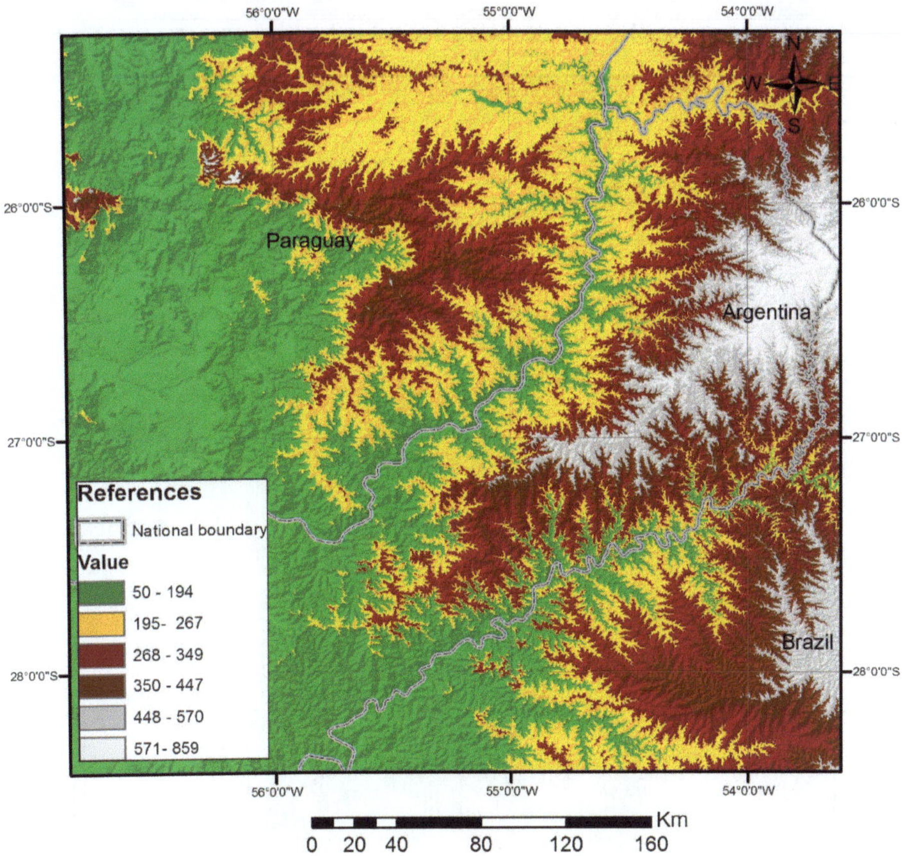

Fig. 2 Basaltic hills of the province of Misiones. Hypsographical map. This region is located southwest of the Paraná Plateau, presenting altitudes between 200 and 860 m a.s.l

units were partially covered afterwards by continental, Oligocene, Miocene, and Pliocene deposits. Where they remained exposed at the surface, they started a period of long-term landscape evolution under tropical climate which was probably sustained during the end of the Mesozoic and most of the Tertiary. Little is known about the paleolandscape of Misiones, where most authors have considered the summit surfaces as just structural surfaces of the basaltic flows. However, Brazilian geomorphologists have identified the "planalto de Vacaria" or Vacaria Surface (Ab'Sáber 1969), a summit surface developed upon the Early Cretaceous basalts, between 950 and 1,100 m a.s.l. The Vacaria Surface is deeply cut by stream canyons, adapted to modern fractures and joints. This surface extends into Argentina and Uruguay, as the landscape slopes towards the S and the SW. The age of the surface is considered to be Late Cretaceous or Paleogene.

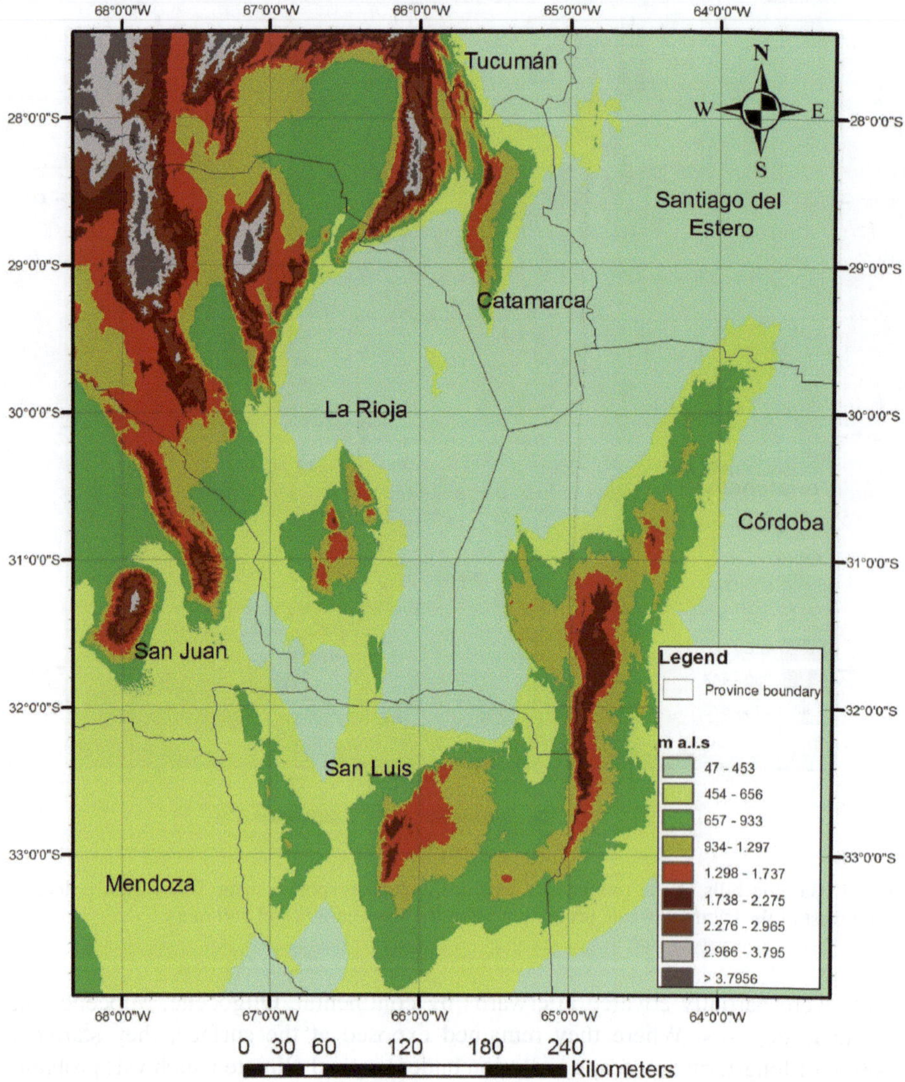

Fig. 3 Sierras Pampeanas of Córdoba, San Luis, La Rioja, San Juan, and Catamarca. Hypsographic map. They are divided into two main regions, the northwestern mountains reaching heights above 3,500 m a.s.l. and the eastern mountains showing heights below 3,000 m a.s.l

Sierras Pampeanas of Córdoba, San Luis, La Rioja, San Juan, and Catamarca

The Pampean Ranges (Figs. 1 and 3) are characterized by extensive planation surfaces at or near their summits, which have usually been considered as a single unit and named a "peneplain" or erosion surface, with assigned ages ranging between the Paleozoic and the end of the Mesozoic, and which would have been exhumed during the Tertiary orogeny. These concepts were strongly challenged by Carignano et al. (1999), who proposed that there is no a single surface and that the existing surfaces are chronologically and genetically different, and they have remained exposed to the atmosphere since the time of their formation (Carignano and Cioccale 2008). A detailed reconstruction of the ancient landscapes of the southernmost extreme of the Sierras de Córdoba has been presented by Andreazzini and Degiovanni (2014).

The ideas about the geomorphological development of the Sierras Pampeanas started with Stelzner (1885), Brackebusch (1879, 1880, 1891), and Bodenbender (1890, 1905, 1907, 1911) who prepared the first stratigraphic schemes of the Sierras Pampeanas and noted the clear coincidence in the elevation of the mountain summits, which they sometimes called "altiplains." When Bodenbender (1905, 1911) suggested that the uplifting of the Sierras Pampeanas took place due to the Andean movements during the Tertiary, he settled the basis for the interpretation and regional correlation that was characterized, mainly, by the extension to these areas of observations done somewhere else in the Andean Cordillera. This was a significant precedent that has conditioned the geomorphological interpretation of the Sierras Pampeanas to the present day. Rovereto (1911) did the first purely geomorphological studies in Argentina, dedicating a chapter to the Sierras de Córdoba, defined by him with an outstanding vision as "a gigantic residual mass of a Paleozoic mountain." This author, also, considered that what he called the "semiplains" of the sierras corresponded to four different erosion surfaces which he named as "peneplains." Though he never mentioned it, the influence of William Davis' concepts was obvious. The three first surfaces would have been developed during the Paleozoic and the fourth during the Mesozoic of pre-Cretaceous age. Rovereto (1911) recognized that, in the Sierras de Córdoba, a superposition of the Andean and Brazilian-Uruguayan structural styles existed. He observed for the first time that the historical geology of these sierras was almost identical to that of SW Brazil. In this sense, he was the first in assigning Cretaceous age to the sedimentary sequences that outcrop east of the Sierras de Córdoba, but, because he was clearly ignored by his colleagues, the idea of a Permian-Triassic age for these beds was sustained until the 1970s, when, thanks to radiometric dating, their Cretaceous age was confirmed.

The observations and deductions by Rovereto (1911) were practically ignored in his time due to the usual criteria of extending the basics of Andean geology to these central regions. Gerth (1914, 1927), Rassmuss (1916), Beder (1916), and Rimann (1926) recognized only one erosion surface in different areas of the Sierras Pampeanas, which had been formed between the Late Paleozoic and the Cretaceous, with a general agreement in a Permian-Carboniferous age. Gerth (1914) proposed also that such erosion surface would have been exhumed.

The following purely geomorphological investigation of this region was done by Schmieder (1921) who, under the influence of other German geologists, confirmed the hypothesis of one single, dismembered Paleozoic surface, uplifted by the Andean movements. In spite of that, this author described in detail the remnants of the surface, noting the presence of "inselbergs" in the highest remnant, the Pampa de Achala. In this work, the first geomorphological map of the region, together with cross sections, was presented, in which the remnants of the surface were represented according to their position and characteristics. The mapped units are in a very clear agreement with those of Rovereto (1911). Schmieder (1921) indicated that he was referring to the erosion surface in the sense of "Rumpffläche" of the German geologists. This term has no genetic meaning (Gross 1948), but in the Spanish version, that word was replaced by "peneplain" though with the note that it was not equivalent to the genetic term in the sense of Davis (1899, 1909). This serious mistake, probably due to the lack of a proper Spanish term and under the influence of the American literature, set conditions to the future interpretation of the papers authored by the German geologists and caused several problems in the present knowledge of the Sierras Pampeanas geomorphology.

One of the most relevant steps forwards of modern geomorphology was when Walther Penck published his theory on the geomorphological evolution and modeling of the Earth landscape (Penck 1924). His theories were in fact firstly conceived following his fieldwork in Argentina (Gross 1948), where he lead reconnaissance work in Northwestern Sierras Pampeanas (Penck 1914, 1920). From his observations in that area and the Sierras de Córdoba, Penck (1924) suggested the existence of four erosion surfaces, generated by gradual slope retreat by a complex mechanism, each one with its own characteristics and different ages, clearly rejecting the idea of one single planation surface. All of Penck's conclusions are founded in a very careful geomorphological reconstruction, supported by a strict stratigraphic and structural control.

The first quarter of the twentieth century was characterized by the development of important geological and geomorphological theories, such as the hypothesis on the connection between the Sierras Pampeanas, the Sierras of Buenos Aires province, the Brazilian-Uruguayan Massif, and South Africa (Frenguelli 1921), originated in the works of Bodenbender (1895, 1911), Walther (1912), and Keidel (1916, 1922), later confirmed by Du Toit and Reed (1927). Contrarily, in the following years, the absence of new geomorphological ideas is surprising, and moreover, a clear recession in the geomorphological investigations took place, drifting away from those brilliant epochs of the precedent decades. Thus, the general belief is that there was only one erosion surface, called a "peneplain," without a clear genetic concept to support it, as a consequence of the mixture of the local literature (based on the work of the German geologists) and the wide global domination of the Davisian concepts. A clear example of this situation is observed in the work of Schlagintweit (1954), who preferred to avoid the term "peneplain" to refer instead to what he called a "coherent original semiplain," which he observed in the Sierras de Córdoba. This author recognized as "monadnocks" the hills rising over the general level of the plain, cited the contribution by Gross (1948), and recommended the analysis of Lester C. King's (1953) paper.

Thus, a sort of academic chaos was generated on the genetic and chronological interpretation of the surfaces which still remains today, as shown by papers in which some authors considered the surface as a "pediment" (González Díaz 1974), and later, closely supporting the Davisian concepts, the erosion surfaces were interpreted as isolated portions of one, single regional "peneplain" of Paleozoic-Tertiary age, developed during one extended cycle of fluvial erosion (González Díaz 1981). Others, such as Sayago (1983, 1986), considered the "primeval plain" as a "peneplain" formed by physical weathering and sheet runoff under semiarid climate, however without ruling out, at a same time, a possible origin by deep chemical weathering under tropical or subtropical climate.

Jordan et al. (1989) estimated by thermochronometry the age of the "basement peneplain" of the Sierras Pampeanas, that is, the major planation surface found at the summit of these ranges. Their results suggested that the surface is not a Paleozoic, neither an Early Mesozoic exhumed "peneplain," and that by Triassic times the rocks now exposed in the planation surface were at 2–4 km down the existing surface at that time. Note that these authors are using the term "peneplain" as any regional surface of low relief created by erosion. This study also revealed that the so-called single "peneplain" is in fact a complex of many different planation surfaces, of varied age and origin, some of which were formed in times as apart as 300–400 million years, but none of them is Late Cenozoic in age. Therefore, their denudation took place long before the Late Cenozoic. Note that these techniques always require a few kilometers of erosion more than geomorphic evidence suggests, but it is a good approximation. There are no significant sedimentary units of Jurassic age in the Sierras Pampeanas, and non-marine Cretaceous sedimentary and volcanic rocks are very scarce, mostly limited to the extreme eastern and in some of the western Triassic depositional centers. It is therefore clear that the Sierras Pampeanas have been a positive element of the landscape as a whole since the Triassic, and locally, perhaps even from the Permian or even before. These authors have established that the denudation rate in the Sierras Pampeanas during the Mesozoic and Cenozoic was very slow, between 0.012 and 0.026 km/Ma, though still much faster than in the Australian craton. The available data suggest that some of the planation surfaces are in fact exhumed Early Paleozoic surfaces, but most of the surfaces were formed by deep weathering after the Middle Triassic and continuing denudation which persisted until the Neogene. The most important results of this study are that the so-called basement peneplain is in fact a diachronic group of erosion surfaces, separated in time by 300–400 Ma, and that the younger unit was formed later than the Middle Triassic.

In recent years, thanks to the collaboration with distinguished South African colleagues such as Tim Partridge and Rodney Maud, it has become clear that it is necessary to return to the ideas of Lester C. King (1950, 1956, 1963, 1967) on the evolution of landscape as a conceptual framework, following the methodology in the work of Partridge and Maud (1987) on the study of equivalent erosion surfaces in Southern Africa.

The Sierra Norte and Sierra de Ambargasta (30° 40′–26° 30′ S; 65° 25′–63° 15′W) have an elongated shape in a N-S direction, with an extension of over 65,000 km^2 and a maximum altitude of 1,140 m a.s.l. (Carignano and Cioccale 2008). These ranges are very useful to understand the genesis of these paleolandscapes. They are composed of Precambrian-Early Paleozoic crystalline basement, and though they were affected also by the Andean orogeny, they do not show the characteristic asymmetrical profile of the other Pampean Ranges. The Sierra Norte is characterized by a slightly convex upwards shape, in which several extensive planation surfaces are found as stepped levels with planed summits. These subhorizontal surfaces may be easily reconstructed. Four major surfaces are found at approximately 900–800, 750–600, 650–550, and 500–350 m a.s.l., distributed in a concentric way around the ranges and having progressive decreasing age. The uppermost level cuts Late Paleozoic sedimentary rocks, whereas the 2nd surface is covered by Cretaceous breccias and fanglomerates and the 3rd one by Cretaceous conglomerates and sandstones. Finally, the 4th planation surface is covered by Miocene marine sediments, a transgressive facies from the Atlantic Ocean (Carignano and Cioccale 2008). In the 1st and 2nd levels, extensive remnants of the weathering profiles have been preserved, with in situ corestones and gruss and huge, dome-shaped inselbergs. These conditions allow the interpretation of these surfaces as etchplains, with later partial removal of the saprolite. The ancient weathering fronts are also observed affecting Permian sedimentary rocks below the Cretaceous sedimentary cover; therefore, regional weathering took place probably during the Late Jurassic or Early Cretaceous. The thickness of the original weathered zone is estimated based upon the local relief of inselbergs (see Linton 1955), of up to several hundred meters. The upper planation surfaces were likely uncovered ever since their formation (Carignano and Cioccale 2008).

The Sierras Chicas and the Sierras Grandes de Córdoba have a very complex and illustrative geomorphological history. Several erosion surfaces have been identified, using geomorphological, geometrical-structural, and sedimentological criteria (Carignano et al. 1999; Figs. 4 and 5). The morphogenetic evolution of these ranges, which represent the most impressive of the Central-Eastern Sierras Pampeanas, has been reconstructed based on such criteria. During Jurassic times, a long period of tectonic quiescence and predominantly humid tropical climate enabled the progressive development of a broad planation surface. Corestones, bornhardts, tors, and deep weathering profiles and their occurrence are interpreted as residual landforms pertaining to that primeval surface. The Late Jurassic-Early Cretaceous interval was marked by continental rifting and the ranges degradation under less humid, semiarid climates. During this rifting interval, each major faulting event generated its own particular erosion cycle. Thus, two additional planation surfaces were developed in a very long and complex denudation cycle. Remnants of these surfaces are still preserved around the nuclei of each of the larger blocks of the Sierras Pampeanas (Figs. 6 and 7). Such surfaces were weathered again during the latest Cretaceous-Paleocene times, and further erosion developed a fourth planation surface.

During the Miocene, a fifth planation surface was developed. Thick and mature calcretes remain as evidence of long-term, climate stability conditions. Due to the faulting and uplifting during the last 10 Ma, almost all of these surfaces have been broken and partially tilted.

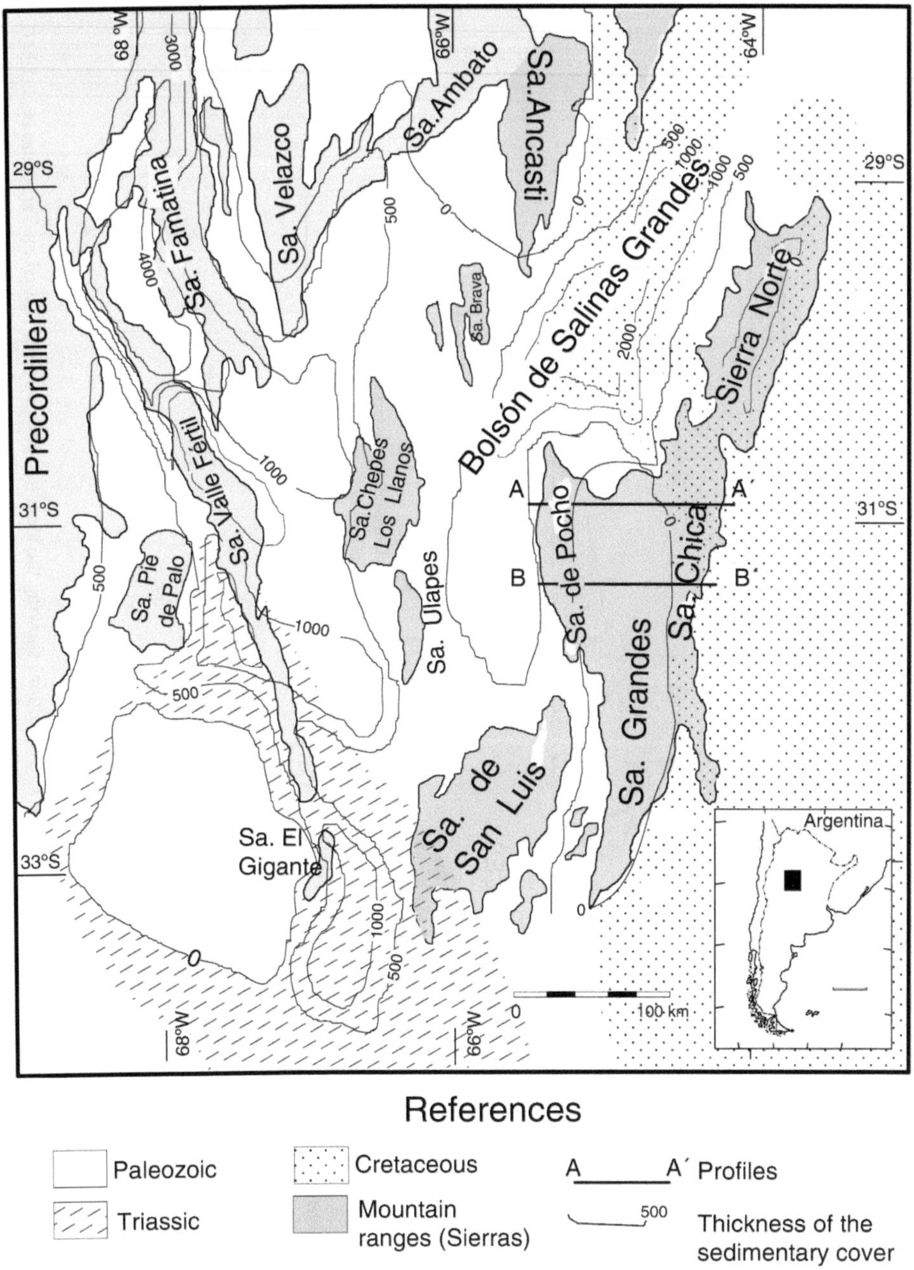

Fig. 4 Geographical distribution of Sierras Pampeanas, with local names for each of the ranges (Modified from Carignano et al. 1999)

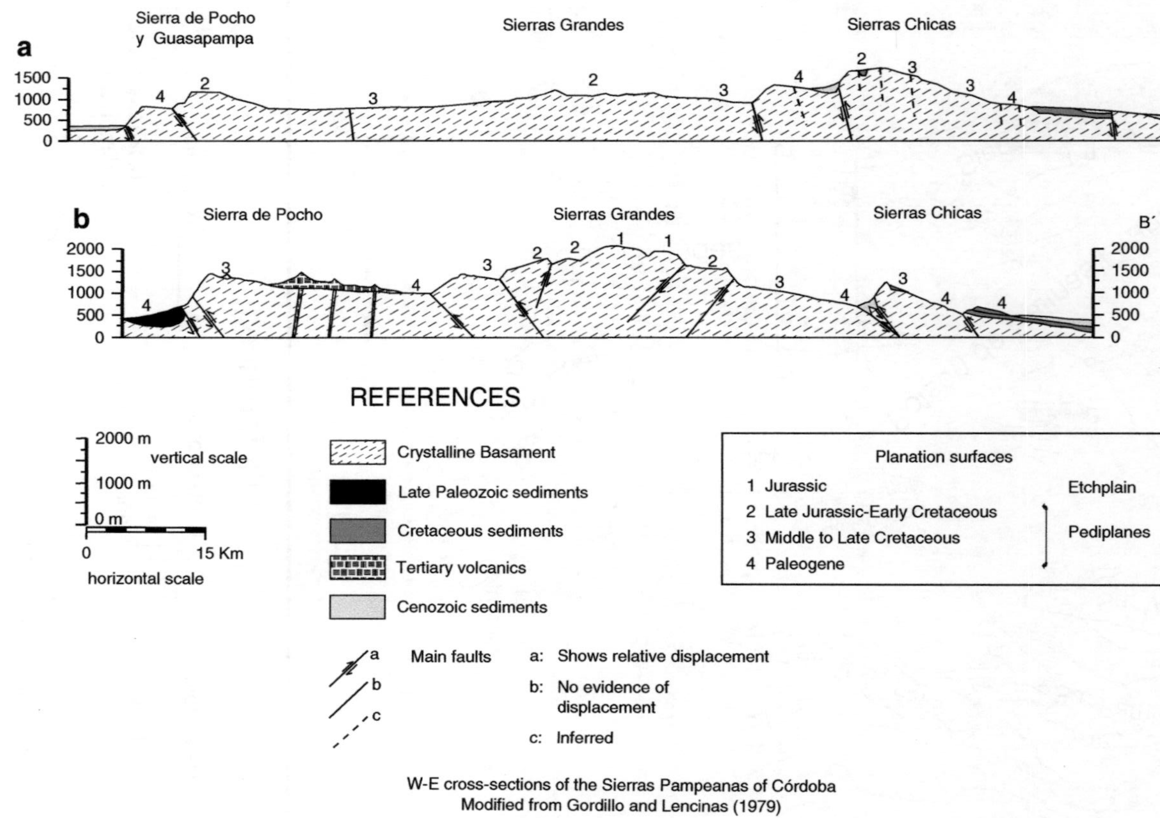

Fig. 5 Sierras Pampeanas of Córdoba, San Luis, and eastern La Rioja. Cross sections showing the identified paleosurfaces (Based on Carignano et al. 1999)

Fig. 6 Pampa de Achala, Sierras Grandes de Córdoba. Uppermost paleosurface showing its ample extension. Granite landscape corresponding to the base of the ancient weathering front, with granite corestones and many other minor features (Photograph by J. Rabassa 2000)

Fig. 7 Sierras Chicas de Córdoba. Paleoweathering, base of a laterite profile. *On top*, Holocene soils

Fig. 8 Sierra de Olta, eastern La Rioja. Carboniferous mountain glacial valley in cross section, carved in Ordovician granites and schists. The valley and the glacial sediments contained within it have been denudated during the Late Cenozoic. In the upper portion, remnants of a Cretaceous planation surface (Photograph by J. Rabassa 2005)

Carignano et al. (1999) proposed that at least the oldest of these surfaces may be tentatively correlated with similar landscapes in Eastern Brazil, Uruguay, and Southern Africa.

The nature and age of these ancient Gondwana paleosurfaces are clearly exceeding the Late Mesozoic tectonic and climatic cycle in the Sierras Pampeanas. In fact, evidence of an ancient planation surface has been found by Socha et al. (2006, 2014) at Sierra de Olta, province of La Rioja (Fig. 8). However, this planation surface is a Devonian or Early Carboniferous (?) etchplain, which was uplifted by the Late Paleozoic orogeny and later deeply eroded by mountain glaciers during the Middle to Late Carboniferous, in the core of the classical Gondwana glaciations. These were mountain glaciers which were located in high, mountainous, coastal ranges, marginally situated in relation with the huge continental ice sheet centered in the present territories of Brazil and South Africa. The whole paleolandscape complex was probably affected by deep chemical weathering in tropical climate during the Middle to Late Mesozoic, forming a younger etchplain at a lower topographical level, which was later denudated during the Cenozoic.

Glaciers covered Gondwana during the Late Paleozoic (Carboniferous and Permian), and then the Sierra de Olta was covered by mountain glaciers, which geomorphological evidence has been preserved (see Socha et al. 2014). Basal and ablation till deposits have been found at the bottom and sides of the ancient glacial valleys, forming terminal valley moraines, with the presence of sequences of advancing and retreating glaciers, and other glaciogenic sediments corresponding to ice-contact glaciofluvial and glaciolacustrine environments. Main and tributary glacial valleys have been recognized and the paleoslopes of the glacial valleys have been reconstructed, as well as their sizes and gradients. The glaciogenic deposits are resting on top of Ordovician granites and metamorphic rocks, which are the preglacial bedrock. The basal contact of the glaciogenic deposits is showing relict preglacial weathering profiles, with the presence of paleoweathering fronts and corestones. The valley moraines are composed of very large Ordovician granite boulders, up to several meters in diameter, very well rounded, strongly equidimensional, and all of very similar dimensions.

These characteristics of high rounding and strong sphericity are very rare in normal glaciogenic systems. It is interpreted that the boulders are in fact preexisting corestones, pre-rounded by deep chemical weathering processes in an etchplain and then incorporated by the glacier to its sedimentary load, very likely transported mainly in supraglacial position and deposited in terminal moraine environments. The presence of the paleoweathering fronts and the corestones is only compatible with the existence of planation surfaces due to deep chemical weathering, developed in continental environments, perhaps under conditions of tropical and hyper-tropical climates during the Late Devonian or the earliest Carboniferous. These paleoclimatic conditions would have been maintained until the regional mountain glaciation of the Carboniferous and, most likely, the Middle Carboniferous, when the preexisting paleosurfaces were strongly eroded by the mountain ice sheet. The mountain glaciations landscape was later buried under Permian marine sandstones and remained so perhaps even up to the Middle Jurassic, when the deep weathering and erosion processes developed new planation surfaces in the Late Jurassic and the Cretaceous, whose remnants are still forming the summit surfaces of the mountains in the region. The following denudation during the Cenozoic completed the exhumation of the Paleozoic paleolandscapes until reaching their present distribution and location.

The Sierras de San Luis (33° S, 66° W) are an extension of the Sierras de Córdoba towards the southwest. Their composition and structural style are similar to the Sierras Pampeanas (Costa et al. 1999). The block mountain ranges were uplifted along regional faults. These authors have noticed the uncertain age of the interior paleosurfaces found in these ranges, which is a serious inconvenience for understanding the Neogene evolution of this region. This paleosurface is widely preserved on the eastern slope (the slope away from the Andes), but it is disrupted by the Neogene tectonics. The paleosurface was studied to understand their contribution to neotectonic activity. It is characterized by a gentle, undulating landscape, slightly tilted to the east. It is composed of denudated crystalline basement, and it lacks any sedimentary cover older than the Pleistocene, except for Miocene-Pliocene volcanics. The surface had been described as a "peneplain" (González Díaz 1981), "shaped by fluvial systems." Costa et al. (1999) describe also a planation surface underlying the Triassic (?)-Cretaceous infilling of adjacent basins, and they considered that all surface remnants are part of the same paleosurface, though they quoted Jordan et al.'s (1989) findings that the Sierras Pampeanas were mountains and not plains during the Late Paleozoic sedimentation. Based on the available information, Costa et al. (1999) concluded that this surface was formed between the Carboniferous and Triassic-Middle Cretaceous times. Basaltic domes of ages between 60 and 70 Ma are lying on top of the surface. Though the published information is still scarce, it may be suggested that the San Luis surfaces are in fact diachronic and perhaps the buried one is of Paleozoic age. The summit surface, in contrast, could be at least partially an etchplain or a pediplain of Late Mesozoic age, as in the Sierras de Córdoba (Carignano et al. 1999), though the possibility that part of it could be a Paleozoic exhumed feature cannot be ruled out. Spectacular weathering features have been observed in the uppermost granite surface (gnammas; Fig. 9). It is clear that deeper studies are needed on these paleolandscapes of the Sierras de San Luis in the future.

The Central Buenos Aires Positive Area, Including the Sierras Septentrionales (Tandilia), the Sierras Australes (Ventania), and the Pampa Interserrana

The Central Buenos Aires Positive area is a large cratonic region which is clearly related to the southern margin of the Brazilian Shield. This cratonic unit extends for more than 300,000 km^2 (Figs. 1 and 10). Charles Darwin (1876, p. 319) was the first author to mention features of ancient surfaces in Buenos Aires province. Note that his actual observations were much earlier, during the 1830s. Darwin observed, on the flanks of the Ventania ranges, remnants of small patches of conglomerates and breccias, at a height of 300–400 ft (100–130 m) above the plain, "firmly cemented by ferruginous matter to the abrupt and battered face of the quartz." Darwin (1876) also described very precisely and for the first time the La Toma Section on the Río Sauce Grande valley, where Miocene and other Late Cenozoic sedimentary deposits are forming the base of the post-Permian sequence. These continental deposits also confirm that the age of the underlying landscape is, at least, clearly pre-Miocene in age. Darwin (1876, p. 353) suggested as the origin of the materials of the Pampean formation, "the enormous area of Brazil consisting on gneissic and other granitic rocks which have suffered decomposition and been converted into a red, gritty, argillaceous mass to a greater depth than in any other country." Though it is known today that only a fraction of the Pampean deposits has such an origin, it is very interesting that Darwin realized at such an early date the importance of the supply of weathered debris coming from the Brazilian Shield into the surrounding sedimentary basins. Undoubtedly, though famous for his theory of Biological Evolution, Charles Darwin was indeed a brilliant geologist, and his writings after his South American trip in the 1830s prove it.

Fig. 9 Paleoweathering features in a granite landscape, Sierras de San Luis. (**a**) Granite landscape, gnammas, and channels; (**b**) gnammas in granite; (**c**) integrated gnammas by runoff erosion; (**d**) channels in granite, with redeposited silica along the margins of the channel as a sort of geochemical levees (Photographs by J. Rabassa 2003)

Du Toit and Reed (1927) and Du Toit (1937) considered the Tandilia ranges as a buried and partially exhumed mountain chain. Also, Du Toit and Reed (1927, p. 26) vividly described the "consolidated gravels and breccias resting on benches and terraces cut along the inner side of the quartzite chain of the Sierra de la Ventana." They were referring to the "Conglomerado Rojo," formally known as the Cerro Colorado Breccia, and pointed that "I (Du Toit) had great difficulty in realizing that this was another continent and not some portion of one of the southern districts of the Cape" (Du Toit 1954). He noted that the "parallelism is so wonderfully close that the geological histories of these two countries must have been all but identical from mid-Paleozoic down to early Tertiary." According to Du Toit and Reed (1927), Keidel (1916, pp. 37–42) described an uppermost terrace that it is incised in the quartzites beneath the crest of the range, at a level of around 800 m a.s.l. The second has an altitude between 450 and 550 m a.s.l. This second surface bears caps of

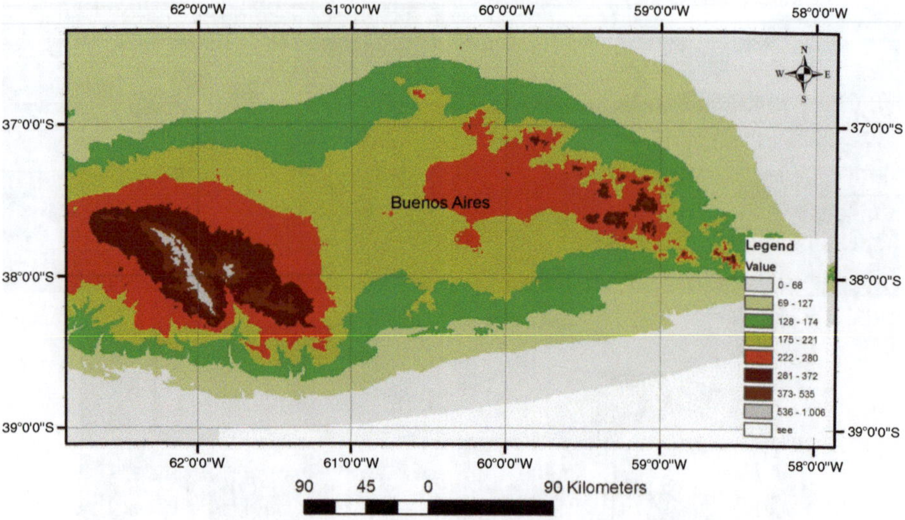

Fig. 10 Central Buenos Aires Positive area. Hypsographic map. This set is composed of the Sierras of Tandil or Tandilia, to the east, and the Sierra de la Ventana or Ventania, to the west. They are low ranges with heights between 300 and 1,000 m a.s.l., with the higher values in the western ranges

hard, bright-red conglomerates and breccias, in sandy cement with iron oxide. The material resembled the so-called High-Level Gravels and the silcretes/ferricretes of the Southern Cape, which are assigned to be formed in planation surfaces "during the early part of the Tertiary" (Du Toit 1954). If the conglomerates in South Africa have an Early Tertiary age (younger than 63 Ma and older than 54.8 Ma, as indicated by Rabassa et al. 1995, 1996, 1997, 1998) and if these units may be correlated as suggested by Du Toit and Reed (1927), then the valleys in which they are found were eroded at least during the Late Cretaceous or earliest Paleogene. It is then possible that these valleys were formed following the well-known, Middle Cretaceous ("Inter-Senonian") uplifting of western Argentina and its corresponding influence in the cratonic areas (the authors are greatly indebted to the late Professor Edgardo Rolleri for these ideas). Later, the landscape dissected on both bedrock and the conglomerates was "in great part buried beneath the mantle of Pampean loess," whereas in South Africa, the general uplift has left it unexposed (Du Toit and Reed 1927). This is perhaps the first scientific explanation of the origin of the Pampa Interserrana buried surface as an extensive planation surface of Early Tertiary age. Du Toit and Reed (1927, p. 107) noted also that there is no evidence of marine Early to Middle Cretaceous along the Pampean coastal region, showing that this area was still united to South Africa. Contrarily, the Late Cretaceous sea penetrated very deeply along the western margin of the Buenos Aires Positive "as far inland as the Sierra Pintada." Therefore, the Buenos Aires Positive had been continuously above sea level perhaps at least since the latest Permian, and all of its landscape was formed subaerially since then.

The Buenos Aires Positive morphostructural unit (Fig. 10; Yrigoyen 1975), whose geomorphological features allow to consider it as of complex landscape, is composed of the following geological provinces: Tandilia or Sierras de Tandil, Ventania or Sierra de la Ventana, and the Llanura or Pampa Interserrana, in the sense of Rolleri (1975) (see Harrington 1980; Zárate and Rabassa 2005). Tandilia is defined as a system of block-shaped mountains, and it is composed of a discontinuous group of hills and low ranges, extending for more than 350 km, with elevations from 50 to 250 m above the surrounding sedimentary plain. The cross section of the system is clearly asymmetrical, with a very neat and abrupt northeastern margin (known as the "Costa de Heusser") and a very smooth southwestern one (named as the "Costa de Claraz"). The first one corresponds to a NW-SE fault escarpment representing the huge fault that bounds the Tandilia system and the Salado tectonic basin. The western border of the Buenos Aires Positive is bounded by the Colorado-Macachín basin. These basins were initiated during the Late Jurassic-Early Cretaceous (Demoulin et al. 2005). The orientation of the Salado basin is inherited from Late Precambrian structures. Both basins display thicknesses of Cretaceous and younger sediments up to 6–7 km. At the base of the Salado basin infill are continental deposits interlayered with volcanic and volcano-clastic rocks associated with the Early Cretaceous rifting phase. Above an angular unconformity, the next sequence corresponds to marine environments, with major phases of marine deposition taking place in the Late Cretaceous-Paleocene and the Miocene-Pliocene and a well-defined Eocene/Oligocene regression.

Large-scale geomorphological units, basically ancient planation surfaces, have been recognized during recent studies (Rabassa et al. 1995, 1998; Demoulin et al. 2005; see Figs. 11 and 12). Along the SE section (Mar del Plata-Balcarce), the summit of the ranges lies as a high surface between 200 and 250 m a.s.l. The relative elevation is variable between 100 and 150 m and rises progressively from the coast towards the hinterland. NW of the city of Balcarce, the surface continues as an erosion feature that cuts the Proterozoic granites and migmatites at elevations of 300–350 m a.s.l. At Tandil, the surface is preserved at the summits of numerous granitic hills. At the regional scale, the surface has a slightly undulating topography, with a small relief of a few tens of meters between 300 and 350 m a.s.l. contour lines. The morphological features at a smaller scale vary with bedrock lithology. In the segment bearing Cambrian-Ordovician quartzites, the mesas and tablelands dominate. They usually show solution features in quartz. In the area of Balcarce, Tandil, and Olavarría, the sections of weathered granitoids are frequent, sometimes with the presence of kaolinite. Other authors have interpreted them as hydrothermal products. However, the surface distribution, their relationship with the surface, and the associated geomorphological features suggest that they are related to continuous weathering profiles instead. NW of Balcarce, on the granitoid basement, the surface presents low elevation inselbergs, whose summits are covered by frequent granitic corestones, occasionally resting on a gruss bed. Around the city of Tandil, the surface is highly dissected, with a few remnants as inselbergs and tors covered by corestones (Fig. 13). The weathering products occur along the slopes. They are corestones with concentric weathering rinds in a matrix of gruss-like, weathered

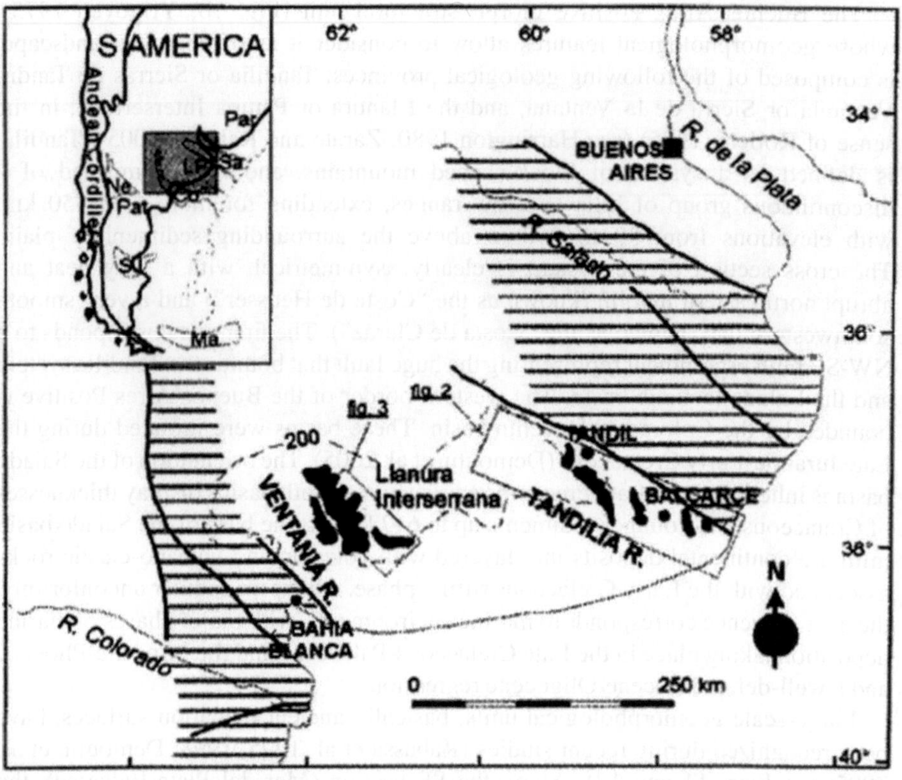

Fig. 11 Buenos Aires Positive tectonic element, including the Ventania and Tandilia ranges, and the intermediate, loess covered Llanura Interserrana (Inter-ranges plains). The Tandilia ranges are geologically and structurally related to similar areas in Uruguay and Southern Brazil. The hatched areas are sedimentary basins developed in Cretaceous times, due to the Gondwana rifting process, which isolated the positive area described. To the east, the Salado basin; to the west, the Colorado-Macachín basin (Modified from Demoulin et al. 2005)

rock (Cerro El Centinela) or fresh granite (Cerro La Movediza; Fig. 14). The "Piedra Movediza" (i.e., the "moving rock") was a rocking stone found on top of a huge inselberg, as part of a partially dismantled tor (see Linton 1955), near the city of Tandil (Fig. 15). Its nature and origin was surprisingly described by Estanislao Zeballos as early as 1876, and quite correctly interpreted it as the result of chemical weathering (Zeballos 1876), although the impact of lighting was also partially invoked. This author also suggested that there were many other rocking boulders at the top of neighboring hills (which was never confirmed later on by us or other authors), wisely observing the relationship between the finding of boulders as remnants on the hill tops. It should be noted that its genesis under humid subtropical climate (a climate type that never existed in the area after the Paleogene) was very early recognized by Walther (1924; in: Fairbridge 1968b). There is no evidence of marked kaolinization, suggesting that either other clays were formed in the

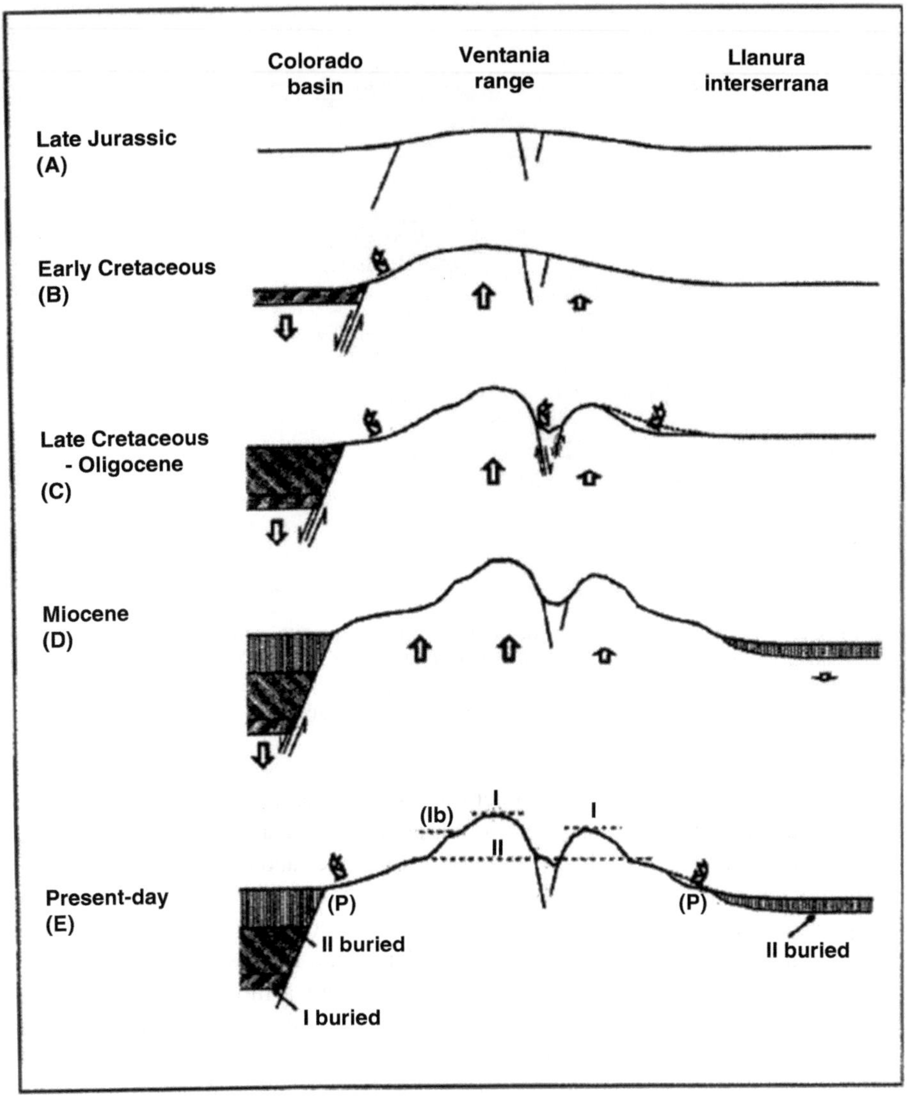

Fig. 12 Structural and geomorphological evolution of the Ventania ranges, with the adjacent Llanura Interserrana and the Colorado basin. The evolution started in Late Jurassic times. (*I*) Pre-Cretaceous landscape; (*Ib*) Intermediate inter-Cretaceous surface; (*II*) Paleogene planation surface; *P* Pliocene pediments. The hatched sections correspond to sediments coming from the Sierras that deposited in the adjacent Colorado basin, as the weathered materials were denudated (Modified from Demoulin et al. 2005)

weathering process or only the roots of the weathering front have been preserved with a very irregular distribution. In Balcarce, isolated remnants of a higher surface occur (Cerro El Sombrero, 420 m a.s.l.). In Tandil, there are several remnants of this

Fig. 13 Cerro La Movediza, Sierras de Tandil. Ancient granite corestones as remnants of paleoweathering under warm-wet tropical climates, at the base of the weathering front. At least one of these corestones evolved into a perched rocking stone (La Movediza, i.e., "the moving one") (We acknowledge this photograph taken by Marcelo Zárate 1998)

surface in the flat summits of the higher ranges (La Juanita, Alta de Vela, Cerro La Blanca), carved in the Proterozoic granitoids at 450–500 m a.s.l. Where no remnants of the weathered cover have been preserved, corestones and tors are the local-scale morphological features associated with these surfaces. The third morphological unit is represented by pediments and alluvial fans, surrounding the ranges, but they belong to the Late Cenozoic (Rabassa 1973; Zárate and Rabassa 2005).

The Ventania ranges, or Sierra de la Ventana, are a mountain system about 180 km by 60 km, composed of subparallel ranges. They reach between 400 and 700 m above the surrounding plains, with maximum elevation of 1,240 m a.s.l. The Curamalal, Bravard, and de la Ventana ranges have preserved remnants of two paleosurfaces located at different topographic heights. Keidel (1916) was the first author to describe and precisely map them. More recently, Pereyra and Ferrer (1995) and Pereyra (1996) pointed out that the higher planation surface of the northeastern ranges of Ventania was probably formed in the time between the Permian collision of Patagonia and central Argentina and the Late Jurassic opening of the Colorado basin. They recognized only one erosion surface and considered that the differences in elevation were due to differential erosion on various lithological types. Rabassa et al. (1995) suggested that the relict landscapes of the Sierras de la Ventana and Tandil, along with the morphology of other cratonic areas of Argentina, should be reinterpreted in a Gondwanic perspective. The sequence of uplifted surfaces would be then linked to the Late Jurassic-Early Cretaceous rifting of South America and

Fig. 14 Cerro La Movediza, Sierras de Tandil. Paleoweathering front with granite corestones in situ, surrounded by chemically altered granite transformed in clays (Photograph by J. Rabassa 1998)

Africa, whereas their intensely weathered bedrock might point to Mesozoic and Paleocene tropical conditions rather than to the cooler and drier Neogene.

Demoulin et al. (2005) indicated that the oldest paleosurface is forming the summit surface of the ranges, with elevations between 800 and 900 m a.s.l. The lower surface, less extensive, has been excavated at elevations of 600–700 m a.s.l., in the southern slope of Sierra de la Ventana. Another extensive surface is located in the intermountain basin of Las Vertientes, developed between 450 and 500 m a.s.l. In this surface, the Río Sauce Grande valley is excavated up to 100 m. In the northern part of the sierras, the surface is found at both sides of the longitudinal depression of Valle de las Grutas; it is also preserved in the outer piedmont of the eastern sierras (Las Tunas, Pillahuincó). Eastwards, the Sierra de Las Tunas presents a pattern of step-like surfaces similar to that of the western ranges, with the sole difference that

Fig. 15 Cerro La Movediza, Sierras de Tandil. The "Piedra Movediza," the rocking stone, as it was in place until the beginnings of the twentieth century. The acrobat, exposing his circus abilities, tries to show that he can keep himself as balanced as the "Piedra Modeviza" itself. This rocking stone was a very early touristic attraction for the area (Photograph kindly provided by Professor Hugo Nario, Tandil, from his collection)

the summit surface is at 700–750 m a.s.l. and the lower one at 550–600 m a.s.l. Along the outer segments of Ventania, pediments have been developed, connecting outwards with the Llanura or Pampa Interserrana plains.

The Ventania paleosurfaces are much less leveled than those of Tandilia, basically due to their lithological differences (Rabassa et al. 1998). Relict sedimentary deposits have been found in some of these surfaces, such as silicified breccias and conglomerates, which occur in different locations of the group of western sierras (Harrington 1936), closely associated to the 450 m a.s.l. surface (Keidel 1916; Zárate et al. 1995). Known as the "Conglomerado Rojo" (Harrington 1936) and later formally named as the Cerro Colorado Breccia (Andreis et al. 1971), these deposits are present in the Valle de las Grutas and both sides of Sierra de la Ventana. They are generally located in front of small valleys draining the sierras, and they are formed by fans or slope deposits, in a whitish to reddish, sandy matrix, cemented with silica and iron oxides and hydroxides (Andreis et al. 1971; Zárate et al. 1998). The hills with silicified deposits are covered by boulder-rich layers. Though these boulders were firstly considered as a separate unit (Las Malvinas Formation, De Francesco, 1971, in Fidalgo et al. 1975), the age and nature of this unit should be revised. The age of the Cerro Colorado Breccia, which traditionally had been considered as of Miocene age, has been reinterpreted, and a Cretaceous or Early Paleocene age has been proposed as the most likely (Zárate et al. 1995, 1998). It is possible that new studies may then reconsider the age of the Las Malvinas Formation, or part of it, which could be in fact a residual, lixiviated breccia of the Cerro Colorado Breccia and therefore part of the latter.

The Llanura or Pampa Interserrana includes the "Pampa Interserrana" as named by Frenguelli (1950) and also the piedmont areas of Ventania and the SW piedmont of Tandilia. It is composed of a plain with elevations of little above 200 m a.s.l. in the central portion in between both mountain systems and lowering gradually southwards towards the Atlantic Ocean. The landscape of the Pampa Interserrana is composed of Late Cenozoic continental deposits which are capped by a thick calcareous duricrusts. The dominant landforms are pediments and loess accumulation plains. These Late Cenozoic units overlie a vast planation surface, which is found a few tens of meters to a couple of hundred meters below the present surface. This ancient surface, clearly pre-Miocene at least, is probably related to the ancient surfaces described for both main ranges. It is in fact developed most on top of the Late Paleozoic deposits of the eastern Ventania ranges. However, further studies are needed to confirm the nature and age of this ancient planation surface, since it is totally buried by the Late Cenozoic sedimentary cover, with the exception of a few, small and isolated, highly weathered outcrops of Carboniferous and Permian rocks.

In summary, the Buenos Aires Positive region was never covered by the sea since at least the Middle to Late Triassic (Uliana and Biddle 1988), but perhaps even since the latest Permian. Thus, this is probably one of the most spectacular examples of long-term landscape evolution in Argentina, with a continued geomorphological history of over 230 Ma of subaerial geomorphological development. The large-scale morphological units and associated weathering products in the Tandilia and Ventania ranges have been described, together with two main planation surfaces, encountered at varying altitudes in different sectors of these ranges. The lower surface is characterized by the roots of kaolinized weathering profiles in Tandilia and by silicified conglomerates around Sierra de la Ventana. In an interpretative model linking the range morphogenesis to the tectono-sedimentary evolution of the bordering Salado and Colorado basins, it has been suggested that the main morphogenetic stages are actually related to the Late Jurassic-Early Cretaceous South Atlantic rifting and, afterwards, to the Miocene tectonic uplift (Demoulin et al. 2005). Thus, the uplifted surfaces appear much older than commonly believed, being respectively of pre-Cretaceous and Late Cretaceous-Paleogene (?) ages. The very low denudation rates that have been established (Demoulin et al. 2005), such as $\sim$4 m per million year, are explained by the very limited (if any) Meso-Cenozoic uplift experienced by the Buenos Aires ranges.

According to Rabassa et al. (1995), the main tectonic event affecting the Pampean long-term morphogenesis has been the Late Jurassic-Early Cretaceous rifting of the South Atlantic Ocean. Therefore, the unification of the pre-Cretaceous surfaces on both sides of the South Atlantic is acceptable since they were part of the same continent and were sharing a similar topography, as Du Toit and Reed (1927) indicated. However, the asymmetric character of the South Atlantic rifting induced different uplift amplitudes in South Africa than those in Southern Brazil-Uruguay-Argentina, opposing the South African Great Escarpment on one side to insignificant along-rift slopes in southern South America. Denudation was consequently much smaller in eastern Argentina than along the South African coast, further north in the rift zone. Finally, the poor preservation of kaolinized bedrock on the Argentine surfaces with respect to the extended weathering mantles covering African or Brazilian surfaces possibly points to a superimposed climatic influence on denudation effectiveness (Demoulin et al. 2005).

The San Rafael or Sierra Pintada Block in Mendoza

The Sierra Pintada or San Rafael Block is located in the central portion of the province of Mendoza, not far from the Andean Cordillera but with a distinctive geological history (see Zárate and Folguera 2014; Figs. 1 and 16). It shares some stratigraphical and structural characteristics with the Precordillera and the Cordillera Frontal, but it was never covered by the Mesozoic seas, and it has probably remained as a positive element in the landscape since the Triassic. Criado Roque (1972a) recognized the existence of two main paleolandscapes. The first one developed after the recession of the sea, in the Early Carboniferous, when a steep relief was generated. Later, after the deposition of Late Triassic volcano-clastic sediments, the entire Jurassic and Cretaceous period is not represented. This huge hiatus is characterized by ample "erosion and peneplanization" (Criado Roque 1972a, p. 295) of this geological unit. These surfaces were covered by Tertiary sediments, mostly related to the Andean uplift. No further information was provided for the nature and genesis of these landforms, but they may be correlated to the Gondwana paleosurfaces of the Sierras Pampeanas.

Later, Criado Roque and Ibáñez (1980) confirmed the existence of these two paleolandscapes, and the persistence of the block as a positive area throughout the Mesozoic, when it underwent intensive erosion. These authors identified also the youngest bedrock unit predating the surface as of Early Triassic age and then Late Cretaceous sediments on top of the paleolandscape, thus providing limiting ages for this ancient surface. These conditions are very similar to those identified in the Sierras Chicas de Córdoba (Carignano et al. 1999).

The San Rafael Block has shown no tectonic deformation whatsoever since the Triassic, in spite of its nearness to the Andean Cordillera, thus making it a highly stable, cratonic area of Western Argentina. This unit may be geologically connected with the Sierras de Lihuel Calel (see below).

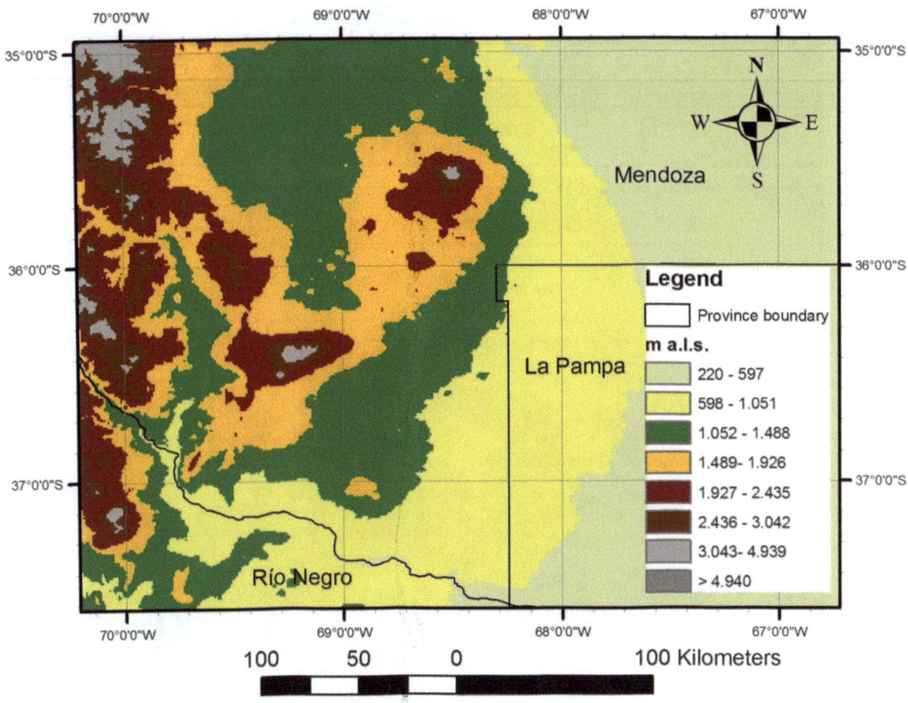

Fig. 16 San Rafael Massif or Sierra Pintada Block. Hypsographic map. The relief of this area is generally bounded by the contours of 450–1,800 m a.s.l. The highest peaks correspond to isolated mountain ranges and volcanoes, among which the Nevado volcano (3,810 m a.s.l.) stands in the southern part. In this sector the basalts have partially covered the ancient relief developed during the Paleozoic, Triassic, and Tertiary. The surrounding hills and a raised range were preserved as relicts

The Sierras de Lihuel Calel in La Pampa

The name Sierras de Lihuel Calel is a general term used to name a belt of remnants of barely outcropping ancient rocks in the province of La Pampa, between lat. 36° and 39° S. These remnants are bounded to the NE by the Macachín Basin and to the SW by the Neuquén Basin, both tectonic depressions related to the late Mesozoic rifting processes.

This unit has been considered by Criado Roque (1972b) and Criado Roque and Ibáñez (1980) as a continuation of the San Rafael Block towards the SE (Figs. 1 and 17). The petrology, structure, and geomorphology of this area have been thoroughly discussed by Zárate and Folguera (2014) and Aguilera et al. (2014a). As in the San Rafael Block, the stratigraphic sequence ends in the Early Triassic, with the tuffs and ignimbrites of the Lihuel Calel Fm. These rocks have been thoroughly eroded during the rest of the Mesozoic and, most likely, even the Early Tertiary. The scarce outcrops of these units are almost totally covered by Tertiary and Quaternary

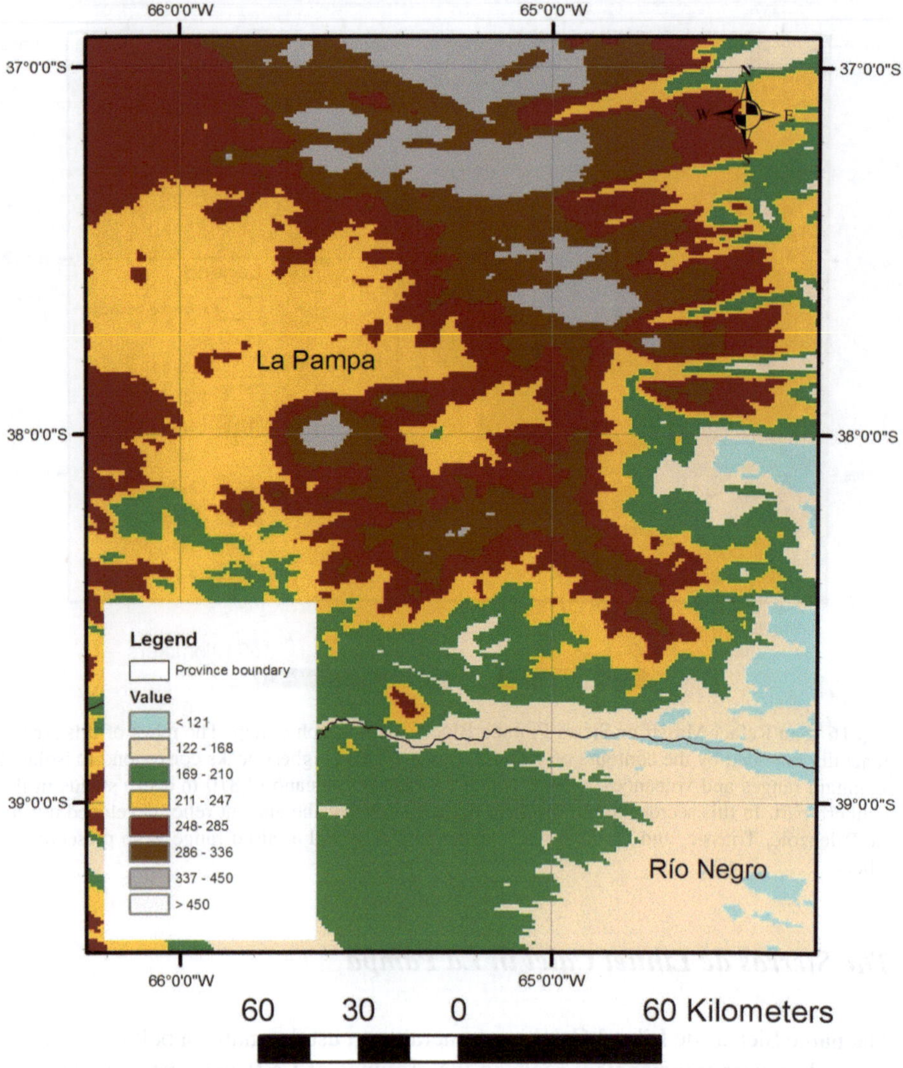

Fig. 17 The Sierras de Lihuel Calel in the province of La Pampa, Central Argentina. Hypsographic map. The ranges are oriented in a northwest-southeast direction, forming several chains with different orientation, covering a square area of approximately 15 km per side. The highest peak reaches 589 m a.s.l

sediments. Therefore, their relationship with the original paleosurfaces is not clear, though the visible portion of it may be related to inselbergs and other residual features. It is highly possible that the existing paleosurface is in fact part of the Gondwana or post-Gondwana surfaces, as recognized in the Sierras de Córdoba by Carignano et al. (1999).

This unit, together with the previous one, is still very poorly known from a paleogeomorphological point of view and clearly deserves further attention.

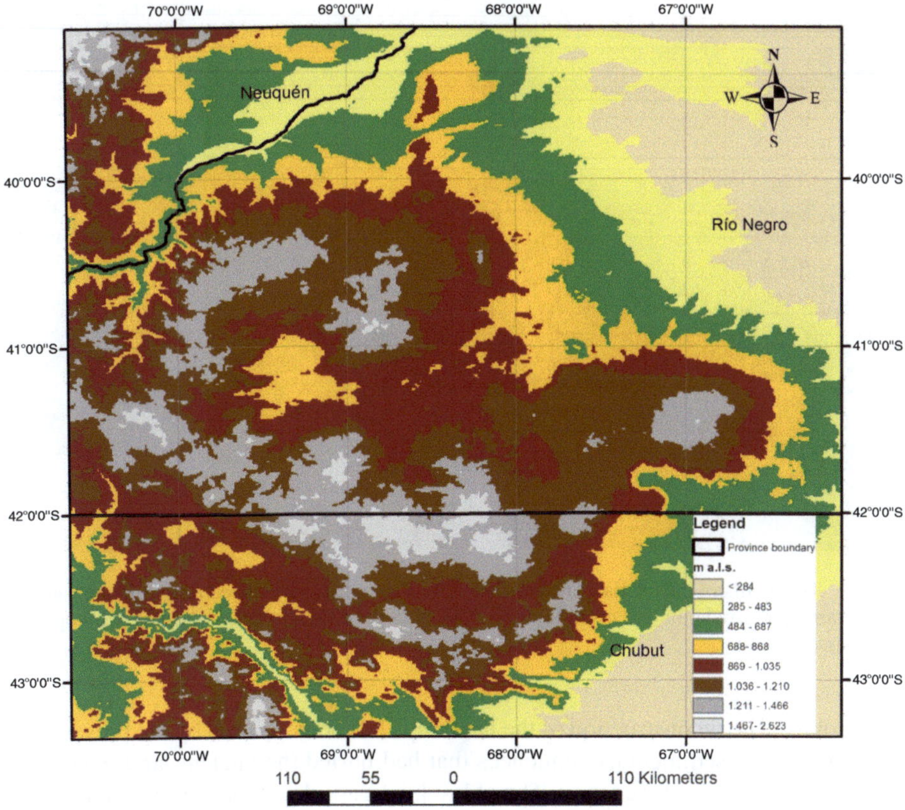

Fig. 18 The Northern Patagonian Massif. Hypsographic map

The Northern Patagonian Massif

The Northern Patagonian Massif is an isolated craton which occupies the northernmost portion of Patagonia (Figs. 1 and 18). It is bounded by Late Mesozoic basins like the Colorado-Negro basin to the north and the San Jorge Gulf basin to the south and the Northern Patagonian Andes, including the Ñirihuau-Ñorquinco basin, to the west, which were uplifted in the Middle to Late Tertiary. This massif has been a positive element of the crust at least since the Permian. The sea flooded its margins during various events in the Jurassic and Cretaceous. Moreover, occasional platform sea transgressions extended locally in several episodes during the Maastrichtian and the Tertiary.

Rabassa (1975, 1978a, b) described the existence of three superposed paleolandscapes in the western portion of the massif, successively developed in the Late Cretaceous-Paleocene, the Oligocene-Early Miocene, and the Pliocene. Some of these paleolandscapes have been exhumed and slightly modified during the

Fig. 19 The Northern Patagonian Massif. Comallo paleosurface, developed as a planation surface on metamorphic rocks of the crystalline basement, in pre-Late Cretaceous times (Photograph: J. Rabassa 2000)

Quaternary. The reconstruction of the landscape has been favored by the presence of several volcano-sedimentary sequences that had buried the ancient landscapes and preserved them from denudation. The oldest landscape developed over the Paleozoic Crystalline Basement, mostly granites, migmatites, and other metamorphic rocks, and Early Mesozoic units, of Triassic and Early Jurassic age. The western area has then drainage towards the Pacific Ocean, before the uplifting of the Patagonian Andes, a fact which had been already mentioned by Groeber (1929). The paleolandscape is a partially exhumed etchplain, formed under wet tropical climate, widely extended across the massif, with the development of inselbergs and tors (Figs. 19, 20 and 21). Humid tropical climates vanished forever from the massif since perhaps the Middle Cretaceous; therefore, the etchplains are preserved remnants of these old climates. This ancient landscape was fluvially eroded and buried by latest Cretaceous continental sediments and the Paleocene-Eocene volcano-sedimentary sequence of the Ventana Fm. (Rabassa 1975, 1978a, b; Fig. 22). Therefore, this landscape was generated between the Middle Jurassic and the Late Cretaceous, although it is possible that it is in fact a palimpsest of several geomorphological cycles, including at least one or two pediplains and a fluvial cycle during that period. Later on, when the Patagonian Andes was uplifted, the drainage direction switched towards the Atlantic Ocean in the Late Oligocene-Early Miocene, and by the Middle Miocene, the basic lines of the present drainage system had been established. A deep, fluvial valley landscape developed in this latter epoch, and it was filled by the ignimbrites and other pyroclastic and sedimentary deposits of

Fig. 20 Inselbergs in a planation surface developed on Permian granites, Pilcaniyeu, province of Río Negro (Photograph: J. Rabassa 2000)

the Collón Curá Fm., of undisputed Middle to Late Miocene age (ca. 15 Ma). The fluvial landscape was buried, and then Pliocene pediments, extending from the rising mountain front to the west, leveled most of the area.

The action of deep chemical weathering in ancient times over the massif is confirmed by the presence of clays and other residual materials accumulated in those basins marginal to the craton, which are of high economic interest. Domínguez (1988) has identified residual kaolins and other clay deposits in the sediments of the Challacó Formation, marine units of the Jurassic of the Central Neuquén Basin deposits. The source for these sediments was located SE and SW of the study area, that is, the western margin of the Northern Patagonian Massif. The clays were the consequence of intense weathering of the Choiyoy Volcanic Group, rhyolites and rhyodacites. Similar origin would have had comparable deposits of the lower Río Chubut valley. In this case, the source area would have been the southern margin of the Massif. Therefore, the entire massif was under similar deep weathering conditions. The kaolinite materials would have been then the result of intensive erosion of a former planation surface, probably of Late Triassic or Early Jurassic age.

The presence of ancient landscapes developed under tropical climates in the Northern Patagonian Massif has been recently recognized by Aguilera (2006), Aguilera and Rabassa (2010), Aguilera et al. (2010, 2014b), Martínez and Rabassa (2014), and Aragón et al. (2010, 2014). The huge planation surfaces described in these papers are one of the most extensive Gondwana Landscapes in Argentina, and they merit the continuation of these studies (Fig. 23).

Fig. 21 Granite corestones developed on Permian granites, Pilcaniyeu, province of Rio Negro. The corestones have piled up as the tors formed as part of a Cretaceous planation surface are dismantled by Cenozoic denudation (Photograph: J. Rabassa 2000)

The Deseado Massif

The Deseado Massif is an isolated craton which was named as "nesocraton" by Harrington (1962), who identified it as a long-term positive, stable, and undeformed structural unit. The Deseado Massif is located in Southern Patagonia, separated from the Northern Patagonian Massif by the San Jorge Gulf Basin and from the Southern Patagonian and Fuegian Andes by the Austral Basin (De Giusto et al. 1980; Figs. 1 and 24). Both these basins are tectonic depressions related to the Late Mesozoic rifting events. It is basically a very large tableland, with very small local relief, located at about 1,000 m a.s.l. Gondwana Landscapes in the Deseado Massif were briefly mentioned by Rabassa et al. (1996), who cited extensive planation

Fig. 22 Late Cretaceous sediments lying on planation surface developed on crystalline basement, metamorphic rocks, and granites, Comallo, province of Río Negro. The surface is possibly Late Jurassic-Early Cretaceous in age (Photograph: J. Rabassa 2000)

Fig. 23 Planation surface developed on granite, El Cuy-Los Menucos, province of Río Negro. Note the abundant granite corestones and the tors under dismantling processes (Photograph: J. Rabassa 2000)

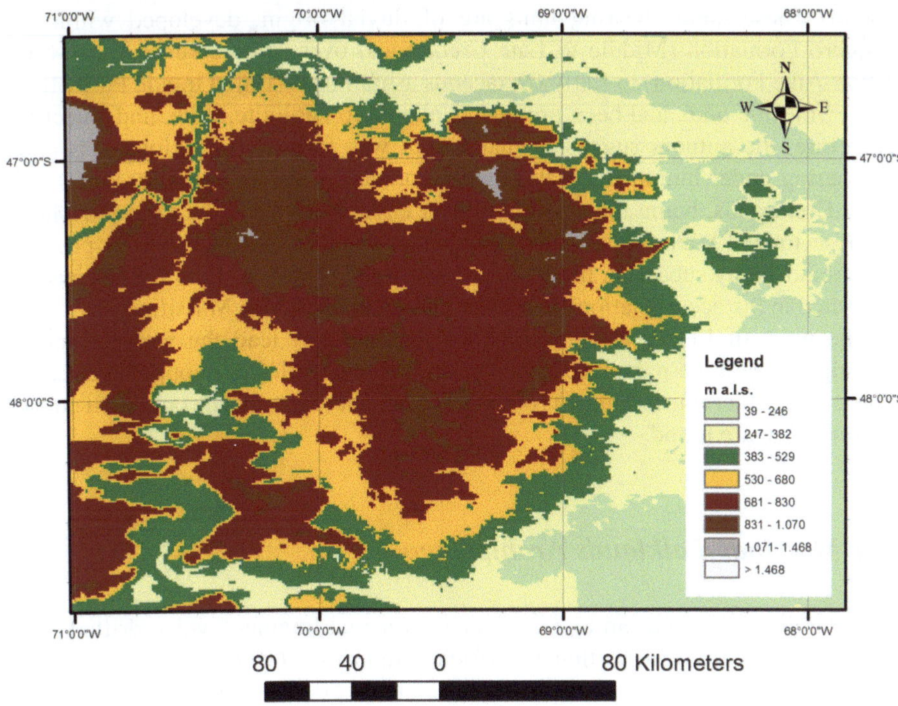

Fig. 24 The Deseado Massif, southern Patagonia. Hypsographic map

surfaces on the Middle to Late Jurassic volcanics and volcano-clastics of the Chon-Aike Formation and other units of the Bahía Laura Group. The region is presently under study, and a preliminary report has been presented by Bétard et al. (2014), based upon much modern methodology and scientific criteria. The Bajo Grande and Baqueró formations of latest Jurassic to Late Cretaceous age are deposited unconformably over this surface, but De Giusto et al. (1980) mentioned also the existence of an angular unconformity and a true paleolandscape in between the latter two units. The possibility of younger planation surfaces, eroded on top of the Cretaceous units should be investigated further. Earliest Tertiary marine deposits of the Salamanca Fm. are found on top of a planation surface, eroded on top of the Cretaceous continental sediments. Above the cited surface, only scattered outcrops of Tertiary volcanics are found, together with erosion remnants, such as inselbergs and tors, and younger volcanic cones. Although no detailed paleogeomorphological studies have been developed yet in this area, it is clear that at least three very well-developed Late Mesozoic paleosurfaces are present, making this unit exceptionally attractive for future investigations.

The existence of residual deposits due to deep chemical weathering in tropical climates was recognized by Cravero and Domínguez (1992, and papers cited there), who described kaolin deposits in Santa Cruz, at the southern portion of the Deseado

Massif. These kaolin-bearing units are of fluvial origin, developed within the Baqueró Formation (Middle to Late Cretaceous) over Middle Jurassic ash-flows (Chon-Aike Formation) to Early Cretaceous ashfall tuffs (Bajo Grande Formation). These volcanic units had been previously altered to kaolinite, illite, and smectites. Therefore, the authors suggested that kaolin was formed by regional, chemical weathering under humid tropical climates sometimes in the latest Jurassic to the Early Cretaceous, being these the dominant environments on the Deseado Massif in those times. Afterwards, when the climate changed, the weathering products were removed by subaerial, fluvial processes in the Middle to Late Cretaceous and deposited in the accumulation areas surrounding the Deseado Massif.

The work of Bétard et al. (2014) will undoubtedly lead the way for further investigations on the paleolandscapes of this remote, isolated, unpopulated region, where the southernmost remnants of Gondwana Landscapes in any continent outside of Antarctica are found.

The Malvinas-Falklands Archipelago

The Malvinas-Falklands archipelago is a continental fragment which drifted away from the southernmost portion of Africa. Clapperton (1993, p. 543) described smoothly rolling uplands, at an average height of 500–600 m a.s.l., with highest summits around 700 m a.s.l., closely adjusted to underlying structure and lithology, which reflect prolonged evolution by subaerial denudation, as expected in a former portion of Gondwanaland (Fig. 25). These topographic levels have been interpreted by Clapperton (1993) as remains of planation surfaces, but their age is still unknown, although they are clearly Triassic or younger.

Concluding Remarks

The analysis of the different studies on the Gondwana Landscapes performed in Argentina since the end of the nineteenth century suggests the following conclusions:

(a) During the Mesozoic and most of the Paleogene, the Sierras Pampeanas had long periods of stability that have favored the development of deep weathering and the formation of etchplains, under hyper-tropical climates.
(b) The Sierras Pampeanas show remnants of exhumed surfaces as well as relict landscapes that were never covered by sediments after their formation and have been denuded since the "hyper-tropical" climates changed towards drier and highly seasonal climates in the Tertiary.

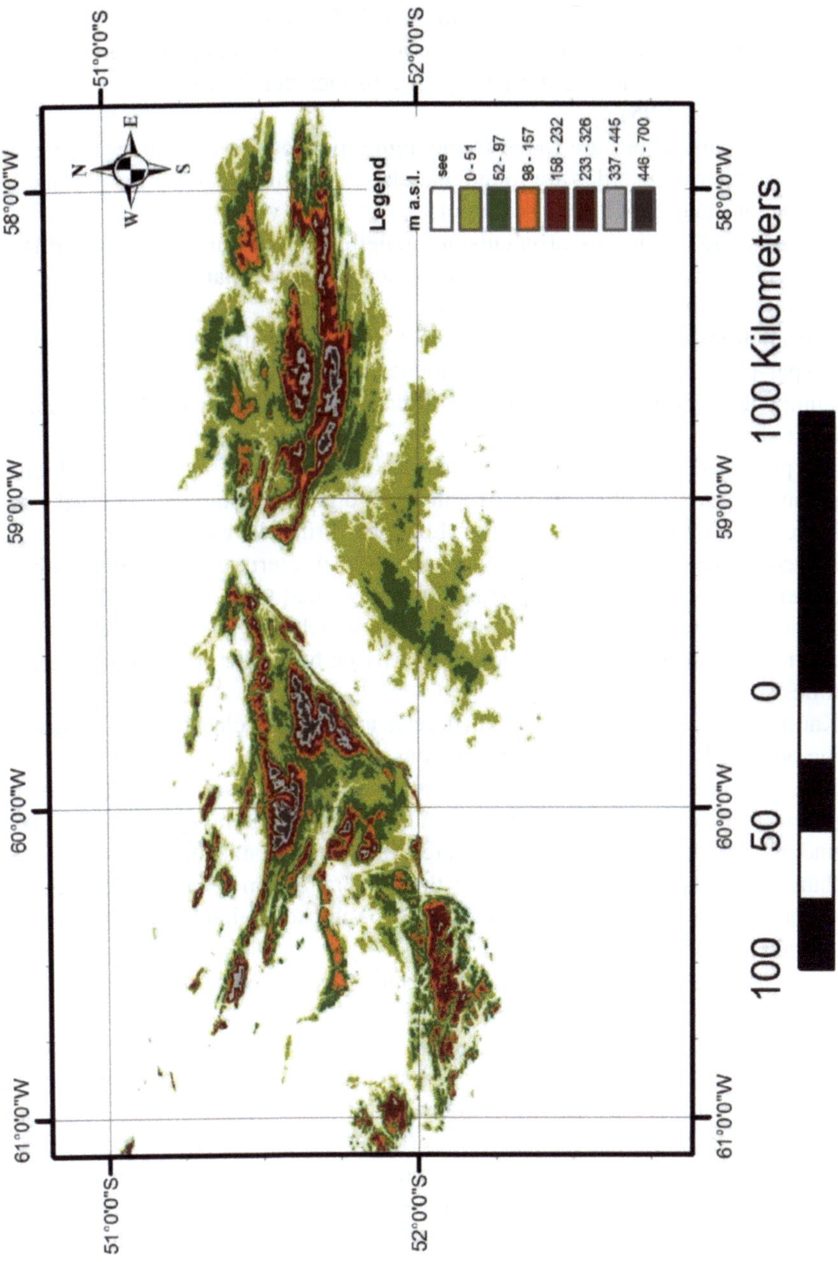

Fig. 25 The Malvinas-Falklands archipelago. Hypsographic map

(c) The relict and exhumed landforms may be related to the surrounding sedimentary record.
(d) The exhumed landscapes are associated with areas of the ranges that were covered by Cretaceous sediments, due to relief inversion due to the process of tectonic inversion following the impact of the Andean orogeny over extensional structures formed during the rifting (for further details on these topics, see Schmidt et al. 1995).
(e) The relict landscapes are found in the higher portions of the ranges where the deep weathering profiles are found and also in the sites that are surrounded by Cretaceous and/or Miocene sediments.
(f) Surfaces and erosion escarpments are genetically and chronologically different, with the oldest ones located in the inner and higher parts of the ranges and the younger ones along their margins.
(g) The relationship between deep chemical weathering and wet tropical or even "hyper-tropical" climate is undisputable, because the generated landforms cannot be generated under drier climatic conditions, as has been shown in many other areas of the world.
(h) The proposed "hyper-tropical" climate (Rabassa 2010, 2014) has no analogues among the present climates of Argentina, and perhaps not even on Earth, and its actual parameters and extension should be further investigated.
(i) Moreover, tropical climates disappeared from western, central, and southern Argentina by the Early Tertiary and never returned to these areas. Therefore, any erosion remnants or residual deposits which may be assigned to these environmental conditions should necessarily be formed in pre-Middle Tertiary times.
(j) Kaolin formation, ferricretes, and silcretes are features clearly associated with the observed etchplains in several regions of Argentina.
(k) Several types of planation surfaces have been recognized, particularly etchplains, formed under warm-wet, tropical conditions, and pediplains, developed under semiarid, seasonally wet climates. In general terms, etchplains are older than the pediplains. It is possible that several geomorphologic cycles are superposed in some areas, forming true palimpsests that should be further studied and unraveled.
(l) Some of the studied landscapes have been exposed to the atmosphere repeatedly for very long periods, in some cases continuously for more than 100 million years.
(m) The co-genetic relationship of the ancient landscapes of Argentina with South Africa, such as the Late Jurassic Gondwana surfaces, the Cretaceous etchplains, the Late Cretaceous and Early Tertiary pediplains, and the "High-Level Gravels" of the Sierras de la Ventana and the Southern Cape, has been strongly supported by field studies. However, the coevolution of the landscapes in both continents was interrupted by the opening of the South Atlantic Ocean. Later on in the Cenozoic, when the ocean at these latitudes became wider, a change in ocean circulation took place and, therefore, profound modifications in the distribution of climate belts occurred.

(n) The pediplains are of a much more local nature and should be analyzed within the framework of regional landscape evolution and may not be developed simultaneously in all areas of Argentina where these Gondwana Landscapes are found.
(o) Inselbergs, bornhardts, perched rocks, rocking boulders, weathering profiles, weathering fronts, granite landforms, granite micro-landforms, quartzite disintegration, and other features and landforms have been very useful in the identification of ancient landscapes.
(p) Old landscapes are indicators of the persistence of most of the studied cratonic regions as positive elements since at least the Jurassic and perhaps even since the Permian.
(q) The presently observed mountains in the Ventania landscape would be of Late Mesozoic age, which were then slowly and barely modified after denudation in the Tertiary, forming an outstanding example of long-term landscape evolution in a cratonic area.
(r) The observed paleolandscapes are very useful in understanding the morphogenetic and environmental conditions of cratonic areas, and the associated marginal sedimentary basins, during the Late Mesozoic and the Paleogene.
(s) These landscapes are an undisputable proof of conditions of long-term stability of both tectonics and climate in southern South America.
(t) The studied landscapes were never again covered by transgressive seas, nor covered by large thicknesses of continental deposits, with the exception of Late Cretaceous-Tertiary shallow seas and Cretaceous and Tertiary pyroclastics.
(u) The existence of a Devonian (?)-Early Carboniferous (?) etchplain, later eroded by the Carboniferous-Permian Gondwana glaciations and exhumed during the Late Cenozoic, has been proven for at least the Sierra de Olta, La Rioja, but it may be also found in other areas in the western Sierras Pampeanas.
(v) Gondwana Landscapes are an important part of the surviving relief in the cratonic areas of Argentina and should be treated appropriately.

Thornbury (1954) presented his personal view about ancient landscapes. In his well-known "fundamental concepts of Geomorphology," he stated that most of the Earth topography has an age that is not older than the Pleistocene, whereas the topography older than the Tertiary is almost negligible. He added also that "if they exist" (showing his profound doubts about these landscapes), it is likely that they are in fact exhumed landforms, which do not correspond to features which would have been exposed to degradation through vast periods of time. Finally, he maintained that "99% of the present Earth surface has an age later than the Middle Miocene."

The evidence presented here for the frequent existence of these ancient landforms in the cratonic regions of Argentina contradicts Thornbury's (1954) ideas. We need to revise the geomorphology of the cratonic areas of Argentina (and other South American countries, such as Brazil, Venezuela, Guyana, Uruguay, and Paraguay) in terms of long-term landscape evolution. This revision should be conducted from a Gondwana viewpoint instead of the present dominant Andean vision.

Acknowledgments To CONICET, the Universidad Nacional de Córdoba, and the Universidad Nacional de Tierra del Fuego, for supporting our work in these fields along many years.

To the National Geographic Society (USA), with many thanks for their generous grant for the study of the Argentine Paleolandscapes, between 1993 and 1998.

To Professor Cliff Ollier (University of Western Australia, Perth), for his strong support for our work in Argentina during his two visits to Argentina in 1999, 2004, and 2012 and a most careful reviewing of an earlier version of this manuscript.

To our Argentine colleagues and friends, Professors Marcelo Zárate and Carlos Costa, for their participation in different fieldwork excursions to different areas of Argentina and sharing with us their ideas, knowledge, and experience.

Professor Nat Rutter (University of Alberta, Edmonton, Canada) shared with us a field trip in 2006 to our study areas in Córdoba and La Rioja provinces and provided firm and serious advice and criticism. Additionally, a field trip to these same areas was developed with professors and colleagues of the Department of Geological Sciences, University of Wisconsin-Madison in 2005, and special research work was performed with Dr. Betty Socha as part of her PhD dissertation (see Socha et al. 2014).

Finally, JR would like to express his deepest thanks to the memory of the late Professor Edgardo Rolleri, who supported our studies with his valuable experience, comments, and opinions and his helpful encouragement about the scientific problems mentioned in this chapter, during unforgettable talks in his house of La Plata along his last years.

References

Ab'Sáber AN (1969) Participação das superficies aplainadas nas paisagens do Rio Grande do Sul. Universidade de São Paulo, Instituto de Geografia 11:1–17

Aguilera EY (2006) Identificación y distribución de distintas superficies geomorfológicas de escaso relieve local, por medio de sensores remotos en el área del Macizo Norpatagónico, provincia de Río Negro. Unpublished doctoral thesis, Facultad de Ciencias Naturales y Museo, Universidad Nacional de La Plata, 285 pp

Aguilera EY, Rabassa J (2010) Origin of the Northern Patagonian Massif regional paleosurface. Geociências 29(4):467–478. São Paulo, UNESP, Brazil

Aguilera EY, Aragón E, Carretero S (2010) The paleosurface on the Paso del Sapo volcanic rocks, Chubut, Argentina. Geociências 29(4):479–486. São Paulo, UNESP, Brazil

Aguilera EY, Sato AM, Llambías E, Tickyj H (2014a) Erosion surface and granitic morphology in the Sierras de Lihuel Calel, province of La Pampa, Argentina. In: Rabassa J, Ollier C (eds) Gondwana landscapes in southern South America. Springer, Dordrecht, pp 393–422

Aguilera EY, Rabassa J, Aragón E (2014b) Paleo-landscapes of the Northern Patagonian Massif, Argentina. In: Rabassa J, Ollier C (eds) Gondwana landscapes in southern South America. Springer, Dordrecht, pp 423–445

Andreazzini MJ, Degiovanni SB (2014) Geomorphology of paleosurfaces in the Sierras de Comechingones, Central Pampean Ranges, Argentina. In: Rabassa J, Ollier C (eds) Gondwana landscapes in southern South America. Springer, Dordrecht, pp 305–330

Andreis R, Spalletti L, Mazzoni M (1971) Sedimentología de la Brecha Cerro Colorado, Sierras Australes de la provincia de Buenos Aires. Reunión sobre Geología Sierras Australes Bonaerenses, Provincia de Buenos Aires (La Plata), pp 65–96

Aragón E, Aguilera EY, Cavarozzi C, Ribot A (2010) The North Patagonian Altiplano and the Somón Curá basaltic plateau. Geociências 29(4):527–532. São Paulo, UNESP, Brazil

Aragón E, Aguilera EY, Cavarozzi CE, Ribot A (2014) The exhumation of the Northern Patagonian Massif Gondwana planation surface due to uprising during the Oligocene. In: Rabassa J, Ollier C (eds) Gondwana landscapes in southern South America. Springer, Dordrecht, pp 517–525

Beder R (1916) Estudios geológicos e hidrológicos en los alrededores de Villa Dolores. Boletín de la Dirección Nacional de Minería Geología e Hidrología 14:1–26
Bétard F, Peulvast J-P, Rabassa J, Aguilera EY (2014) Meso-Cenozoic paleotopographies and paleolandscapes in the Deseado Massif (Santa Cruz province, Argentina). In: Rabassa J, Ollier C (eds) Gondwana landscapes in southern South America. Springer, Dordrecht, pp 477–501
Bodenbender G (1890) La cuenca del valle del Río Primero en Córdoba. Descripción geológica del valle del Río Primero desde la Sierra de Córdoba hasta la Mar Chiquita. Boletín de la Academia Nacional de Ciencias, Córdoba- Argentina 12:5–55
Bodenbender G (1895) Devónico y Gondwana en la República Argentina. Boletín de la Academia Nacional de Ciencias, Córdoba-Argentina 15(2–3):201–252
Bodenbender G (1905) La Sierra de Córdoba. Constitución geológica y productos minerales de aplicación. Anales del Ministerio de Agricultura de la Nación, Sección Geología 1(2):1–146. Buenos Aires
Bodenbender G (1907) Petrografía: Meláfidos, basaltos y andesitas en la sierra de Córdoba. Diabasas, porfiritos augíticos y meláfidos entre la Sierra de Córdoba y la Precordillera. Anales del Ministerio de Agricultura de la Nación, Sección Geología 2(3):1–34, Buenos Aires
Bodenbender G (1911) Constitución geológica de la parte meridional de La Rioja y regiones limítrofes. Boletín de la Academia Nacional de Ciencias, Córdoba-Argentina 19:1–220
Brackebush L (1879) Informe. Boletín de la Academia Nacional Ciencias Córdoba-Argentina 3:251–262
Brackebush L (1880) Informe sobre la marcha del Museo Mineralógico de la Universidad Nacional de Córdoba. Boletín de la Academia Nacional de Ciencias, Córdoba- Argentina 3:135–163
Brackebush L (1891) Mapa Geológico del Interior de la República Argentina. Escala 1:1.000.000, Gotha. Academia Nacional de Ciencias, Córdoba-Argentina
Carignano C, Cioccale M (1997) Landscapes antiquity of the Central Sierras Pampeanas (Argentina): geomorphic evolution since the Gondwana times. In: 4th international conference on geomorphology, abstracts, Supplementi di Geografia Fisica e Dinamica Quaternaria, Supplement 3, 1, 104. Torino, Italy
Carignano C, Cioccale M (2008) Geomorfología de la Sierra Norte-Ambargasta, provincias de Córdoba y Santiago del Estero, Argentina. XVII Congreso Geológico Argentino, Actas, pp 1189–1190. San Salvador de Jujuy
Carignano C, Cioccale M, Rabassa J (1999) Landscape antiquity of the Central Eastern Sierras Pampeanas (Argentina): Geomorphological evolution since Gondwanic times. Zeitschrift für Geomorphologie, NF, Supplement Band 118:245–268. Berlin-Stuttgart
Clapperton CM (1993) Quaternary geology and geomorphology of South America. Elsevier, Amsterdam, 779 pp
Costa CH, Giaccardi AD, González Díaz EF (1999) Palaeolandsurfaces and neotectonic analysis in the Southern Sierras Pampeanas. In: Smith BJ, Whalley WB, Warke PA (eds) Uplift, erosion and stability: perspectives on long-term landscape development. Geological Society, London, Special Publication, 162, pp 229–238. London
Cravero MF, Domínguez EA (1992) Kaolin deposits in the Lower Cretaceous Baqueró Formation (Santa Cruz province, Patagonia, Argentina). J S Am Earth Sci 6(4):223–235
Criado Roque P (1972a) Bloque de San Rafael. In: Leanza AL (ed) Geología Regional Argentina. Academia Nacional de Ciencias, Córdoba, pp 288–296
Criado Roque P (1972b) Cinturón Móvil Mendocino-Pampeano. In: Leanza AL (ed) Geología Regional Argentina. Academia Nacional de Ciencias, Córdoba, pp 297–304
Criado Roque P, Ibáñez G (1980) Provincia Geológica Sanrafaelino-Pampeana. In: Turner JC (coordinator) Geología Regional Argentina, vol 1. Academia Nacional de Ciencias, Córdoba, pp 837–869
Darwin C (1876) Geological observations on the volcanic islands and parts of South America visited during the voyage of H.M.S. Beagle, 2nd edn. Smith, Elder and Co., London
Davis WM (1899) The geographical cycle. Geogr J 14:481–504
Davis WM (1909) Geographical essay. Ginn and Company, New York, 777 p

De Giusto JM, Di Persia CA, Pezzi E (1980) Nesocratón del Deseado. In: Turner JC (coordinator) Geología Regional Argentina, vol 2. Academia Nacional de Ciencias, Córdoba, pp 1389–1430

Demoulin A, Zárate M, Rabassa J (2005) Long-term landscape development: a perspective from the southern Buenos Aires ranges of east central Argentina. J S Am Earth Sci 19:193–204

Domínguez E (1988) Posición estratigráfica y mineralogía de las capas arcillosas de la Formación Challacó (Jurásico) en Rincón del Aguila, provincia del Neuquén. Revista Asociación Geológica Argentina 43(3):343–355

Du Toit AL (1937) Our wandering continents. An hypothesis of continental drifting. Oliver & Boyd, London

Du Toit AL (1954) The geology of South Africa. Oliver & Boyd, Edinburgh, 611 pp

Du Toit AL, Reed FRC (1927) A geological comparison of South America with South Africa. Carnegie Institution, Washington, DC

Fairbridge RW (ed) (1968a) The encyclopedia of geomorphology. Reinhold Book Corporation, New York

Fairbridge RW (1968b) Perched Rock/Boulder. In: Fairbridge RW (ed) Encyclopedia of geomorphology. Reinhold, New York, pp 823–824

Fidalgo F, De Francesco F, Pascual R (1975) Geología superficial de la llanura bonaerense. In: Relatorio de la Geología de la provincia de Buenos Aires, 6º Congreso Geológico Argentino. Bahía Blanca, pp 103–138

Frenguelli J (1921) Los terrenos de la costa Atlántica en los alrededores de Miramar (Provincia de Buenos Aires) y sus correlaciones. Boletín de la Academia Nacional de Ciencias, Córdoba-Argentina 24:325–485

Frenguelli J (1950) Rasgos generales de la morfología y la geología de la provincia de Buenos Aires. LEMIT, serie II, 33. La Plata

Gerth E (1914) Constitución geológica, hidrogeológica y minerales de aplicación de la provincia de San Luis. Anales de la Dirección General de Minas, Geología e Hidrología 10(2):1–64

Gerth E (1927) El Morro de San Luis. Un Cráter de elevación. Boletín de la Academia Nacional de Ciencias de Córdoba 30:171–180

González Díaz EF (1974) Superficies de erosión (abanicos rocosos) exhumadas en el flanco occidental de la Sierra de Ambato al sur de la Quebrada de La Cébila (La Rioja). Rev Asoc Geol Argent 29:5–22

González Díaz EF (1981) Geomorfología. In: Yrigoyen M (ed) Geología de la Provincia de San Luis. Relatorio 8º Congreso Geológico Argentino. Buenos Aires, pp 193–236

Groeber PF (1929) Líneas fundamentales de la geología del Neuquén. Dirección General de Minas, Geología e Hidrología, Publicación Nº 58, Buenos Aires

Gross W (1948) Cuadro morfológico del Valle de Punilla. Revista Facultad de Ciencias Exactas, Físicas y Naturales, Universidad Nacional de Córdoba 11:1–14

Harrington HJ (1936) El Conglomerado Rojo de las sierras australes de la provincia de Buenos Aires y sus relaciones con el relieve de montaña. Obra del Cincuentenario, Museo de la Plata 2:145–184

Harrington HJ (1962) Paleogeographic development of South America. Bull Am Assoc Petroleum Geologists 46(10):1773–1814

Harrington HJ (1980) Sierras Australes de la provincia de Buenos Aires. Segundo Simposio de Geología Regional Argentina 2:967–984

Jordan TE, Zeitler P, Ramos V, Gleadow AJW (1989) Thermochronometric data on the development of the basement peneplain in the Sierras Pampeanas, Argentina. J S Am Earth Sci 2(3):207–222

Keidel J (1916) La geología de las Sierras de la Provincia de Buenos Aires y sus relaciones con las montañas de Sud África y los Andes. Anales Ministerio Agricultura Argentina, 11, 3. Buenos Aires

Keidel J (1922) Sobre la influencia de los cambios climatéricos cuatemarios en el relieve de la región seca de los Andes centrales y septentrionales de la Argentina. Boletín de la Dirección General de Minas, Geología e Hidrología 5:3–19

King LC (1950) The study of the World's plainlands: a new approach to geomorphology. Geol Soc Lond Quart J 106(1):101–131
King LC (1953) Canons of landscape evolution. Geol Soc Am Bull 64:721–752
King LC (1956) A geomorphological comparison between Brazil and South Africa. Quart J Geol Soc Lond 112:445–474
King LC (1963) South African scenery, 3rd edn. Oliver & Boyd, Edinburgh, 308 pp
King LC (1967) The morphology of the earth, 2nd edn. Oliver & Boyd, Edinburgh
Kröhling D, Brunetto E, Galina G, Zalazar MC, Iriondo M (2014) Planation surfaces on the Paraná basaltic plateau, South America. In: Rabassa J, Ollier C (eds) Gondwana landscapes in southern South America. Springer, Dordrecht, pp 247–303
Linton DL (1955) The problem of tors. Geogr J 121:470
Martínez OA, Rabassa J (2014) The Rhyolitic Plateau of the Marifil Formation (Jurassic): a Gondwana paleosurface in the southeastern portion of the Northern Patagonia Massif. In: Rabassa J, Ollier C (eds) Gondwana landscapes in southern South America. Springer, Dordrecht, pp 447–476
Ollier C (2014a) Some principles in the study of planation surfaces. In: Rabassa J, Ollier C (eds) Gondwana landscapes in southern South America. Springer, Dordrecht, pp 47–59
Ollier C (2014b) Planation surfaces of the Gondwana continents: synthesis and problems. In: Rabassa J, Ollier C (eds) Gondwana landscapes in southern South America. Springer, Dordrecht, pp 61–96
Partridge T, Maud R (1987) Geomorphic evolution of Southern Africa since the Mesozoic. S Afr J Geol 90:179–208
Penck W (1914) La estructura geológica del Valle de Fiambalá y las cordilleras limítrofes al norte de Tinogasta. Ministerio de Agricultura de la Nación, Boletín 17, Buenos Aires
Penck W (1920) Der Südrand der Puna de Atacama (NWArgentinien). Ein Beitrag zur Kenntnis des Andinen Gebirgstypus und der Frage der Gebirgsbildung. Der Abhandlungen der Sachsischen Akademie der Wissenschaften, Leipzig 1:3–420
Penck W (1924) Die Morphologische Analyse. Geographisches Abh., 2, 2. Stuttgart, J. Engelhron, 283 pp
Pereyra FX (1996) Caracterización geomórfica y evolución del paisaje de un sector de las Sierras Australes, provincia de Buenos Aires. Revista Asociación Geológica Argentina 51(3):248–260. Buenos Aires
Pereyra FX, Ferrer J (1995) Geomorfología del flanco nororiental de las Sierras Australes, provincia de Buenos Aires. IV Jornadas Geológicas y Geofísicas Bonaerenses, Actas 1:239–247. Junín
Rabassa J (1973) Geología Superficial en la hoja "Sierras de Tandil", provincia de Buenos Aires. LEMIT, Anales, Serie II 240:115–160. La Plata
Rabassa J (1975) Geología de la región de Pilcaniyeu-Comallo. Unpublished doctoral thesis, Facultad de Ciencias Naturales y Museo, Universidad Nacional de La Plata, and Publicaciones del Departamento de Recursos Naturales y Energía, Fundación Bariloche 17, pp 1–128, San Carlos de Bariloche
Rabassa J (1978a) Estratigrafía de la región Pilcaniyeu-Comallo, provincia de Río Negro. 7° Congreso Geológico Argentino, Actas 1:731–746, Neuquén
Rabassa J (1978b) Paleorrelieves cenozoicos en la región Pilcaniyeu-Comallo, provincia de Río Negro, Argentina. 7° Congreso Geológico Argentino, Actas 2:77–87, Neuquén
Rabassa J (2010) Gondwanic Paleolandscapes: long-term landscape evolution, genesis, distribution, and age. Geociências 29(4):541–570. São Paulo, UNESP, Brazil
Rabassa J (2014) Some concepts on Gondwana landscapes: long-term landscape evolution, genesis, distribution and age. In: Rabassa J, Ollier C (eds) Gondwana landscapes in southern South America. Springer, Dordrecht, pp 9–46
Rabassa J, Zárate MA, Camilión MC, Partridge T, Maud R (1995) Relieves relictuales de Tandilia y Ventania. IV Jornadas Geológicas Bonaerenses. Actas:249–256. Noviembre de 1995, La Plata

Rabassa J, Zárate M, Cioccale M, Carignano C, Partridge TC, Maud R (1996) Paisajes relictuales de edad Gondwánica en áreas cratónicas de Argentina. 13º Congreso Geológico Argentino and 3rd. Congreso de Exploración de Hidrocarburos, Actas 4, 219. Buenos Aires

Rabassa J, Zárate M, Partridge TC, Maud R, Cioccale M, Carignano C (1997) Gondwanic relict paleolandscapes in cratonic areas of Argentina. In: 4th international conference on geomorphology, abstracts, Supplementi di Geografia Fisica e Dinamica Quaternaria, Supplement 3, 1, 321. Torino, Italy

Rabassa J, Zárate MA, Demoulin A, Camilión MC, Partridge T, Maud R (1998) Superficies de erosión y morfogénesis de Tandilia y Ventania. V Jornadas Geológicas Bonaerenses, Actas:111–118. Mar del Plata

Rabassa J, Carignano C, Cioccale M (2010) Gondwana paleosurfaces in Argentina: an introduction. Geociências 29(4):439–466. São Paulo, UNESP

Rassmuss E (1916) Rasgos geológicos generales de las Sierras Pampeanas. Dirección General de Minas, Geología e Hidrología, Boletín 13 B, Buenos Aires

Renne PR, Ernesto M, Pacca IG, Coe RS, Glen JM, Prévot M, Perrin M (1992) The age of Paraná Flood Volcanism, rifting of Gondwanaland, and the Jurassic-Cretaceous boundary. Science 258:975–979

Rimann E (1926) Estudio geológico de la Sierra Chica entre Ongamira y Dolores. Córdoba-Argentina Boletín de la Academia Nacional de Ciencias 23:9–191

Rolleri EO (1975) Provincias geológicas bonaerenses. In: 6º Congreso Geológico Argentina, Relatorio de la Geología de la Provincia de Buenos Aires, pp 29–54

Rovereto G (1911) Studi di geomorfologia argentina. I. La Sierra di Córdova. Tipogr. Della Pace e Cuggiani, 19 p

Sayago JM (1983) Geomorfología de la Sierra de Ancasti. La Geología de la Sierra de Ancasti. Münstersche Forschungen zur Geologie und Paläontologie, Münster, pp 265–284

Sayago JM (1986) Morfoclimas y paleoformas en la evolución geomorfológica de la Sierra de Ancasti (Cuaternario Superior) Catamarca. Revista de la Asociación Geológica Argentina 41(1–2):155–164

Schlagintweit O (1954) Una interesante dislocación en Potrero de Garay. Revista de la Asociación Geológica Argentina 9(3):135–154

Schmidt CJ, Astini R, Costa C, Gardini C, Kraemer P (1995) Cretaceous rifting, alluvial fan sedimentation and Neogene inversion, Southern Sierras Pampeanas, Argentina. In: Tankard AJ, Suárez RS, Welsink H (eds) Petroleum Basins of South America. American Association of Petroleum Geologists, Memoir 62, pp 341–358

Schmieder O (1921) Apuntes geomorfológicos de la Sierra Grande de Córdoba. Córdoba-Argentina. Boletín de la Academia Nacional de Ciencias 25:181–204

Socha BJ, Carignano C, Rabassa J, Mickelson DM (2006) Sedimentological record of Carboniferous mountain glaciation in an exhumed paleovalley, eastern Paganzo Basin, La Rioja Province, Argentina. Geological Society of America (Philadelphia Annual Meeting), Abstracts, 38, 62

Socha BJ, Carignano C, Rabassa J, Mickelson DM (2014) Gondwana glacial paleolandscape, diamictite record of Carboniferous valley glaciation, and preglacial remnants of an ancient weathering front in northwestern Argentina. In: Rabassa J, Ollier C (eds) Gondwana landscapes in southern South America. Springer, Dordrecht, pp 331–363

Stelzner A (1885) Comunicaciones al Profesor H.B. Geinitz: Primera carta. In: Homenaje al 150 aniversario de la Independencia Argentina. Boletín de la Academia Nacional de Ciencias de Córdoba 45:115–120

Thornbury WC (1954) Principles of geomorphology. Wiley, New York, 618 pp

Uliana MA, Biddle KT (1988) Mesozoic-Cenozoic paleogeographic and geodynamic evolution of southern South America. Revista Brasileira de Geociências 18(2):172–190

Walther J (1912) Über transgressionen der oberen "Gondwana Formation" in Sud-brasilien und Uruguay. Centralbl.Mineralogie und Palaeontologie, pp 385–403

Yrigoyen MR (1975) Geología del subsuelo y plataforma continental. In: 6º Congreso Geológico Argentino, Bahía Blanca-Buenos Aires, Relatorio Geología de la provincia de Buenos Aires, pp 139–168

Zárate M, Folguera A (2014) Planation surfaces of Central Western Argentina. In: Rabassa J, Ollier C (eds) Gondwana landscapes in southern South America. Springer, Dordrecht, pp 365–392

Zárate M, Rabassa J (2005) Geomorfología de la Provincia de Buenos Aires. In: 16º Congreso Geológico Argentino, Buenos Aires. Relatorio de la Geología y Recursos Minerales de la Provincia de Buenos Aires, pp 119–138

Zárate M, Rabassa J, Partridge TC, Maud R (1995) La Brecha Cerro Colorado, es Miocena? IV Jornadas Geológicas Bonaerenses, Actas 1:159–168

Zárate M, Rabassa J, Maud R, Partridge TC (1998) La silicificación de la Brecha Cerro Colorado: clasificación, génesis e implicancias ambientales. V Jornadas Geológicas Bonaerenses, Actas 1:165–173

Zeballos ES (1876) Estudio geológico sobre la provincia de Buenos Aires. Anales de la Sociedad Científica Argentina 1876:5–66

Planation Surfaces on the Paraná Basaltic Plateau, South America

Daniela Kröhling, Ernesto Brunetto, Gabriel Galina, M. Cecilia Zalazar, and Martín Iriondo

Abstract SRTM data constituted a good resource for morphometrical analyses of the extensive Paraná Basaltic Plateau (southern Brazil and northeastern Argentina, South America). The plateau is a stepped system of high-level surfaces separated by escarpments and with incised fluvial valleys, mainly belonging to the Upper Paraná and the Upper Uruguay River basins. Palaeosurface remnants of such basins preserve attributes that have been identified in digital elevation models. Generation of hypsometric curves in six representative tributary basins of the Uruguay River basin and also in one tributary basin of the Paraná River in the region permitted to identify, classify and map the main Cenozoic planation surfaces of the basaltic plateau. Other morphometric parameters such as longitudinal profiles and isobase lines were produced also to delimit such surfaces. Field geomorphological analyses were performed, also including the description of toposequences. Three groups of hypothetical hypsometric curves are deduced from proximal value sets for predicted base levels. Each mapped surface was considered between the minimum height for correlative surfaces in all of the subbasins and the minimum height of the next higher surface. Using that approach, which is based on the clustering from the modelled base level in the river mouth of the subbasins, three main palaeosurfaces were defined in northeastern Argentina. Complementarily, three intermediate or

D. Kröhling (✉) • M. Iriondo
CONICET, Facultad de Ingeniería y Ciencias Hídricas (FICH), Universidad Nacional del Litoral, CC. 217, 3000 Santa Fe, Argentina
e-mail: dkrohli@gmail.com

E. Brunetto
CICYTTP-CONICET, 3105 Diamante, Argentina

G. Galina
FICH, Universidad Nacional del Litoral, CC. 217, 3000 Santa Fe, Argentina

M.C. Zalazar
Facultad de Ingeniería Química, Universidad Nacional del Litoral, 3000 Santa Fe, Argentina

J. Rabassa and C. Ollier (eds.), *Gondwana Landscapes in southern South America*, Springer Earth System Sciences, DOI 10.1007/978-94-007-7702-6_10,

secondary surfaces also were identified based on morphometric analyses, taking into account that small flat remnants at the same level suggest that they could be remnants of a formerly extensive plain.

Remnants of the Late Cretaceous-Palaeogene King's *Sul-American Surface* are well developed in the headwaters of the Uruguay River (on the subbasins of the Pelotas and Canoas Rivers, Brazil), located immediately westwards from the Serra Geral erosion scarp. *Sul-Americana 1* (1,277–1,080 m a.s.l.) and *Sul-Americana 2* (1,080–880 m a.s.l.) *Surfaces* in this work are equivalent to the King's *Sul-Americana Pediplain,* represented by an extended low-relief surface. If the Upper Uruguay River basin corresponded to the lower segment of such surface, a predicted base-level value would be estimated around 880 m a.s.l. The higher relict surface of the plateau in Argentina corresponds to a secondary planation surface named *Bernardo de Irigoyen Erosion Surface* (675–880 m a.s.l.). The King's *Velhas Surface* (Palaeogene) is mainly preserved on water divides as remnants in the Upper Uruguay basin in Brazil, and it is represented by a landscape of gentle and well-rounded hills. It is correlated in this chapter with *Aristóbulo del Valle Erosion Surface*, which in the Argentine Misiones province comprises the plane-top watershed between the Paraná and Uruguay basins (490–675 m a.s.l.). Because of the proximity of these large fluvial collectors, pediplanation and pedimentation processes leaved a narrow remnant that locally are restricted to ridges of planed tops. The general landscape of this surface is represented by rounded hills and flat-bottomed fluvial valleys. A predicted base-level value for this surface could be estimated around 310–425 m a.s.l. The main lower surface identified in this morphometric analysis corresponds to the *Apóstoles Surface* (110–165 m a.s.l.), which would be generated by the Plio-Pleistocene King's *Paraguaçu cycle* of deep fluvial incision on the *Velhas/Aristóbulo Surface*. The base level for the lower surface remnant of this erosive cycle is above 100 m a.s.l. The parallel retreat of slopes along the major valleys to their coalescence generated an extended erosion surface in southeastern Upper Uruguay basin, named *Apóstoles Pediplain*. It is formed by convex, low and long hills with very gentle and simple slopes, alternating with wide and very shallow fluvial valleys that have consumed much of the interfluves bearing remnants of the Velhas cycle.

The formation of the Late Cretaceous–Cenozoic regional erosion surfaces mapped in a relatively distant area of the passive margin is mainly related to fluvial erosion and slope retreat process. Most fluvial erosion was concentrated in tributary valleys of the Paraná and Uruguay Rivers destroying the flat basaltic surface and generating a new erosion surface below. Field data also indicate that geomorphic processes like fluvial, surface wash and mass movement (rock fall, debris flow) were important. Uplift and rejuvenation of valleys before the new surface can be graded to the lower level were required. For the planation surfaces identified in the study region, the base level to which the erosion process developed was the Uruguay River or the Paraná River. The steeper longitudinal profile observed in a tributary of the Paraná, compared to the closest Uruguay tributary in northeastern Argentina, points out that erosion is more active on the Paraná system. Planation surfaces exert control on the active headward erosion by valley development and rock landslides.

Keywords Gondwana • Argentina • Misiones • Parana basalts • Planation surfaces

Introduction to Landscape Evolution

Between the theories that explain long-term regional landscape evolution, the concepts of *planation surfaces* or *erosion surfaces* are exposed in relation to nearly flat topographical surfaces that are over longer distances and that cut across geological structures and different rock types. Planation surfaces have much significance in morphotectonic studies, being used to infer magnitudes of surface uplift and subsidence histories (Ollier and Pain 2000). Fluvial erosion is the commonest type of erosion in planation surfaces. *Plateaus* indicate vertical uplift of a former low-lying plain. In such upland areas, there may be identified extensive and elevated plains that sometimes bear signs of an erosional origin followed by subsequent dissection (planation surfaces) separated by escarpments.

There are at least a few points persistently disputed, including the nature of process, or processes, leading to planation, the meaning of planation surfaces in long-term landscape evolution, and the possibility of producing and maintaining flat surfaces without recourse to erosional processes acting over protracted time spans. Several mechanisms have been proposed to account for the origin of near-level surfaces. Accordingly, specific types of planation surfaces are distinguished (Migón 2004). The term *palaeoplain*, proposed by Ollier (1991) and related to descriptive definition linked to an old plain without genetic connotations, would be preferable in some cases.

A landscape would suffer repeated "cycles of erosion" involving an initial rapid uplift followed by a slow wearing down. In the known Davisian model (Davis 1899, 1909; working in the humid–temperate areas of northeast USA), it was assumed that uplift takes place quickly and that uplift and planation take place alternately. The land is gradually worn down by the operation of geomorphological processes, without further complications being produced by tectonic movements. Furthermore, slopes within landscapes decline in steepness through time until an extensive flat region was produced close to the base level. Isolated residual hills (monadnocks) along the divides between the broad-floored rivers might rise above it, limited by gentle slopes; otherwise, slope gradients should be very low and drainage lines should not be incised. This surface of low relief formed through erosion (hillslope processes, mainly fluvial and gravity driven; weathering also could be important) over protracted spans of time being at the penultimate stage of development is a *peneplain*. Also, Davis suggested the term "pastplain" to describe a peneplain which has been uplifted and now shows the initial stage of dissection.

From Davis, many theories to explain the landscape evolution have been expounded. A variation on Davis's scheme was offered by Penck (1953), who worked in a much more tectonically active area of the Andes, considering that in many landscapes, uplift and denudation occur at the same time. The continuous and gradual interaction of tectonic processes and denudation leads to a different model of landscape evolution, in which the evolution of individual slopes is thought to determine the evolution of the entire landscape. Three main slope forms evolve with different combinations of uplift and denudation rates. This theory couples evolution of the hillslope to conditions at the slope base. Penck's complicated analysis predicted both slope recession and slope decline, a result that extends Davis's simple idea of slope downwearing and consequent relief reduction (see details in Ollier and Pain 2000; Goudie 2004; Migón 2004; Huggett 2007; among others).

An alternative model of slope development by parallel retreat over long distances leading to *pediplanation* was put forward by King (1953, 1957). Initially working in the semiarid climate of South Africa, King (1953) proposed a morphologically intermediate conceptual model, in which the steep and rocky upper slopes retreat parallel to themselves and undergo replacement by a pediment, which is a lower gradient surface with some regolith cover. Pediment-forming processes develop under surface and subsurface weathering and fluvial processes. Pediments gradually coalesce and eventually consume any residual mountain masses reducing the relief to form a regional, ever-growing planation surface that is a *pediplain*. The combination of rocky and regolith-covered slope elements is significant in this model, while the tectonic assumptions return to the stability advocated by Davis (Kirkby 2004). According to King (1953), the most important processes are sheet wash on pediments at the foot of scarps and mass failure and gully erosion on the steep slopes. Scarps retreat virtually as fast as knickpoints move upstreams, so that the distribution of successive erosion cycles bears no relationship to the drainage pattern whatsoever. He argued that retreating scarps resulted in isostatic uplift, generally in the form of large-scale warping, initiating a new cycle of scarp retreat.

According to Huggett (2007), the Kingian model of repeated pediplanation envisaged long-term cycles, too. Indeed, remnants of erosion surfaces can be identified globally (King 1983). However, following Huggett (2007), King's views are not widely accepted and have been challenged (e.g. Summerfield 1984; Ollier 1991, 1993). A popular theme is that the landscape alternates between stages of relative stability and stages of relative instability. An early version of this idea, which still has considerable currency, is the theory of biostasy and rhexistasy (Erhart 1955). According to this model, landscape change involves long periods of biostasy, associated with stability and soil development, broken in upon by short periods of rhexistasy, marked by instability and soil erosion. During biostasy (the normal state), streams carry small loads of suspended sediments but large loads of dissolved materials are removed to the oceans, leaving deep ferrallitic soils and weathering profiles on the continents. Rhexistatic conditions are triggered by bouts of tectonic uplift and lead to the stripping of the ferrallitic soil cover, the headward erosion of streams and the flushing out of residual quartz during entrenchment (Huggett 2007).

Following the synthesis of Ollier and Pain (2000) and Migon (2004), the principal difference between the two models resides in the way of slope development. *Peneplains* are formed by downcutting of rivers followed by valley widening and slope decline (downwearing), whereas *pediplains* are produced by slope (scarp) retreat (backwearing) after a period of downcutting producing a new plain near the new base level. Another difference is that pediplains are not necessarily graded towards base level and they may form stepped landscapes and develop simultaneously at different altitudes. So the pediplain is a diachronic erosion surface getting younger towards the scarp. Landscapes may consist of pediplain remnants of various ages (a stepped landscape). The escarpment retreat rates are depending on lithological resistance, tectonic activity and the intensity of erosion (largely controlled by climate).

Etchplanation was initially seen (Wayland 1933; Büdel 1957) as a specific variant of peneplanation, applicable to low latitudes (tropical and subtropical environments), characterised by deep chemical weathering. It is shown later that *etchplains* form in the subsurface, through rock decay which is intense enough to overcome local differences in rock resistance against weathering. Büdel (1957) proposed the theory of *double planation*, where the surface erosion lowers the land surface at the same rate that chemical etching lowers the weathering front. It envisages land surfaces of low relief being maintained during prolonged, slow uplift by the continuous lowering of double planation surfaces—the wash surface and the basal weathering surface (Büdel 1957, 1982; Thomas 1966). Subsequent removal of weathering products would expose the planar etched topography, which now forms an etchplain. Recently, it was deduced that long-term etching leads to diversification rather than to planation of relief. Many low-latitude surfaces of low relief are cut across the weathering mantle, but the hidden topography of the weathering front is much more varied (Migón 2004).

Taking into account the concepts of Fairbridge and Finkl (1980), it is likely that plains of long geomorphic history have been shaped by various processes, changing over time; hence, they would be polygenetic surfaces rather than any monogenetic peneplains, pediplains or etchplains. In this sense, Fairbridge and Finkl (1980) proposed to return to *peneplains*, preferring to give the term a nongenetic meaning, simply to describe a near flat surface regardless of its origin, setting and evolutionary stage. This point is accepted by Twidale (1983), who describes peneplains as rolling or undulating surfaces of low relief, without referring to their position in respect to base level. Moreover, he argues that there is no means to decipher the mode of past slope evolution leading to the present-day peneplain; hence, the argument focused on the backwearing-or-downwearing issue is by and large pointless (Migón 2004).

The Paraná Basaltic Plateau

The identification of palaeosurfaces at a regional scale is useful on the analysis of the landscape evolution of basaltic plateaus, on the application of morphogenetic analyses of passive margins and cratonic areas and particularly on the studies of fluvial processes related to bedrock streams. Research on the Southern Brazilian Plateau has a great potential in the geomorphological study of erosion surfaces of global significance.

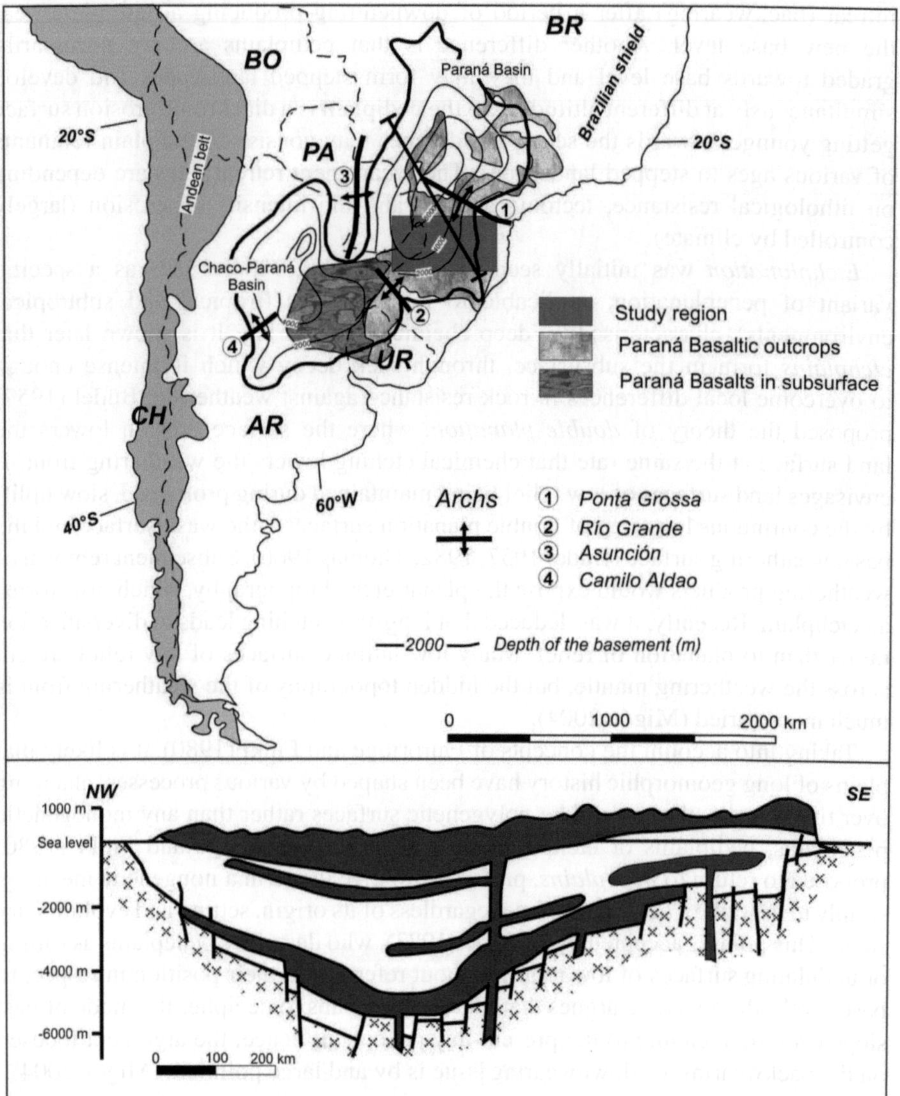

Fig. 1 Location map of Palaeozoic Paraná and Chaco–Paraná intracratonic sag basins in central South America (Taken from Milani and Zalán 1999) and Cretaceous–Palaeogene tholeiitic flood basalts (Taken from Lagorio 2008). In the *lower part* of the figure is a NW–SE cross section of the Paraná Basin

The regional geomorphology is characterised by lowlands in the central-western sector of the Paraná basin correlative to the main depositional centre, a plateau in the central-eastern basin and range relief in eastern border, near the Atlantic coast (Fig. 1, modified from Milani and Zalán 1999 and Lagorio 2008). This landscape has been explained as a result of a complex geological history.

Mesozoic volcanic rocks of the Paraná basin in southeastern South America and the Etendeka region of northwestern Namibia form one of the world's largest flood volcanic provinces, occurred in the 135–130 Ma. time span (Milner et al. 1995; Peate et al. 1992; Torsvik et al. 2009). The Paraná–Etendeka continental flood basalt province is related to the opening of the South Atlantic Ocean and the break-up of western Gondwana in the Early Cretaceous (magnetic anomaly M0 in 120.6 Ma., Torsvik et al. 2009). The Paraná basin covers more than 1.2 million km^2 in southern Brazil, northeastern Argentina, eastern Paraguay and northern Uruguay, with maximum registered thickness of lava flow sequences of 1,700 m (Almeida 1986). Much of the Serra Geral Formation is composed of tholeiitic basalts and basaltic andesites, but in the southern part of the basin, adjacent to the continental margin, silicic rocks form the upper part of the volcanic succession (Melfi et al. 1988). It forms the extensive Paraná Basaltic Plateau, a stepped system of high-level (with maximum elevations of 1,400 m a.s.l.) low-relief surfaces separated by escarpments and with incised fluvial valleys, mainly belonging to the Upper Paraná and the Upper Uruguay River basins.

Basaltic lava flows cover the northeastern domain of the Chaco–Paraná basin, mainly in Argentina, and two-thirds of the Paraná intracratonic basin, which has a NNE–SSW-trending elliptical shape. The main depositional centre of the basin exceeds 7,000 m in thickness. It was proposed the initial subsidence mechanism for the basin was crustal stretching resulting in formation of rift basins, overlaid by intracratonic sags whose accommodation space was subsequently generated by thermal subsidence. Additional subsidence could have been achieved either by crustal stretching (Milani and Zalán 1999). Structurally, the southern portion of the Paraná basin is limited by the N–S-trending Asunción arch in the west and the NW–SE-trending Rio Grande arch in the south. Depositional centres, in central-eastern Paraná basin, are crossed by the NW–SE-trending Ponta Grossa arc. During the Mesozoic, the eastern flank of the Paraná basin had suffered a deep deformation by the South Atlantic rifting and opening of the ocean (Milani and Zalán 1999).

The pure stretching model and simple shear and multiple shear models were invoked to explain the formation of the Brazilian–Angolan margins during the Mesozoic (Aslanian et al. 2009). The presence of passive margins limited by mountain systems and their conjugated borders characterised by wide and gentle coastal plains and extended well-developed rift basins (Ramos 1996) led to explain such an asymmetry by a simple shear stretching model (Lister et al. 1991). The eastern border of Brazil, characterised by range systems, would correspond to an upper plate passive margin, while the African margin, north of the Walvis Ridge–Rio Grande Rise, which acted as a transfer zone, would represent the lower plate passive margin (Ramos 1996). Asymmetrical conjugated margins are formed by simple shear (Wernicke 1985) or multiple detachment surfaces (Lister et al. 1986), generally controlled by old suture zones. Particularly, simple shear and multiple shear models were utilised to interpret the stretching in the Sergipe–Alagoas and in Gabon basins, in sections essentially based on gravity data (Ussami et al. 1986 and Castro 1987, respectively). Later, uplift and erosion were linked to Late Cretaceous or Cenozoic reactivation by hot-spot activity or contractional stresses (Cobbold et al.

2001). Additionally, high local stresses are operating in the eastern craton margin due to density contrast in the continental/oceanic lithospheric transition zone or flexural effects by sedimentary loads in the continental shelf (Assumpçao 1992, 1998; Assumpçao et al. 2004).

King (1956) interpreted the landscape evolution of the eastern South American passive margin by the interplay between long-term denudation and regional uplift. Differences in tectonic styles along that margin are now explained by Bezerra et al. (2008) and others as different responses of previous structures to post-rift stress fields. Also, post-break-up denudation does not present a similar pattern in Africa and South America; denudation rates varied significantly along both passive margins. For many authors, King's model has played a significant role in explaining tectonic activity (or their absence) and climate as both driving mechanisms of morphogenesis in passive margins (Bezerra et al. 2008).

Geomorphological and geodynamic studies of the development of these margins assume either that the elevated plateau represents a pre- or syn-break-up surface that underwent very little erosion or that it remained high despite significant post-rift denudation (Watts 2001, among others). Japsen et al. (2012a) indicated instead that the present elevation of marine incursions in post-rift sediments across the interior of Brazil testifies to post-rift subsidence followed by significant uplift. A synthesis of geological data, landscape analysis and palaeothermal and palaeoburial data by Japsen et al. (2012b) reveals that the present landscape of the Atlantic margin of northeastern Brazil is due to multiple post-rift episodes of burial, uplift and exhumation. The reconstructed four-stage history is the following: (1) after Early Cretaceous break-up, the margin underwent kilometre-scale burial beneath a thick sedimentary cover; (2) uplift episodes in the Campanian and Eocene led to almost complete removal of these deposits by river erosion to base level and to formation of a peneplain (the higher surface) with a deeply weathered surface; (3) the resulting large-scale, low-relief erosion surface was deeply weathered and finally reburied at the Oligocene–Miocene transition; and (4) Miocene uplift and erosion produced a new peneplain (the lower surface) by river incision below the uplifted and reexposed higher surface. Minor uplift in the Quaternary led to the incision below the lower surface and to the formation of the coastal plain.

A reconstruction of the deformational history of the obliquely rifted margin of southeastern Brazil was performed by Cobbold et al. (2001) and Meisling et al. (2001). They interpreted and documented the Late Cretaceous and Cenozoic reactivation of older structures, attributing them to the combined effects of far-field stresses and hot-spot activity. Two episodes of post-rift exhumation of mounting ranges during the Late Cretaceous and Eocene were reported, based on fission-track data (Cobbold et al. 2001). Late Cretaceous–Palaeogene alkaline intrusions were attributed to the Trindade hot spot and interpreted as emplaced along reactivated strike-slip faults and transfer zones. The structures originated in a transtensional stress regimen became inverted during Neogene transpression. According to those data, it is possible to recognise two episodes of thermal reactivation associated to extensional regimen in the Late Cretaceous (90–80 Ma.) and the Palaeogene (50–40 Ma.) and a phase of Neogene uplift (25–0 Ma.) linked to an Andean compressive context (Gallagher et al. 1994; Cobbold et al. 2001; Meisling et al. 2001) in an on-shore area, relatively distant of the passive margin.

First studies on palaeosurfaces in Brazil begun with the Davisian model (Davis 1899) of concordant mountaintops for defining levels of erosion. Stepped systems of planation surfaces as a result of cyclic evolution related to regional uplift were investigated by King (1956, 1967) in Brazil. King (1956) extended the parallel retreat of hillslopes as a model for hillslope development of southern Brazil. He identified the *Gondwana Surface* (*Upper Cretaceous*), as a result of generalised planation. It is considered as the oldest surface of Brazil, occupying the highest position and exposed from the Late Cretaceous to the present times. All of the ancient surfaces were exhumed or are fossil. Following King (1956), the general landscape of Brazil is represented by a vast pediplain, generated by denudation between the Lower Cretaceous and the Middle Tertiary. It was later reduced to a plateau dissected by polycyclic stream incision that excavated valleys almost on the entire surface, with coincident heights on the interfluves.

The *Sul-American erosion cycle* of King (*Lower Tertiary*) is represented by remnants of the *Sul-American Pediplain*. According to King (1956), the erosion cycles, after the post-Cretaceous uplift, are marked by the opening of valleys that destroyed a large part of that pediplain. Locally these subsequent cycles included an advanced stage of planation. The *Velhas erosive cycle* of the *Upper Tertiary* is represented by valleys that dissected the Sul-American Surface. Locally, this cycle includes a phase of generalised planation that generated a *Velhas Surface*, represented by an undulating landscape, with pediments. Velhas surface penetrated along the main rivers, destroying most of the Sul-American Surface. The typical landscape is a surface on which the two cycles are involved. The Pleistocene erosion cycle (*Paraguaçu cycle*) is characterised by the opening of fluvial valleys (King 1956). Following this author, both planation surfaces juxtaposed at different levels are separated by relatively steep erosion scarps. The scarps are undergoing recession by headward erosion. This is evident for King (1956) from the existence of pediments, also showing the way in which the erosion surface was generated (by development and coalescence of isolated pediments). The stepped landscape then shows that it evolved from the regression of scarps and pedimentation. In King's (1956) opinion, each planation surface remains virtually unchanged until it is destroyed by the scarp of the subsequent erosion cycle.

The ages of the main erosion surfaces studied recently in northeastern Brazil (Japsen et al. 2012b) are coincident with the ages suggested by, e.g. King (1967) and Bigarella (1975). Two regional peneplains in northeastern Brazil identified by Japsen et al. (2012b), also taking into account the results of Peulvast et al. (2008) and corresponding to a lower surface (*Paraguaçu/Sertaneja*) and remnants of a higher surface (*Sul-Americana*), were formed by erosion during Neogene and Late Cretaceous–Palaeogene times, respectively. Japsen et al. (2012b) concluded that fluvial erosion down to the base level is the key to understanding the formation of extensive, low-relief surfaces mapped in northeastern Brazil. The process of valley widening by river erosion eventually results in a peneplain. Valley incision below such a surface is evidence for a lowering of base level (uplift of the landmass or drop in sea level), with subsequent formation of new valley floors grading to sea level and thus possibly resulting in the formation of a new peneplain. The height difference between the valley floor and the overlying surface therefore indicates the amount of uplift or fall in base level. Those authors inferred in northeastern Brazil a recent change in base level that fits well with the estimate of Bigarella (1975), who suggested that this change took place since ca. 2 Ma. Both of the regional erosion surfaces developed during the Cenozoic, consequently sea level, are the most likely base level to which the surfaces graded (Japsen et al. 2012b).

Ollier (2003), Ollier and Pain (2000) and Ollier (2012) claim that margins with a marginal and asymmetrical swell (or bulge parallel to the continental edge and the coast, with the steeper slope to the coast) are the dominant type of passive margin, as the Brazilian margin. The marginal swell (or "rim highlands", by L. King) appears to have formed after a period of planation and it is separated from the Atlantic coast by a great escarpment. In southern Brazil this escarpment is in general known as Serra Geral, a local name that refers as how it is seen from the coast (like a mountain range). It is represented by a steep erosional scarp up to 1,000 m high and hundreds of km long, locally controlled by the lithology of the eroded formations but without structural control. It runs roughly parallel to the coast (with a NNW–SSW orientation, locally named "Aparados da Serra") and separates the interior basaltic plateau from the contrasting Atlantic coastal plain, including topographically irregular areas indicating that the old plateau surface has been destroyed and leaving also shallow valleys. The mechanism of parallel slope retreat seems to be responsible for creating the escarpments, following drainage lines to the coast which first makes great gorges and then valley widening, and coalescence creates a continuous escarpment. The top of the escarpment is the drainage divide between coastal and inland drainage; all major streams like the tributaries of the Uruguay River basin drain westwards on the gentle inland slope to the Paraná. The creation of the escarpment was related to the formation of a new continental margin with the break-up of the Gondwana and the opening of the South Atlantic Sea.

Taking into account the analysis of Bezerra et al. (2008), the foundation of King's model is based on the following premises: (1) Continental areas uplift episodically, almost synchronously, and uniformly; (2) parallel-scarp retreat prevails over downwearing, which is regarded as a minimum (this retreat propagates inland as denudation makes progress); and (3) knickpoints retreat inland over long distances along rivers and slopes. The approach used in this model favours the spatial correlation of planation surfaces, usually widely scattered, which would be based on their elevation and relative position in the landscape. This correlation would eventually lead to an estimated age of erosion surfaces, especially if a link could be established between the erosion surface and a sedimentary deposit of known age.

Other investigations in the Southern Brazilian Plateau proposed four periods of morphogenesis related to uniform uplift, following planation process proposed by King. Bigarella and Becker (1975) suggested important internal morphological differentiations, which are associated to the influence of local factors in the regional evolution. As Maack (1947) observed in southern Brazil, some of the planation surfaces, identified in the plateaus, are in fact structural surfaces. Structural platforms are common in basaltic terrains because of the flatness of the top surface of the flow. In case of multiple flows, denudation may expose top parts of each major lava flow and produce stepped topography, reminiscent of a generation of planation surfaces of various ages. In fact, all benches have structural foundation and may all have similar ages (Migón 2004).

Almeida (1956) generated the first geomorphological studies on the Paraná Basaltic Plateau (Paraná and Santa Catarina states, Brazil). According to Bigarella and Becker (1975), erosion processes on the Brazilian Plateau gave origin to a stepped morphology with development of numerous "cuestas", but a tabular morphology predominates. Waterfalls are structurally controlled by sub-horizontal rocks. The large pediplanation surface (Palaeogene) is known in Brazil as *Pd3* (Ab'Sáber 1969; Bigarella et al. 1965). *Pd2* and *Pd1* pediplanes (Bigarella and Andrade 1965) are equivalent to the surfaces generated by the Velhas and Paraguaçu cycles of King, respectively.

Japsen et al. (2012b) and Vasconcelos and Carmo (2008) found a distinct relation between Ar–Ar ages for lateritic weathering profiles and three main landscape types in Brazil. Weathering ages are 70–30 Ma. for the main plateau surfaces (*Sul-Americana*), 15–6 Ma. for intermediate surfaces (*Velhas*) and less than 4 Ma. for lower surfaces (*Paraguaçu*). Taking into account the major gaps in the record, Japsen et al. (2012b) concluded for northeastern Brazil that the reburial of the higher surface by an Oligocene–Miocene cover provides a straightforward explanation for the gap in the weathering record between 30 and 15 Ma., prior to the removal of this cover during the Miocene uplift event (event also suggested by King 1967 and Cobbold et al. 2001, among others). The presence of an extensive Cretaceous cover prior to Campanian uplift and exhumation may similarly explain the lack of pre-70 Ma. weathering ages.

In part of the region comprising the Upper Uruguay River basin in Brazil, Justus et al. (1986) have recognised the three palaeosurfaces of the landscape identified by King (1956) and designated them as *Região Geomorfológica Planalto das Araucárias (Sul-Americana), Região Geomorfológica Planalto das Missões (Velhas) and Região Geomorfológica Planalto da Campanha (Paraguaçu).* Recently, Paisani et al. (2008) proposed a hypothetic evolutionary model for the planation surfaces of an area of the basaltic plateau (southwestern Paraná and northwestern Santa Catarina States, Brazil), from field data and from a hypsometric classification obtained by radar images. These authors suggested, conversely to the authors cited previously, that the surfaces were formed simultaneously. Also they deduced that (1) the planation surfaces have developed mainly by etch processes, (2) the type of lava flow has little influence on the development of planation surfaces, (3) the tectonic factors have been very important in the formation of landforms in layers and (4) the kind of subtropical climate had a very important role on the planation morphology of the surfaces.

From a morphotectonic point of view, the Argentine sector of the Paraná Basaltic Plateau is part of the southern Paraná Basin, bounded in the south by the Asunción and Rio Grande archs. This region is considered by Popolizio (1972) as a transitional area between the uplift of the Southern Brazilian Plateau to the northeast and the subsiding style of the middle-lower Paraná basin southwestward of the province of Corrientes (Argentina). For this reason, table-like landscape has remained unchanged in the Misiones province, with appreciable uplift, compared with the morphology of the province of Corrientes, characterised by a structural style represented by vertical faults that generated differential tectonic blocks movements (Popolizio 1972). This author suggests that the fracture tectonics that affected the basaltic plateau in the Misiones province (represented by fracture systems without vertical displacements—NE-SW main direction and NW–SE and E–W systems with minor frequencies) played a secondary role with respect to the morphoclimatic processes that modelled the region.

The studied region is under subtropical humid conditions today, with mean annual temperatures between 19 and 21 °C and rainfall ranges between 1,700 and 2,200 mm/year. The present vegetation is represented by savannah and tropical forest. Then, some of the landscape development processes can be linked to tropical geomorphology. This thematic area was initiated in South America by Tricart (1956, 1959), who focused his attention on fluvial processes and the impacts of climate change. Between the Brazilian landscape and macroecological domains differentiated by Ab'Sáber (2000), the Upper Uruguay River basin in Brazil is included in the *Domain of the Araucaria Plateau.* It is characterised by medium altitude plateaus, 800–1,300 m a.s.l., covered with Araucaria forests of diverse density, including mosaics of mixed prairies. Depth of weathering is very variable, with imperfectly developed convex hills. There is eventual colluvium on slopes, covering sub-recent topography, with large microrelief irregularities, corresponding to a drier climate. Vegetation stocks closely related to the present inter- and subtropical ecosystems were developed after Mid-Tertiary times. During the Quaternary, such floras fluctuated in space, controlled by successive climatic changes (Ab'Sáber 2000).

Previous Morphometric Analysis of the Basaltic Plateau in the Upper Uruguay River Basin

Palaeosurfaces represent time intervals long enough for distinctive correlated features to develop and must be distinguished among themselves by their descriptive attributes (Widdowson 1997). Hence, the morphometric analysis method is used as an auxiliary tool in many geomorphological investigations in different regions of the world, including that comprising the study of erosion surfaces. Golts and Rosenthal (1993), Zuchiewicz and Oaks (1993), Grohmann (2004) and Grohmann et al. (2007) discussed the development of that method. The availability of digital elevation data and processing capacity today stimulates the use of hypsometric curves and digital elevation to represent relief forms and models as basic mapping tools (Soares and Riffel 2006). These authors demonstrated for three neighbouring hydrographical basins in southern Brazil that hypsometric curves can be used to identify palaeosurfaces in multi-history landscapes by manipulating their attributes.

Palaeosurface remnants are large areas with relatively similar elevations that appear in hypsometric curves as rather flat segments. Hypsometric curves generated for each drainage basin permit to compare the landscape evolution for basins that differ in extent and steepness, for evaluating the geomorphic maturity of catchments and landforms (Strahler 1952a). The shape of a hypsometric integral (Strahler 1952b) changes from concave–convex to concave as the basin reaches the equilibrium (mature) stage; concave curves indicate planations. Reinterpretation of the stepped topography, combined with the analysis of palaeolandforms, drainage anomalies and structural controls, was made by many authors in different areas of Brazil.

First results of morphometric analyses of the Paraná Basaltic Plateau were previously presented by Kröhling et al. (2009, 2011). Remnants of the King's *Sul-American Surface* are well developed in the headwaters of the Uruguay River (on the subbasins of the Pelotas and Canoas Rivers, Brazil) at around 1,000–1,200 m.a.s.l. and located immediately westwards from the Serra Geral erosion scarp. The *Velhas Surface* is widespread in the Upper Uruguay River basin in southern Brazil and it is represented by a landscape of gentle and well-rounded hills (locally named as "coxilhas").

Three erosion surfaces have been recognised and described in the basaltic plateau of the province of Misiones (Argentina) by Iriondo and Kröhling (2008), mainly on basis of field data. The higher remnant surface exposed of the plateau in Argentina is restricted to the area of Bernardo de Irigoyen (26°16′S and 53°39′W), northeastern Misiones province, and found at altitudes around 800 m a.s.l. This area would have undergone intense physical and chemical weathering and persistent erosion action; small circular depressions in the basaltic rocks (200–300 m in diameter and 5–10 m depth) presently occupied by peat bogs are frequent. These depressions could be relicts of fluvial erosive forms.

According to Iriondo and Kröhling (2004), the landscape of the province of Misiones is formed by the intermediate surface, mostly close to 500 m a.s.l., forming a narrow plateau that is geographically known as "Sierra de Misiones". The surface is represented by rounded hills and flat-bottomed fluvial valleys, and it is correlated with the surface generated by the Velhas erosive cycle of King (1956) in Brazil. The *Velhas Surface* of Misiones (or Pd2 of Bigarella et al. 1965) was preserved on divides as remnants, being mainly represented by the flat-topped watershed between the Paraná and Uruguay basins, with a NE–SW direction in the central part of the province (Fig. 2). Because of the proximity of these large fluvial collectors, pediplanation and pedimentation processes left narrow remnants that are restricted to ridges with planed tops.

The lower level of the general landscape (close to 200 m a.s.l.) forms a peripheral area in the southeastern of Misiones province that is locally named Apóstoles Surface. It would have been generated by the King's *Paraguaçu cycle* (or Bigarella's

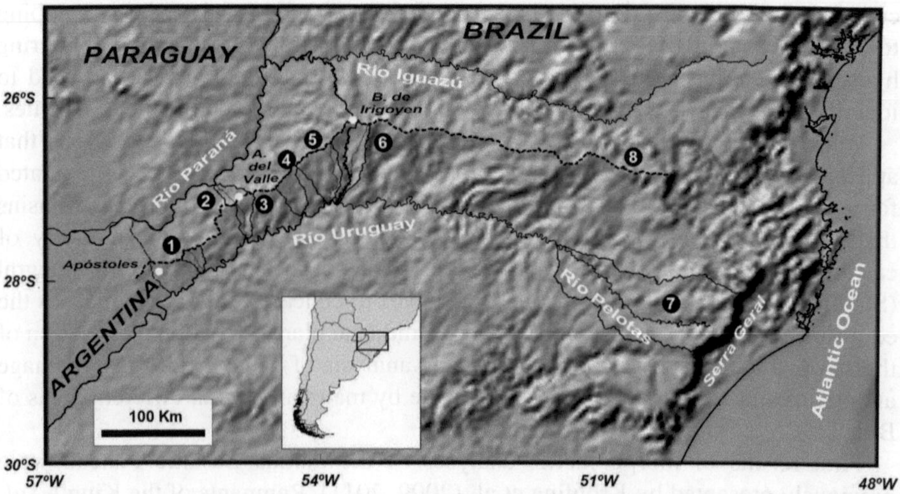

Fig. 2 Location of drainage subbasins in the Upper Uruguay River basin (southern Brazil–northeastern Argentina) on a shadow relief map derived from a DEM. Subbasins and main collector rivers: *1* Itacaruaré; *2* Cuña Pirú River; *3* Acaraguá; *4* El Soberbio; *5* Yabotí; *6* Pepirí Guazú; *7* Pelotas River basin; *8* divide of Paraná–Uruguay fluvial systems

Pd1), destroying the intermediate surface (*Velhas Surface*) by the regression of erosion scarps and a deep incision and widening of fluvial valleys.

The main purpose of this chapter is to apply morphometric analysis techniques to a large area of the Paraná Basaltic Plateau in the Upper Uruguay River basin (26°30′–28°45′ lat. S and 49°15′–56° long. W), covering a large part of the States of Rio Grande do Sul and Santa Catarina, southern Brazil, and an area comprising the province of Misiones and the northeastern of Corrientes province in northeastern Argentina (Fig. 2). Detailed works on the analysis of the erosion surfaces of Misiones province (29,800 km^2) are presented. The investigation began with the identification of palaeosurfaces in selected tributary stream basins of the larger Uruguay and Paraná River basins. The analysis was extended on watershed in order to identify palaeosurface remnants and to correlate them to those described by other authors in southern Brazil. Geomorphological and stratigraphical field data of the Uruguay River basin are integrated in the work of Iriondo and Kröhling (2008).

Materials and Methods

In order to understand the hillslope form evolution and the characterisation of hillslope sequences in the study region according to dominant process regimes, field observations and indirect measurements from maps, satellite images and digital elevation models (DEMs) were applied. Geographical Information Systems (GISs), with algorithms to calculate morphometric properties and statistics, provided the base for morphometric analysis. Raster-based DEMs from the Shuttle Radar Topographic Mission (SRTM, distributed at horizontal resolution of 3 arcsec—approximately a 90 × 90 m grid—Jarvis et al. 2008) were used.

Elevation data for selected hydrographical subbasins were extracted from the generated DEM for the region comprising the Upper Uruguay River basin. The perimeters of the drainage basins were calculated by free-software GRASS GIS (GRASS Development Team 2005). Different steps involved previously the obtainment of flow-accumulation maps (r.watershed) and the simulation of the stream networks. Once the outlet coordinates were obtained, drainage basin perimeters were delineated by "r.basin" module. Morphometric properties of a drainage basin are quantitative attributes of the landscape that are derived from the elevation surface and fluvial network within a basin. The method of morphometric analysis was applied to the following tributary river basins in Misiones province: Itacaruaré, Acaraguá, El Soberbio, Yabotí and Pepirí Guazú. Also, a tributary basin of the Paraná River (Cuña Pirú River basin) was considered. The Pelotas River basin (the main tributary of the Upper Uruguay River), with its watersheds near the Serra Geral scarp in southern Brazil, was also analysed for a comparative study (Fig. 2).

An isobase surface is a hypothetical surface determined by the intersection of drainages of similar order with an erosion surface, associated with the reorganisation of the drainage networks. It can be considered as a simplified surface of the original topography, where the elevations above the isobase surface are discarded (Filosofov 1960). The isobase map was generated by GRASS-GIS, following the methodology proposed by Grohmann et al. (2007). Steiner (2007) compared different GIS software platforms (GRASS GIS and ArcGIS) for production and manipulation of morphometric data, slope, aspect and surface roughness maps and concluded that the DEM-derived products are equivalent. The author also concluded that isobase surface maps generated by GRASS present smoother shapes and contours than maps produced by ArcGIS.

The DEM was smoothed with a 7 × 7 filter to minimise the effects of possible noise. An isobase represents the line that marks an erosion surface, by stream incision. According to Golts and Rosenthal (1993), the resulting surface is related to similar stages of erosion and can be considered as a product of recent tectonic–erosive events. These hypothetical surfaces can also be associated to future stages of erosion once the backwearing and/or downwearing thrives.

Stream channel network was extracted from a 90 × 90 SRTM DEM using the Single Flow Direction (SFD-D8) algorithm (r. watershed command). The stream Strahler order (1952a) has been assigned to polylines of the streams in the attribute table. The elevation isobase map was made from the intersection of contours with 1st-, 2nd-, 3rd- and 4th–5th-order stream channels. The resulting raster was interpolated in order to obtain the hypothetical map (Wobbe 2007).

According to Soares and Riffel (2006), the region under a hypsometric curve (being a summation of area intervals per altitude) represents the amount of rock between a river mouth and the erosion surface. Altitude frequency histograms for each tributary basin were produced from the extracted SRTM data, using the ENVI program. Hypsometric curves for these subbasins were generated also from SRTM data; respective curves represent the relationship between altitude and cumulative areas under given contour intervals around watersheds. The procedure followed in this work is presented in detail by Soares and Riffel (2006): "The histograms for the several classes of H are constructed and the cumulative frequency is converted to area (A) between the highest point and the H level. A hypsometric curve represents the cumulative frequency of points with altitude H within class intervals, from the highest to the lowest point. In order to compare different basins, the H and A values are converted to fractions of their respective maximum values to give comparable hypsometric curves. The experimental hypsometric curves are segmented between minimum (Homin) and maximum (Homax) observed values of altitude in histograms; the minimum (Homin) is taken from the modal lower class boundary, because this modal class represents the larger area of altitude interval; this class therefore represents the lower mean slope interval and as a consequence, the closest preserved level to the early (eroded) base level of the hydrographical basin. The operation is taken in order not to consider altimetry data related to erosion events governed by successive base levels".

The relief gradient in a hydrologic basin dominated by fluvial erosion base level can be modelled by a logarithmic function. This model is fitted to each segment by regression analysis using the least-squares method:

$$H = a - b\ln(A)$$

where H is the minimum altitude to which area A is circumscribed and a and b are the estimated parameters (a is the maximum predicted altitude and b is a coefficient that is proportional to the average declivity in the altitude interval). This allows calculating the minimum palaeosurface altitude. For each investigated basin, the simulation of a stream network was carried out by using DEM data. Also longitudinal profiles of each collector of selected fluvial subbasins were extracted through ArcGIS program.

Field data taken during many previous years of studies on geomorphology and Quaternary geology of the upper Uruguay River basin (NE Argentina and SW Brazil) supported the necessary base (Iriondo and Kröhling 2004, 2008). Field geomorphological analyses on valleys and watershed divides were performed, also including the description of toposequences. Uruguay and Yabotí River valleys near Moconá waterfalls and south watershed divide of the El Soberbio River subbasin were specially surveyed.

Generated Hypsometric Curve Analysis

The main surfaces of the basaltic plateau in the Upper Uruguay River basin (*Sul-Americana, Velhas and Paraguaçu*) can be distinguished from a polymodal distribution observed in the frequency histogram of elevations to regional scale

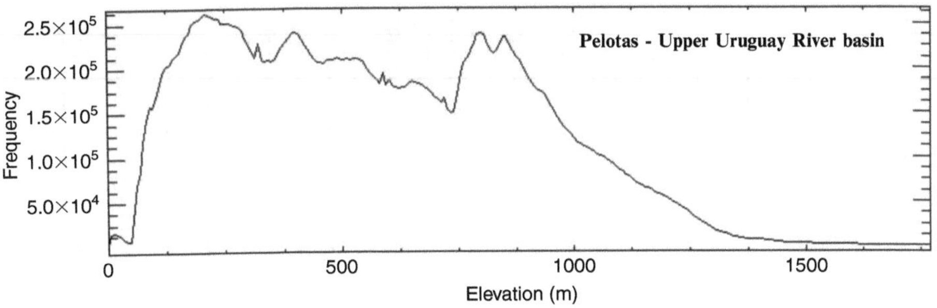

Fig. 3 Frequency histogram of regional elevations for the Pelotas and the Upper Uruguay River basins derived from SRTM DEM (90 m)

(Fig. 3). Table 1 summarises the main modes of the statistical distribution of altitudes for each studied tributary basins, taken from the respective frequency histograms. The histograms of Fig. 4 correspond to the three basins with main headwaters in the upper level of the landscape in Misiones province. The main mode of the Pepirí Guazú River basin, at the Argentine–Brazilian border, is at 512 m a.s.l.; secondary modes are at 692 and 806 m a.s.l. The Yabotí River basin shows a main mode at 524 m a.s.l., close to the main mode of the Pepirí area. El Soberbio River histogram shows a main peak at 384 m a.s.l. and two secondary modes at 276 and 476 m a.s.l.

Histograms generated for the adjacent Cuña Pirú and Acaraguá River basins (tributaries of the Paraná and Uruguay basins, respectively) are presented in Fig. 5. The Cuña Pirú basin presents two well-defined clusters of elevations frequencies. The main mode is at 184 m a.s.l., with a secondary mode around 453 m a.s.l. The Acaraguá basin shows a polymodal distribution (Mo: 451, 316 and 181 m a.s.l.). The Itacaruaré basin distribution is bimodal, with the main mode close to 144 m a.s.l. and secondary mode around 225 m a.s.l. (Fig. 6). The main mode in the highest altimetry value considering the six tributary basin investigated was obtained in the Pelotas River basin (Mo: 922 m a.s.l.), with the secondary mode at 1,139 m a.s.l. (Fig. 6).

Hypsometric curves generated for each basin allow the identification of significant low-gradient surfaces or terraces (see an example in Fig. 7). Such relatively flat segments are indicative of palaeosurfaces that could be related to ancient base levels. From regression analysis (see Table 1), the adjustment of the low-gradient segments permits to reconstruct two hypothetical hypsometric curves and consequently to estimate the elevation for the former base levels. In order to compare the different basins, each area is converted to fractions of their respective maximum values to obtain normalised hypsometric curves (Fig. 8 and Table 1). Three groups of hypothetical hypsometric curves are deduced from proximal value sets for predicted base levels. Using this approach, which is based on the clustering from the modelled base level in the river mouth of the basins, three main palaeosurfaces are defined in the Argentine area of the Upper Uruguay River basin

Table 1 Analysis of hypsometric curves of seven fluvial subbasins of the studied region in order to identify and describe preserved levels (erosion surfaces) of early base levels of the respective basins. Logarithmic functions describe fluvial erosion base level for each hydrologic subbasin

Hydrographic basin	Modal elevation (m.a.s.l.)	Minimum elevation (m.a.s.l.)	Maximum elevation (m.a.s.l)	Predicted base level (m.a.s.l)	Log. model	R^2
Pelotas						
Sul-Americana 1 palaeosurface	1,138.9	1,079.8	1,276.9	903.6	$y = -181.3\ln(x) + 2{,}626.1$	0.9910
Sul-Americana 2 palaeosurface	922.1	882.6	981.6	840.9	$y = -223.81\ln(x)$	0.9915
Pepirí Guazú						
Bernardo de Irigoyen palaeosurface	692	675.68	708.35	548.8	$y = -72.44\ln(x) + 1{,}108.1$	0.9854
Aristóbulo 1 palaeosurface	512.4	479.70	594.02	425.1	$y = -150\ln(x) + 1{,}583.2$	0.9913
Yabotí						
Aristóbulo 1 palaeosurface	523.8	492.34	555.15	424.1	$y = -76.63\ln(x) + 1{,}010$	0.9833
Paraíso						
Aristóbulo 2 palaeosurface	475.55	429.6	506.18	356.7	$y = -81.894\ln(x) + 927.98$	0.9851
Aristóbulo 2 palaeosurface	383.66	368.34	414.29	310.2	$y = -130.2\ln(x) + 1{,}218.4$	0.9975
Apóstoles 1 palaeosurface	276.45	245.82	322.39	216.8	$y = -372.12\ln(x) + 2{,}812.5$	0.9992
Acaraguá						
Aristóbulo 2 palaeosurface	451.1	436.05	481.14	377.4	$y = -37.92\ln(x) + 627.54$	0.9849
Apóstoles 1 palaeosurface	315.8	255.67	360.89	206.8	$y = -175\ln(x) + 1{,}361$	0.9878
Apóstoles 2 palaeosurface	180.5	165.48	225.61	134.7	$y = -401.1\ln(x) + 2{,}780.2$	0.9995
Cuña Pirú						
Aristóbulo 2 palaeosurface	453.2	453.22	498.16	402.7	$y = -28.11\ln(x) + 577.33$	0.979
Apóstoles 2 palaeosurface	161.1	213.55	168.61	183.6	$y = -129.4\ln(x) + 965.04$	0.9895
Itacaruaré						
Apóstoles 2 palaeosurface	225.3	211.04	282.20	157.4	$y = -62.47\ln(x) + 541.62$	0.9811
Apóstoles 3 palaeosurface	143.8	111.39	182.57	105.8	$y = -119.2\ln(x) + 838.88$	0.9963

Fig. 4 Frequency histograms of elevations for the Pepirí Guazú, Yabotí and El Soberbio River basins, obtained from SRTM DEM (90 m)

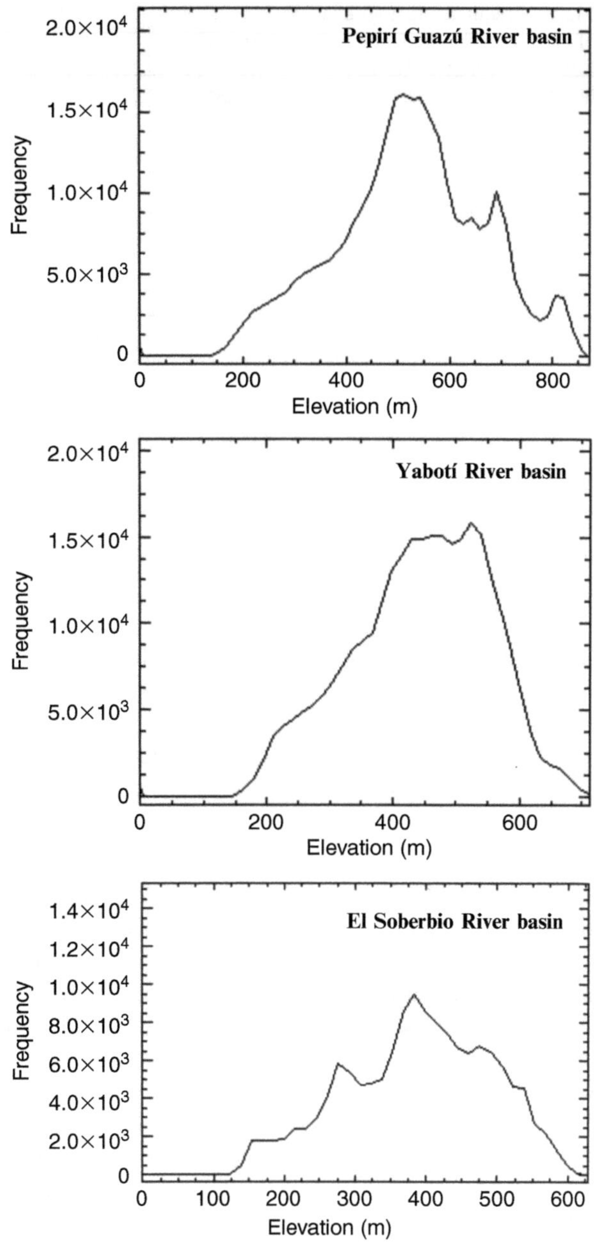

(named *Erosion Surfaces Bernardo de Irigoyen, Aristóbulo del Valle and Apóstoles*). Complementary intermediate or secondary palaeosurfaces also appear from the analysis (*Aristóbulo 2; Apóstoles 2 and 3*). Therefore, each mapped palaeosurface

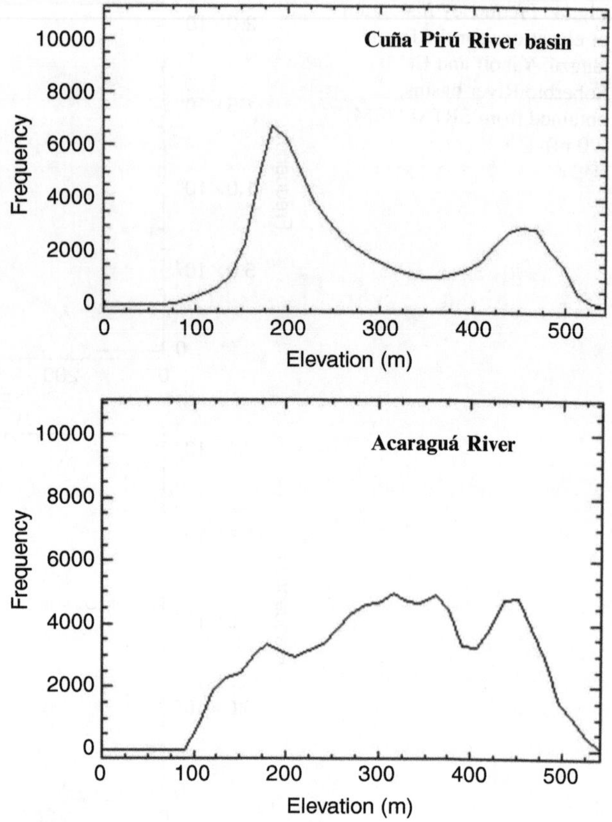

Fig. 5 Frequency histograms of elevations for the Cuña Pirú and Acaraguá River basins

was considered between the minimum H_{min} for correlative surfaces in all of the basins and the H_{min} of the next higher surface (Table 1 and Fig. 9).

A more detailed view for the resulting surfaces in the Misiones province and the analysed drainage basins is presented in Fig. 10. In the Pepirí Guazú River basin, the *Aristóbulo del Valle Surface* is dominant (Fig. 11). Steps or knickpoints are easily seen in the longitudinal profile of the collector in coincidence with the lower limits defined for *Bernardo de Irigoyen, Aristóbulo 2 and Apóstoles 1 Erosion Surfaces.* Similarly, the *Aristóbulo Surface* covers the largest area in the Río Yabotí basin, although the *Aristóbulo 2 and Apóstoles 1 Surfaces* would influence around 550 and 280 m a.s.l., respectively, on the longitudinal profile of the collector (Fig. 12). On the El Soberbio River Basin, the *Aristóbulo del Valle Surface* dominates. The knickpoint at about 275 m a.s.l. could indicate the *Apóstoles 1 Surface* influence on the stream longitudinal profile (Fig. 13).

Figure 14 shows that *Apóstoles 2* has a larger representation in the Cuña Pirú River basin; *Apóstoles Surface 3* and *Aristóbulo 2* occupy most of the Acaraguá River basin. The main controls on the respective longitudinal profiles of the collectors are *Apóstoles Surfaces 3 and* 2, respectively (around 180 and 120 m a.s.l.;

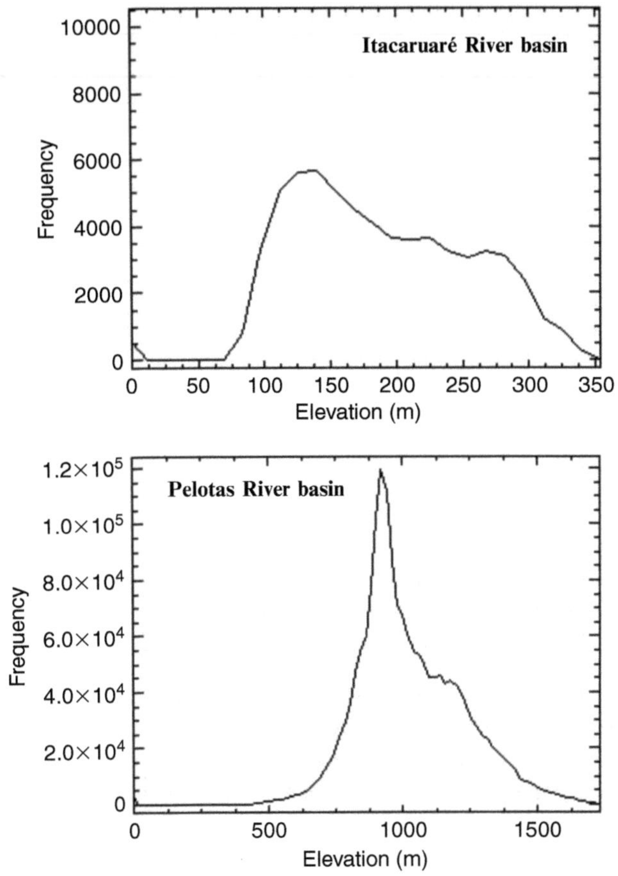

Fig. 6 Elevations frequency histograms for the Itacaruaré and Pelotas River basins

Fig. 14). *Apóstoles Surface 3* is mainly preserved in the Itacaruaré River basin (Fig. 15) and the stream longitudinal profile is also controlled by *Apóstoles Surfaces 3 and 2* (see the inflections at 120 and 180 m a.s.l.).

The Pelotas River basin (southern Brazil) is almost totally developed on the *Sul-Americana Erosion Surface* (Fig. 16); an important control, represented by a knickpoint in the longitudinal profile of the river, corresponds to *Sul-Americana Surface 1* (around 1,080 m a.s.l.).

The extension of each mapped palaeosurface into narrow valleys carved on the next higher surfaces (Figs. 11, 12, 13, 14, and 15) is not entirely representative, because they were eroded, partially by current fluvial networks. However, it is clear that the erosion surfaces exert control on the streams, observed from the knickpoints in longitudinal profiles.

The scarp that limits *Aristóbulo del Valle Surface* is highlighted by a noticeable anomaly in the distribution of the isobase lines, elaborated from the intersection of

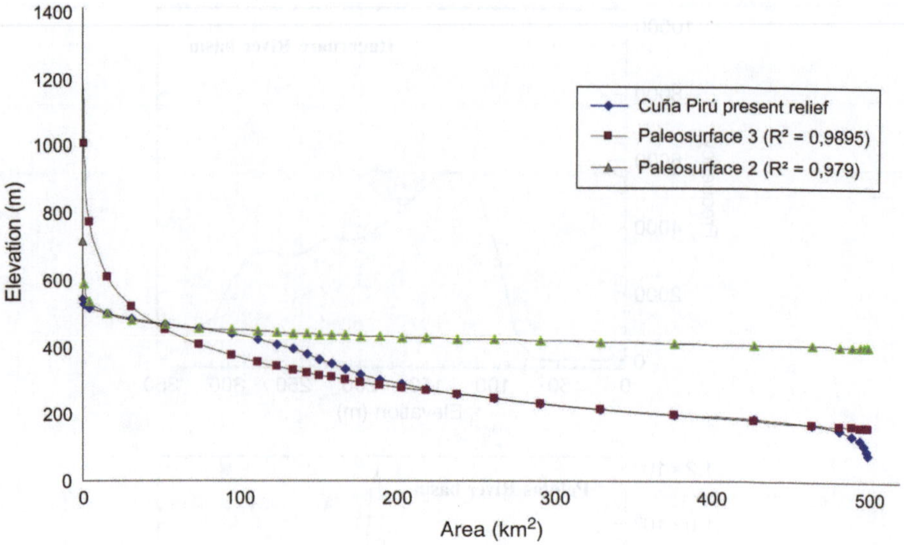

Fig. 7 Example of an analysis of hypsometric curves of the Cuña Pirú River basin. They represent the present relief and modelled base levels for relict surfaces

contour lines with the 2nd- and 3rd-order stream channels belonging to the Cuña Pirú River basin (Fig. 17b). The strong control on the headwaters of this tributary basin is also observed in the longitudinal profile of the collector (Fig. 14, profile B-B′). The isobase map constructed on the 1st-order streams allows the delimitation of the area that remains unaffected by the current fluvial erosion (Fig. 17a) and that corresponds to the remnant of the *Aristóbulo Surface 2.* Slopes that connect both levels of the landscape (*Aristóbulo and Apóstoles Surfaces*) are sharp with frequent debris composed of boulders of high angularity, which suggests that the process of areal erosion was produced mainly by rock landslides. The sequence of isobase maps drawn for different hierarchy orders indicates a noticeable smoothing in the largest stream order hypothetical map (4th–5th) (Fig. 17c). This could indicate how the basaltic hillslopes will evolve by slope retreat (pediplanation), generating smooth convex hills as observed today in the *Velhas Surface* at the Pelotas River Basin (Fig. 18c, d).

According to the obtained results, the *Sul-Americana Surfaces 1 and 2* in this work (Fig. 9) are equivalent to the King's *Sul-Americana Erosion Surface* (or Pd3 of Bigarella et al. 1965). If the Upper Uruguay River basin corresponded to the lower segment of such surface, a predicted base-level value would be estimated around 880 m a.s.l. (Table 1). Both are well represented in the Pelotas River basin (Fig. 16). The higher remnant surface of the area of Bernardo de Irigoyen (northeastern Argentina) was considered a "sensu lato" equivalent of the Palaeogene King's *Sul-American Surface* in morphological and stratigraphical sense by Iriondo and Kröhling (2008), but it is now reinterpreted as a different palaeosurface. It

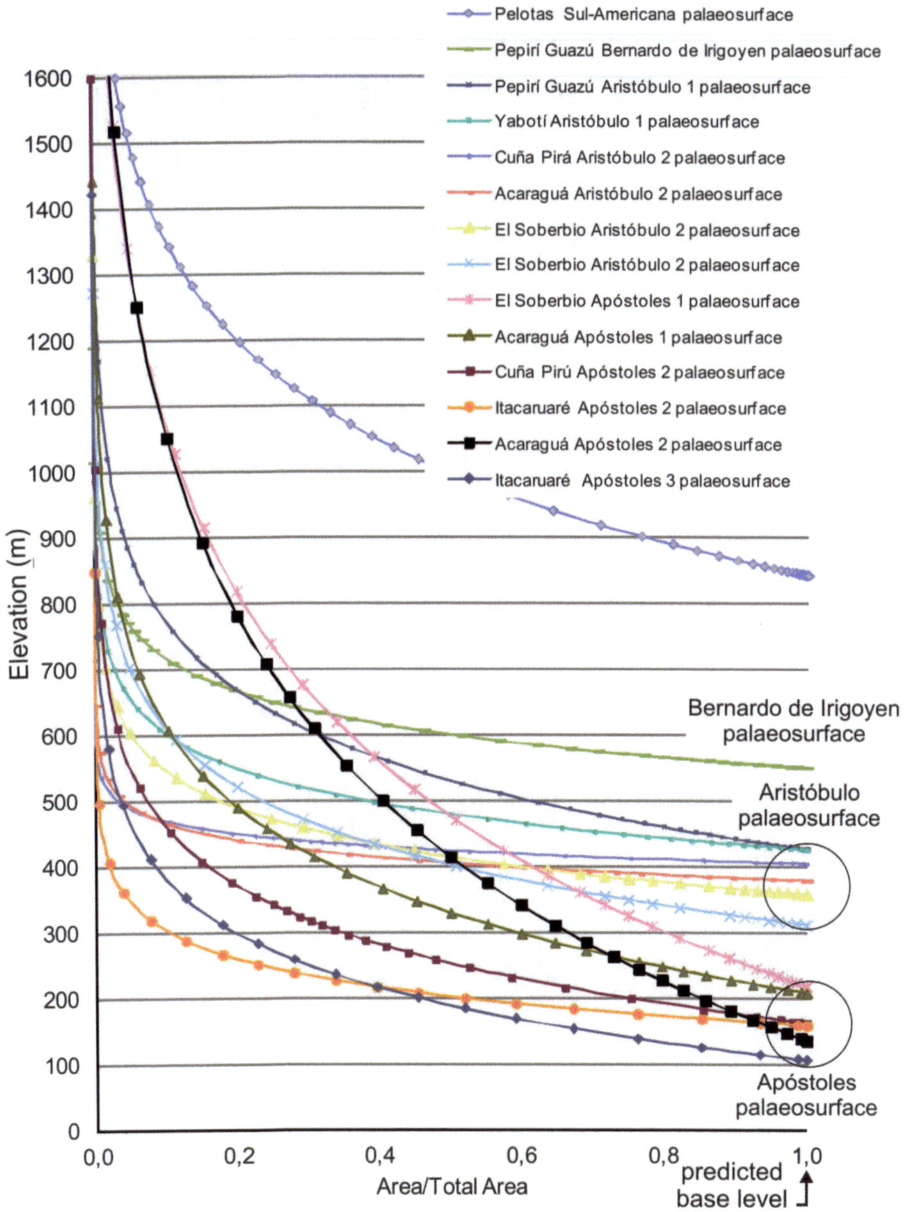

Fig. 8 Modelled hypsometric curves for the seven studied drainage subbasins. Each curve is labelled as an interpreted erosion surface. The inferred base levels at the mouth of the lower basins may be clustered in two groups, which surfaces could be interpreted as belonging to two well-defined stages

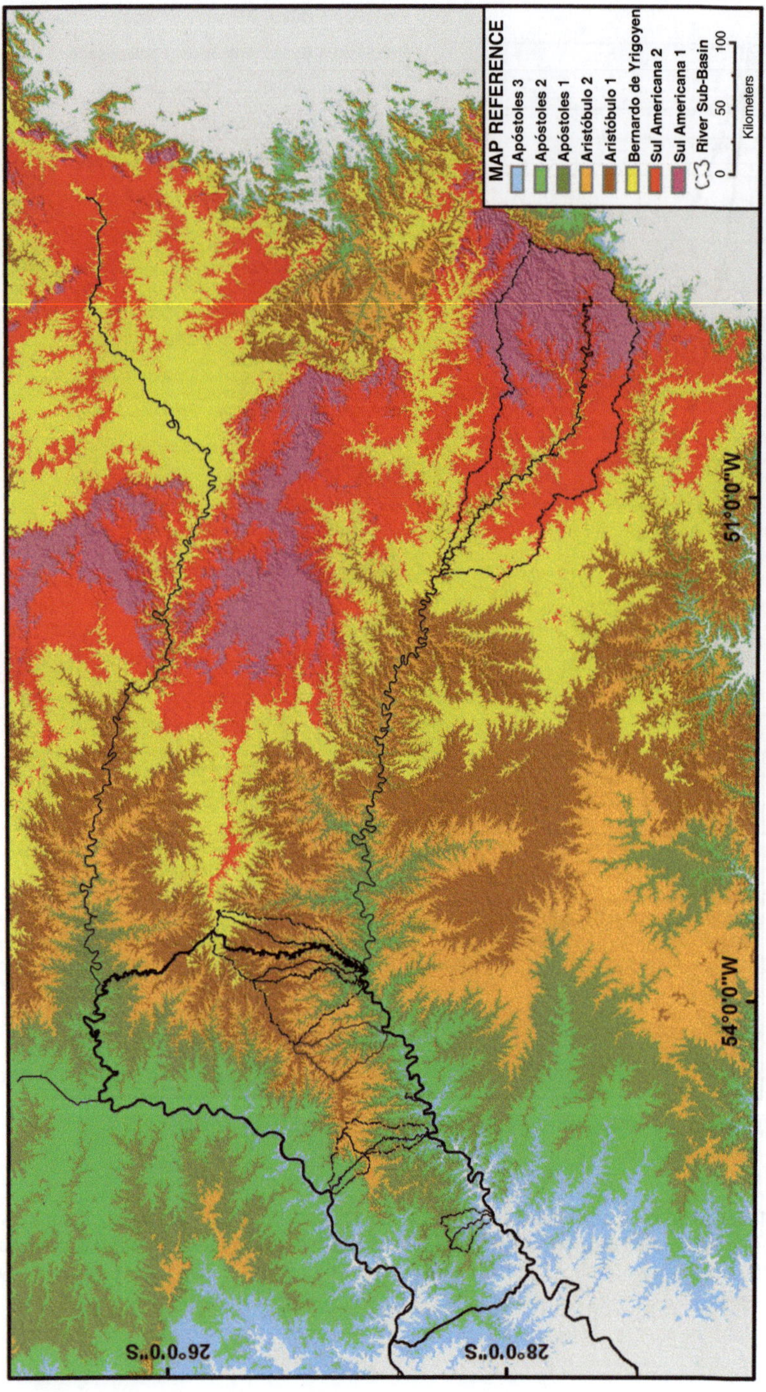

Fig. 9 Regional erosion surfaces map for the Upper Uruguay River basin defined from morphological analyses

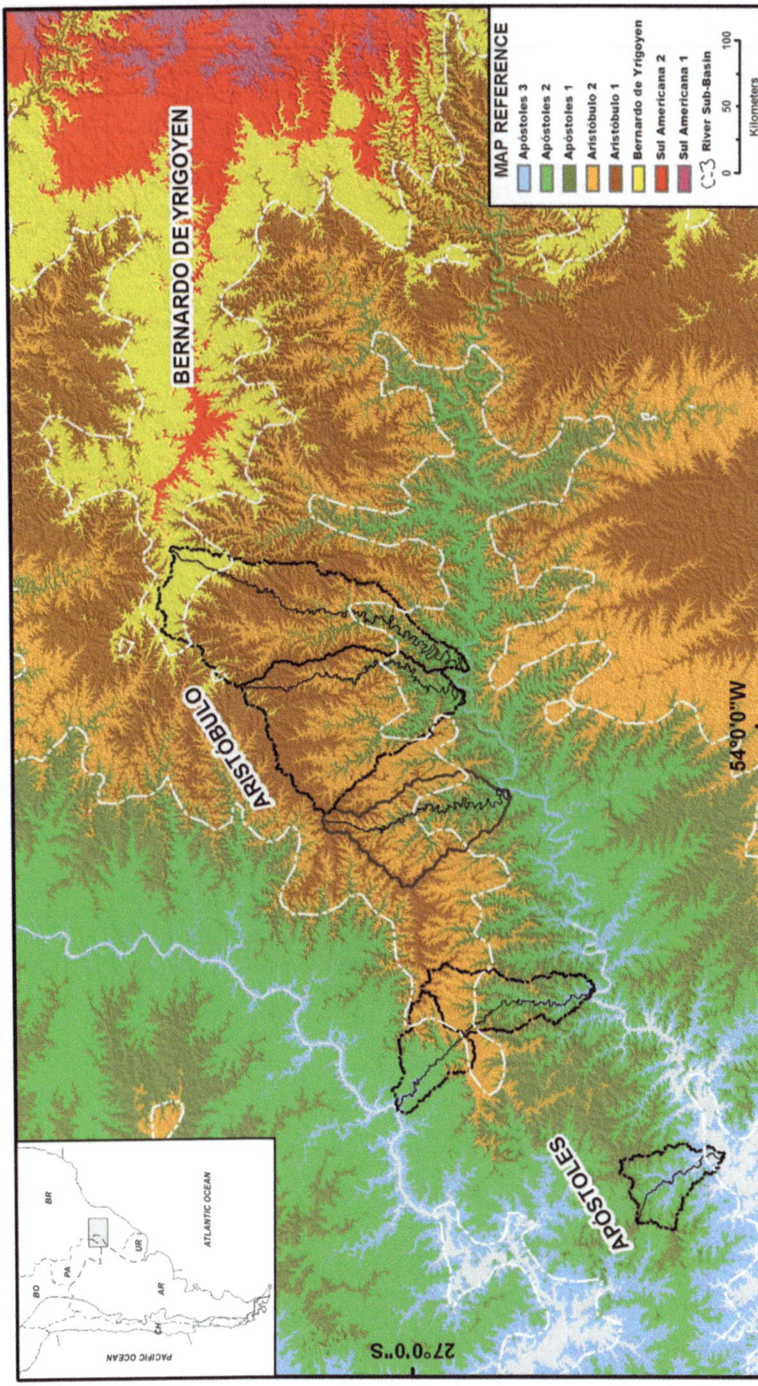

Fig. 10 Erosion surfaces map of Misiones Province (northeastern Argentina), at the southwestern area of the Paraná Basaltic Plateau

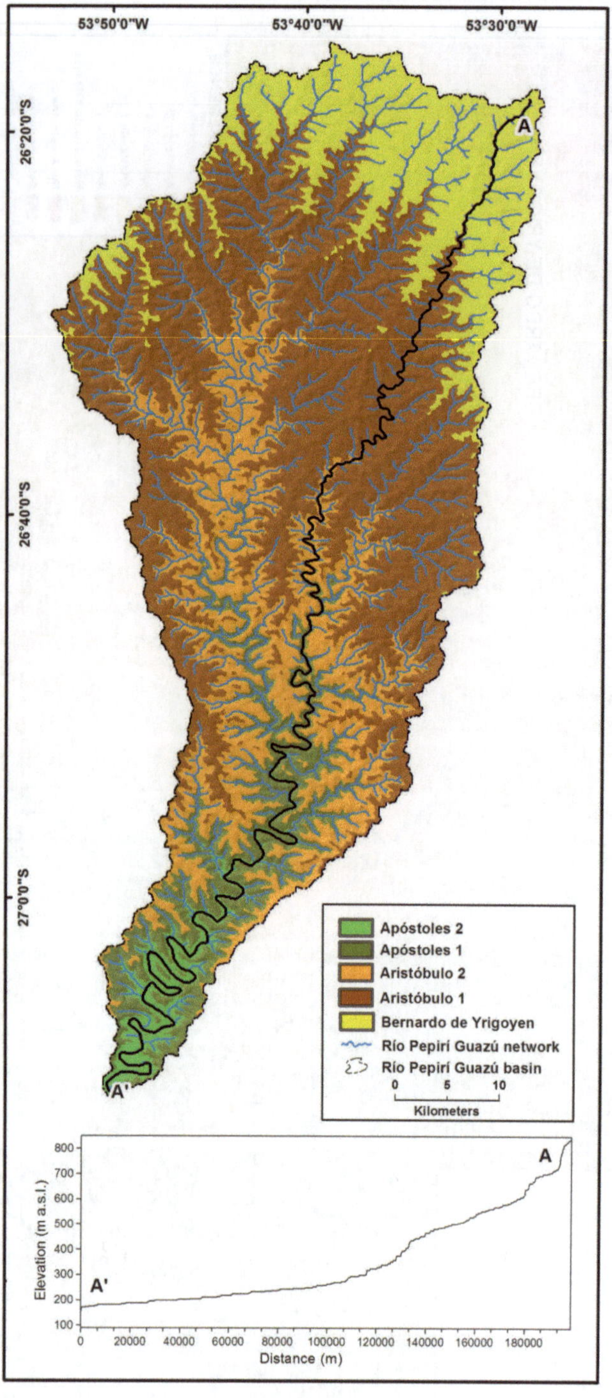

Fig. 11 SRTM-DEM-based map of erosion surfaces in the Pepirí Guazú River basin

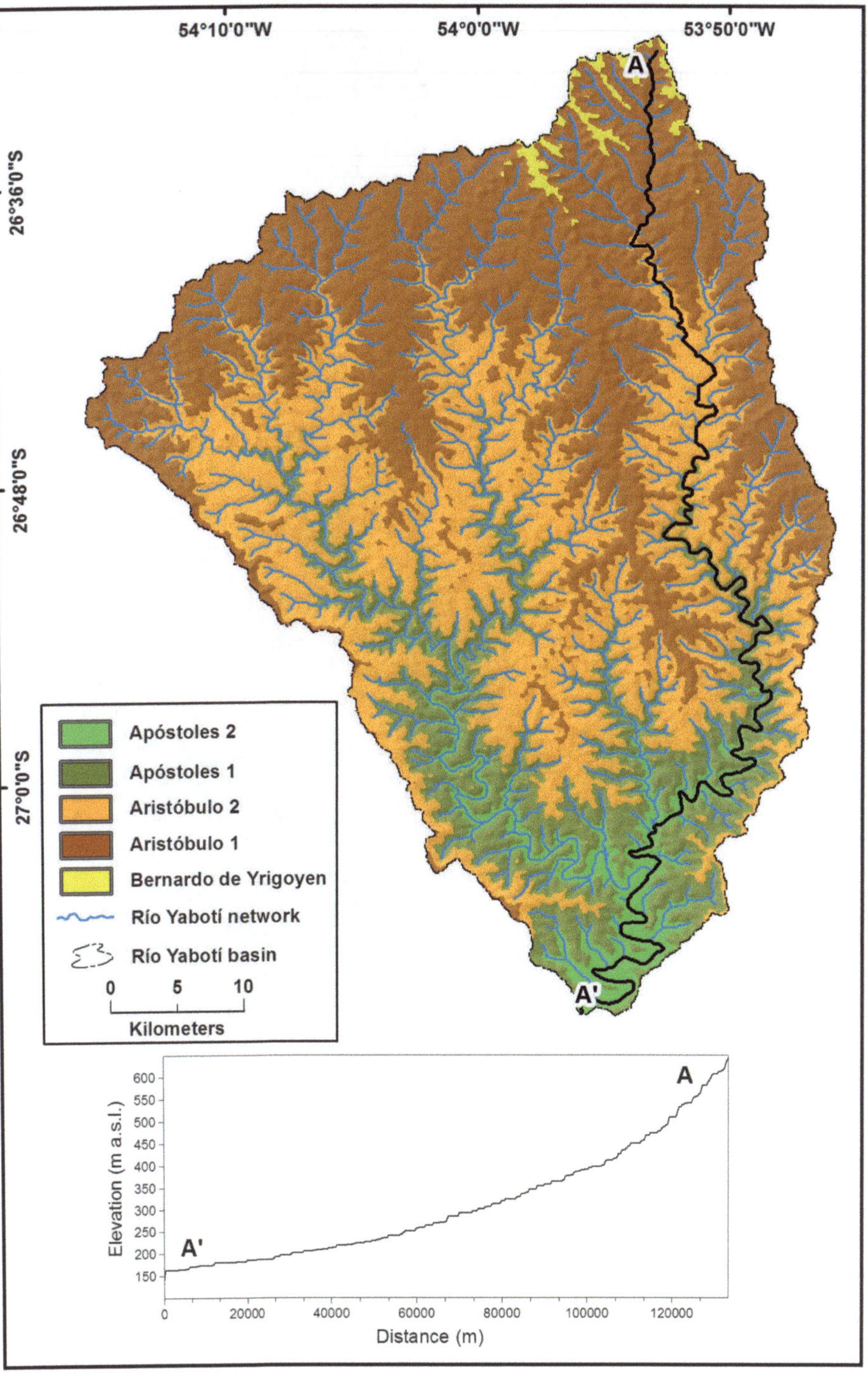

Fig. 12 SRTM-DEM-based map of erosion surfaces in the Yabotí River basin

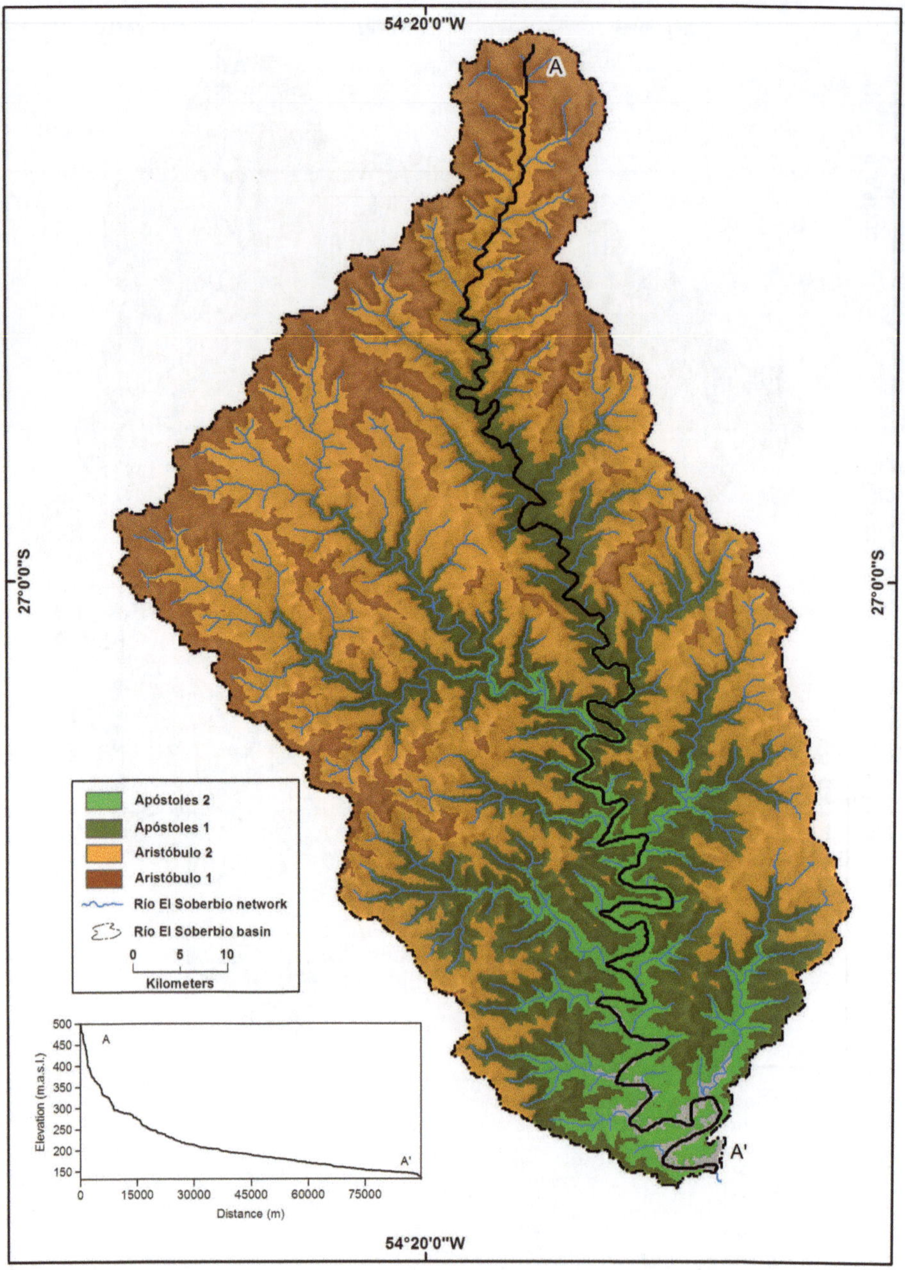

Fig. 13 SRTM-DEM-based map of erosion surfaces in the El Soberbio River basin

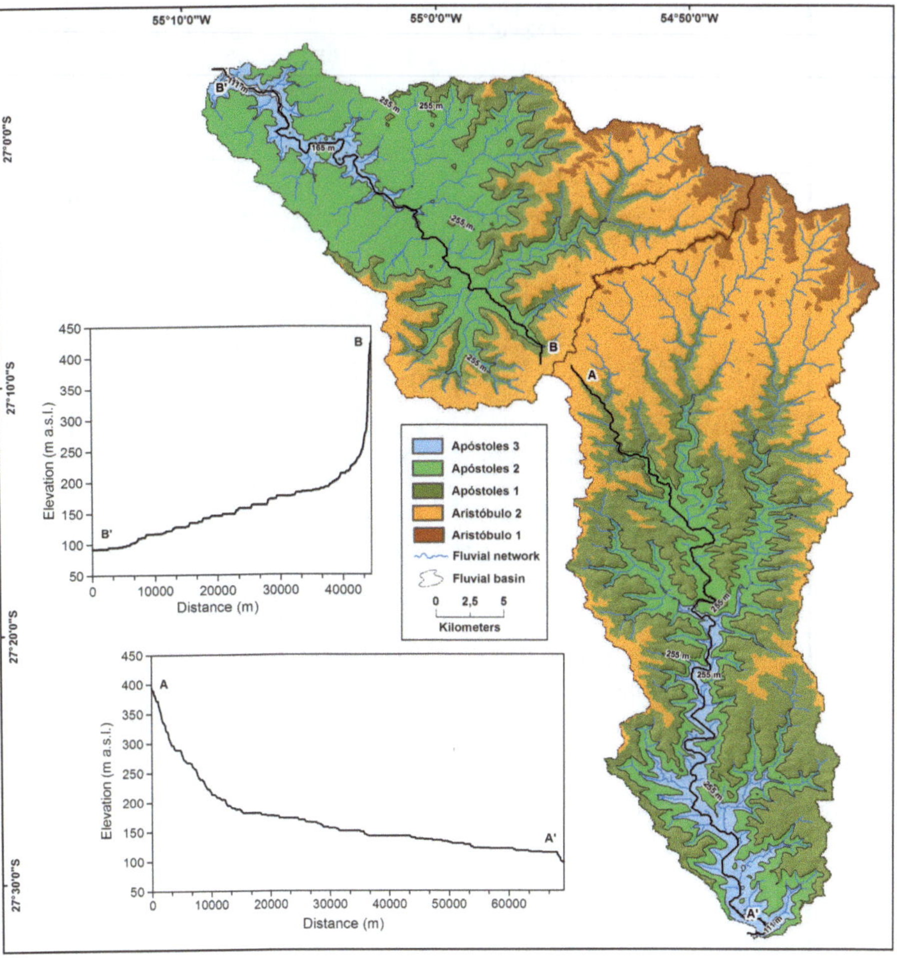

Fig. 14 SRTM-DEM-based map of erosion surfaces in the Cuña Pirú (B-B′ profile) and Acaraguá River basins (A-A′ profile)

is characterised by the *Bernardo de Irigoyen Erosion Surface* (675–880 m a.s.l.; Fig. 10 and Tables 1 and 2).

King's *Velhas Surface* (or Pd2 of Bigarella et al. 1965) is correlated in this study with *Aristóbulo del Valle Erosion Surface*, which in Misiones province comprises the plane-top watershed between the Paraná and Uruguay basins, with a typical area near the town of Aristóbulo del Valle (370–675 m a.s.l.; Fig. 10). A predicted base-level value for this surface could be estimated around 310–425 m a.s.l. (Table 1). The lower *Apóstoles 1, 2 and 3 Erosion Surfaces* identified in this morphometric analysis (Fig. 10) correspond to the locally named *Apóstoles Surface*, which would be generated by the King's *Paraguaçu cycle* (or Bigarella's Pd1). The base level for these lower surfaces is above 100 m a.s.l. (Tables 1 and 2).

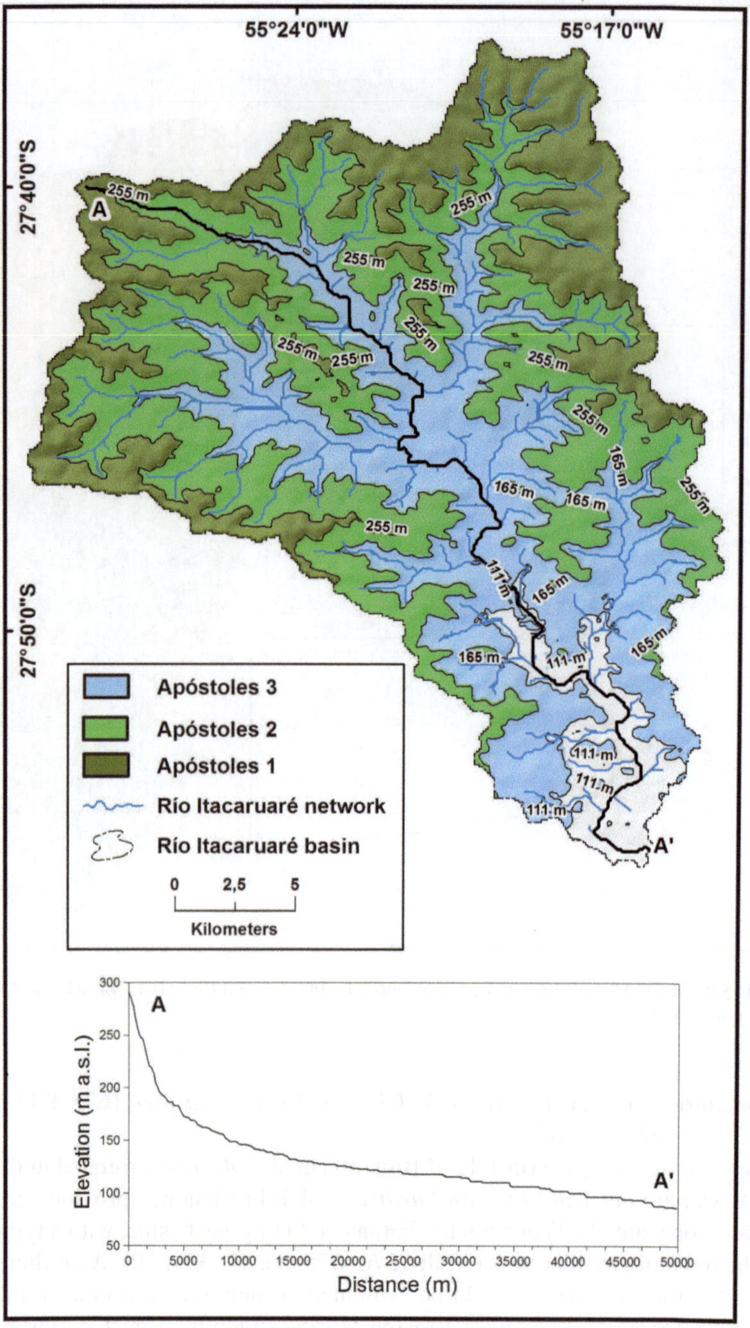

Fig. 15 SRTM-DEM-based map of erosion surfaces in the Itacaruaré River basin

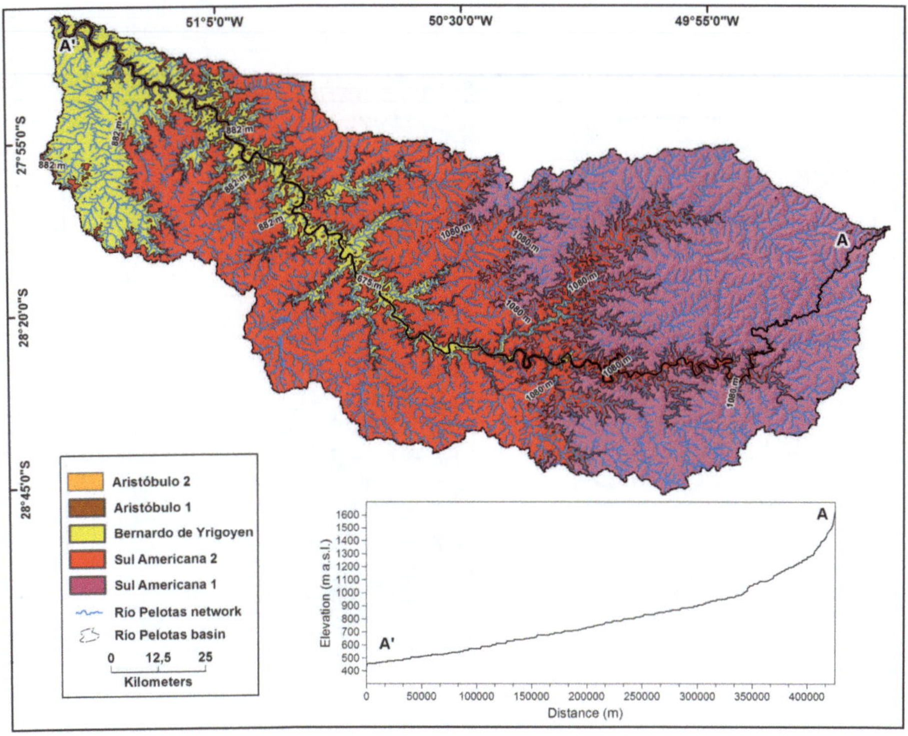

Fig. 16 SRTM-DEM-based map of erosion surfaces in the Pelotas River basin

Field Geological Data

The Sul-Americana Erosion Surface *(Brazil)*

The upper basin of the Pelotas River covers a large area of the *Sul-Americana Surface* (Fig. 9) that is known as *Planalto dos Campos Gerais* (Justus et al. 1986). In general it is an extended sub-horizontal basaltic surface (with a relief of few tens of metres), with the presence of small circular depressions (locally named "dales") and shallow valleys. Bigarella et al. (1965) interpreted the origin of dales as a product of suffusion during less humid periods, provoked by water runoff along pre-existing weakness tectonic lineaments.

Locally, there are dispersed basaltic blocks on the surface (that in general is formed by weathered basalt); the accumulation of coarser materials can form screes, especially at the tableland hills resulting from fluvial erosion of this surface (Fig. 18a, b). In some sites, deep fluvial valleys (160 up to 350 m deep) are incised

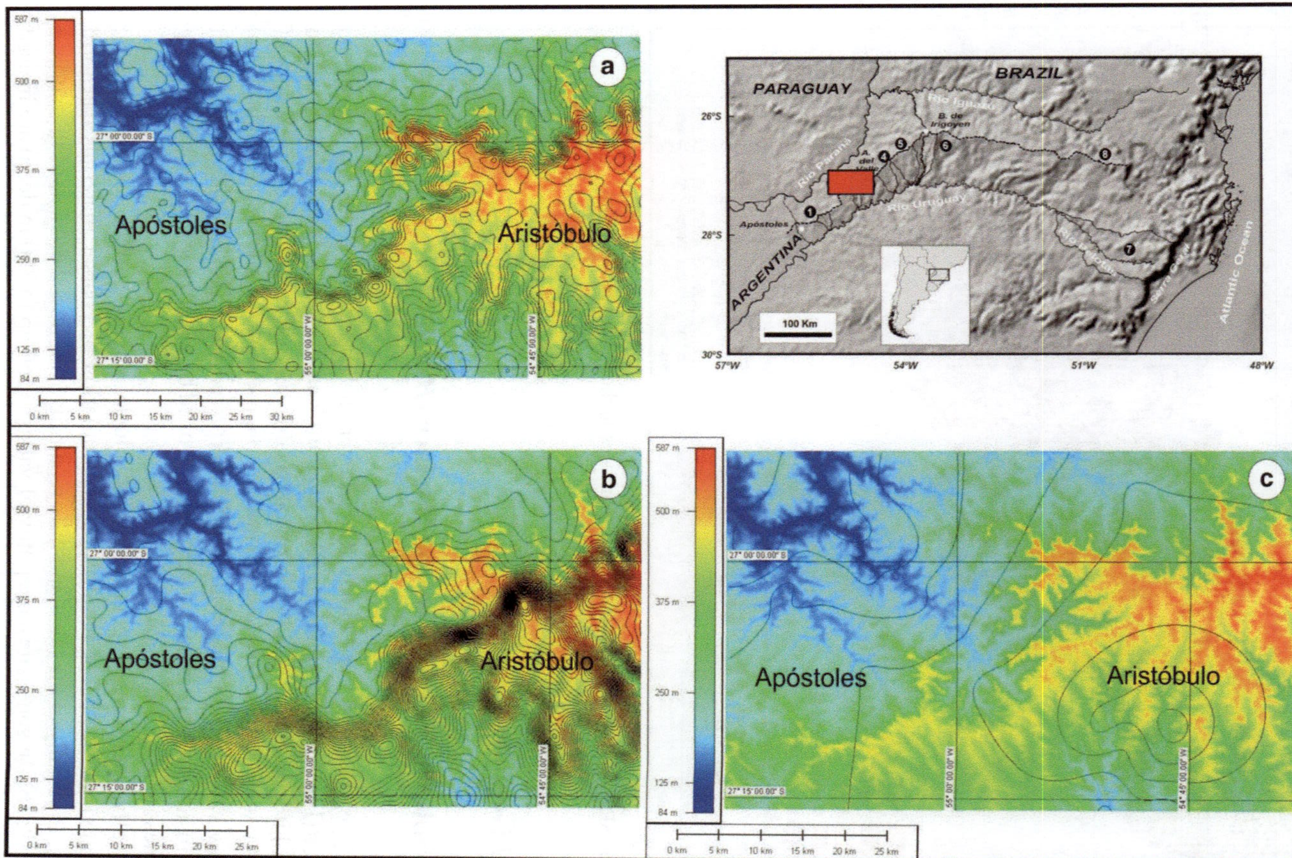

Fig. 17 Isobase maps in the Aristóbulo del Valle area (Misiones, northeastern Argentina). (**a**) Isolines drawn from 1st-order streams, (**b**) isolines drawn from 2nd- and 3rd-order streams and (**c**) isolines from 4th- and 5th-order streams

Fig. 18 Geomorphology of the Upper Pelotas–Uruguay River basin (southern Brazil). (**a**) A view of the Serra Geral Erosional Scarp, at ca. 1,700 m a.s.l. in the headwaters of the Pelotas basin; (**b**) knickpoint affecting a 1st-order tributary profile of the Pelotas River; (**c**) typical landscape with convex hills belonging to the *Velhas Surface* at the Canoas River and near the confluence with the Pelotas River; (**d**) a smooth convex hill representative of the *Velhas Surface* in the Campos Novos area; (**e**) the lower fluvial terrace on the Pelotas River valley; (**f**) view of both fluvial terraces of the Upper Uruguay River valley, in the Chapecó area

Table 2 Ranges of elevations defined for each erosion surface identified in the studied region on the basis of the morphological analyses

	Elevation (m.a.s.l.)	
Palaeosurface	Minimum	Maximum
Sul-Americana 1	1,080	no defined
Sul-Americana 2	880	1,080
Bernardo de Irigoyen	675	880
Aristóbulo 1	490	675
Aristóbulo 2	370	490
Apóstoles 1	255	370
Apóstoles 2	165	255
Apóstoles	110	165

in the surface, reducing the surface to isolated tableland hills or to a discontinuous line of crests. Paisani et al. (2008) consider this surface as the only pediplain of their study region.

Another large area of the *Sul-Americana Surface* is the Serra do Espigão that forms the divides between the Iguazú and the Uruguay River basins (Brazil; Fig. 9). A narrow relict segment of the *Sul-Americana Surface 2* extends to the southwest, near the Argentine–Brazilian border. It is represented by narrow tableland hills subordinated to the lower erosive surface (*Bernardo de Irigoyen Surface*).

The Bernardo de Irigoyen Erosion Surface *(Brazil, Argentina)*

It is an intermediate surface between the *Sul-Americana* and the *Velhas/Aristóbulo Erosive Surfaces* (Fig. 9), generated by slope retreat and pedimentation. Its contact with the Sul-Americana is frequently characterised by pediments. Fluvial erosive processes have been contributing to pediment formation; the tributaries of the upland fluvial subbasins eroded back the *Sul-Americana Surface* by lateral channel migration leaving a sub-horizontal level that locally presents a microrelief. This surface forms the divide between the Pelotas/Uruguay River basin and the Jacuí River basin. This surface is the base level of the Pelotas and Canoas River subbasins (Upper Uruguay basin) (Fig. 9). A large remnant of this surface extends westward, bordering the divide between the Iguazú and the Uruguay River basins, also entering to Argentina. The area of Bernardo de Irigoyen (Argentina)/Barracão (Brazil) represents a typical landscape of this erosion surface (around 800 m a.s.l.). It is formed by weathered basalt with small circular depressions (200–300 m in diameter and 5–10 m depth) presently occupied by peat bogs. It forms the headwaters of the Pepirí Guazú River basin at the Argentine–Brazilian border (Figs. 10 and 11). Scarce small remnants of this surface appear at the headwaters of the Yabotí River basin (Fig. 12).

The Velhas/Aristóbulo del Valle Erosion Surface *(Brazil, Argentina)*

The Velhas Surface (Brazil)

A highly dissected surface formed by a concordance of crests and valleys of the Velhas erosive cycle is differentiated in large areas of the Upper Uruguay basin. The general landscape is defined by convex hills with gentle slopes ("coxilhas"; Fig. 18c, d) and similar relative highs, alternating with wide alluvial plains of the Velhas cycle (Fig. 19a, b). In the headwaters of drainage subbasins, the presence of circular or elliptical depressions is common, with flat bottom ("dales"). A typical convex hill of the Velhas landscape is elliptic, with a symmetrical profile (600–1,200 m long and up to 50 m relative high respect to the closest fluvial depression; Fig. 18d). The linear segments of the hillslopes are debris slopes composed of basaltic blocks in a fine brown matrix.

The Aristóbulo del Valle Surface (Misiones)

This is represented typically by the *Aristóbulo 1 Surface* (Fig. 10), a narrow top surface that forms the divides between the Paraná and Uruguay River basins (Aristóbulo del Valle–San Pedro localities). It consists of rounded hills, separated by shallow fluvial valleys that locally correspond to dry valleys, with flat bottom and without a channel. The pattern of that drainage net is different to the present one. The view of the water-divide lines in the field is formed by those convex hills intercalated by depressions, all formed by gentle slopes. That suggests the importance of the fluvial process on landscape modelling.

In the Oberá area, the *Aristóbulo 2 Surface* is formed by convex hills of 1–3 km long and 20 up to 30 m high, with respect to the general level of the landscape (ca. 370 m a.s.l.). There, the fluvial erosion was controlled by the joints and fractures of the basaltic rock (NW–SE main direction), evolved to valleys 5 up to 7 km long and 0.8 up to 2 km wide. Isolated tableland hills also appear (400–425 m a.s.l.).

The *Aristóbulo del Valle Surface* dominates on the following tributaries basins of the Uruguay River in Misiones province: the Pepirí Guazú, the Yabotí and the El Soberbio (Figs. 11, 12 and 13). A representative view of the *Aristóbulo del Valle Surface* is obtained from different sites of the Provincial Araucaria Reserve, at the headwaters of the Yabotí River basin (San Pedro locality). In the hillslope profile that limits this surface with the lower erosion surface, different unit profiles were identified, according to the definitions of Wood (1942). An active scarp segment of steep slope angle is represented by a gravity (or derivation) slope. The downslope talus (or debris) slope segment has a lower gradient than the upper segment of the slope, and it is composed of sedimentary materials from upslope by mass wasting

Fig. 19 Geomorphology of the Upper Uruguay River basin (Argentine–Brazilian border). (**a**) Relict islands generated by erosion of the lower fluvial terrace of the Uruguay River, in the area of Iraí. General view of the landscape formed by the *Paraguaçu cycle;* (**b**) lower fluvial terrace of the Uruguay valley, connected to the hills of the *Paraguaçu cycle* in the Mondai–Itapiranga area; (**c**) panoramic view of Yabotí River valley, incised (*Paraguaçu cycle)* into the *Aristóbulo/Velhas Surface;* (**d**) the Moconá waterfalls along the Uruguay bedrock channel. (**e**) View of the hills of the *Apóstoles 1 Surface*. Incised meander of the Uruguay River into the *Apóstoles 1 Surface* near El Soberbio; (**f**) view of the Encantado waterfall, formed by a 1st-order tributary of the Cuña Pirú River on the erosional scarp that connects the *Aristóbulo del Valle 1 Surface with the Aristóbulo 2 Surface*, in central Misiones province

processes, surface wash and subsurface water movement. The deposit that composes this linear slope is mainly formed by an accumulation of basaltic blocks. The waning slope (pediment) is a rectilinear-concave upward erosional element, produced by surface wash that connects to the lower slope segment, represented by the valley floor and dominated by alluvial processes.

A typical stratigraphical profile outcropping at the hillslopes of fluvial valleys of the area begins from the base to the top of the profile with (1) a fresh basaltic rock, (2) a saprolite (weathered basalt), (3) an agglomerate formed by angular basaltic blocks in a fine clastic matrix and (4) a fine aeolian material (filling topographic depressions). A typical depression of the *Aristóbulo/Velhas erosive surface,* situated near the Yabotí River headwaters, is 6 m deep and 300 m in diameter. Its bottom is formed by basaltic rock and it is partially occupied by temporary swamps (Fig. 20c).

In general, a more fluvial integration is observed on the *Aristóbulo 2 Surface* in comparison to the *Aristóbulo 1.* There, 1st-order channels are in general 2 m wide and 1 m deep, with high gradients (with pebbles as bed load). The 2nd-order channels occupy deep valleys (20–30 m, 10 m wide and lower longitudinal gradients), with fluvial bars including fallen trees from the forest covering the valley slopes. Field data indicate that each 2nd-order channel of the tributary subbasins of the Uruguay (Argentine margin) has a waterfall (10–16 m high) generated by the knickpoint at the *Aristóbulo 2/Apóstoles 1 Surfaces* (Fig. 20d).

A 52 km geological transect along the southwestern border of the El Soberbio River basin, following the provincial route 13 (San Vicente–El Soberbio; Fig. 22), is presented in Fig. 23. Both *Aristóbulo Surfaces* dominate on this basin (Fig. 10). It is notable in the stream longitudinal profile the knickpoint at about 275 m a.s.l., coinciding with the scarp between the *Apóstoles 1* and *Aristóbulo 2 Surfaces* (Fig. 13). The outcropping profiles of the *Aristóbulo 1 Surface* (around 500–550 m a.s.l.) at the El Soberbio basin divides are represented by weathered basalt covered by fine aeolian material, except in an area of the divide, where the saprolite of basaltic rocks forms the surface, also reaching higher altitudes. The *Aristóbulo 2 Surface*, with a mean altitude of 400 m a.s.l. and an extension of 10 km, is composed of fresh basaltic rocks (Fig. 20d), except for some sites where weathered basalt outcrops (Fig. 24e, f).

Representative Fluvial Basins of the Aristóbulo Del Valle Surface

The representative tributary basins of the right margin of the Uruguay River in Misiones with an extended *Aristóbulo Surface* are the El Soberbio, the Yabotí and the Pepirí Guazú River basins (Figs. 2 and 10). Their bedrock channels are incised meanders occupying fluvial valleys (50–70 km long, deep and narrow with high gradients), covered by the forest. The bedrock bottoms are in general occupied by the channel partially covered by basaltic blocks and gravels and including rapids and waterfalls. The sinuous channels are locally controlled by the fractures and joints of the basaltic rocks.

Fig. 20 Geomorphology of the Misiones province (northeastern Argentina). (**a**) View of the Yabotí River valley near the mouth in the Uruguay River; (**b**) the Moconá waterfalls and a view of the Argentine *Apóstoles 1 Surface;* (**c**) a swampy environment occupying a circular depression into the inherited relief on *Aristóbulo Surface*, at ca. 550 m a.s.l. and near San Pedro locality (Provincial Araucaria reserve); (**d**) weathered basaltic slope of a knickpoint affecting the profile of a lower order channel. Slope retreating by disruption of the rock and alteration of joints; (**e**) view of the Uruguay River valley incised into the *Apóstoles 1 Surface;* (**f**) another view of the Uruguay River valley. Tableland hills correspond to *Apóstoles 1 Surface*. The steps at the middle of the hillslopes would be associated to the *Apóstoles 2 Surface*

Fig. 21 (**a–f**) Different views of the stepped landscape of southeastern Misiones province near the Uruguay River valley (northeastern Argentina). Narrow butte tops and planned top hills represent the *Apóstoles 1 and 2 Surfaces*. Look at U-palaeovalleys carved on these surfaces, whose base levels are in higher elevations than the present Uruguay River channel (**a**). (**g–h**) View of the typical profile of hillslopes

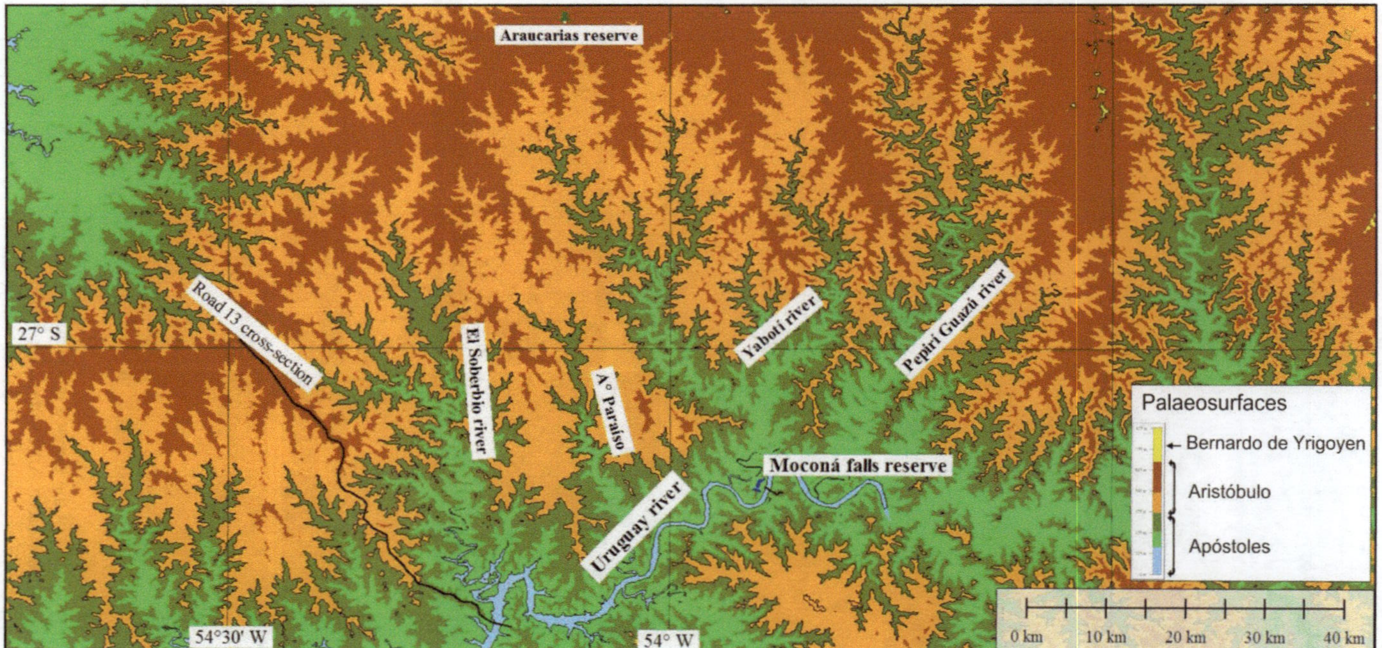

Fig. 22 Representative area dominated by the presence of the *Aristóbulo Surface*, in El Soberbio, Yabotí and Pepirí Guazú River basins (northeastern Misiones province). Points and tracks of field observations are also represented

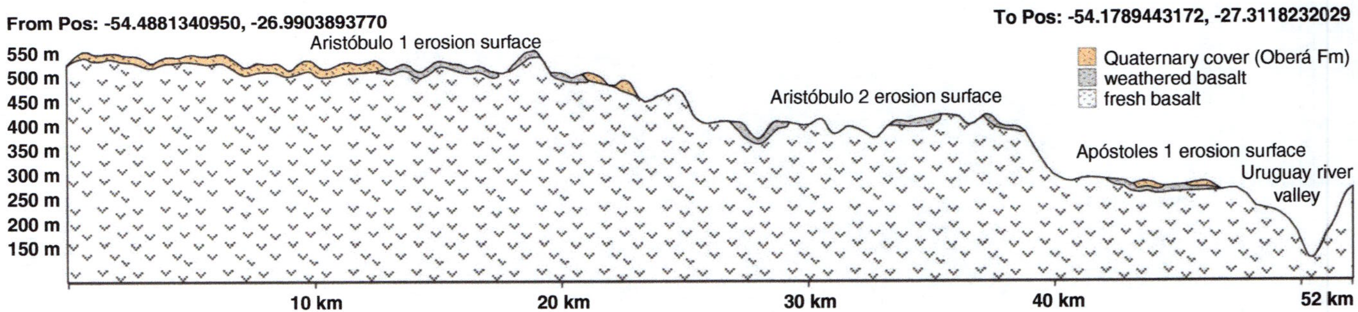

Fig. 23 NW–SE cross section along the route 13 (see Fig. 22). The section corresponds to the west divide of the El Soberbio River basin. Discrete segments were interpreted as different erosion surfaces

Fig. 24 Views of the landscape of southeastern Misiones province (Argentina). (**a**) *Apóstoles 3 Surface* at the Uruguay River valley; (**b**) the surface of the lower terrace of the Uruguay River connected to the slopes of the *Apóstoles 3 Surface;* (**c**) the Itacaruaré River valley near the mouth in the Uruguay River; (**d**) the Acaraguá River valley near the mouth in the Uruguay; (**e–f**) detailed views of the outcropping profiles formed by weathered basalt, controlled by internal anisotropies of the rock

The Apóstoles Surface/Paraguaçu (Argentina, Brazil)

The *Paraguaçu* cycle of deep fluvial incision on the *Velhas/Aristóbulo Surface* and the parallel retreat of slopes along the major valleys to their coalescence generated an extended erosion surface at the southeastern of the province of Misiones and the southwestern of the Rio Grande do Sul State (RGS, Brazil), defined as *Apóstoles Surface 3*. The area of Apóstoles in Argentina and the area between São José and São Borja (RGS) represent the typical landscape of this surface. It is formed by convex, low and long hills with very gentle and simple slopes, alternating with wide and very shallow fluvial valleys that have consumed much of the interfluves bearing remnants of the *Aristóbulo Surface*. Relative small fluvial drainage subbasins, formed by minor river channels of low gradient and low morphogenetic potential, are typical in the present landscape. The *Apóstoles Surface* is generally covered by fine aeolian material. The base level for this lower surface in the Upper Uruguay basin (around 100 m a.s.l.; Table 1, Fig. 8) was the Uruguay River.

Northeastward of the locality of San Javier (Misiones)–Porto Xavier (Brazil), the *Paraguaçu cycle* is expressed by narrow valleys of direct tributaries of the Uruguay River, near their mouths (e.g. El Soberbio River basin; Fig. 13). The *Apóstoles Surface 3* forms the present bottom of the subbasins of the southeastern of Misiones, of low gradients (Fig. 10; Acaraguá, Fig. 14, and Itacaruaré River basin, Fig. 15). A typical landscape of the Apóstoles 1 and 2 Surfaces is seen at the Uruguay River valley in that area (Fig. 21a–f). In the southwestern area of the Brazilian Uruguay River basin, the *Paraguaçu* cycle is also represented by the bottom of the tributary fluvial basins near the collector (*Apóstoles 3*). In those areas, the remnant of the *Aristóbulo/Velhas Surface* constitutes long and narrow crests along the interfluves of the rivers subbasins, mainly represented by the intermediate *Apóstoles Surfaces 1 and 2* (Fig. 20e, f). In the province of Misiones, between the L. Alem–Cerro Azul areas, a relict long tableland hill, locally known as Sierra del Imán (325 m a.s.l.), corresponds to the *Apóstoles 1 Surface*. At the southwestern external border of the Apóstoles Surface, discontinuous ferruginous crusts were formed.

The Valley of the Uruguay River Incised into the Basaltic Plateau

The fluvial collector of the Upper Uruguay basin presents a typical pattern of incised meanders into the basaltic plateau, with structural control related to the join and fractures patterns of the rock. Their headwaters are at the large Serral Geral erosive scarp (1,500–1,600 m a.s.l.), where the collector is named Pelotas River (Brazil) (Fig. 18a). There, the river occupies a deep valley of high gradient and with lateral slopes of high gradient (35–50°), with a bottom occupied by a river channel of 150 up to 250 m wide with local rapids (Fig. 18b) and few rock islands generated by lateral fluvial erosion. The incised meander valley of the Pelotas River (with a meander wavelength of 3.3 km in the first 100 km to 9.3 km wavelength in the downstream segment of the river, 250 km long) was generated on the *Sul-Americana Surface* (Pd1), before the formation of the *Velhas/Aristóbulo Surface* (Figs. 16 and 18e). At the confluence of the Pelotas and Canoas River (near the locality of Barracão, Brazil), the Uruguay River (with a river channel up to 450 m wide and a wavelength of 9.7 km) occupies a deep valley that is gradually widening downstream. It has excavated tens of metres in the *Paraguaçu erosive cycle* in the *Velhas/Aristóbulo Surface* (Pd2; Fig. 9). The valley slopes are mainly represented by the gravity slope (basalt, with slope gradients of 40–50° favouring the fall processes) and the talus or debris slope, composed of coarse basaltic blocks in a brown clayey silt matrix, today stabilised by the forest. Locally, both fluvial terraces (up to 200 m wide in the Chapecó area) appear in the valley floor of the Uruguay (Fig. 18f). The river channel upstream of the Argentine–Brazilian border has a variable wide (between 200 and 950 m) and forms composite meanders (characterised by a mean wavelength of 13.3 km, with curvature radius of 2–3 and 5 km, respectively). Fluvial islands (100 up to 1 km long and 5 up to 25 m wide) appear as an erosive rest of the lower fluvial terrace (some relicts of the terrace have 200 m wide) (Fig. 19a).

Along the international border, the fluvial valley is mainly excavated into the *Velhas/Aristóbulo Surface* (Fig. 19a, b, c). The fluvial erosive process of lateral widening is at present very active; because of that the lower terrace (100 m wide and at 7 m above the river during normal periods) on both river margins is under erosion. At El Soberbio locality (near the mouth of the El Soberbio River), the Uruguay River valley is 3 km wide and 80 m deep (Fig. 19e). Their hillslopes are mainly formed by the free face (40°), the talus or debris slope and the waning slope or pediment. It is a rectilinear-concave upward erosional slope, produced by surface wash that connects in general to the *Apóstoles 3 Surface* (Figs. 9 and 10). This erosion surface is located ca. 15 m above the river. The valley-floor basement contains the lower fluvial terrace. Above El Soberbio, the fluvial channel is 200 m wide; downstream of it, the river channel reaches up to 800–1,000 m wide (locally with extreme values of 1.6 km). The meander wavelength along this segment is between 8.2 and 9 km. Some islands, up to 4,000 km long and 200–600 m wide, appear as relict areas of the lower fluvial terrace. Levees are common in some segments of the lower fluvial terrace.

The Hillslopes of the Uruguay River Valley (Apóstoles Surface) in the Area of the Moconá Ecological Reserve (Misiones)

A geomorphic catena (a sequence of linked slope units) comprising a typical segment of the Uruguay River valley slope, in a minor elevated tectonic block (Moconá area; Figs. 19d, 20b and 25), provided evidences on landscape processes. In general, the arrangement of units within the hillslope profile of the river valley is the upper wash slope or convex slope (the *Apóstoles 1 Surface)*, the gravity (or derivation) slope (the *Apóstoles 2 Surface)*, the talus (or debris) slope and the footslope ramp (Fig. 26). Morpho-sedimentological characteristics of the footslope ramp are the following: the gradient of the ramp-like slope is ca. 15°. It is ca. 2 km long and it is covered by a discontinuous regolith deposit (from 0.20 up to 2 m in thickness). The slope deposit is stratigraphically and texturally heterogeneous; it is composed of weathered basaltic rock fragments (fine to medium residual blocks) in a fine matrix (red silt). This deposit could be considered as a "taluvium", a term that in the sense of Ward (2004) is bridging the gap between talus (composed of rock fragments) and colluvium (fine material only). The development of taluvium from talus has been related to weathering of the rock, although its formation by

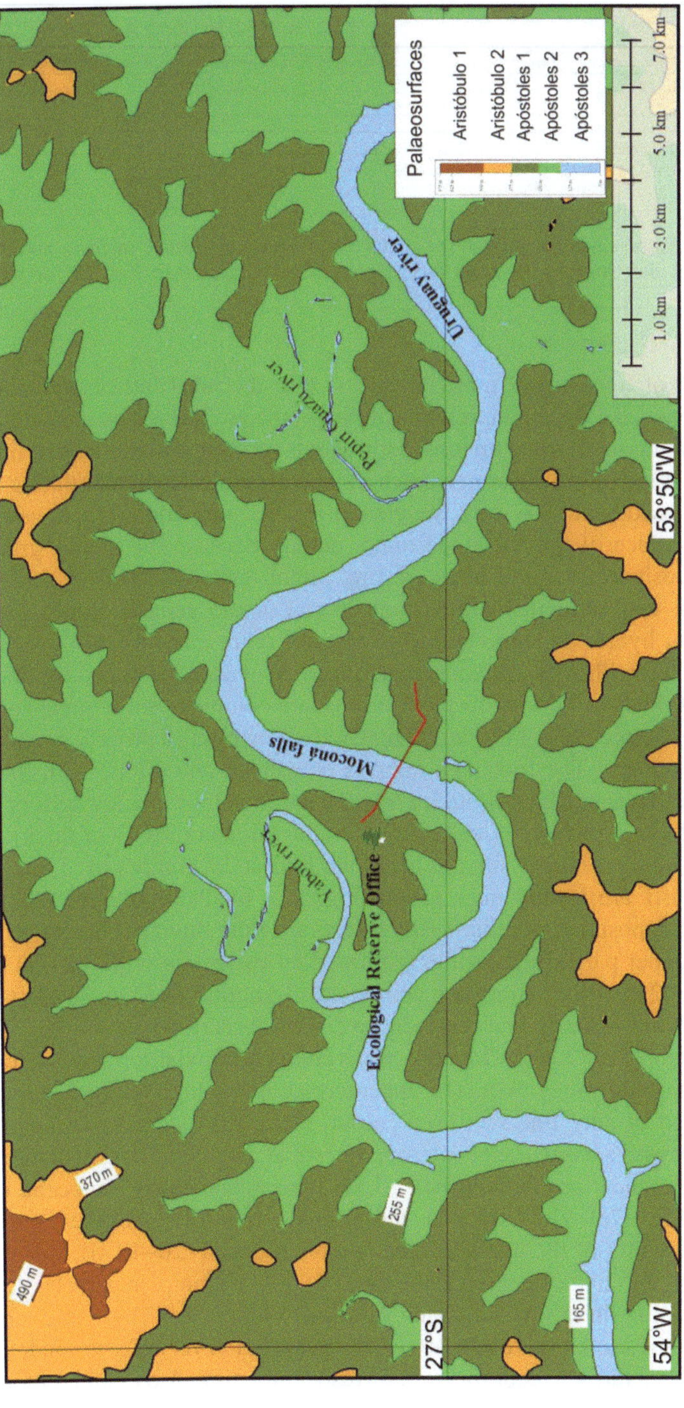

Fig. 25 Detailed DEM-based map of erosion surfaces discriminated in the area of the Moconá–Yabotí Biosphere Reserve/Parque Estadual do Turvo (at the northeastern Argentina–southwestern Brazilian border). The landscape is dominated by the *Apóstoles surfaces*

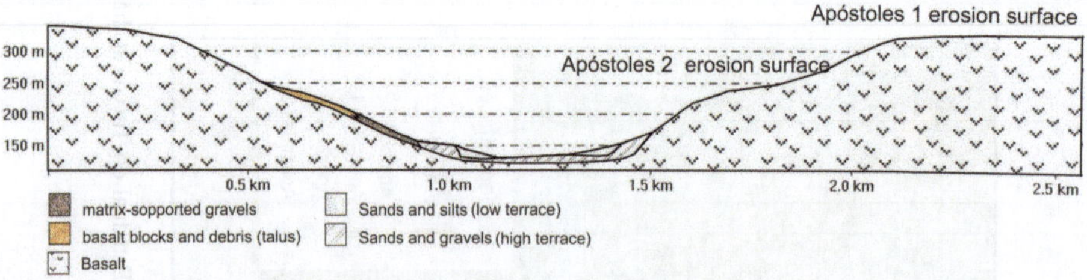

Fig. 26 NW–SE cross section through the contact between the *Apóstoles surfaces* and the slopes of the valley of the Uruguay River in the Moconá Reserve area (see Fig. 25 for location of toposequences)

incorporation of finer airfall material (loess) into the talus lattice, has also been proposed. Hence, the upper centimetres of the profile are formed by a discontinuous aeolian silty material. It filled up local depressions forming patches from 5 up to 7 m in diameter. This aeolian cover generates a slope profile with an undulating surface (0.10–0.30 m wide and 3–6 m in wavelength). Soil creep and surface wash affect the present surface of the footslope ramp, mainly because the slopes gradient is below the threshold for rapid mass wasting. Small grooves (up to 3 m wide and 0.20 m deep; with smooth margins, transitional to the general slope) generated by the present hydric dynamic and a microrelief represented by hollows and big tree roots mixed with rock fragments and the silty matrix, produced by the fall of some trees of the forest during tropical storms, are common. Subsequent soil creep produces a rearrangement of the coarse materials, generating a process of smoothing of the ramp surface. The footslope ramp is connected downslope with the upper terrace of the Uruguay River through a transitional footslope of a few tens of metres wide. This slope is formed by colluvial and alluvial materials (3 m thick). At the base appears massive clayey silt, red in colour. Fine strata composed of fine gravels in a silty matrix are locally intercalated (Fig. 27a, b). Palaeocurrent data of this fluvial gravel deposit are parallel to the direction of the fluvial collector (Uruguay River). The upper part of the profile is formed by two strata with internal erosive contacts (Fig. 27b). Both strata are composed of angular and coarse clasts (poor selected) in a sandy matrix that represents a distal debris slope.

Downstream of the known Moconá waterfalls, two fluvial terraces were developed on the basaltic valley floor. The upper terrace is 6 m high on the lower terrace. It is composed of silty fine sand, red colour, massive to laminated (modified by bioturbation), with a basal gravel outcropping. An archaeological level at the middle of the stratigraphical profile suggests a Late Pleistocene–Early Holocene age for this terrace (Fig. 27c). The terrace is well formed on both sides of the river. 600 m upstream, the width of the upper terrace is more reduced (20–40 m). It is directly contacted with the coarse material at the base of the slope ramp, in part characterised by coarse basaltic blocks accumulated by gravity processes. There, the terrace is composed of fine sand with few lenses of fine gravels and very coarse sand, without sedimentary structures, except for local diffuse lamination. A dynamic of torrential

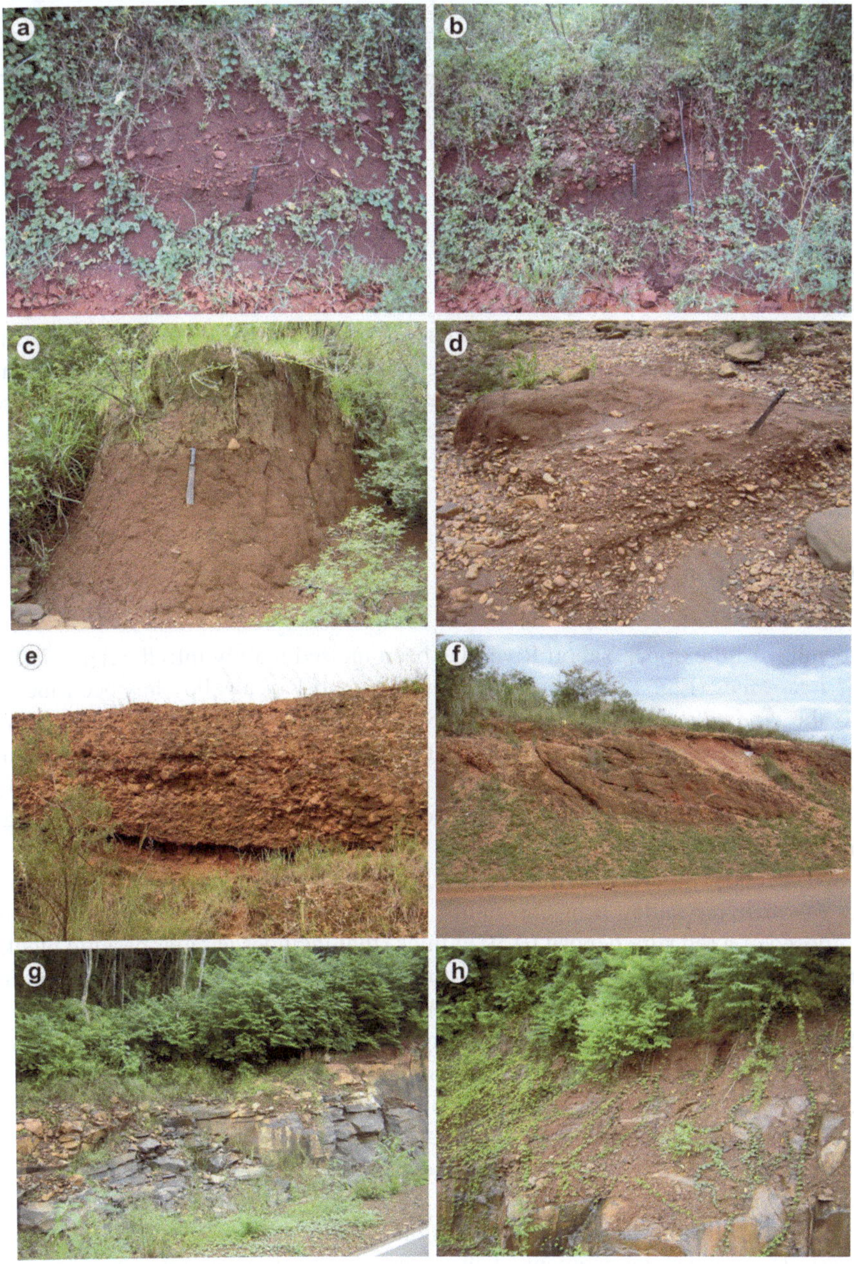

Fig. 27 Views of the linked slope units described in a hillslope of the Uruguay valley in the Moconá area (northeastern Misiones). (**a**–**b**) Footslope ramp deposits. A contact between the gravitational and water flow accumulation is observed; (**c**) fine-grained sediments forming the upper fluvial terrace of the Uruguay River; (**d**) coarse-grained materials of the Uruguay River lower terrace; (**e**–**f**) outcropping profile of the Uruguay River upper terrace in the San Javier–Apóstoles area (southeastern Misiones); (**g**) rockfall talus forming part of the slopes of the Yabotí valley; (**h**) hanging tributary palaeovalleys along the slopes of the Yabotí valley

sedimentary accumulation is interpreted from the upper terrace deposit. The bedrock channel is partially covered by very probable pre-Quaternary fluvial conglomerate. The basaltic composition of this deposit is similar to that forming the scarce present fluvial banks. On the Argentine margin of the Uruguay River, the lower terrace is 100 m wide and consists of fine sand, reddish brown in colour, including lenticular strata formed by coarse gravels and well-rounded (with low sphericity) boulders (Fig. 27d).

Downstream, in general the valley of the Uruguay River is composed of the same toposequences: the intermediate Apóstoles 1 and 2 Surfaces, the talus (or debris) slope, and the footslope ramp, connected to the fluvial terraces.

At the locality of San Javier, characterised by the extended *Apóstoles 3* surface (southwestern Misiones) close to the Uruguay River appears a level of fluvial terrace (108 m a.s.l.). It outcrops 20 m above the present level of the river. It is composed of basaltic and siliceous gravels in scarce sandy matrix (Fig. 27e), fining upwards and organised in strata (dipping 10–19° to the river channel) (Fig. 27f).

The Yabotí River Valley

The valley of the upper Yabotí River has been incised mainly into the *Aristóbulo 2 and 1 Surfaces* (Fig. 12), whereas the lower Yabotí River valley has been incised into the *Apóstoles 3 and 2 Surfaces* (Figs. 10, 19c and 25). The lower valley was formed by widening of the immediate area of the channel (pedimentation and pediplanation). The valley, representative of the Paraguaçu cycle, constitutes a 60 m deep canyon (Fig. 20a). The river bed (ca. 50 m wide) contains rapids and bars formed by basaltic pebbles and gravels.

Sedimentary outcrops at the divide between the Uruguay River valley and the Yabotí River valley near the mouth, formed by the *Apóstoles Surface 1*, are composed of a red silty deposit (0.50 m thick) that includes angular basaltic blocks. An aeolian contribution is deduced from this material. It covers the weathering basalt. The hillslope characterising the *Apóstoles Surface 2* (the gravity slope) is affected by episodic hydric erosion (1st-order channels), resulted in deep and narrow valleys. Horizontal levels are observed on the hillslope surfaces, controlled by the weathered basalt and the petrologic structure of the original volcanic rocks (alveolar, amygdaloid and fluidal structures) (Fig. 24e) and also the joint and fracture patterns. The regolith (3–5 m thick) resulted in blocks and pebbles, mainly as a product of rock exfoliation. These coarse materials are a source of rounded clasts (Fig. 24f) for the mass wasting processes affecting the slopes (and locally forming a debris slope). The downslope segment of the valley is the waning slope, a rectilinear-concave upward slope, produced by surface wash that connects to the lower slope segment (the footslope ramp, with a gradient of 9° in its upper part), formed by a scree deposit (Fig. 27g). The concave segment of the ramp is in general connected to a narrow fluvial terrace of the Yabotí, located 6 m above the channel. The terrace deposit is in general formed by subangular basaltic gravels and blocks and it is connected by a subvertical slope. Downslope of it is the lower fluvial terrace (at 1.5 m above the present river), composed of a silty deposit. Locally, the outcropping profile of the basaltic slope of the valley shows minor fluvial channels (10–20 m wide and 2–6 m deep) filled with massive fine sediments (Fig. 27h). These were interpreted as hanging tributary valleys (at 60 m above the present level of the river); below of them the Yabotí River has excavated a deep channel (canyon 35 m deep). On the right margin of the Yabotí, the lower fluvial terrace is connected with a red silty deposit, including in some points fine gravel lenses, that forms a 4 m high vertical profile. Upslope this slope segment is linked to the subvertical/vertical slope segment of the canyon, formed by weathered basalt. A debris slope with a steep gradient (50°) locally appears downslope of the rocky hillslopes. This rockfall talus is formed by coarse angular blocks and gravels in a silty matrix (agglomerate with more than 15 % matrix) that downslope connects to the ramp footslope (with a gradient of 9°), composed of coarse alluvial materials, 0.50–1.5 m thick (subangular or sub-rounded coarse to fine gravels and pebbles in a silty matrix).

An evolutive frame from the interpretation of the morpho-stratigraphical characteristics of the lower Yabotí River valley may be as follows: (1) development of the river basin under humid climate, with a base level located 60 m above the present bedrock channel; (2) base level change and modification of the drainage network, possibly related to tectonics and excavation of the river canyon, leaving hanging and inactive tributary valleys; (3) installation of a dry subtropical climate that generated the mass deposits of the valley hillslopes and locally it triggered sheet flow processes; (4) Aeolian dust accumulation and subsequent formation of mixed materials on the ramps and infilling of the tributary valleys by the loessic material; and (5) installation of a humid tropical climate. The incision of the canyon continued, in consequence, the steepened valley walls below the knickpoint favoring the formation of waterfalls in some of the present fluvial tributaries of the Yabotí.

At the confluence between the Yabotí River and the Uruguay appears the lower terrace of the Uruguay (25 m above the present cannel), a loamy deposit, including pebbles and gravels. The terrace is affected there by rotational landslides, generating a stepped morphology of the river margin.

The Los Muertos River Valley

A minor tributary subbasin of the Uruguay (Los Muertos River Basin, 30 km long) has its upper basin mainly developed on the *Aristóbulo 2 Surface* and its lower basin on the *Apóstoles Surface* (Fig. 10). The valley slopes are nearly totally covered by the forest. At 3 km downstream of the water divides, the valley of Los Muertos begins with a vertical waterfall, 30 m high (at the *Aristóbulo 1/Aristóbulo 2* scarp). Its associated plunge pool is partially covered by coarse basaltic blocks. Upstream the knickpoint, the river (7–10 m wide and 0.40 m water proof, with a discharge of 0.33–0.55 m^3/s; the suspended sediment suspension concentration is less than 1 ppm) cuts a debris deposit, 1 m thick, formed by middle to coarse angular blocks.

The *Aristóbulo 1 Surface* there (580 m a.s.l.) is formed by a regolith on basalt. In a section of the river valley located 7 km downstream, each footslope ramp on both sides of the river is covering a pediment elaborated on basalt. These ramps are formed by fine reddish brown sediment, with torrential strata at the base, formed by pebbles and gravels. The ramp deposit, 2–2.5 m thick, was dated by TL in ca. 3.5 ka B.P. The next section, 7 km downstream, is represented by a wider valley, with a level of fluvial terrace (100 m wide on the right margin and 300 m wide on the left margin), dated by TL in ca. 10 ka. B.P. (Iriondo et al. 2001), and a channel flanked by levees, 1.5–8 m high and 10–20 m wide (TL dated in ca. 1.5 ka B.P. At the river mouth, the river reaches 20 m wide, with 1 m water proof during normal periods). There, the fluvial terrace deposit is 5.5 m thick. This bedrock river shallow channel contains erosive forms (e.g. potholes up to 1.5 m deep) and point bars formed by fine (upstream) to medium-coarse (downstream) sub-rounded to rounded basaltic pebbles and fine-medium blocks. The mouth bar is formed of clayey sand, with high concentration of vegetal remains (TL dated in ca. 1.1 ka B.P.)

The main characteristics of the Los Muertos valley evolution are as follows: (1) the upper valley is generated by the evolution of pediments through a central line (channel), and meanwhile the lower valley was elaborated by lateral fluvial erosion; (2) the Holocene deposits reproduce the previous landforms elaborated on the basalt (mainly a ramp footslope on the upper basin and a fluvial terrace on the lower basin); (3) the inorganic colloids are an important component in the sedimentary process and the consequent landscape evolution; (4) the bedrock channel is excavated on the basalt, with fluvial levees; (5) the river bedload is formed by pebbles, that mobilises during the floods; and (6) the suspended load is very low.

The Acaraguá River Valley

The *Aristóbulo 2 Surface* is dominant on the Upper Acaraguá River basin. *The Apóstoles 3* and *Apóstoles 1 Surfaces* occupy most of the lower basin (Figs. 10, 14 and 24d). There, both *Apóstoles Surfaces* are represented by low hills, with gentle slopes; the basin divide corresponds to the higher planned top hills (*Aristóbulo 2;* 450 m a.s.l.).

The Cuña Pirú River Valley

This tributary basin of the Paraná River is mainly developed on the *Apóstoles 2 Surface* (Fig. 14). In general, its 1st-order channels have a waterfall at the very narrow surface that connects the *Aristóbulo 2 Surface* with the *Apóstoles 2 Surface* (elevation difference: 260–270 m). The Encantado waterfall is the most scenic of them (Fig. 19f). The knickpoint has a complex morphology that from upstream to downstream is composed of a series of rapids (slope gradient: 30–40°), a vertical waterfall 70 m high connected to a plunge pool (200 m in diameter) and a lower segment with high gradient forming rapids. The present control of the erosion by the *Aristóbulo 2 Surface* is markedly showed by isobase maps (Fig. 17a, b).

The Itacaruaré River Valley

The *Apóstoles Surface 3* is mainly preserved in the Itacaruaré River basin (Fig. 15). Its headwaters are on a narrow relict (ridge) of the *Aristóbulo 2 Surface* that corresponds to the Sierra del Imán (325 m a.s.l.). Near its mouth, the river occupies a wide valley (800 m), with a flat valley floor and a channel (15 m wide) including bars and islands covered by the forest and locally a fluvial terrace (Fig. 24c). The valley is elaborated on a general landscape of long and low basaltic hills, smoothened by a Quaternary mantle.

Discussion

The planation surfaces of the Paraná Basaltic Plateau of southern Brazil have been classically interpreted as the result of pedimentation processes, although Paisani et al. (2008) proposed that the planation surfaces of Paraná and Santa Catarina States (southern Brazil) would be developed by etchplanation processes. They concluded that the type of lava flow has little influence on the development of planation surfaces but that the subtropical climate had a very important role on the morphology of the surfaces. According to Ollier and Pain (2000), it is unlikely that weathering causes general surface lowering, in the most of landscapes. Where weathering profiles are deep, it is found that constant volume alteration features are present within a metre or so of the top of the weathered rock. If the volume remains the same, there can be no general surface lowering due to weathering. Our field data suggest that the general surface lowering was mainly by erosion, not by weathering. Most fluvial erosion was concentrated in tributary valleys of the Paraná and Uruguay Rivers destroying flat surfaces and generating a new erosion surface below. Local base levels, represented by these large river collectors, controlled the fluvial erosion. Uplift and rejuvenation of valleys before the new surface can be graded to the lower level were required. Then, the pediplanation slope retreat away from drainage lines is more applicable to understand the landscape evolution of the region, also considering that lateral planation and floodplain development by mainstreams is common in humid areas like the study region. Field data also indicate that superficial geomorphic processes like surface wash and mass movement were also important. Where the regolith or saprolite was thin and the valley slopes were steep, slope-eroding processes were driven mainly by rock fall, debris flow or washload transport.

According to the data of Japsen et al. (2012b), related to the absolute timing and magnitude of discrete burial and exhumation events that conditioned the modelling of erosion surfaces in the margin of northeastern Brazil, the Campanian, Eocene and Miocene uplift phases coincide with three main phases of Andean orogeny, which occurred during periods of relatively rapid convergence at the Andean margin of South America—Peruvian (90–75 Ma), Incaic (50–40 Ma) and Quechuan (25–0 Ma). Those authors suggested that all these uplift events have a common cause, which is lateral resistance to plate motion. On the other hand, considering the reconstruction of the deformational history of the obliquely rifted margin of southeastern Brazil by Cobbold et al. (2001) and Meisling et al. (2001), Late Cretaceous and Cenozoic reactivation of older structures could be most important. Cobbold et al. (2001) attributed them to the combined effects of far-field stresses and hot-spot activity, defining two episodes of post-rift exhumation of mounting ranges during the Late Cretaceous and Eocene, based on fission-track data. Alkaline intrusions were attributed to the Trindade hot spot and interpreted as emplaced along reactivated strike-slip faults and transfer zones. The structures originated in a transtensional stress regimen became inverted during Neogene transpression. Hence, in an on-shore area, relatively distant of the passive margin, Gallagher et al. (1994), Cobbold et al. (2001) and Meisling et al. (2001) recognised two episodes of thermal reactivation associated to extensional regimen in the Late Cretaceous (90–80 Ma.) and the Palaeogene (50–40 Ma.) and a phase of Neogene uplift (25–0 Ma.) linked to the Andean compressive context.

Taking into account the interpretations of Assumpçao (1992, 1998) and Lima et al. (1997), away from the local flexural disturbances of the continental margin, the observed stress regimen in southeastern Brazil is in good agreement with theoretical stress model of the South American plate, derived from finite element modelling (Meijer and Wortel 1992; Coblentz and Richardson 1996). This regional regime is a compressive stress field oriented roughly E–W to ESE–WNW (Ricomini and Assumpçao 1999). The authors hold that although the major positive geoid anomalies in Brazil are mainly related to thermal reactivation associated with alkaline magmatic activity during the Late Cretaceous–Tertiary, they correlated well with uplifted areas by neotectonics and seismic activity in several regions of Brazil. Hence, the last uplift phase seems to be associated to far-field stresses, probably restricted to the Pliocene–Pleistocene, as Bigarella (1975) suggested for de Paraguaçu erosion cycle.

Data supplied by Torsvik et al. (2009; see Fig. 2 in that work) suggest the magmatic and hot-spot activity that generated thermal high gradient was the main responsible of uplift during the Late Cretaceous and the Palaeogene. These data correlate with the timing proposed by Cobbold et al. (2001) for the alkaline magmatic activity.

Independently of the invoked causes for the uplift events, there would be an agreement in the existence of three principal uplift episodes post-break-up of Gondwana, which could correlate with the three main King's erosion cycles, making an adjustment in the chronological sequence. We propose here that the main recognised discrete erosion surfaces in the Paraná Basaltic Plateau (northeastern Argentina and southern Brazil) from quantitative geomorphologic data can be correlated to these uplift events, establishing a chronological frame based on numerical data. A more complex history of uplift-subsidence events could account for the stepped landscape in a hydrographical basin distant of the coastal area, like the Uruguay River basin. Many other discrete planation surfaces observed in this work and in Paisani et al. (2008) in the neighbouring Iguazú River basin reinforce that deduction.

More knowledge about variations in tectonic forcing, the base-level history and the Quaternary climate changes of the region will give important information for the interpretation of landscape evolution and especially taken into account a genetic connotation. Also, some difficulties of separating local base-level controls from structural controls could be solved. The studied sequence of planation surfaces may provide a history of uplifting, once accurate dating of surfaces in the studied region can be generated. It will permit to place each of the identified planation surfaces in a detailed chronological sequence.

Conclusions

The morphometric analysis is an appropriate method for the reconstruction of long-term landscape development of the large Paraná Basaltic Plateau. SRTM data constitute a good resource for this analysis. Planation surfaces in the present-day landscape of the Upper Uruguay River basin (southern Brazil and northeastern Argentina) may be identified in digital elevation models. Remnants of planation surfaces in general appear on water divides after the landscape had been dissected. The generation of hypsometric curves in six representative tributary basins of the Uruguay basin and also in one small basin tributary of the Paraná River permitted to identify, classify by height range and map the main Late–Cretaceous–Cenozoic planation surfaces. The altitude range of the occurrence of each surface and its correlation over wider areas was obtained. Other morphometric parameters such as longitudinal profiles and isobase lines, complementary with field data, were utilised also to corroborate such erosion surfaces.

The modelling of different stages of evolution of the erosion surfaces in a key area of the province of Misiones (Argentina) by using isobase maps allowed making predictions about how pediplanation processes will continue in the future. It constitutes a present analogous example to understand the formation of ancient erosion surfaces of the Paraná Basaltic Plateau.

The planation surfaces recognised in this research indicate the number of erosional cycles experienced by the landscape. There is a good correlation between our recognised erosion surfaces and the King's planation surfaces: *Sul-American Surface* (Late Cretaceous-Palaeogene), the *Velhas Surface* (Palaeogene) and the King's *Paraguaçu cycle* (Plio-Pleistocene). The higher erosion surface (880–1,080 m a.s.l.) has been affected by several climatic periods and by a wide range of surface processes. In the Upper Uruguay River, remnants of this old surface appear in the Pelotas River area in southern Brazil.

The higher remnant surface exposed of the basaltic plateau in Argentina is restricted to a narrow area at the northeastern of Misiones province and found at altitudes between 675 and 880 m a.s.l. It corresponds to a secondary planation surface named *Bernardo de Irigoyen Surface*. The *Velhas Surface* is mainly preserved on divides as remnants, being represented by the plane-top watershed between the Paraná and Uruguay River basins, with a NE–SW direction in the central part of the province of Misiones. It was designed as *Aristóbulo 1 Surface* (490–675 m a.s.l.). Because of the proximity of these large fluvial collectors, pediplanation and pedimentation processes leaved a narrow remnant that locally are restricted to ridges of planed tops. The *Apóstoles 3 Surface* (110–165 m a.s.l.) is an extended erosion surface at the southeastern of Misiones and the southwestern of the Rio Grande do Sul State of Brazil. Secondary surfaces (*Aristóbulo 2 Surface and Apóstoles 1 and 2 Surfaces)* were identified based on morphometric analyses, taking into account that small relicts at the same level suggest that they could be remnants of a formerly extensive plain. For those planation surfaces identified in the study region, the base level for erosion was provided by the Uruguay or the Paraná Rivers. The steeper longitudinal profile observed in a tributary of the Paraná River (the Cuña Pirú River), compared to the closest Uruguay tributary (the Acaraguá), points out that erosion is more active on the Paraná fluvial system. The steeper sections, which show knickpoints, result from outcrops of the basalt and/or the active headward erosion by valley development and rock landslides. The presence of the erosion surfaces also appears to exert control on these processes.

References

Ab'Sáber AN (1969) Participação das superficies aplainadas nas paisagens do Nordeste Brasileiro, vol 19, Geomorfologia. Instituto de Geografia, Universidade de São Paulo, São Paulo, pp 1–38

Ab'Sáber AN (2000) The natural organization of Brazilian inter and subtropical landscapes. Rev Inst Geol 21(1/2):57–70

Almeida F (1956) O Planalto basáltico do Bacia do Paraná. Bol Paul Geogr 24:3–34

Almeida F (1986) Distribução regional e relações tectónicas do magmatismo pós-Paleozóico no Brasil. Rev Bras Geociênc 15:325–349

Aslanian D, Moulin M, Olivet J-L, Unternehr P, Matias L, Bache F, Rabineau M, Nouze H, Klingelhoefer F, Contrucci I, Labails C (2009) Brazilian and African passive margins of the Central Segment of the South Atlantic Ocean: kinematic constrains. Tectonophys (Special Issue: Role of Magmatism) 468:98–112

Assumpçao M (1992) The regional intraplate stress field in South America. J Geophys Res 97(8):11889–11903

Assumpçao M (1998) Focal mechanisms of small earthquakes in the southeastern Brazilian shield: a test of stress models of South American Plate. J Geophys J Int 133:490–498

Assumpçao M, Schimmel M, Escalante C, Barbosa JR, Rocha M, Barros LV (2004) Intraplate seismicity in SE Brazil: stress concentration in lithospheric thin spots. J Geophys J Int 159:390–399

Bezerra F, Brito Neves B, Corrêa A, Barreto A, Suguio K (2008) Late Pleistocene tectonic-geomorphological development within a passive margin – the Cariatá trough, northeastern Brazil. Geophys J Roy Astron Soc 97:555–582

Bigarella J (1975) The Barreiras Group in northeast Brazil. An Acad Bras Cienc 47:365–393
Bigarella J, Andrade G (1965) Contribution to the study of the Brazilian Quaternary. Geol Soc Am Spec Pap 84:433–451
Bigarella J, Becker R (1975) Topics for discussion, international symposium on the quaternary (Southern Brazil). Bol Parana Geociênc 33:169–272
Bigarella J, Mousinho M, Silva J (1965) Pediplanos, pedimentos e seus depósitos correlativos no Brasil. Bol Parana Geogr 16(17):117–151
Büdel J (1957) Die 'Doppelten Einebnungsflachen' in den feuchten Tropen. Z Geomorphol NF 1:201–228
Büdel J (1982) Climatic geomorphology (trans: Fischer L, Busche D). Princeton University Press, Princeton, 443 pp
Castro ACM Jr (1987) The northeastern Brazil and Gabon Basins; a double rifting system associated with multiple crustal detachment surfaces. Tectonics 6(6):727–738
Cobbold PR, Meisling KE, Mount VS (2001) Reactivation of an obliquely rifted margin, Campos and Santos Basins, southeastern Brazil. AAPG Bull 85(11):1925–1944
Coblentz DD, Richardson RM (1996) Analysis of the South American intraplate stress field. J Geophys Res 101(B4):8643–8657
Davis WM (1899) The geographical cycle. Geogr J 14(5):481–504
Davis WM (1909) Geographical essays. Ginn and Co., Boston, 777 pp
Erhart H (1955) Biostasie et rhexistasie: esquise d'une théorie sur le rôle de la pédongenèse en tant que phénomène géologique. Comptes Rendus Academie des Sciences française 241:1218–1220
Fairbridge R, Finkl C (1980) Cratonic regime, unconformities and peneplains. J Geol 88:69–86
Filosofov V (1960) Brief guide to morphometric methods in search of tectonic structures. Saratov University Public House, Saratov (Russian language)
Gallagher K, Hawkesworth C, Mantovani M (1994) The denudation history of the onshore continental margin of S.E. Brazil inferred from fission track data. J Geophys Res 99:18117–18145
Golts S, Rosenthal E (1993) A morphotectonic map of the northern Arava in Israel derived from isobase lines. Geomorphology 7:305–315
Goudie AS (2004) Cycle of erosion. In: Goudie AS (ed) Encyclopedia of geomorphology, vol 1, A–I. Routledge, London/New York, pp 213–214
GRASS Development Team (2005) Geographic Resources Analysis Support System (GRASS), GNU General Public License. Electronic document, http://grass.itc.it
Grohmann C (2004) Morphometric analysis in Geographic Information Systems: application of free software GRASS and R. Comput Geosci 30:1055–1067
Grohmann C, Riccomini C, Machado Alves F (2007) SRTM-based morphotectonic analysis of the Poços de Caldas Alkaline Massif, southeastern Brazil. Comput Geosci 33(1):10–19
Huggett RJ (2007) Fundamentals of geomorphology. Routledge, London/New York, p 463
Iriondo M, Kröhling D (2004) Cenozoic geomorphological evolution of the Uruguay River Basin, South America. In: International Geological Congress (IUGS), 32, Florence. Abstracts, 1 pp
Iriondo M, Kröhling D (2008) Cambios ambientales en la cuenca del Uruguay-desde dos millones de años hasta el Presente. Colección Ciencia y Técnica, Ediciones UNL (Universidad Nacional del Litoral), Santa Fe, Argentina, 360 pp
Japsen P, Chalmers J, Green P, Bonow J (2012a) Elevated, passive continental margins: not rift shoulders, but expressions of episodic, post-rift burial and exhumation. Global Planet Change 90–91:73–86
Japsen P et al (2012b) Episodic burial and exhumation in NE Brazil after opening of the South Atlantic. Geol Soc Am Bull 124(5–6):800–816. doi:10.1130/B30515.1
Jarvis A, Reuter H, Nelson A, Guevara E (2008) Holefilled seamless SRTM data V4. International Centre for Tropical Agriculture (CIAT). Available from http://srtm.csi.cgiar.org
Justus J, Machado M, Franco M (1986) Geomorfología. Levantamento de Recursos Naturais. Folha SH.22 Porto Alegre e parte das folhas SH.21 Uruguaiana e SI.22 Lagoa Mirim, 33, 2. Fundação Instituto Brasileiro de Geografia e Estatística (IBGE), Rio de Janeiro, pp 313–392
King LC (1953) Canons of landscape evolution. Geol Soc Am Bull 64:721–752
King LC (1956) A Geomorfologia do Brasil Oriental. Rev Bras Geogr, IBGE 2:147–265

King LC (1957) The uniformitarian nature of hillslopes. Trans Edinb Geol Soc 17:81–102
King LC (1967) The morphology of the earth, 2nd edn. Oliver & Boyd, Edinburgh/London, 726 pp
King LC (1983) Wandering continents and spreading sea floors on an expanding earth. Wiley, Chichester, 232 pp
Kirkby M (2004) Slope, evolution. In: Goudie AS (ed) Encyclopedia of geomorphology, vol 2, J–Z. Routledge, London/New York, pp 963–968
Kröhling D, Iriondo M, Brunetto E, Galina G (2009) Evolución del paisaje de la meseta basáltica en la alta cuenca del rio Uruguay. In: IV Congreso Argentino De Cuaternario y Geomorfología, 4, Congresso Da Associação Brasileira De Estudos Do Quaternário, 12 and Reunión sobre el Cuaternario de América del Sur, 2, La Plata. Actas de Resúmenes, La Plata, Argentina, 1 pp
Kröhling D, Brunetto E, Galina G, Zalazar MC (2011) Palaeosurfaces analysis on the Cretaceous basaltic plateau on the Upper Uruguay River basin (NE Argentina and Southern Brazil). Geociências 30(1):31–46
Lagorio SL (2008) Early Cretaceous alkaline volcanism of the Sierra Chica of Córdoba (Argentina): mineralogy, geochemistry and petrogenesis. J S Am Earth Sci 26:152–171
Lima CC, Nascimento E, Assumpção M (1997) Stress orientations in Brazilian sedimentary basins from breakout analysis: implications for force models in the South American plate. Geophys J Int 130:112–124
Lister GS, Etheridge MA, Symonds PA (1986) Detachment faulting and the evolution of passive continental margins. Geology 14(3):246–250
Lister GS, Etheridge MA, Symonds PA (1991) Detachment models for the formation of passive continental margins. Tectonics 10(5):1038–1064
Maack R (1947) Breves noticias sobre a geología dos Estados do Paraná e Santa Catarina. Arq Biol Tecnol 2:63–154
Meijer PTH, Wortel MJR (1992) The dynamics of motion of the South American plate. J Geophys Res 97(8):11915–11931
Meisling KE, Cobbold PR, Mount VS (2001) Segmentation of an obliquely rifted margin, Campos and Santos basins, southeastern Brazil. AAPG Bull 85(11):1903–1924
Melfi AJ, Piccirillo EM, Nardy AJR (1988) Geological and magmatic aspects of the Paraná basin – an introduction. In: Piccirillo EM, Melfi AJ (eds) The Mesozoic flood volcanism of the Paraná Basin: petrogenetic and geophysical aspects. Inst. Astr. Bn. Geotis. Publ, São Paulo, pp 1–13
Migón P (2004) Planation surface. In: Goudie AS (ed) Encyclopedia of geomorphology, vol 2, J–Z. Routledge, London/New York, pp 788–792
Milani EJ, Zalán PV (1999) An outline of the geology and petroleum systems of Paleozoic interior basins of South America. Episodes 22(3):199–205
Milner SC, Duncan AR, Whittingham AM, Ewart A (1995) Trans-Atlantic correlation of eruptive sequences and individual silicic units within the Paraná-Etendeka igneous province. J Volcanol Geotherm Res 69:137–157
Ollier CD (1991) Ancient landforms. Belhaven Press, London/New York, 233 pp
Ollier CD (1993) Age of soils and landforms in Uganda. Israel J Earth Sci 41:227–231
Ollier CD (2003) The origin of mountains on an expanding earth, and other hypotheses. In: Scalera G, Jacob KH (eds) Why expanding earth? A book in honour of Ott Christoph Hilgenberg. INGV (InstitutoNazionale di Geofisica e Vulcanologia), Rome, pp 129–160
Ollier CD (2012) Extension everywhere – rifts, continental margins and island arcs. In: Scalera G, Boschi E, Cwojdzinski S (eds) The Earth expansion evidence – a challenge for geology, geophysics and astronomy selected contributions to the interdisciplinary workshop of the 37th international School of Geophysics, EMFCSC, Erice, 4–9 Oct 2011, pp 61–76
Ollier CD, Pain CF (2000) The origin of mountains. Routledge, London, 345 pp
Paisani JC, Pontelli ME, Andres J (2008) Superfícies aplainadas em zona morfoclimática subtropical úmida no planalto basáltico da bacia do Paraná (SW Paraná/NW Santa Catarina): primeira aproximação. Geociências 27(4):541–553
Peate DW, Hawkesworth CJ, Mantovani MSM (1992) Chemical stratigraphy of the Paraná lavas (South America): classification of magma types and their spatial distribution. Bull Volcanol 55:119–139

Penck W (1953) Morphological analysis of landforms (trans: Czech H, Boswell KC). Macmillan, London, pp 1–18
Peulvast JP, Sales VC, Bétard F, Gunnell Y (2008) Low post-Cenomanian denudation depths across the Brazilian northeast: implications for long-term landscape evolution at a transform continental margin. Global Planet Change 62:39–60
Popolizio E (1972) Geomorfología del relieve de plataforma de la provincia de Misiones y zonas aledañas. An Soc Argent Estud Geogr XV:17–82
Ramos VA (1996) Evolución tectónica de la plataforma continental. XIII° Congreso Geológico Argentino y IIII° Congreso de Exploración de Hidrocarburos. In: Ramos VA, Turic MA (eds) Geología y Recursos Naturales de la Plataforma Continental Argentina. Relatorio 21, pp 385–404
Riccomini C, Assumpção M (1999) Quaternary tectonics in Brazil. Episodes 22:221–225
Soares P, Riffel BS (2006) Hypsometric curves as a tool for paleosurface mapping. Math Geol 38(6):679–695
Steiner S (2007) Aquisição e processamento de dados morfométricos derivados do modelo digital de elevação SRTM. Dissertação (Mestrado) – Instituto de Geociências, Universidade de São Paulo, São Paulo, 38 p
Strahler A (1952a) Dynamic basis of geomorphology. Geol Soc Am Bull 63:923–938
Strahler A (1952b) Hypsometric (areal-altitude) analysis of erosional topography. Geol Soc Am Bull 63:1117–1142
Summerfield MA (1984) Plate tectonics and landscape development on the African continent. In: Morisawa M, Hack JT (eds) Tectonic geomorphology, The symposia on geomorphology, international series no. 15. George Allen & Unwin, Boston, pp 27–51
Thomas MF (1966) Some geomorphological implications of deep weathering patterns in crystalline rocks in Nigeria. Trans Inst Br Geogr 40:73–193
Torsvik TH, Rousse S, Labails C, Smethurst MA (2009) A new scheme for the opening of the South Atlantic Ocean and the dissection of an Aptian salt basin. Geophys J Int 177:1315–1333
Tricart J (1956) Comparaison entre les conditions de façonnement des lits fluviaux en zone temperée et. zone intertropicale. C R Acad Sci 245:555–557
Tricart J (1959) Observations sur le façonnement des rapides des rivières intertropicales. Bull Sect Geogr Com Trav Hist Sci 68:333–343
Twidale CR (1983) Pediments, peneplains and ultiplains. Rev Geomorphol Dyn 32:1–35
Ussami N, Karner GD, Bott MHP (1986) Crustal detachment during South Atlantic rifting and formation of Tucano–Gabon Basin System. Nature 322(6080):629–632
Vasconcelos P, Carmo ID (2008) Calibrating denudation chronology through 40Ar/39Ar weathering geochronology. In: 33rd international geological conference, Abstract CD-ROM, Oslo, 1 p
Ward S (2004) Taluvium. In: Goudie AS (ed) Encyclopedia of geomorphology, vol 2, J–Z. Routledge, London/New York, pp 1035–1036
Watts A (2001) Isostasy and flexure of the lithosphere. Cambridge University Press, Cambridge
Wayland EJ (1933) Peneplains and some other erosional platforms. Annual report bulletin, Protectorate of Uganda Geological Survival Department of Mines, Note 1, pp 77–79
Wernicke B (1985) Uniform sense normal simple shear of continental lithosphere. Can J Earth Sci 22:108–125
Widdowson M (1997) The geomorphological and geological importance of paleosurfaces. In: Widdowson M (ed) Paleosurfaces: recognition, reconstruction and paleoenvironmental interpretation, Special publication, 120. Geological Society of London, London, pp 1–12
Wobbe F (2007) Geomorphological parameter extraction in GRASS GIS-Isobase map calculation. TU Bergakademie Freiberg, Institut für Geologie. http://www.rsg.tu-freiberg.de/twiki/bin/view/Main/GeomorphologyGrassGIS
Wood A (1942) The development of hillside slopes. Proc Geol Assoc Lond 53:128–140
Zuchiewicz W, Oaks JRR (1993) Geomorphology and structure of the Bear River Range, NE Utah: a morphometric approach. Z Geomorphol Suppl Bd. 94:41–55

Geomorphology of Paleosurfaces in the Sierras de Comechingones, Central Pampean Ranges, Argentina

M. Jimena Andreazzini and Susana B. Degiovanni

Abstract Remnants of pre-Andean erosion paleosurfaces have been already described by several authors in different sectors of the cratonic area of the Sierras Pampeanas of Argentina. Their origin and age are still under discussion, but in general, it is assumed that they correspond to etchplains and pediplains developed between the Middle-Late Jurassic and the Paleogene. The objective of this chapter is to present a morphological-morphometric characterization of the planation surfaces that are preserved in the summit portions of the Sierra de Comechingones (Sierras Grandes, Córdoba), between 32° 22′–32° 52′ S and 64° 45′–64° 57′ W, and propose genetic-evolutionary models that would consider the incidence of lithology and fracturing degree on the configuration and preservation of these paleolandforms. These features are located between 2,150 and 1,500 m a.s.l. and they are developed upon the Precambrian gneissic-migmatitic rocks (the Monte Guazú Complex), mylonites and ultramylonites (the Guacha Corral shear zone), and Devonian granites (the Cerro Áspero Batholith). The latter are accompanied by fluorite epithermal mineralizations (Early Cretaceous). The paleosurfaces were affected by the Andean orogeny and modeled by the Neogene erosion cycle, being partially covered by Quaternary loess units. To complete this analysis, petrographic, structural, stratigraphic, sedimentological, metalogenetic, and geochronological studies were performed both at the local and regional scales, thus generating our own information in the field and in the lab. Based upon topographic sheets, satellite images, and digital elevation models (DEM), the paleosurfaces were identified and mapped, with special attention to the different lithology on which they have been carved, the drainage network and the depth of fluvial incision were defined, the morphometric parameters were obtained (such as slope, area, elevation, drainage density, and mean length of first-order stream channels [Lm1]), and six topographic-geological

M.J. Andreazzini (✉) • S.B. Degiovanni
Departamento de Geología, Universidad Nacional de Río Cuarto, Ruta Nacional 36, Km. 601, X5804BYA Río Cuarto, Córdoba, Argentina
e-mail: jandreazzini@yahoo.com.ar

J. Rabassa and C. Ollier (eds.), *Gondwana Landscapes in southern South America*, Springer Earth System Sciences, DOI 10.1007/978-94-007-7702-6_11,

sections, considered as representative of the regional conditions, were prepared. In the field, the existing landforms and morphology aspects, active processes, and lithology sections, among other parameters, were described. In the metamorphic environment, the erosion surfaces show a homogeneous relief (generally crest-like ridges) with low topographic and morphological variability (slope, 2–4 %; drainage density, Dd < 4; Lm1 = 150–250 m; incision depth, 10–20 m), and, in general, they are better preserved than those corresponding to the granite environment. The latter are more heterogeneous (slope varying between 4.5 and 7.5 %); they exhibit landforms of varied size and type (crests, rounded boulders, tafoni, tors, domes, among other features) and show greater fluvial dissection (Dd, 6.5–7.5; Lm1, 70–110 m; incision depth, 100–110 m). The geomorphological analysis particularly integrated with metalogenetic and geochronological studies of the fluorite deposits and the stratigraphic and sedimentological investigation of the adjacent Cretaceous sedimentary basins allow the following considerations: (a) the studied cratonic area was a positive element of the landscape at least since the Carboniferous-Permian, and since then, it has been the subject of different denudation cycles; (b) previous to the Andean orogeny, there was a very long period of stability in which the tectonic and erosion processes, generally of low activity, favored the development of these erosion surfaces, which began to be denudated much later than these movements, with their remnants located today in the summit or water-divide areas; (c) at least since the Late Cretaceous, the Cerro Áspero Batholith was not being exhumed and, thus, the studied paleosurfaces may be assigned to a Late Cretaceous-Paleogene age; (d) the studied erosion surfaces belong to one single paleosurface and its altitudinal variations are due to the tilting during the Andean events; (e) the paleosurfaces are polygenetic, a result of the oscillation between wetter periods (dominant chemical weathering, channeled fluvial action) and more arid ones (pedimentation), thus conforming spatial arrangement of the palimpsest type; and (e) a marked lithological control on the morphology and degree of preservation of coeval paleosurfaces do exist.

Keywords Gondwana • Argentina • Sierras Pampeanas • Planation surfaces • Granite morphology

Introduction

The presence of planation surfaces in the Sierras Pampeanas region was recognized towards the end of the nineteenth century and the beginnings of the twentieth century, in the pioneering works of Stelzner (1885), Brackebush (1879, 1880, 1891), Bodenbender (1895, 1905, 1907, 1911, 1929), Rovereto (1911), Beder (1916), Rassmuss (1916), Rimann (1926), Penck (1914, 1920), and Schmieder (1921). In these first investigations, the origin, evolution, and age of these paleosurfaces were already considered controversial. Despite the time elapsed, the discussion is still alive today.

Some authors interpreted these surfaces as part of a single, unique "peneplain," in the sense of William M. Davis, which developed sometime between the Late Paleozoic and the Miocene and which was later faulted, tilted, and exhumed during the Andean orogeny (González Díaz 1981; Jordan et al. 1989; Costa et al. 1999; Beltramone 2007), whereas others have proposed a polygenetic model, strongly based on the concepts of Lester C. King (1963) and recognized several planation levels of different origin and age (Carignano et al. 1999; Carignano and Cioccale 1997, 2008; Cioccale and Carignano 2009; Rabassa et al. 1996, 1997, 2010; Rabassa 2010, 2014). The better examples of these paleosurfaces are preserved as remnants found in the higher areas (locally known as "pampas de altura") of the main mountain ranges which form these Sierras. In the Sierras Pampeanas of the province of Córdoba, several levels have been recognized between 2,000 and 500 m a.s.l., among which the Pampa de Achala and the Pampa de San Luis (2,000 and 1,800 m a.s.l., respectively), located in the Sierras Grandes, are noted for to their extension and elevation, followed in order of importance by the Pampa de Olaen (Sierras Grandes), Pampa de Pocho (Sierra de Pocho), and the summit surface of the Sierra Chica (1,500–1,000 m a.s.l.). These surfaces are generally covered by Late Cenozoic, but mainly Quaternary, colluvial, and loess sediments, with an average thickness of less than 1 m, with a few exceptions where the cover just exceeds 5 m.

According to Carignano et al. (1999) and Rabassa et al. (2010), the highest paleosurface is an "etchplain," mainly generated by deep chemical weathering under very warm and wet environmental conditions, during the Late Triassic-Middle Jurassic interval, whereas the paleosurfaces that occur in a step-like pattern around the nucleus of the larger Sierras blocks are the result of the activity of erosion agents related to mostly semiarid climates, basically by pedimentation processes, in this case corresponding to extensive denudation periods during the Late Jurassic-Cretaceous and Late Cretaceous-Miocene (Carignano et al. 1999; Cioccale 1999; Rabassa et al. 2010). Contrarily, Beltramone (2007) interpreted the different levels of erosion surfaces as corresponding to a unique pre-Jurassic peneplain, buried and later exhumed, which was dismembered by the Tertiary Andean tectonics.

The cited references show that the scientific works dealing with genetic-evolutionary aspects of the different paleosurfaces of the Sierras de Córdoba have greatly increased in recent decades, both at the local and regional scales, but much has still to be done. In the particular case of the Sierra de Comechingones, which comprises the southern part of the Sierras Grandes, there are very few studies about these relict paleosurfaces. Degiovanni et al. (2003) described the surfaces found in the southern sector of this mountain range, which are considered polygenetic (in the sense of Klein 1985). Coniglio et al. (2000) and Coniglio (2006) studied the fluorite mineralization in the Cerro Áspero Batholith (Sierra de Comechingones) and indicated that the formation of these epithermal deposits would have taken place in a scenario coeval with the development of erosion surfaces of regional extent. Finally, some papers have analyzed the Quaternary sequence found in the "pampas de altura" of these ranges, among them Andreazzini et al. (2012) and Krapovickas and Tauber (2012a, b).

To improve the knowledge of these geomorphological relicts, this chapter analyzes the planation surfaces of the north-central portion of the Sierras de Comechingones. These units are mainly preserved in the summit areas and, secondarily, in lower blocks of the eastern slope of the Sierras. The surfaces were developed on metamorphic and granitic rocks. We present a morphological and morphometric characterization of the surfaces and propose genetic-evolutionary models which discuss the influence of the lithology and degree of fracturing in the configuration and preservation of the paleolandforms.

Location of the Study Area

The observations presented here were made on the relict surfaces, located in the Sierra de Comechingones (Sierras Pampeanas de Córdoba, Argentina) between 32° 22′ and 32° 52′ south latitude and between 64° 45′ and 64° 57′ west longitude (Fig. 1). The Sierra de Comechingones constitutes the southern end of the Sierras Grandes, with an approximate length of 150 km, forming part of the boundary between the provinces of Córdoba and San Luis. Its elevation decreases towards the south, from 2,884 m a.s.l. (Cerro Champaquí) to 650 m a.s.l. at its terminal outcrops.

Regional Geological Setting

The Sierras Pampeanas are mountain ranges of roughly N-S orientation of approximately 500 km long and 200 km wide, which are mega-blocks of Precambrian-Early Paleozoic crystalline basement, faulted and tilted towards the east during the Andean (Tertiary) orogeny. They exhibit a markedly asymmetrical cross section. The western fault scarps have a steeper gradient, with variable displacements (200–1,000 m), whereas the eastern structural slopes are smoother and show well-defined steps, which have been associated with tectonics and/or erosion processes. In either case, these different levels are remnants of paleosurfaces affected by the Late Cenozoic geomorphological processes.

During Carboniferous and Permian times the Sierras were already a positive element under denudation conditions, and a thick sequence of silicoclastic sediments were deposited in the eastern area, adjacent to the ancient Pampean Arch, whose outcrops form today a narrow belt in the NW margin and the southern end of the Sierras (Hünicken and Pensa 1980a, b).

In Late Permian-Early Triassic times, an extensional regime created the Triassic rifts. Deposits of this age have not yet been identified either in the Mesozoic sedimentary column of the Chaco-Pampean Plains (Chebli et al. 1999) or in the ranges themselves, which are considered as of a positive nature. According to Carignano et al. (1999), the clastic continental sequences accumulated in the adjacent basins

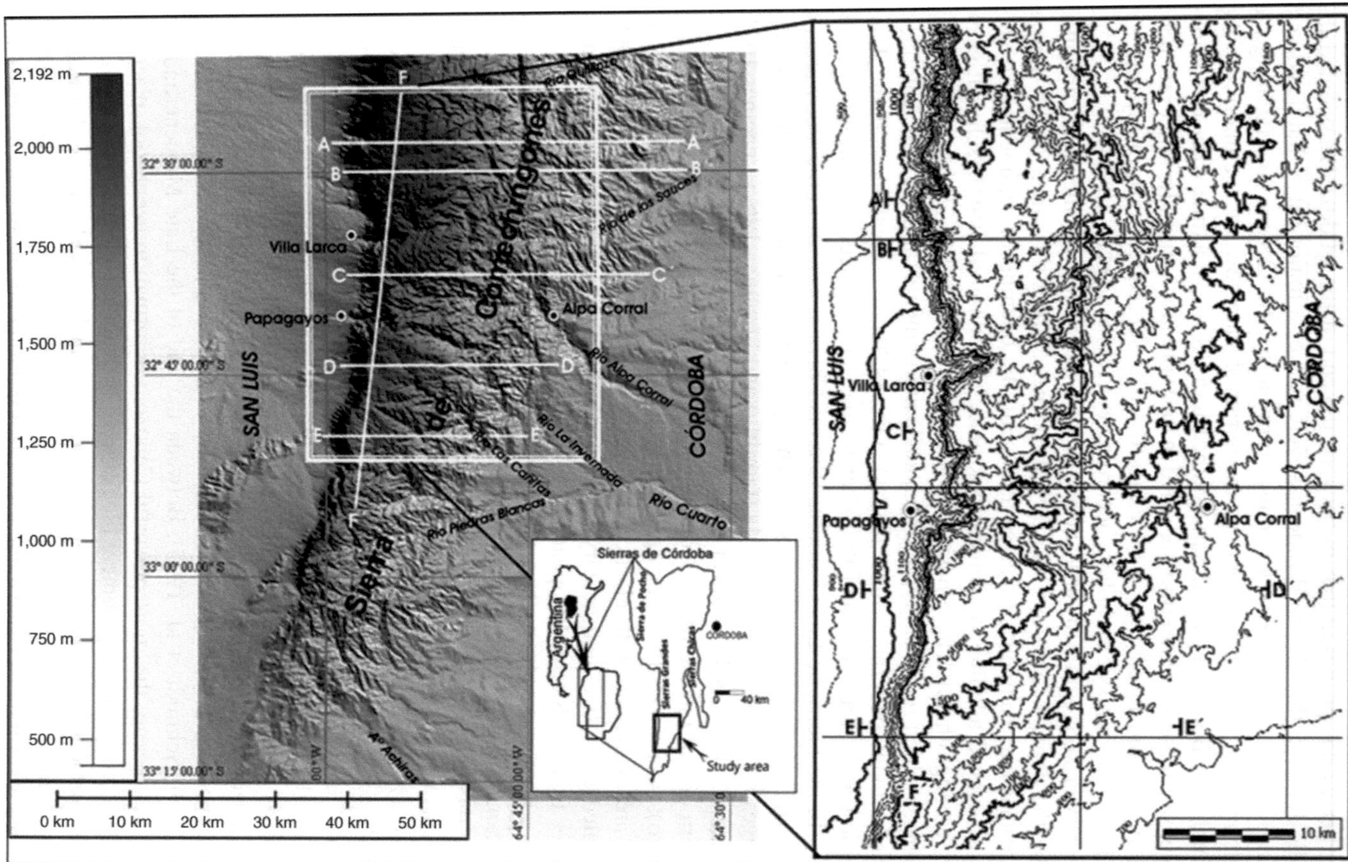

Fig. 1 Location of the study area. A DEM (digital elevation model) and details of the topography of the studied zone. The location of the analyzed sections is also shown in both figures

indicate that warm-wet environmental conditions had been relatively uniform and persistent, with some semiarid intercalated periods. These same authors, on the basis of earlier works, proposed that in between the end of the Middle Triassic and the Middle Jurassic, the Sierras Pampeanas were part of a major ridge that acted as a barrier to the moist air masses coming from the Pacific Ocean, thus favoring the development of a desert environment in the rain shadow towards the interior of the continent (the Botucatú paleodesert, southeastern Brazil), at a time when conditions of intense chemical weathering prevailed in the mountain ranges. Its relict products, such as tors and inselbergs, were exposed under subsequent erosion cycles (the Middle to Late Jurassic etchplain, according to Carignano et al. 1999).

The extension regime towards the end of the Gondwana orogenic cycle continued during the entire Cretaceous and the Paleocene. During this time, in the Sierras Pampeanas and Pampean Plains environment, the opening of sedimentary basins with associated alkaline volcanism took place (Uliana et al. 1990; Schmidt et al. 1995; among others). These basins are generally filled by two sedimentary mega-sequences which represent typical "bajada" environments (alluvial fans, calcretes, braided fluvial patterns, saline playas). One of them corresponds to the Late Jurassic-Early Cretaceous and the other to the Late Cretaceous (Carignano et al. 1999). In the Sierra de Los Cóndores (Sierras Chicas), east from the study area, a volcanoclastic sequence is exposed which represents the first cycle, where basalts have ages between 154 and 112 Ma (Gordillo and Lencinas 1967; Lencinas 1971; González 1971; Sánchez 2001). The second cycle has been recognized in the southern end of the Sierras de Comechingones (Cerro La Garrapata, La Leoncita, Madero), where basaltic rocks aged between 55 and 75 $\pm$ 5 Ma (López and Solá 1981) are associated with clastic continental sequences. Carignano et al. (1999) correlated these two sedimentary cycles with the development of two pediment surfaces found in Sierras Pampeanas.

Since the Eocene, the Andean movements determined the evolution of the Sierras Pampeanas, raising the mountain blocks and defining intermontane basins which, during the whole Paleogene and the Early Miocene, were filled by continental sediments corresponding to alluvial fans, braided channels, and subordinate sub-aqueous environments. During the great Atlantic transgression of the Middle to Late Miocene, the Sierras Pampeanas were surrounded by a shallow epicontinental sea and, even though they remained as a positive element, the erosion rate diminished significantly. In the Miocene-Pliocene transition, the segmentation of the Andes took place and the Early Paleozoic sutures and the normal faulting of the Paleozoic and Mesozoic were tectonically inverted, thus being in most cases transformed into high-angle, reverse faults, with east-dipping fault planes (Chebli et al. 1999; Schmidt et al. 1995). Between the end of the Late Miocene and the Early Pleistocene, the most powerful Cenozoic deformation event took place. In this period, as the large basement blocks were fragmented, elevated, and tilted, the erosion surfaces were dismembered and underwent long and intense denudation cycles (González Díaz 1981; Carignano et al. 1999; Degiovanni et al. 2003). There were two main episodes: one in the Pliocene and another in the Pleistocene heralded by a volcanic period that covered in some areas the third level of pedimentation (Carignano et al. 1999).

Finally, the Quaternary of the Pampean region is represented by an alternation of loess and alluvium, and to a smaller extent, lacustrine and colluvial deposits, in which paleosols and calcretes are interbedded. These sequences are in general associated with climatic variations (mostly glacial and interglacial cycles) which characterized this period (Cantú 1992; Iriondo 1997, 1999; among others) and also with neotectonic events (Sagripanti 2006; Degiovanni et al. 2003; Costa et al. 1999).

Local Geological Setting

The geological knowledge of the Sierra de Comechingones was for a long time restricted to the work of Sosic (1964) and Gordillo and Lencinas (1979). Later works discussed petrologic, geochemical, and structural conditions of the region, both in the metamorphic and granitoid rocks (Nullo et al. 1991; Fagiano et al. 1993; Feliú 1994; Otamendi 1995; Otamendi et al. 1996; Stuart-Smith and Skirrow 1997; Coniglio and Esparza 1988; Porta 1992; Pinotti et al. 1996, 1997, 2002; Martino 2003; Fagiano and Martino 2004; Fagiano 2007; Coniglio 2006; Coniglio et al. 2010; among others).

A geological map of the study area (Fig. 2) shows the distribution of metamorphic rocks of the Monte Guazú Complex, mylonitic rocks of the Guacha Corral shearing zone, and granitic rocks of the Cerro Áspero-Alpa Corral Batholith (Fagiano 2007; Rey Ripoll 2008; Coniglio et al. 2010).

The Monte Guazú Complex is recognized as the lithostratigraphic unit with wider distribution in the Sierra de Comechingones, and it is composed of rocks which are the product of regional metamorphism. Migmatites and paragneisses dominate, of homogeneous nature and poorly developed schistosity, and oval to lens-like bodies of tonalite and granodiorite orthogneisses are interposed. Second in importance are the amphibolites forming elongated, tabular, and elliptical outcrops. Small to medium granitic, simple pegmatite bodies, concordant with the metamorphic rocks, are common in the area. Veins of hydrothermal quartz, either concordant or discordant with bedrock structure, have been identified (Fagiano 2007).

The Guacha Corral shear zone is one of the more extensive found in the Sierras de Córdoba. It extends from the southern portion of the Achala Batholith to the southern end of the Sierra de Comechingones, with an approximate length of 120 km. It has a varying width with a maximum of 20 km in the northern part, but in its medial section is segmented by the intrusion of the Cerro Áspero Batholith. In the southern portion, the shear zone is in numerous narrow bands 100–200 m thick (Fagiano and Martino 2004). It is a wide zone of ductile deformation of N-S orientation, which provides a penetrating, mylonitic foliation superimposed to a previous one of regional development. The contacts between the shear zone and the migmatitic complex are gradual, and it is common to find sequences composed of, from the outer to the inner part of the belt, deformed stromatites, protomylonites, mylonites, and ultramylonites.

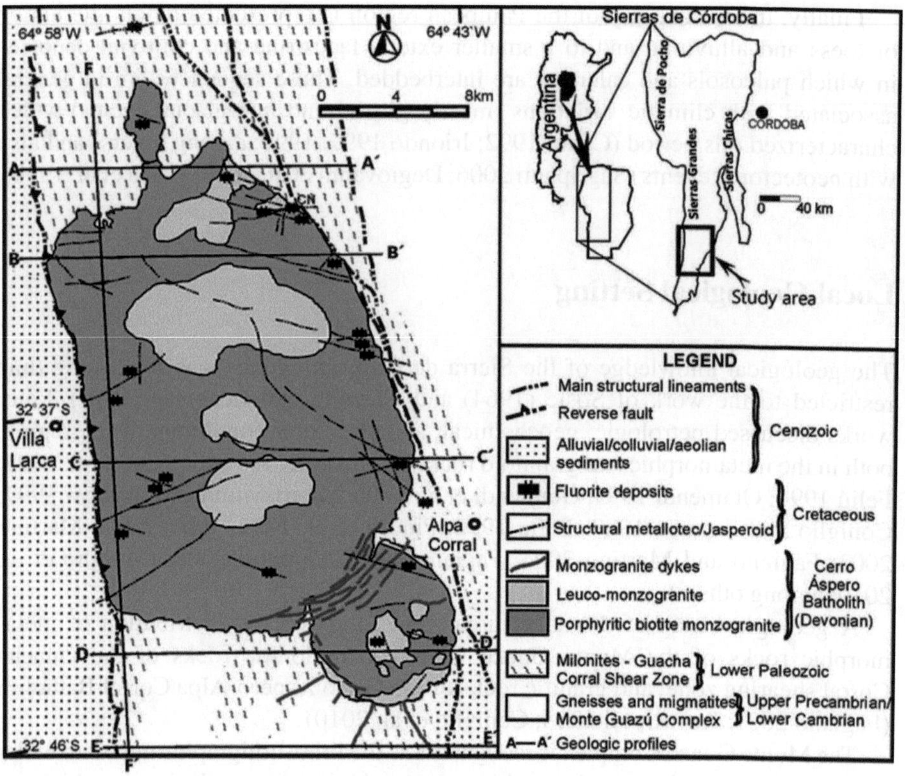

Fig. 2 Geological map of the study area (Modified from Coniglio et al. 2010). The location of six geological sections discussed in this chapter is also shown

The Cerro Áspero Batholith intrudes the medial section of the Monte Guazú Complex. It has an area of 440 km^2 and is the most important granitic complex of the Sierra de Comechingones. It is of Middle to Late Devonian age (Pinotti et al. 2002). According to these authors, this batholith was built up by the successive emplacement of three plutonic bodies (Los Cerros, El Talita, and Alpa Corral, located from N to S, respectively) which have individually developed facies variations (inner, outer, and summit units and a suite of associated dykes). As shown in Fig. 2, the inner units comprise most of 60 % of the batholith surface, and they are represented by very coarse-grained, porphyritic biotite monzogranites (Pinotti et al. 2002). In the outermost and summit units, as well as in the dykes, two-mica or exclusively muscovite leucogranites are dominant, whose composition varies between monzogranites and alkaline feldspar granites (Coniglio 2006).

Closer to the El Talita and Alpa Corral plutons (in the S and SE contacts), a discontinuous thermal contact aureole has been described, with an approximate thickness of 500 m, formed by rocks of hornblende hornfels and albite-epidote facies (Fagiano et al. 2006).

The contacts with the metamorphic rocks and the observed field relationships indicate a post-kinematic intrusive emplacement and, associated to this magmatism, wolfram and molybdenum-bearing hydrothermal deposits and epithermal fluorite veins are found.

Galindo et al. (1997) obtained two Sm-Nd ages of 117 ± 26 Ma and 131 ± 22 Ma, from fluorite found in the Cerro Áspero Batholith (Bubú Mine) and of the Achala Batholith (La Nueva Mine), respectively. The veins are emplaced in subvertical structures, which record several stages of opening and infilling with fluorite-chalcedony-dominated epithermal mineralization. Based on mineralogical, geochemical, and stable isotope studies, Coniglio et al. (2000, 2004, 2006) proposed as responsible for their genesis the action of heated meteoric fluids which, due to fluid-rock interaction processes, interchanged isotopically with the host granites and leached part of the fluorine content. According to Coniglio et al. (2000, 2004), changes in fluorite texture, from massive towards richer in open spaces, from I to IV stages, are accompanied by a gradual cooling of the hydrothermal fluid and isotopic restrictions that suggest depositional conditions for the minerals in successively shallower levels with respect to the paleosurface (from approximately 2 km), with the present exposition of the four stages at the same level. These authors have suggested the occurrence of erosion processes during the lifetime of the hydrothermal system as a possible scenario to explain the textural evolution of this mineralization, which was probably related to the development of erosion surfaces in the Sierras Pampeanas from the Late Jurassic to the Late Cretaceous (Coniglio 2006).

Over a clear erosional unconformity, the metamorphic rocks and granites are covered by loess and loess-like Quaternary deposits (Fig. 3), with thickness of a few centimeters up to more than 5 m, which are exposed in eroded channels (Fig. 3a, b). In some places, a discontinuous level of calcrete is interbedded in the sequence; the calcrete has been dated in $4{,}180 \pm 80$ ^{14}C yr BP (Andreazzini et al. 2012). The soils developed on the Quaternary cover show a significant degree of evolution, with abundant organic matter and clayey horizons (Bt).

It must be noted that the different rock types exposed in the Sierra de Comechingones, particularly in the paleosurfaces, do not exhibit a significant weathering profile and, in general, they occur with little or no alteration, both in the valleys and on the divides (Fig. 3a–c). In most cases regolith is absent, whereas in others a partial alteration is observed (mechanical disaggregation, with minimum oxidation and clay formation) of a few decimeters.

Methodology

The methodology used in this chapter is centered upon the synthesis of information coming from different fields of the Earth Sciences, with the purpose of adding evidence which would allow the discussion of several aspects of the genesis, age, evolution, and preservation of the studied paleolandforms. Numerous previous

Fig. 3 Different contact zones between the metamorphic crystalline basement (Late Precambrian-Early Paleozoic) and the sedimentary cover (Quaternary). (**a**) and (**b**) location in valleys; (**c**) location on the divides

papers (in geology, geomorphology, geochronology, metalogenesis, stratigraphy, etc.) were used together with our own field and laboratory information to characterize the paleosurfaces developed on different rocks (basically, metamorphic rocks and granites).

By means of the analysis of topographic sheets (scale 1:50.000, Instituto Geográfico Nacional of Argentina), Landsat and Google Earth satellite images, and digital elevation models (DEM) such as SRTM (90 m resolution), the relicts of the planation surfaces present in the study area were identified and delimited. Some morphometric parameters such as slope, area drainage density, and stream length were obtained. These last two indicators were obtained from the analysis of order 3 basins, considered to be representative of the lithological environments.

In the field, the morphology, stratigraphic profiles, and active processes were described.

Based upon this information, five E-W profiles transverse to the Sierra de Comechingones, and one N-S profile was constructed. These profiles were selected considering the presence of planation surface relicts on the different lithology types observed in the study zone.

Using the SAGA GIS software, the drainage network and the map of fluvial incision depth were compiled.

Finally, an integrated analysis of the available information was completed.

Results

The results presented here concern erosion surface remnants developed on the granitic rocks of the Cerro Áspero Batholith and on older metamorphic and mylonitic rocks located both N and S of the intrusive body. The paleolandforms are located between 2,150 and 1,500 m a.s.l., occupying an approximate area of 300 km^2. These landforms are tilted towards the E and S-SE, as is observed in the topographic-geological profiles of Fig. 4.

Figure 5 shows aerial views of these ancient surfaces in different geological contexts. The paleosurface relief within the metamorphic region shows great homogeneity (Fig. 5a), whereas those of the granitic environment present a much more irregular aspect (Fig. 5b, c). Among them, an erosion bench in the order of 100–110 m (up to 200 m in some places) may be observed, as it is shown in the topographic map of Fig. 1 and in the F–F′ section (Fig. 4), although exaggerated by fluvial incision of the Arroyo Papagayos, and in the aerial view of Fig. 5b.

Due to the differences that these surfaces present in both environments in relation to topography, landforms, and drainage network characteristics, among other variables, they will be treated separately.

Planation Surfaces in the Metamorphic Rock Environment

The planation surface occupies an approximate area of 60 km^2 over the metamorphic rocks located north of the Cerro Áspero Batholith. It is found between the 2,100 and 1,900 m a.s.l. contour lines and has a variable slope between 2 and 4 % (A–A′ section; Figs. 1, 2 and 4).

South of the batholith, the paleosurface is located between 1,800 and 1,400 m a.s.l., it covers an approximate area of 90 km^2, and its slope is in the order of 3–5 % (D–D′ and F–F′ sections; Figs. 1, 2 and 4).

These surfaces show a very homogeneous relief, smoothly undulating, with minimal internal elevation benches associated with fluvial valleys, which in general do not exceed 10–15 m. They have a Quaternary sediment cover, formed by loess and loess-like materials, which is preserved over most of the area (Figs. 6a, b).

Rocky outcrops are mainly restricted to the fluvial channels, and they exhibit low relief of crest-like landforms controlled by the orientation of the metamorphic structures. In some places where orthogneisses are exposed, the local relief is more heterogeneous and, with larger steps, composed of rounded blocks, tafoni, and pedestal rocks, among others (Fig. 6d), similar to those developed on granitoid rocks.

In the drainage network, the trunk streams of higher importance (order 3 and 4) are controlled by different structures of varying strike, where the NW-SE and NE-SW systems are dominant, the drainage pattern is angular, but the lower order streams present a dendritic pattern (Fig. 7). Drainage density is generally smaller than 4 and the mean length of the order 1 channels varies between 150 and 250 m.

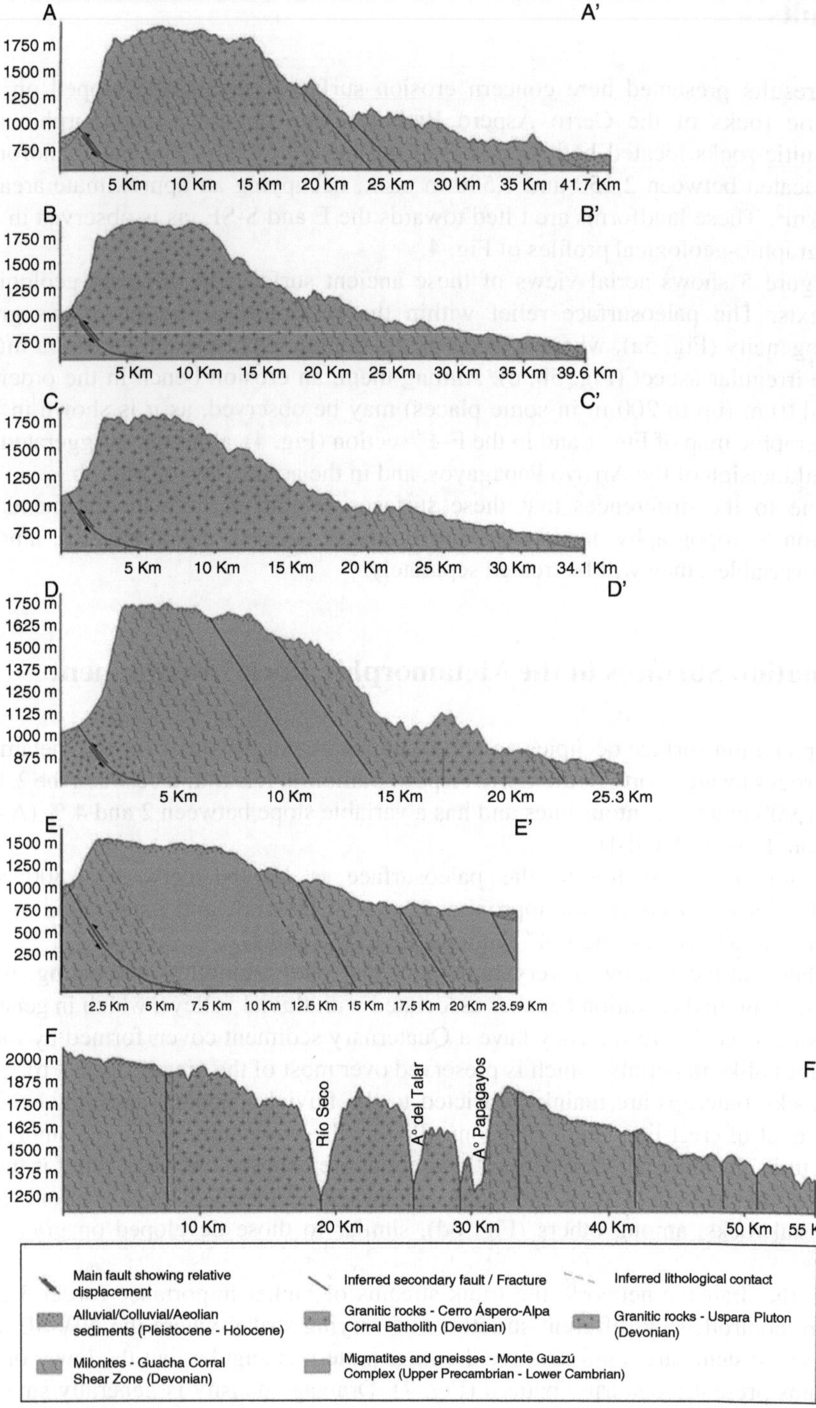

Fig. 4 Topographic-geological profiles of the central sector of the Sierra de Comechingones. The location of the profiles is presented in Figs. 1 and 2

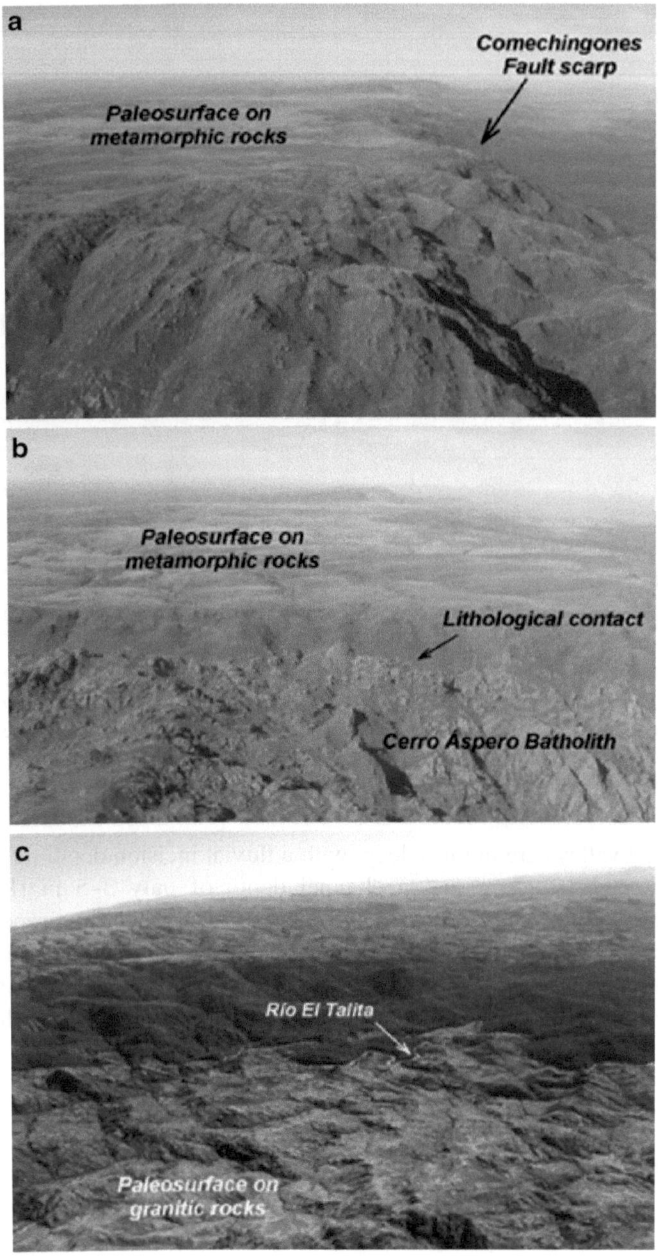

Fig. 5 (**a**) N-S view of the paleosurface on metamorphic rocks (Monte Guazú Complex, Río Las Cañitas upper basin). (**b**) N-S view of the contact between the Cerro Áspero Batholith and the paleosurface carved on metamorphic rocks. (**c**) S-N view of the paleosurface on granitic rocks (Cerro Áspero Batholith)

Fig. 6 (**a**) View of the paleosurface elaborated on the metamorphic rocks of the Monte Guazú Complex, where in the central sector channels developed on the Quaternary sediment sequence, covering the metamorphic rocks, are developed. (**b**) View of the paleosurface; at the background, the Comechingones fault scarp. (**c**) Fluvial valley carved on the paleosurface, Arroyo Los Comederos. (**d**) Mesoscale boulder type landforms, with partially developed tafoni

The fluvial valleys are not too deep, with a fluvial incision depth that in general does not exceed 10–15 m and a channel depth of only 3–5 m (Fig. 6c). In the drainage network and channel depth maps (Fig. 8), it is observed that these paleosurfaces correspond to the origin of the La Tapa, Guacha Corral, and Quillinzo stream basins and that, in these environments, the incision is normally smaller than 40 m, with dominant values of the order of 10–20 m. Locally, active rill formation processes are observed associated with the Quaternary sediments.

Planation Surfaces in the Granitic Environment

Within the area occupied by the Cerro Áspero Batholith, the remnants of the summit paleosurface occupy an approximate area of 150 km^2, with a variable slope between 4.5 and 7.5 % and an elevation between 2,000 and 1,800 m a.s.l. in the northern portion of the batholith and between 1,800 and 1,600 m a.s.l. in its southern sector (B–B′ and C–C′ profiles; Fig. 4).

Fig. 7 Google Earth satellite image showing the characteristic drainage network of the paleosurface when developed on metamorphic rocks

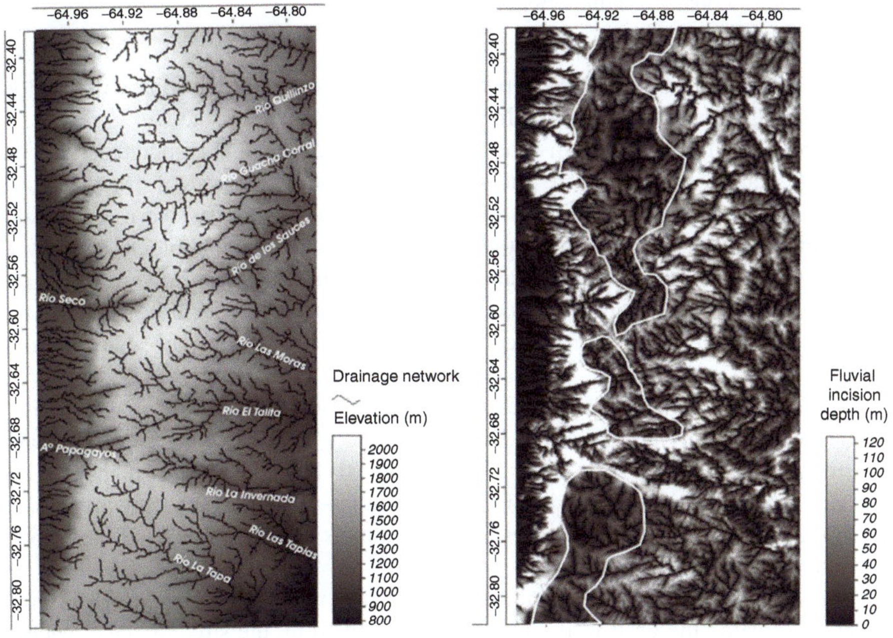

Fig. 8 Drainage network and fluvial incision depth maps, obtained from the SRTM DEM. The *whitish solid line* depicts the extent of the remnants of the paleosurface

Fig. 9 Bell-like morphology (**a** and **b**) and smaller landforms such as tors, rounded boulders, and tafoni (**c** and **d**), within the Cerro Áspero Batholith domain

The landscape of the granitic pampas combines rocky outcrops with sectors covered by Quaternary sediments. In the latter, rill formation is active.

On rocky outcrops medium-size landforms are dominant, associated with the batholith margin facies, where the rock shows a high degree of fracturing. In those portions where the granite exhibits high jointing density (western margin and the Comechingones fault scarp front), relatively small angular blocks dominate. To a lesser extent in the pampas and somewhat more common towards the east, large bell-like granitic landforms occur, like those presented in Fig. 9a, b. Minor granitic landforms such as rounded boulders, tafoni, and tors are also common (Fig. 9b–d).

This surface, unlike the surfaces on metamorphic rocks, is characterized by a larger degree of fluvial dissection. Drainage density reaches values of 6.5–7.5 and the mean length of the first-order channels is 70–110 m. As shown in Fig. 10, the drainage network pattern is of the rectangular-angular type, showing a high structural control that for the larger order streams is given by NNW-SSE and NE-SW conjugate system structures.

The valleys are only slightly incised and are very wide on the pampas (Fig. 11a), whereas towards the east, in the Cenozoic erosion front, the streams have reached a high level of incision in several sections (Fig. 11b). In these areas, the fluvial divides correspond to the level of the regional planation surface. In Fig. 8 and in the

Fig. 10 Google Earth image showing the characteristic drainage network in granitic environments

Fig. 11 (**a**) Fluvial valley in the granitic pampas; (**b**) deep stream incision section east of the Cerro Áspero Batholith

F–F′ profile (Fig. 4), the important erosion work performed by the La Invernada, El Talita, Las Moras, Papagayos, Seco, and other streams has been recognized, reaching incision depths of up to 100–110 m.

Discussion

The integrated analysis of the geomorphological, petrographic, structural, stratigraphic, sedimentological, metalogenetic, and geochronological information, both of the study area and of the regional setting, has allowed the discussion of various aspects related to the origin, age, and evolution of the studied paleosurfaces.

The mineralogical, geochemical, and stable isotope studies performed by Coniglio et al. (2004) and Coniglio (2006) on epithermal fluorite deposits found in the granitic paleosurfaces are herein of special interest. These papers indicated the large supply of meteoric waters to the hydrothermal system and the depositional conditions of these minerals which were particularly shallow (less than 2 km depth). Therefore, these contributions suggested that:

(a) During the development of the hydrothermal system, the climatic conditions of the area would have been wet enough to sustain such large water supply.
(b) The coexistence of erosion processes simultaneously with the mineralization period.

Taking into account the radiometric ages obtained by Galindo et al. (1997) on the fluorite deposits of the Sierras Pampeanas de Córdoba (Early Cretaceous) and the duration of the mineralization period dated by Jelinek et al. (1999), of approximately 50 Ma (Early Cretaceous-Early Paleogene), in similar fluorite deposits of Santa Catarina State (Brazil), Coniglio (2006) postulated that after the exhumation of the Sierras Pampeanas in the Early Carboniferous, the rate of uplifting and erosion should have been very slow, considering that the mineralization affected the summit units of the Cerro Áspero Batholith. According to this author, "although the possibility that part of the evolution of the epithermal system would have happened during the uplifting of the Sierras Pampeanas at the beginning of the Andean cycle cannot be ruled out, the banded-crust like dominium and the infilling of spaces generated in transtensive domain of the mineralization are not primarily coincident with clearly compressive kinematics."

This evidence allows us to point out that the studied paleosurfaces were being formed during the cited mineralization epoch and that, at least until the Late Cretaceous-Early Paleogene, the present levels that exhibit veins of fluorite in the Cerro Áspero Batholith which were formed at depths shallower than 2 km were not exposed yet.

The analysis of the sedimentary sequences that infilled the adjacent Mesozoic basins contributes to reinforce these interpretations. In the Sierra de los Cóndores Group (Late Jurassic-Early Cretaceous), the compositional and morphometric analyses of gravel and sandstones by Sánchez and Minudri (1993) and Minudri and Sánchez (1993) indicate that in the basal layers, the clasts are predominantly of metamorphic rocks, with some participation of pegmatite types, but towards the top of the sequence, the percentage of the pegmatites increases. This compositional change could be attributed to a gradual thickness diminution of the roof rocks of the Cerro Áspero Batholith, where increased content of such rocks is expected, suggesting that during the Early Cretaceous, the erosion surface has not reached yet the mineral-bearing plutonic rocks.

Besides, Sánchez and Minudri (1993) and Minudri and Sánchez (1993) pointed out that, in general, the studied clasts of conglomerates and sandstones pertaining to the Sierra de los Cóndores Group present a low grade of alteration and a highly variable degree of sphericity and roundness, depending upon their composition, which would be indicating moderate (or at least not extreme) weathering conditions in the source areas (i.e., the Sierras de Comechingones) for the Late Jurassic-Early Cretaceous period.

In relation to this, it should be noted that neither regolith levels have been found yet in the studied paleosurfaces nor observations about them in any of the accessed petrographic papers for this area of the Sierras Pampeanas (Fagiano 2007; Coniglio 2006; Otamendi 1995; among others). None of these papers report the presence of significant alteration levels in the outcropping crystalline basement rocks, while Sánchez (2001) described a weakly altered regolith bed underlying the basal sequence of the Sierra de los Cóndores Group.

Several authors studied the Cretaceous basins of the Sierras Pampeanas of Córdoba (Gordillo and Lencinas 1967; Sánchez et al. 1990; Astini et al. 1993; Piovano 1996; Sánchez 2001), and they interpreted the paleoenvironmental conditions for that period. Particularly, in the Sierras de los Cóndores Group, Sánchez et al. (1990) and Sánchez (2001) recognized facies intercalations of breccias and conglomerates, sandstones and conglomerates, sandstones and mudstones, mudstones and evaporates, which they interpreted as belonging to an alluvial fan paleoenvironment, with local variations in the production of sediments related to the activity of the main faults, volcanic activity, and, secondarily, climatic variations from subhumid to arid or semiarid. Therefore, the dominant erosion processes in such period would be associated to well-defined fluvial systems, developed in the mountain sectors, with the formation of alluvial/colluvial fans, and also to the penetration in sedimentary basins located farther east, where the intercalation of calcretes and evaporates is frequent. Similar conditions are described for the Late Cretaceous and Tertiary (López and Solá 1981) in volcano-clastic sequences of the southernmost part of Sierra de Comechingones and in the Tertiary deposits described in the Sierras Pampeanas environment.

The investigations steered by Linares et al. (1961) in the outcropping Tertiary sequences of the Valle de Punilla (the localities of Santa María de Punilla, Cosquín, Bialet Massé), farther north from the study area, in addition to confirm the recently described paleoenvironmental conditions, provided compositional information which is of great interest to establish the source areas. These authors indicated that in the basal levels (Early Eocene in age), the present lithology corresponds to the degradation of the granitic areas of Sierras Grandes, such as those comprising the Achala Batholith, whereas the upper levels incorporated metamorphic rock clasts which are pertaining to the Sierras Chicas, indicating the uplifting of this mountain range but only after the Eocene.

In the mineralogical descriptions, and particularly those of the basal sequence, Linares et al. (1961) did not mention the presence of fluorite-chalcedony clasts, with provenance from the Achala Batholith and the erosion of its mineral deposits, which are coeval with those of the Cerro Áspero Batholith. Towards the southern end of the Sierras Pampeanas of Córdoba, closer to the study area, there are neither records nor scientific studies concerning the sedimentary sequences of the Early Eocene, with the exception of Cantú (1992) when he assigned the basal sequences of the middle-outer fan of the Alpa Corral River to the Pliocene, although his work did not include rock compositional analyses. Thus, with the available evidence, it is assumed that the erosion processes had not reached yet the mineral-rich layers in the Early to Middle Tertiary (i.e., Eocene-Miocene), previous to the main Andean uplift in the Pliocene-Pleistocene.

As a consequence of such important tectonic event, the crystalline basement blocks were fragmented, tilted, and displaced, up to 1,000 m of vertical displacement in the main fault scarps, according to Gordillo and Lencinas (1979). The Cretaceous sequences were also uplifted as a result of the tectonic inversion, thus becoming exposed to subsequent erosion processes. The main fluvial systems, preceding the Andean uplift (González Díaz 1981; Carignano et al. 1999; Degiovanni et al. 2003, among others), started a marked process of incision and headwater erosion, which continues until present times.

The papers by Degiovanni et al. (2003) and Villegas et al. (2006) show that the longitudinal profiles of some stream channels which drain the eastern and southeastern slopes of the Sierra de Comechingones (Río La Tapa, Piedras Blancas, Cuenca Río Cuarto, and the Achiras and Las Lajas creeks, among others) are clearly unadjusted and the tributaries of lower order (1 and 2, basically) exhibit certain gradient values which are lower (1.5–3.0 %) than those of the collecting streams of higher order (essentially 3 and 4). This local anomaly to the gradient law, according to Horton (1945), is related to the presence of relict paleosurfaces which are still almost not incised in the headwater areas.

There are still very few studies in the Sierras Pampeanas environment that have taken into consideration the development, morphology, and preservation of the relict paleosurfaces, with a few references which basically described granitic landforms (Carignano et al. 1999; Cioccale and Carignano 2009). In this sense, it is understood that this chapter provides a valuable tool when analyzing in detail the existing morphology in different lithological environments. The paleosurfaces developed over the gneissic-migmatitic rocks of the Monte Guazú Complex and the mylonites of the Guacha Corral shear zone (north and south of the Cerro Áspero Batholith, respectively) show a very homogeneous morphology, with internal relief of quite small magnitude and equivalent elevations, probing a likely behavior of these lithological types and poor spatial variability. On the contrary, the degree of fracturing and the compositional and textural variations in granitic rocks, associated to the various facies present, favored the development of a more heterogeneous surface, where it is common to find residual landforms with higher topographic break than in the previous case. The comparative analysis of drainage density indicates that the dissection degree is twofold in the granitic environment, whereas the length of stream channels of order 1 diminishes to one half, showing a greater susceptibility for this lithological type though facing the same erosion processes.

These same lithological and structural characteristics are valid to explain the degree of preservation of these paleolandforms. Although the relict surfaces are located in the main water divides, the intensity of the erosion processes was larger in the southern portion of the batholith where, as it may be observed in the fluvial incision map, the La Invernada, El Talita, Las Moras, Seco, and Papagayos streams, controlled by larger structures (some of them with more recent activity), have generated an important elimination of materials with an incision depth of around 100–120 m in rocks corresponding to the porphyritic biotite monzogranite facies. The local relief that it is observed in the bedrock-intrusion contact is also attributed to differential erosion, and it is basically related to the presence of contact metamorphism rocks, which provide a larger resistance to bedrock in this area.

Concluding Remarks

Based upon the preceding discussion, the following conclusions may be forwarded:

1. The studied cratonic area was a positive element of the region at least since the Carboniferous and Permian, and since then, it has been subject to different denudation cycles, both associated to the distension tectonic environments of the Mesozoic and also the Andean compressive actions during the Cenozoic.
2. Previous to the Andean orogeny, during the Middle to Late Tertiary, there was a long period of relative stability, where the tectonic and erosion processes, generally of low energy and magnitude, favored the development of erosion surfaces, which began to be denudated after these movements, being preserved their remnants along the higher water divides.
3. At least until the Late Cretaceous, the Cerro Áspero Batholith was not exhumed; therefore, the age assigned to the studied paleosurfaces is comprised between the Late Cretaceous and the Paleogene.
4. The remnants of the studied erosion surfaces correspond to one single paleosurface and the altitudinal variations observed, both in N-S and W-E directions, correspond to the tilting of the basement blocks during the Andean episodes.
5. The paleosurfaces would have a polygenetic origin, a result of the alternation of wetter periods (humid/subhumid), in which chemical weathering processes and channel erosion would have been dominant, with others of more arid conditions, when mechanical weathering and disaggregation and overland and/or poorly organized surficial runoff would have been prevailing, with the development of pediment formation. Due to this, it could be explained that landforms corresponding to different landscape evolution cycles would be coexisting, with palimpsest spatial arrangements.
6. A marked lithological control over the geomorphology and the degree of preservation of coeval paleosurfaces exist, whether they are developed over high-grade metamorphic rocks and/or mylonites (Monte Guazú Complex, Guacha Corral shear zone) or over granitic rocks (Cerro Áspero Batholith). In the metamorphic rock environment, the erosion surfaces present a lesser topographic and morphological variation, and, in general, they are better preserved than those developed over granitic rocks, thus suggesting the great influence that compositional, textural, and structural variations exert over geomechanic properties.

Acknowledgments This chapter has been completed within the framework of research projects funded by the Secretaría de Ciencia y Técnica of the Universidad Nacional de Río Cuarto (SECyT-UNRC), the Agencia Nacional de Promoción Científica y Tecnológica of Argentina (ANPCyT), and the Ministerio de Ciencia y Tecnología of the province of Córdoba (MINCyT). The authors would like to express their gratitude to Professor Cliff Ollier for the critical review of a first draft of the manuscript and his most valuable observations and recommendations and to Dr. Jorge Rabassa for his invitation to publish this chapter, his multiple suggestions, and his help with the English text.

References

Andreazzini MJ, Degiovanni S, Cantú M, Grumelli M del T, Schiavo H (2012) Estudio preliminar de secuencias cuaternarias en pampas de altura de la Sierra de Comechingones (cuenca alta del Río Cuarto-Córdoba). V Congreso Argentino de Cuaternario y Geomorfología, Río Cuarto, Abstracts volume, 29

Astini RA, Pezzi LI, Massei GA (1993) Paleogeografía y paleoambientes del Cretácico de la Sierra de Pajarillo-Copacabana-Maza, noroeste de Córdoba. XII Argentine Geological Congress, Actas 1, pp 170–176. Buenos Aires

Beder R (1916) Estudios geológicos e hidrológicos en los alrededores de Villa Dolores. Boletín de la Dirección Nacional de Minería Geología e Hidrología 14:1–26. Buenos Aires

Beltramone CA (2007) Las superficies de erosión en las sierras Pampeanas de Córdoba: algunas consideraciones sobre su génesis. Revista de la Asociación Geológica Argentina 62(3):478–482

Bodenbender G (1895) Devónico y Gondwana en la República Argentina. Boletín de la Academia Nacional de Ciencias, Córdoba, Argentina 15(2-3):201–252

Bodenbender G (1905) La Sierra de Córdoba. Constitución geológica y productos minerales de aplicación. Anales del Ministerio de Agricultura de la Nación, Sección Geología 1(2):1–146. Buenos Aires

Bodenbender G (1907) Petrografía: Meláfidos, basaltos y andesitas en la sierra de Córdoba. Diabasas, porfiritos augíticos y meláfidos entre la Sierra de Córdoba y la Precordillera. Anales del Ministerio de Agricultura de la Nación, Sección Geología 2(3):1–34

Bodenbender G (1911) Constitución geológica de la parte meridional de La Rioja y regiones limítrofes. Boletín de la Academia Nacional de Ciencias, Córdoba, Argentina 19:1–220

Bodenbender G (1929) Triásico y Terciario de la falda oriental de la Sierra de Córdoba. Boletín de la Academia Nacional de Ciencias 31:73–139. Córdoba

Brackebush L (1879) Informe. Boletín de la Academia Nacional Ciencias Córdoba, Argentina 3:251–262

Brackebush L (1880) Informe sobre la marcha del Museo Mineralógico de la Universidad Nacional de Córdoba. Boletín de la Academia Nacional de Ciencias, Córdoba, Argentina 3:135–163

Brackebush L (1891) Mapa Geológico del Interior de la República Argentina. Escala 1:1.000.000, Gotha. Academia Nacional de Ciencia, Córdoba

Cantú MP (1992) Holoceno de la provincia de Córdoba. Manual: Holoceno de la República Argentina. Tomo I. Simposio Internacional sobre el Holoceno en América del Sur, 24 pp, Paraná, Argentina

Carignano C, Cioccale M (1997) Landscapes antiquity of the Central Sierras Pampeanas (Argentina): geomorphic evolution since the Gondwana times. In: 4th international conference on geomorphology, Italy, Abstracts volume, Supplementi di Geografia Fisica e Dinamica Quaternaria Torino, Supplement, 1997, 3(1), p 104

Carignano C, Cioccale M (2008) Geomorfología de la Sierra Norte-Ambargasta, provincias de Córdoba y Santiago del Estero, Argentina. In: Congreso Geológico Argentino, 17, 2008, San Salvador de Jujuy, Actas, pp 1189–1190

Carignano C, Cioccale MA, Rabassa J (1999) Landscapes antiquity of the Central Sierras Pampeanas (Argentina): geomorphic evolution since the Gondwana times. J Zeitschrift für Geomorphologie: Ann Geomorphol Suppl Band 118:245–268

Chebli G, Mozetic M, Rossello C, Bühler M (1999) Cuencas Sedimentarias de la Llanura Chacopampeana. Instituto de Geología y Recursos Minerales. Geología Argentina. Anales. Buenos Aires

Cioccale M (1999) Investigación geomorfológica de cuencas serranas, estudio geomorfológico integral: morfodinámica, morfometría y morfogénesis del flanco oriental de las Sierras Chicas de Córdoba. Unpublished doctoral thesis, Universidad Nacional de Córdoba, Córdoba, Argentina

Cioccale M, Carignano C (2009) Evolución geomorfológica de las Sierras Chicas, Córdoba, Argentina: una visión Gondwánica. IV Congreso Argentino de Cuaternario y Geomorfología, Abstracts volumen, N° 6. La Plata. Argentina. – cioccale, m. y c. carignano, 2009

Coniglio J (2006) Evolución petrológica y metalogenética del batolito Cerro Áspero en relación con el ciclo geoquímico endógeno del flúor, Sierra de Comechingones, Córdoba, Argentina. Unpublished doctoral thesis, Universidad Nacional de Río Cuarto, Río Cuarto, Argentina, 163 p

Coniglio J, Esparza A (1988) Geología del sector sur del Batolito Cerro Áspero-Alpa Corral, Córdoba, Argentina. V Congreso Geológico Chileno, 2, E1–E15

Coniglio JE, Pérez Xavier RP, Pinotti LP, D'eramo FJ (2000) Ore forming fluid of vein-type fluorite deposits of Cerro Áspero batholith, southern Córdoba Province, Argentina. Int Geol Rev 42(4):368–383

Coniglio JE, Pérez Xavier RP, Pinotti LP, D'eramo FJ (2004) Estudio de isótopos estables aplicado a los depósitos de fluorita del batolito Cerro Áspero. Córdoba. 7° Congreso de Mineralogía y Metalogenia, Actas 1, pp 171–176. Río Cuarto

Coniglio JE, D'eramo FJ, Pinotti LP, Demartis M, Petrelli HA (2006) Magmatismo devónico de las Sierras de Córdoba: fuente posible de flúor de las mineralizaciones mesozoicas. El ejemplo del batolito Cerro Áspero. 8° Congreso de Mineralogía y Metalogenia, Actas 1, pp 227–232. Buenos Aires

Coniglio JE, D'eramo FJ, Pinotti LP, Demartis M, Agulleiro Insúa LA, Petrelli HA (2010) Control estructural de las mineralizaciones de fluorita del Batolito Cerro Áspero, Sierras Pampeanas de Córdoba. Revista de la Asociación Geológica Argentina 67(4):507–520

Costa CH, Giaccardi AD, González Díaz EF (1999) Palaeolandsurfaces and neotectonic analysis in the Southern Sierras Pampeanas. In: Smith BJ, Whalley WB, Warke PA (eds) Uplift, erosion and stability: Perspectives on long-term landscape development. Geological Society, London, Special publication 162, pp 229–238

Degiovanni S, Villegas M, Doffo N (2003) Geomorfología del extremo Sur de la Sierra de Comechingones. II Congreso Nacional de Cuaternario y Geomorfología (San Miguel de Tucumán). Cuaternario y Geomorfología. ISBN 987-9390-47-4, pp 257–266

Fagiano MR (2007) Geología y petrología del basamento cristalino de Las Albahacas, Sur de la Sierra de Comechingones, Córdoba. Unpublished doctoral thesis, Universidad Nacional de Río Cuarto, 375 p

Fagiano MR, Martino R (2004) Cinemática y petrología de la faja de cizalla Guacha Corral en el extremo austral de la Sierra de Comechingones, Provincia de Córdoba. Avances en Microtectónica y Geología Estructural. Asociación Geológica Argentina Serie D: Publicación Especial 47:45–50

Fagiano M, Otamendi J, Nullo F (1993) Geología y Petrología del granito Los Nogales. XII Congreso Geológico Argentino and II Congreso de Exploración de Hidrocarburos 4, pp 33–41. Mendoza

Fagiano M, Pinotti L, Tramannoni P, Esparza A (2006) Petrología de la aureola de contacto del batolito Cerro Áspero, Sierra de Comechingones, Córdoba, Argentina. Avances en Mineralogía, Metalogenia y Petrología. 8° Congreso de Mineralogía y Metalogenia, Buenos Aires, pp 339–346

Feliú G (1994) Geología del granito Inti Huasi, Córdoba. Unpublished graduation thesis, Universidad Nacional de Río Cuarto

Galindo C, Pankhurst R, Casquet C, Coniglio JE, Baldo E, Rapela C, Saavedra J (1997) Age, Sr and Nd isotope systematics, and origin of two fluorite lodes, Sierras Pampeanas, Argentina. Int Geol Rev 39:948–954

González RR (1971) Edades radimétricas de algunos cuerpos eruptivos de Argentina. Revista Asociación Geológica Argentina 26:411–412

González Díaz EF (1981) Geomorfología. In: Irigoyen M (ed) Geología de la Provincia de San Luis. Relatorio VIII Congreso Geológico Argentino, Buenos Aires, pp 193–236

Gordillo C, Lencinas A (1967) Geología y petrología del extremo oeste de la Sierra de Los Cóndores, Córdoba. Boletín Academia Nacional de Ciencias, Córdoba 46(1):73–108

Gordillo C, Lencinas A (1979) Sierras Pampeanas de Córdoba y San Luis. Segundo Simposio de Geología Regional Argentina 1:577–650. Academia Nacional de Ciencias, Córdoba

Horton RE (1945) Erosional development of streams and their drainage basins: hydrophysical approach to Quantitative Morphology. Geol Soc Am Bull 56:275–370

Hünicken R, Pensa C (1980a) Formación Los Ranqueles (Carbónico) de Sampacho y Cerro Suco en el Sudoeste de Córdoba, Argentina. IUGS-UNESCO Proyecto 211, Boletín 3:5–6

Hünicken R, Pensa C (1980b) Estratigrafía y tectónica de las sedimentitas neopaleozoicas (Formación Chancaní) y de las filitas (Formación La Mermela) del borde occidental de las Sierras de Pocho y Guasapampa, Provincia de Córdoba, República Argentina. Boletín Academia Nacional de Ciencias de Córdoba 53(3–4):255–285

Iriondo M (1997) Models of deposition of loess and loessoids in the Upper Quaternary of South America. J S Am Earth Sci 10(1):71–79

Iriondo M (1999) Climatic changes in the South American plains: records of a continent –scale oscillation. Quat Int 57–58:93–112

Jelinek AR, Bastos Neto AC, Lelarge MLV, Soliani jr E (1999) Apatite fission track dating of fluorite ore veins from Santa Catarina State, Brazil: a complex hydrothermal evolution. J S Am Earth Sci 12:367–377

Jordan TE, Zeitler P, Ramos V, Gleadow AJW (1989) Thermochronometric data on the development of the basement peneplain in the Sierras Pampeanas, Argentina. J S Am Earth Sci 2(3):207–222

King LC (1963) Canons of landscape evolution. Geol Soc Am Bull 64:721–752

Klein C (1985) La notion de cycle en Géomorphologie. Revue Geólogie Dynamique & Geographie Physique 26(2):95–107

Krapovickas JM, Tauber AA (2012a) Atos Pampa, Área Fosilífera Cuaternaria de las Sierras de Córdoba: Estudios Preliminares. V Congreso Argentino de Cuaternario y Geomorfología, Río Cuarto

Krapovickas JM, Tauber AA (2012b) La Estratigrafía en las Pampas de Altura de las Sierras Pampeanas de Córdoba: Modelo Regional. V Congreso Argentino de Cuaternario y Geomorfología, Río Cuarto

Lencinas A (1971) Geología del valle de Punilla entre Bialet Massé y La Cumbre, Provincia de Córdoba. Boletín de la Asociación Geológica de Córdoba 1:61–70

Linares E, Timonieri AJ, Pascual R (1961) La edad de los sedimentos terciarios del valle de Punilla, provincia de Córdoba y la presencia de "Eohyrax rusticus" Ameghino en los mismos. Revista de la Asociación Geológica Argentina 15:191–212

López MG, Solá P (1981) Manifestaciones volcánicas alcalinas de los alrededores de Las Chacras y de la región de Villa-Mercedes-Chaján, provincias de San Luis y Córdoba. VIII Congreso Geológico Argentino, San Luis. Actas 4, pp 967–978

Martino R (2003) Las fajas de deformación dúctil de las Sierras Pampeanas de Córdoba. Revista Asociación Geológica Argentina 58(4):549–571

Minudri C, Sánchez ML (1993) Petrografía de areniscas del Grupo Sierra de Los Cóndores (Cretácico inferior) y su utilidad en la determinación del ambiente geotectónico, Sierra de Los Cóndores, Provincia de Córdoba, República Argentina. V Jornadas Pampeanas de Ciencias Naturales, Santa Rosa de La Pampa, Actas 2, pp 34–35

Nullo F, Fagiano M, Otamendi J, Brein C (1991) Estados de deformación estructural en el basamento cristalino de la Sierra de Comechingones, Achiras, Córdoba. VII R. Microtectónica, 25, Bahía Blanca, Argentina

Otamendi JE (1995) Petrología, geoquímica y estructura del basamento Pre-Carbonífero del extremo sur de la sierra de Comechingones, Córdoba-San Luis, Argentina. Unpublished doctoral thesis, Universidad Nacional de Río Cuarto

Otamendi J, Nullo F, Fagiano M, Aragón E (1996) Dos terrenos metamórficos y estructurales en el extremo sur de la Sierra de Comechingones, Córdoba-San Luis: Algunas implicancias tectónicas. XIII Congreso Geológico Argentino and III Congreso Exploración de Hidrocarburos, 2, pp 249–266. Buenos Aires

Penck W (1914) La estructura geológica del Valle de Fiambalá y las cordilleras limítrofes al norte de Tinogasta. Ministerio de Agricultura de la Nación, Boletín, 17. Buenos Aires

Penck W (1920) Der Südrand der Puna de Atacama (NWArgentinien). Ein Beitrag zur Kenntnis des Andinen Gebirgstypus und der Frage der Gebirgsbildung. Der Abhandlungen der Sachsischen Akademie der Wissenchaften, Leipzig 1, pp 3–420

Pinotti L, Coniglio JE, Llambías EJ (1996) Características geológico-estructurales del Plutón Alpa Corral. Sierras Pampeanas de Córdoba, Argentina. XIII Congreso Geológico Argentino and III Congreso de Exploración de Hidrocarburos, Actas 3, pp 477–483

Pinotti L, Llambías EJ, Coniglio JE (1997) Stopping as a main mechanism of intrusion in post-orogenic granite from the southern part of the Sierra de Comechingones, Sierras Pampeanas de Córdoba, Argentina. In: Second international symposium on granites and associated mineralizations. Abstracts with Programs, Salvador, Bahia, Brazil, pp 323–325

Pinotti L, Coniglio J, Esparza A, D'eramo F, Llambías EJ (2002) Nearly circular plutons emplaced by stoping at shallow crustal levels, Cerro Áspero batholiths, Sierras Pampeanas de Córdoba, Argentina. J S Am Earth Sci 15:251–265

Piovano EL (1996) Correlación de la Formación Saldán (Cretácico temprano) con otras secuencias de las Sierras Pampeanas y de las cuencas Chaco-Paranaense y de Paraná. Rev Asoc Geol Arg 51:29–36

Porta R (1992) Caracterización petrográfica y geoquímica del Batolito Cerro Áspero-Alpa Corral (32°34′-32° 42′ LS y 64°43′-64°52′ LO), Córdoba, Argentina. Revista Estudios Geológicos, Museo Nacional de Ciencias, Madrid 48:237–245

Rabassa J (2010) Gondwana paleolandscapes: long-term landscape evolution, genesis, distribution and age. Geociências 29(4):541–570. São Paulo, UNESP

Rabassa J (2014) Some concepts on Gondwana landscapes: long-term landscape evolution, genesis, distribution and age. In: Rabassa J, Ollier C (eds) Gondwana landscapes in southern South America. Springer, Dordrecht, pp 9–46

Rabassa J, Zárate M, Cioccale M, Carignano C, Partridge TC, Maud R (1996) Paisajes relictuales de edad Gondwánica en áreas cratónicas de Argentina. In: 13° Congreso Geológico Argentino & III Congreso de Exploración de Hidrocarburos 3, Buenos Aires. Actas, Buenos Aires 4, p 219

Rabassa J, Zárate M, Partridge TC, Maud R, Cioccale M, Carignano C (1997) Gondwanic relict paleolandscapes in cratonic areas of Argentina. In: 4th international conference on geomorphology, Torino, Italy. Abstracts volume, Supplementi di Geografia Fisica e Dinamica Quaternaria, Supplement 3(1), p 321

Rabassa J, Carignano C, Cioccale M (2010) Gondwana paleosurfaces in Argentina: an introduction. Geociências 29(4):439–466. São Paulo, UNESP

Rassmuss E (1916) Rasgos geológicos generales de las Sierras Pampeanas. Dirección General de Minas, Geología e Hidrología, Boletín 13 B. Buenos Aires

Rey Ripoll M (2008) Perfil geológico-estructural de la faja cizalla Guacha Corral, Sa. Comechingones, 32° 45′S. Evaluación del potencial metalogenético. Unpublished graduation thesis, Universidad Nacional de Río Cuarto, 87 p

Rimann E (1926) Estudio geológico de la Sierra Chica, entre Ongamira y Dolores, Córdoba, Argentina. Boletín de la Academia Nacional de Ciencias 23:9–191

Rovereto G (1911) Studi di geomorfologia argentina. I. La sierra di Córdova. Tipogr. Della Pace E Cuggiani, 19 p

Sagripanti G (2006) Neotectónica y peligro sísmico de la región de Sampacho, Departamento Río Cuarto, Provincia de Córdoba. Unpublished doctoral thesis, Universidad Nacional de Río Cuarto, 300 p

Sánchez ML (2001) Geología de la cuenca cretácica entre los 32° y 33°50′ de Latitud Sur y entre los 64° y 65°15′ de Longitud Oeste, provincia de Córdoba, República Argentina. Unpublished doctoral thesis, Universidad Nacional de Río Cuarto

Sánchez ML, Minudri C (1993) Variaciones de las propiedades morfométricas y sus significado sedimentológico en las psefitas del Grupo Sierra de Los Cóndores, Provincia de Córdoba, República Argentina. V Jornadas Pampeanas de Ciencias Naturales, Santa Rosa de La Pampa. Actas 2, pp 67–74

Sánchez ML, Villegas MB, Poiré DG (1990) Paleogeografía del Cretácico inferior en el área de la Sierra de los Cóndores, provincia de Córdoba, Rep. Argentina. 3rd Argentine meeting of sedimentology, Actas, pp 235–246. San Juan

Schmidt CJ, Astini RA, Costa CH, Gardini CE, Kaemer PE (1995) Cretaceous rifting, alluvial fan sedimentation and Neogene inversion, Southern Sierras Pampeanas, Argentina. In: Tankard AJ, Suárez RS, Welsink H (eds) Petroleum basins of South America. American Association Petroleum Geology, Memoir 62. American Association of Petroleum Geologists (AAPG), Tulsa, Oklahoma, USA, pp 341–358

Schmieder O (1921) Apunte Geomorfológico sobre la Sierra Grande de Córdoba. Boletín Academia Nacional de Ciencias de Córdoba 25:181–204. Córdoba

Sosic M (1964) Descripción geológica de la hoja 24h, Sierra del Morro (San Luis-Córdoba). Carta Geológico-Económica de la República Argentina. Boletín 95, Dirección Nacional de Geología y Minería, 44 p. Buenos Aires

Stelzner A (1885) Comunicaciones al Profesor H.B. Geinitz: Primera carta. In: Homenaje al 150° aniversario de la Independencia Argentina. Boletín de la Academia Nacional de Ciencias de Córdoba 45:115–120

Stuart-Smith PR, Skirrow RG (1997) Report on 1:100000 Scale Geological and Metallogenic Maps Sheet 3366-24. Provinces of San Luis and Córdoba. Programa Nacional de Cartas Geológicas. Proyecto de Cooperación Argentino-Australiano, 43 p. Buenos Aires

Uliana M, Biddle K, Cerdán J (1990) Mesozoic extension and the formation of Argentine sedimentary basins. In: Tankard A, Blackwill H (eds) Extensional tectonics and stratigraphy of the North Atlantic Margins. American Association Petroleum Geology, Memoir 46. American Association of Petroleum Geologists (AAPG), Tulsa, Oklahoma, USA, pp 599–611

Villegas MB, Degiovanni S, Ulla JP (2006) Tipificación del comportamiento del sector sur de la cuenca del Río Cuarto a través del análisis de variables morfológicas y sedimentológicas. III Congreso Argentino de Cuaternario y Geomorfología, Actas, 1, pp 221–231

Gondwana Glacial Paleolandscape, Diamictite Record of Carboniferous Valley Glaciation, and Preglacial Remnants of an Ancient Weathering Front in Northwestern Argentina

Betty Socha, Claudio Carignano, Jorge Rabassa, and Dave Mickelson

Abstract A record of glacier advance and retreat is preserved in Carboniferous strata exposed in an exhumed glacial paleovalley on the eastern side of the Paganzo basin. Previous investigations have focused on the sandstones in the paleovalley and inferred a glacial lacustrine history. New observations have demonstrated that remnants of a preglacial, ancient weathering front, developed under wet tropical conditions and composed of corestones, are found underneath the glaciogenic deposits. Delta and alluvial fan deposits were also recognized, but no inferences were made from the diamictites in the paleovalley regarding glacial events (Andreis et al., Bol Acad Nac Cienc Cordoba 57:3–119, 1986; Buatois and Mángano, J Paleolimnol 14:1–22, 1995; Sterren and Martínez, El Paleovalle de Olta (Carbonífero): Paleoambiente y Paleogeografía. 13° Congreso Geológico Argentino and 3° Congreso de Exploración de Hidrocarburos, Actas, 2, 89–103, 1996). This chapter focuses on the diamictites and provides a link between the sediment infill and the glacial origin of the paleovalley. We describe diamictites and associated sediments at three main locations: at La Chimenea, near the mouth of the paleovalley; at Mid-Valley, near the middle of the paleovalley; and at the Campsite,

B. Socha
SCS Engineers, 2830 Dairy Drive, Madison, WI, USA
e-mail: bsocha@scsengineers.com

C. Carignano
CONICET, Córdoba, Argentina

Universidad Nacional de Córdoba, Córdoba, Argentina

J. Rabassa (✉)
Laboratorio de Geomorfología y Cuaternario, CADIC-CONICET, Ushuaia, Tierra del Fuego, Argentina

Universidad Nacional de Tierra del Fuego, Ushuaia, Tierra del Fuego, Argentina
e-mail: jrabassa@gmail.com

D. Mickelson
Department of Geoscience, University of Wisconsin-Madison, Madison, WI, USA

J. Rabassa and C. Ollier (eds.), *Gondwana Landscapes in southern South America*, Springer Earth System Sciences, DOI 10.1007/978-94-007-7702-6_12,

near the head of the valley. We interpret some of the diamictites exposed at La Chimenea and at Mid-Valley to be subglacial tillite. Deformation in the sandstone underlying the tillite indicates warm-based conditions as the glacier advanced over soft deformable sediment. At the Campsite location, a diamictite bed, which is about 1.5 m thick, lies within a sequence of alternating sandstone and siltstone beds. The diamictite bed is interpreted to represent an ice-front readvance during a period of ice retreat. The diamictite may be a debrite originating off the ice front, or a subglacial deposit, i.e., a tillite, or a combination of both. Two additional diamictite beds, exposed higher in this sequence of alternating sandstone and siltstone beds, may also record minor ice-front advances into the flooded valley.

Evidence of an ancient, preglacial weathering front (Late Devonian?–Earliest Carboniferous?) has been found in the granitic basement rocks which underlie the glaciogenic deposits, as large corestones included in a weathered regolith. This weathering front was developed under wet tropical conditions, before the onset of Carboniferous glaciations. The tillite and other diamictites overlying the corestones are composed largely of locally derived granitic basement rock. Features observed in the tillite and other diamictites are attributed to rapid rates of deposition, depositional processes, and the susceptibility of pre-weathered granitic basement rock to glacial and other erosional processes. Processes other than glacial erosion and deposition, including mass transport (slumping, rafting, sliding, and debris flow), also operated in the steep-sided valley and contributed large amounts of diamictite and other sediment to the valley fill. Corestones, weathered from the basement rock during a pre-Carboniferous period of intense weathering, constitute the larger clasts in the diamictite and associated deposits. The glacial paleolandscape is very well preserved in detail, after being buried during the Permian and later exhumed in the Cenozoic. The glacial valley was likely a transitional (fjord) environment, as micropaleontological material (Gutiérrez and Limarino, Ameghiniana 38:99–118, 2001) and clay mineral assemblages (Net et al., Sediment Geol 152:183–199, 2002) indicate a marine transgression into the area during the Middle Carboniferous.

Keywords Gondwana • NW Argentina • Carboniferous glaciation • Glacial sediments • Etchplains

Introduction

Glaciers covered much of Gondwana during the Carboniferous and Permian, and although glacial and deglacial sediments are present in large areas of western and northwestern Argentina, the geographic extent and characteristics of the glaciation that occurred are not well resolved. The record of Late Paleozoic glaciation is preserved in depositional basins formed along the active western margin of the Gondwana supercontinent (Fig. 1). Strata found in the westernmost basins in Argentina, the Calingasta–Uspallata and Paganzo basins, are characterized as the products of Middle Carboniferous mountain or valley glaciers (López Gamundi 1987; Limarino and Gutiérrez 1990).

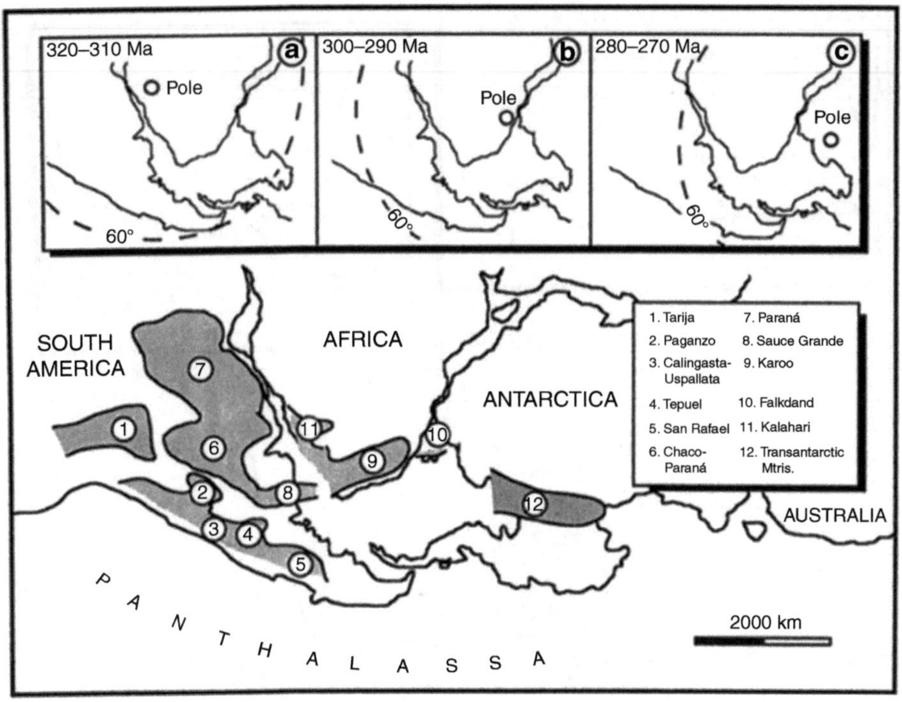

Fig. 1 Generalized paleogeography of the western portion of the Gondwana supercontinent and the Panthalassa Ocean during the Late Paleozoic (From López Gamundi 1997, fig. 8.1, p. 147, by permission of Oxford University Press, USA)

This chapter is a summary of a chapter from the Doctoral Dissertation of one of us (Socha 2007) and other scientific communications originated in such work and other regional studies (see, for instance, Socha et al. 2006; Rabassa et al. 2007).

Along the southwestern and eastern rim of the Paganzo basin (Fig. 2), deep valleys were cut into basement rock. These paleovalleys, which appear infilled with thick sequences of glacial and deglacial sediments, have since been exhumed by modern fluvial erosion exposing a detailed depositional record of changing environmental conditions. Paleovalleys along the southwestern rim are interpreted as paleofjords whose sedimentary successions record the marine flooding of glacial valleys (Dykstra et al. 2006; Kneller et al. 2004).

This study presents the record of glacier advance and retreat preserved in Carboniferous strata found in a paleovalley on the eastern side of the Paganzo basin. The morphology of the paleovalley is strikingly glacial, and the thick sedimentary sequence exposed in the paleovalley includes several facies of sediments. Previously, the paleovalley deposits were interpreted as entirely glacial lacustrine with some deltaic and alluvial fan sediments (Andreis et al. 1986; Buatois and Mángano 1995; Sterren and Martínez 1996). However, ice-rafted dropstones and sediment gravity-flow deposits indicate near-ice deposition.

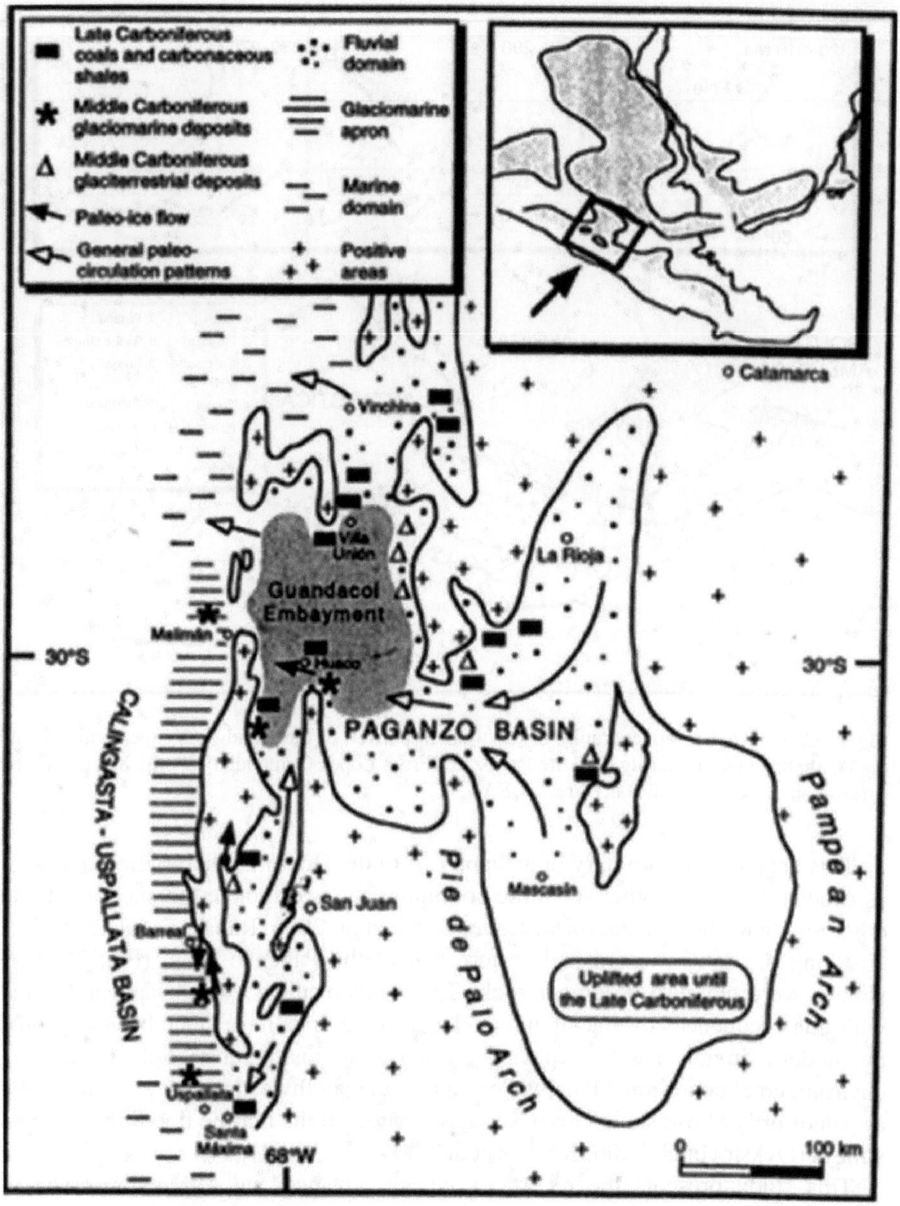

Fig. 2 Paleogeography of the Paganzo and Calingasta–Uspallata basins during the Middle- to Late Carboniferous (From López Gamundi 1997, fig. 9.5, p. 153, by permission of Oxford University Press, USA)

This chapter reports observations of an ancient, preglacial weathering front (Late Devonian?–Earliest Carboniferous?), a Carboniferous glacial paleolandscape, and glacial diamictite and other glaciogenic sediments at three locations in the paleovalley and intends to reconstruct the depositional environments in the relict glacial valley. From the diamictite characteristics, the general conditions at the base of the ice and the mode of sediment transport and deposition are inferred. Direct evidence of warm-based valley glaciation is preserved at the base of a diamictite bed that we interpreted to be subglacial tillite.

Regional Paleogeographic Setting

The Late Paleozoic tectonic setting of the Calingasta–Uspallata and Paganzo basins of Western Argentina was a back-arc to foreland basin formed in association with the subduction of the Pacific Plate beneath the western margin of Gondwana (Ramos 1988). Early Carboniferous sediments are not found in the Paganzo basin, but based on the sedimentary and paleontological record from the adjacent Calingasta–Uspallata basin, the Early Carboniferous (Tournaisian–Visean) was a time of humid, temperate climate (López Gamundi et al. 1992). Sedimentation took place mostly in small intermontane basins. The sediments consist mainly of clast-supported conglomerates and interfingering sandstones and are interpreted as alluvial fan and fluvial deposits. The paleoecology of plant remains found in these deposits also indicates a more temperate and humid climate than later when climate was dominated by glacial events (López Gamundi et al. 1992).

During the Middle Carboniferous (Namurian–Early Westphalian), widespread glacial deposition occurred in basins along the western margin of Gondwana (López Gamundi et al. 1992). The Calingasta–Uspallata and Paganzo basins were at a paleolatitude of about 60° south (Fig. 1), and a combination of high latitude and high altitude contributed to the formation of ice centers along the margins of the basins (López Gamundi 1997). Striations carved in the basement rock and boulder pavements within the Carboniferous strata indicate that ice flow in the Paganzo basin was generally toward the northwest, with local ice-flow directions determined by the local paleoslopes (López Gamundi 1997; Fig. 2). Glacial modification of the topography, following uplift of the region in the Late Devonian and Early Carboniferous, produced significant topographic relief (López Gamundi 1997).

Several models of the Carboniferous glaciation suggest that the Paganzo basin, bounded on the west by the paleo-Precordillera, on the east and south by the Pampean and Pie de Palo Arches, and the Puna high on the north (Buatois and Mángano 1995), was cut off from the marine environment by mountain ranges and that the basin received mainly continental sediments, while to the west the Calingasta–Uspallata basin received mainly marine sediments (López Gamundi 1987; López Gamundi et al. 1992).

Local Setting

The Malanzán subbasin (Fig. 3), located in the eastern Paganzo basin, is commonly referred to as the Malanzán paleovalley (Andreis and Bossi 1981). The Malanzán paleovalley trends south–southwest to north–northeast from Malanzán to Solca

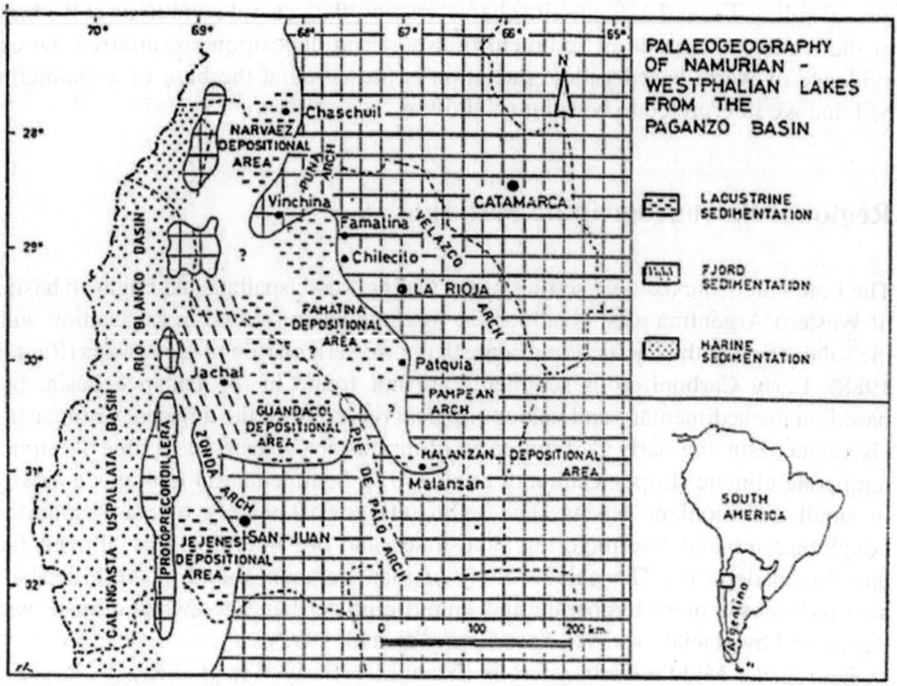

Fig. 3 Paleogeographic map of late Namurian–early Westphalian postglacial lacustrine deposits of the Paganzo Basin (After Buatois et al. 1994). Note location of the Malanzan depositional area in the eastern part of the Paganzo Basin. *Square-crossed* pattern indicates basin basement. (From Buatois and Mangano 1995, *Journal of Paleolimnology*, Kluwer Academic Publishers, v. 14, fig. 1, p. 3, with kind permission from Springer Science and Business Media)

(Fig. 4) and is part of a larger paleovalley system cut into the basement rock. The segment of the paleovalley extending from La Chimenea to Olta (Fig. 4) is referred to as the Olta paleovalley and is the study area for this chapter. Sediments in the Malanzán and Olta paleovalleys have been interpreted as predominantly glaciolacustrine (Andreis et al. 1986; Sterren and Martínez 1996). However, recent findings of Namurian–Early Westphalian marine microfossils in the Malanzán Formation at the Rio Olta extend the marine transgression into the eastern part of the Paganzo basin and imply a paleofjord environment for some of the Carboniferous deposits in the paleovalleys (Gutiérrez and Limarino 2001).

The Olta paleovalley is exposed on the northern end of the Sierra de Los Llanos. The Sierra de Los Llanos is an uplifted, fault-bounded block of Ordovician basement rock. There is a small structural dip to the northeast (Page et al. 2002), oblique to the valley axis, which curves gently from the northeast to the southwest (Fig. 4).

The local subcrop to the Carboniferous deposits consists of Ordovician granite and granodiorite of the Chepes Group, granite and migmatite of the Pacatala Group, and phyllite of the Olta Formation (Fig. 4) (Page et al. 2002). Stratigraphically, the Carboniferous units – the Malanzán and Loma Largo Formations– are overlain to

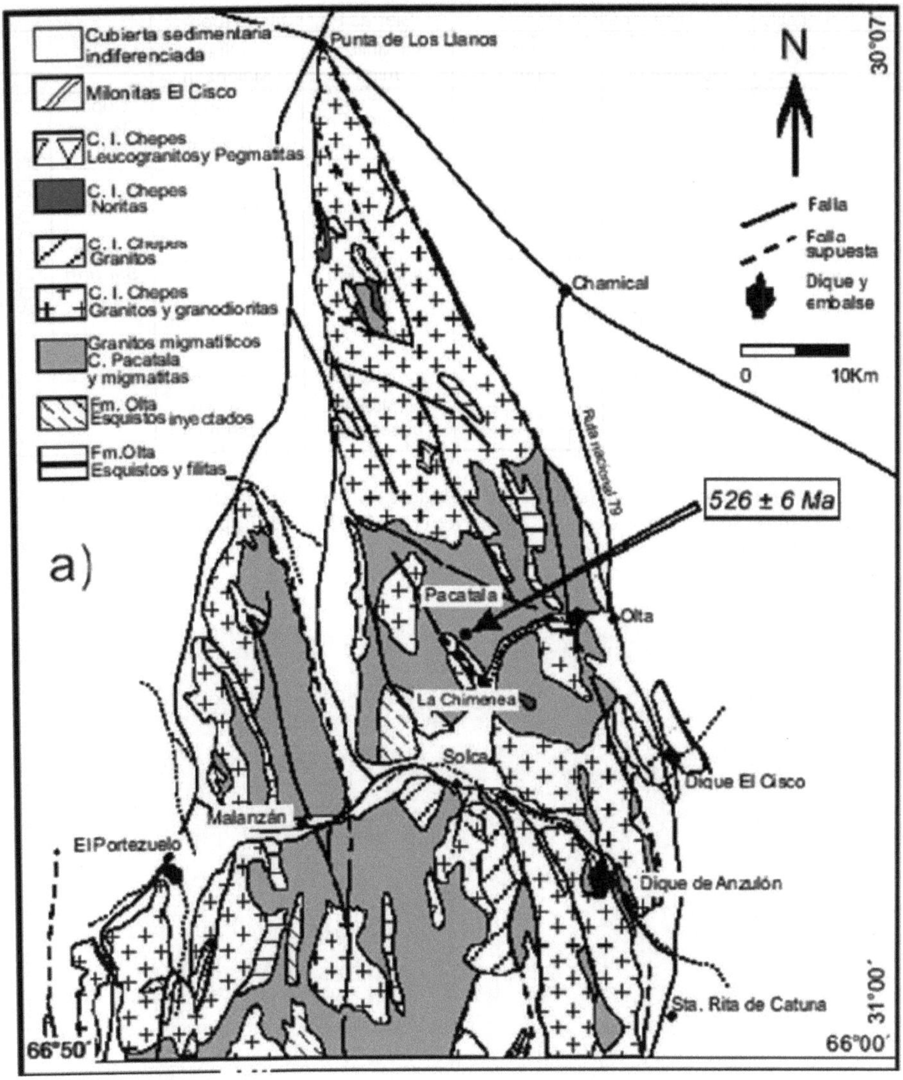

Fig. 4 Geologic map of the Sierra de Los Llanos, Sierra de Malanzán, and the northern part of the Sierra de Chepes (From Page et al. 2002, with permission by the Asociación Geológica Argentina)

the east by the Permian Solca and La Colina Formations (Fig. 5), which consist largely of red, cross-bedded sandstone.

The outline of the paleovalley (apparent on the satellite image, Fig. 6, and shown schematically in Fig. 7) is defined by mapping the valley floor and sides. The paleovalley was partially infilled by glaciogenic deposits, then by Permian sedimentary rocks, and likely exhumed primarily by fluvial erosion during the Cenozoic. The modern stream, the Rio Olta, flows to the northeast and presumably runs parallel to the course of the original valley. Based on the direction of progradation

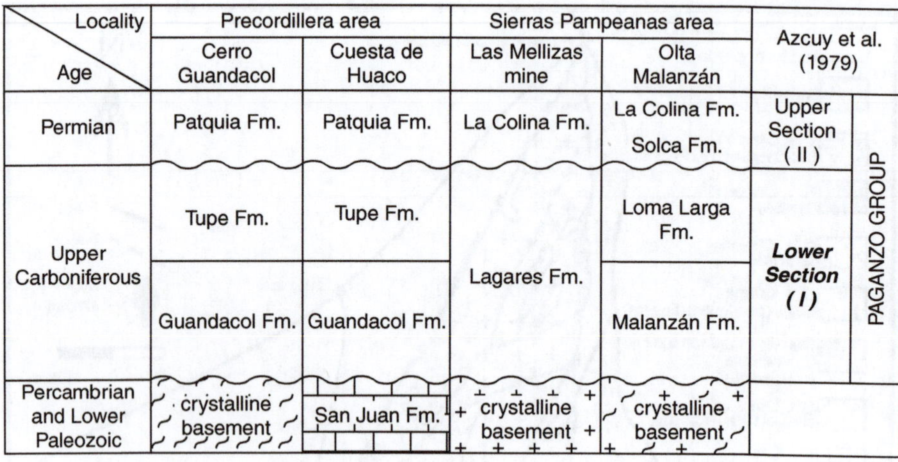

Fig. 5 Stratigraphic chart for units in the lower Paganzo Group (Carboniferous) in the study area (Reprinted from Sedimentary Geology 2002; Net et al. 2002, v. 152, pp. 183–199, fig. 2, with permission from Elsevier)

Fig. 6 Satellite image of the Olta paleovalley. (MDA EarthSat, natural view color; 15 m resolution.) See Fig. 7 for an approximate scale and geographic orientation

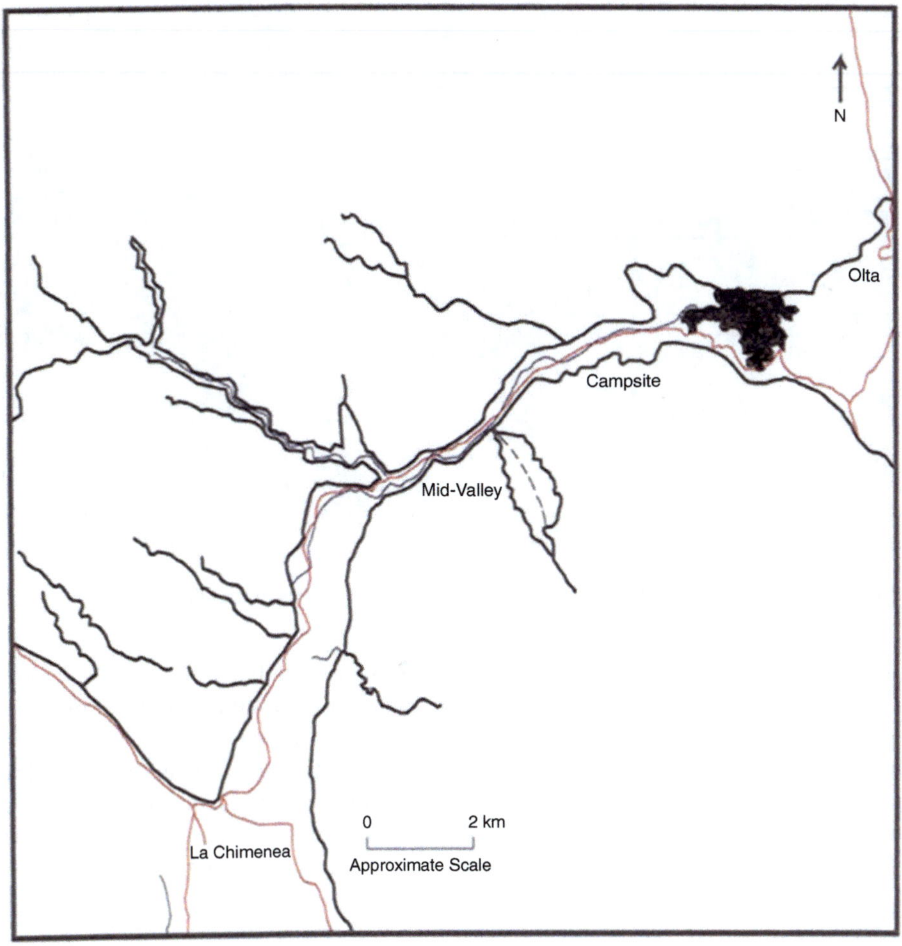

Fig. 7 Section locations in the Olta paleovalley. Approximate locations of roads and streams are shown in *red* and *blue*, respectively. Schematic map is derived from MDA EarthSat image (see **Fig. 6**). La Chimenea – S30° 41′ 00.2″ W66° 23′ 40.3″; Mid-Valley – S30° 38′ 32.7″ W66° 20′ 46.5″; Campsite – S30° 37′ 65.0″ W66° 19′ 17.0″

of delta fronts, the paleoflow direction in the valley was to the southwest (Sterren and Martínez 1996). The stream in the broader, southwestern part of the valley is ephemeral. The paleovalley had very steep sides (50–70°) and was at least 300 m deep, based on the height of the surrounding peaks of the basement rock above the lowest points on the unconformity. The U-shaped cross section of the paleovalley is characteristic of a valley produced by glacial erosion (Fig. 8). Carboniferous deposits are preserved over a distance of more than 12 km along the valley from La Chimenea to about 4 km west of Olta, with a total fill thickness of about 300 m. The valley varies in width from about 2 km at the southwest end, to about 500 m

Fig. 8 U-shaped cross section of the Olta paleovalley. View to west from Olta

at the narrowest part near its midpoint. Approximately halfway up the paleovalley, on the southwestern side, a small tributary valley infilled with clastic sediment, is present (Fig. 6). The tributary valley is interpreted as a hanging valley, carved when the ice surface in the valley was very high, which was then left high above the valley floor after deglaciation. Faults, trending northwest to southeast, cut across the valley mouth at both ends of the valley (Fig. 4). No other major tectonic structures have been observed in or across the paleovalley, suggesting a simple structural setting.

Sedimentary Facies

Several sections in the Olta paleovalley were described and measured by Sterren and Martínez (1996), who identified lithofacies associations, facies variability, and paleocurrent directions. They interpreted the succession, dominated by coarse-grained delta deposits (Gilbert deltas) and thin-bedded lacustrine deposits with rafted clasts, as infill of a glacial paleovalley. Overlying the deltaic sequence are deposits of a braided fluvial system.

A significant portion of the valley infill consists of diamictites, and even though a glacial origin is inferred for the paleovalley, a direct glaciogenic origin has not been inferred for any of the diamictites. This chapter focuses on the diamictites exposed in the Olta paleovalley and provides a link between the sediment infill and the glacial origin of the paleovalley. Diamictites and associated deposits at three main locations in the paleovalley – La Chimenea, Mid-Valley, and the Campsite – are described (Fig. 7).

The following lithofacies code, adapted from Eyles et al. (1983), is used to describe the sections. Inferences made from observations of characteristics of Pleistocene and modern glacier deposits regarding depositional environment are used to interpret the sections of Carboniferous diamictites and associated sediments exposed at the three locations in the Olta paleovalley.

Lithofacies Code Dmm – Diamictite, matrix-supported, massive **Dms** – Diamictite, matrix-supported, stratified **Dcm** – Diamictite clast-supported, massive **Dcs** – Diamictite clast-supported, stratified
Gm – Gravel, massive **Sm** – Sand, massive **Sh** – Sand, horizontally laminated **Sr** – Sand, ripple cross-bedded **Fm** – Fines (silt & clay) massive **Fl** – Fines (silt & clay) laminated

Till and Tillite

Characteristics of till are well documented from Pleistocene and modern glacier environments. Presumably soft sedimentary features of till and associated sediment can be preserved in the rock record as lithified structures in tillite. Therefore, the inferences made from modern and Pleistocene sediment characteristics regarding depositional environment also apply to tillite. In the remainder of this chapter, "till" and "diamicton" are used when referring to unlithified (Pleistocene and modern) sediment, and "tillite" and "diamictite" are used when referring to lithified (pre-Pleistocene) deposits.

The general conditions at the base of a glacier or the base of a lobe of glacier ice, the mode of sediment transport, and the environment of deposition can often be inferred from till characteristics. It is generally accepted that sediment can retain features it obtained while being transported by moving glacier ice.

Glacier ice moves by internal deformation in the ice itself, by sliding at the base of the ice, by deformation in the underlying sediment, or by some combination of these mechanisms (Boulton 1996). Ice motion results from permanent strain of the ice and the glacier bed in response to stress (Benn and Evans 1996). Stress, or force per unit area, can be resolved into stress acting at right angles to the surface (normal stress) and stress acting parallel to the surface (shear stress). Strain (the change in shape and size of a material due to stress) may occur as deformation of the ice, deformation of the bed under the ice, or by sliding at the interface between the ice and the glacier bed. The surface motion or surface velocity of a glacier is the sum of these processes (Fig. 9). Bed deformation can only be significant when sediments are saturated with water at a pressure that is almost as high as the ice overburden pressure (i.e., effective pressure is low), and basal sliding is only expected when the ice is at the pressure melting point (i.e., the glacier is temperate) (Paterson 1994).

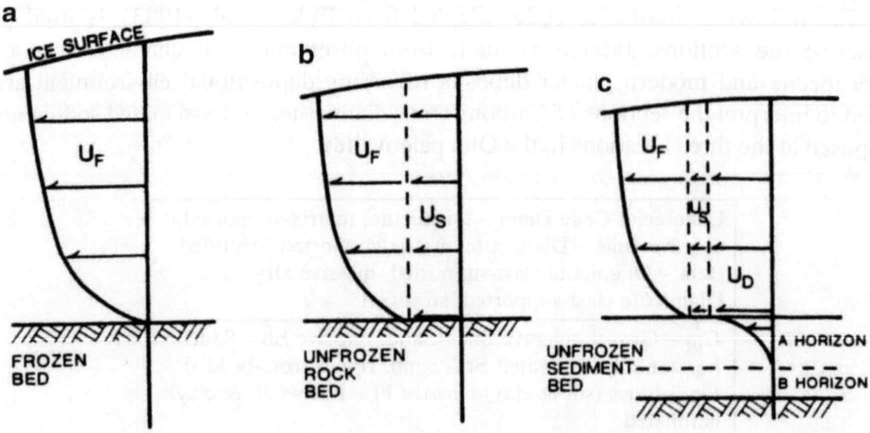

Fig. 9 Schematic diagram of glacier motion (Reprinted from Boulton 1996, Journal of Glaciology, with permission from the International Glaciological Society and G.S. Boulton). At (**a**) glacier rests on a frozen bed and moves by ice deformation alone, at (**b**) glacier rests on an unfrozen rock bed and moves by ice deformation and basal sliding, and at (**c**) glacier rests on an unfrozen sediment bed and moves by ice deformation, basal sliding, and deformation of subglacial sediment. U_F horizontal velocity component due to ice deformation, *Us* horizontal velocity component due to sliding, U_D horizontal velocity component due to sediment deformation

The sedimentary characteristics acquired from moving ice such as evidence of shear, beveled bullet boulders, and strongly developed clast fabric can be preserved in till (Ham and Mickelson 1994). The three main theories of till formation, melt-out, lodgement, and deformation (described further in the following paragraphs) account for the formation and preservation of features acquired from moving ice. However, these characteristics are wholly or partially destroyed or overprinted if the till is later remobilized by sediment gravity flow, fluvial, or other processes.

Clast fabric is a feature of glaciogenic sediment acquired from shearing induced by moving ice. Elongated clasts tend to become aligned parallel to the direction of strain produced by the shear stress of the flowing ice. The alignment of the clasts can be preserved under certain conditions such as deposition by lodgement of particles directly from moving ice (Boulton 1971) or by melt-out from debris-rich ice (Lawson 1979; Ham and Mickelson 1994). Preferred clast orientations can also develop in saturated, unfrozen sediment that is deformed beneath the glacier (Benn 1995; Hart 1998).

Lodgement occurs through interference at the interface between moving ice and rock or older sediment at the glacier bed. Deposition takes place by pressure melting against obstacles (Dreimanis 1976). Melt-out is a process by which sediment is released by melting from stagnant ice often beneath a layer of overburden. The resulting melt-out till will retain some features it acquired in the moving ice, if the meltwater can drain, if the surface slope is low, and if the layer of overburden is thick enough to prevent slump or flow (Boulton 1971, 1972). If sediment is deposited from a debris-rich ice layer with high sediment content (Lawson 1981; Ham and Mickelson 1994), the clasts in the sediment may have a strong, or well-developed, preferred orientation that is parallel to the direction of ice flow. A weaker orientation may indicate deposition from a debris-rich ice layer with low sediment content (Lawson 1981; Ham and Mickelson 1994).

Sediment transported as a saturated, unfrozen, deforming bed beneath the glacier also acquires features resulting from ice flow. Wet sediment is mobilized by the shearing force of the moving glacier, and strain results in change in the geometric relationship of grains to produce new sediment. Elson (1988), Benn (1995), and Hart (1998) suggested that a thin deforming layer would have a strongly developed preferred orientation of clasts and that a thick deforming layer will have clasts with a weakly developed preferred orientation parallel to the direction of ice flow. As shown by laboratory experiments using a thin layer of sediment in a ring shear device, strong clast fabric develops in the direction of shear at small strains (1 to 2) and remains strong through higher strains (Hooyer and Iverson 2000).

Sediment gravity flow is a common process in a supraglacial or proglacial environment. Sediment deposited by sediment gravity flow may have the following characteristics: a basal zone of traction gravel, variable matrix texture, lack of fabric consistently oriented parallel to the direction of ice flow, dipping of undulating basal contacts, and a denser clast concentration (Lawson and Kemmis 1983). Flow noses, rafts of fine-textured laminae, silt and clay stringers, rip-up clasts, basal grooves, clast imbrication or clusters, random clast fabric or fabric parallel to bedding, and erosion and inclusion of underlying material are distinctive features of sediment flows noted by Eyles et al. (1983). Common, prominent distinguishing features of sediment flow deposits are a weak clast fabric that is not parallel to the direction of ice flow, variable texture, presence of stringers of sorted sediment, and clast concentrations.

Grain-size distribution is an important factor affecting how sediment responds to subglacial stress and in determining the processes dominating at the time of deposition. For example, saturated clay-rich sediment is expected to shear under low effective stress; the proportion of large clasts to fine-grained sediment affects deformation as boulders can support part of the shear stress and provide instruments for plowing (Paterson 1994).

Roundness, lithology, and surface textures (striations and polish) of clasts are notable diagnostic features for diamicton formation. Roundness can be an indication of the degree of wear against other clasts, matrix material, or bedrock or can be produced by weathering such as is commonly produced by exfoliation of granite. Striations and polish are evidence of wear due to interaction with matrix material, other clasts, or the bedrock surface (Boulton 1978; Benn 1995).

Subglacial Processes

Traditionally, classifications of till (Dreimanis 1988) focused on tills as genetic end members, i.e., lodgment till, melt-out till, and deformation till. However, recent research focuses on the recognition that the processes of deformation, sliding, lodgement, and plowing coexist in the subglacial environment and produce sedimentary sequences that record a composite signature of transport and deposition (Evans et al. 2006).

The end member classifications do not appear to be adequate genetic designations for till. Observations from Pleistocene sedimentary sequences support a model wherein the ice moves by a combination of deformation of a thin bed of sediment, plowing, lodgement, and sliding (Clayton et al. 1989; Mickelson et al. 1992; Piotrowski and Tulaczyk 1999; Piotrowski et al. 2001, 2004, 2006). In Pleistocene sedimentary sequences, generally undisturbed beds interspersed with highly deformed beds have been observed, and the soft bed of temperate glaciers is envisioned as a mosaic of stable and deforming spots whose distribution varies through time and space (Piotrowski et al. 2004, 2006). Piotrowski et al. (2006) summarized field indicators of bed deformation (intermixed sediment layers, overturned and attenuated folds, boudinage structures, and strong alignment of particles), field indicators of lodgement (grooved surfaces within till layers, plow marks, and consistently striated clasts), and field indicators of basal decoupling (sand stringers, scours under pebbles that are filled with sand, and partly armored till pellets). Till formation can occur under high basal water pressure which periodically caused decoupling of the ice and till (Piotrowski et al. 2006). High basal water pressure beneath Pleistocene ice sheets is also inferred by many other workers including Brown et al. (1987), Cutler et al. (2000), and Wysota (2007).

Recent observations from modern temperate glaciers also indicate that both deformation and sliding occur and that the processes change spatially and temporally (Engelhardt and Kamb 1998; Truffer et al. 1999, 2000; Truffer and Harrison 2006; Porter and Murray 2001).

Inferences made from Pleistocene and modern environments are likely also applicable to ancient depositional environments of tillite. The subglacial environment for till formation beneath Carboniferous glaciers is also expected to have included coexisting processes of deformation, sliding, lodgement, and plowing, and the sedimentary sequences produced likely record a composite signature of the processes. Because the sedimentary sequences were likely produced by a continuum of subglacial processes, we distinguish the sediment only as subglacial tillite and do not apply the specific genetic classification end member names of lodgement, melt-out, or deformation to the sediment. The most critical distinction to be made in reconstruction of a glacial paleoenvironment is the recognition of subglacial tillite, because the presence of subglacial tillite confirms the presence of glacier ice in the environment.

Deposits in the Paleovalley

Diamictites and associated deposits at three locations in the paleovalley are described in the following sections. At the first location (La Chimenea) near the mouth of the paleovalley, and at the second location (Mid-Valley) near the middle of the paleovalley, predominantly diamictite is exposed. At both locations, the diamictite overlies a few meters of sandstone, which lies on the Ordovician basement rock. At the third location (Campsite) near the head of the paleovalley, mostly alternating beds of sandstone and siltstone are exposed. The sequence of alternating sandstone and siltstone beds contains three thin diamictite beds. We focus on these three locations; however, other outcrops of especially sandstone and siltstone are extensive elsewhere in the paleovalley.

Description of Diamictite at La Chimenea

Diamictite, exposed in the roadcut about 1 km east of the village of La Chimenea (Fig. 7), is mostly clast-supported massive (Dcm) and stratified (Dcs) and is composed predominantly of cobbles and boulders of weathered granite. The matrix is fairly well cemented and consists predominantly of feldspar, quartz, biotite, and muscovite. Individual grains vary from silt and fine sand to granules. A weak preferential alignment of clast long axes is present in some of the diamictite. Clasts have long axes with an apparent predominant orientation of about 85° to about 110° and dip 20° to 30°. The axis of this segment of the valley trends about 185°.

West of the road, a total thickness of about 15 m of diamictite is exposed along the northwest side of the valley. The upper approximately 2 m of the diamictite is clast-supported massive (Dcm) and overlies about 1.5 m of matrix-supported massive diamictite (Dmm). The matrix-supported massive diamictite (Dmm) has apparent fabric due to the alignment of the long axes of pebbles, cobbles, and boulders (Fig. 10). The apparent predominant orientation of the clast long axes is about 100° with a dip of about 10–20°. Cobbles, boulders, and pebbles are generally well rounded, and some have an elongated bullet shape. The clasts are exclusively coarse-grained rocks with granitic composition. The surfaces of the clast are rough and weathered. Striations and polish are not present on the clast surfaces. The diamictite matrix is fairly well cemented and consists of predominantly feldspar, quartz, biotite, and muscovite. Individual grains vary from silt and fine sand to granules. Locally, stringers of greenish sandstone are present in the matrix, and small lenses of greenish sandstone underlie some of the cobbles and boulders. The greenish sandstone has abundant weathered biotite.

Underlying the matrix-supported massive diamictite (Dmm) are several meters of mostly clast-supported diamictites (Dcm) similar to that observed in the roadcut east of La Chimenea. Near the base of the outcrop, the diamictite (Dcm) grades laterally to matrix-supported stratified diamictite (Dms) and overlies thin-bedded, horizontally laminated, and ripple cross-bedded sandstone (Sl and Sr) and horizontally laminated siltstone and mudstone (Fl) (Fig. 11).

Fig. 10 Fabric in boulder-rich, matrix-supported diamictite exposed at La Chimenea

Fig. 11 Clast-supported, massive (Dcm), and matrix-supported, stratified diamictite (Dms), overlying green and red sandstone (Sl, Sr), and laminated siltstone and mudstone (Fl), near La Chimenea. Hammer near the center of photo is 26 cm long

Description of Diamictite and Sandstone at Mid-Valley

Extensive areas of diamictite are exposed along the valley walls about halfway between La Chimenea and Olta (Mid-Valley Section; Fig. 7). Although some layers of the diamictite have more matrix, the diamictite is mostly clast-supported massive (Dcm) and stratified (Dcs) and is composed predominantly of cobbles and boulders of weathered granite (Fig. 12). The small tributary valley on the south side of the main valley is cut into the bouldery diamictite, which has a total thickness of more than 40 m in this part of the valley.

Exposed along the stream bank cut into the southeast valley wall is the contact of the basement rock and the Carboniferous deposits. One to two meters of thin-bedded, horizontally laminated and ripple cross-bedded sandstone (Sl and Sr) overlie the basement rock and underlie the diamictite (Figs. 13, 14, and 15). Ripple cross-beds in the sandstone indicate a paleocurrent direction toward about 240°. At the contact with the diamictite, the sandstone beds fold under and over boulders (Fig. 14). The axes of the folds in the sandstone are oriented from about 120° to about 140° (approximately perpendicular to the valley axis, which trends approximately 240°). Some of the diamictite is matrix supported (as seen to the right of the hammer on Figs. 13 and 14) and contains clasts with near-vertical orientations (as seen on the far right side of Fig. 14). However, except for about a meter of diamictite directly overlying the sandstone, the diamictite is dominated by rounded

Fig. 12 Boulder-rich diamictite exposed on south valley wall at the Mid-Valley section. The diamictite varies from clast-supported to matrix-supported and from massive to stratified (Dcm, Dmm, Dcs, and Dms). Approximate height of outcrop in center of photo is 40 m. Note the subhorizontal bedding in the diamictite near the center of the photo. Boulders range in size up to 3 m in diameter and are of predominantly granitic composition

Fig. 13 Boulder-rich diamictite (Dmm, Dcm) overlying folded sandstone (Sr, Sl) layers at Mid-Valley. Hammer in the center of photo is 26 cm long

Fig. 14 Closer view of folds in sandstone underlying boulder-rich diamictite at Mid-Valley. Note weathering rind on cobble to upper left of hammer. Hammer in center of photo is 26 cm long

to subangular large cobbles and boulders (Fig. 13). Some of the clasts have an elongated bullet shape, but approximately equi-dimensional clasts predominate. The surfaces of the cobbles and boulders are rough, and weathering rinds are prominent on some clasts (Fig. 14).

Fig. 15 Folded sandstone layers at Mid-Valley. Hammer is 26 cm long

The diamictite varies from clast supported to matrix supported and from massive to stratified (Dcm, Dmm, Dcs, and Dms). Approximate height of outcrop in the center of the photo (Fig. 12) is 40 m. The scale of the photo is too small to distinguish individual diamictite beds; however, subhorizontal bedding is apparent in the diamictite shown near the center of the photo. The diamictite has abundant large boulders (some up to 3 m in diameter) of predominantly granitic composition rocks. The small tributary valley on the left is filled in with mostly diamictite. The diamictite is underlain by sandstone in some places as shown in Fig. 13.

Striations and polish are not present on the clast surfaces. The matrix of the diamictite is fairly well cemented and consists predominantly of feldspar, quartz, biotite, and muscovite.

Interpretation of Diamictites at La Chimenea and Mid-Valley

The matrix-supported massive diamictite (Dmm) at La Chimenea (Fig. 10) is interpreted to be subglacial tillite based on its uniform appearance, clast fabric and shape, and the presence of stringers of sorted sediment in the matrix. The fabric likely developed as the substrate (highly weathered Ordovician basement rock) was incorporated into the base of the ice by freezing on or moved in traction below the ice in an unfrozen state, as a bed of sediment advected along by the shear stress of the overriding ice. In subglacial till, the preferred orientation of the clast long axes is generally parallel to the direction of ice flow. The assumed general direction of ice flow was parallel to the valley axis which trends to the southwest (about 240°).

However, locally ice flow may have been transverse to the axis of the valley as the ice flow was likely divergent at the valley mouth. Strong fabric is obvious in the diamictite, but fabric direction is only apparent, as the diamictite is well indurated, and clasts cannot be removed to measure the true orientation of their long axes.

The stringers of greenish sandstone may be proglacial fluvial or lacustrine sediment that was frozen onto the base of the ice, or incorporated, in an unfrozen state, as an inclusion in a thicker bed of soft sediment advected along as a traction layer beneath the moving ice. An alternative interpretation is that the thin layers of sorted sediment resulted when high water pressure separated the ice from the bed, and formed a cavity. Meltwater may have flowed between the ice and the bed and produced washed and sorted sediment, which subsequently infilled the cavity. A similar process may have formed the small lenses of sand found locally beneath some of the cobbles and boulders. The sand lenses may have resulted from the infilling of scours that were eroded by meltwater flowing under cobbles and boulders protruding from the base of the ice (Munro-Stasiuk 2000) or meltwater flowing around cobbles and boulders protruding from the bed. Similar sand stringers and sand pockets in Pleistocene sediments have been taken as indicators of pressurized meltwater flow at the glacier bed (Piotrowski et al. 2006).

The clasts in the tillite are not striated or polished because the parent material is very coarse-grained, weathered, somewhat friable, granitic rock. The tillite does not overlie striated and polished basement rocks because the basement rock is also coarse-grained, somewhat friable, weathered granitic rock. The lithology of the tillite is dominated by the local basement rock, though nonlocal lithologies are reported for diamictite at other locations in the paleovalley (Sterren and Martínez 1996). The predominance of the local lithology in the tillite and other diamictites is likely due to the overwhelming abundance of easily eroded, pre-weathered, basement rock.

The diamictite at Mid-Valley (Fig. 13) is more variable than the matrix-supported massive diamictite (Dmm) at La Chimenea, but based on the folding and other deformation in the underlying sandstone, a subglacial origin is also suggested for this sediment. The fold axes in the sandstone are approximately perpendicular to the assumed ice-flow direction down the valley. The folds in the sandstone may indicate that grounded ice advanced over a soft sediment bed and deformed the soft substrate in the direction of ice flow. Effective pressure was likely low as the ice margin advanced into the flooded valley. The plumose form and associated soft sediment structures in the fold shown on Fig. 15 may indicate concurrent dewatering and high water pressure in the sediment as it was overridden and deformed by the advancing ice. The abundance of large boulders in the sedimentary sequence and the overall coarse texture of the tillite and other diamictites are likely due to the erosion and incorporation locally of intensely weathered basement rock. See below for discussion of the origin and interpretation of these weathered granitic boulders.

Glacial erosion produced subglacial tillite, and glacial conditions also likely enhanced the environment for mass wasting, slumping, and avalanching to occur. Most of the diamictite exposed along the valley wall, such as shown on Fig. 12, does not have attributes characteristic of glaciogenic sediment. Although crude stratification is apparent in some of the diamictite (Fig. 12), the majority of the diamictite is extremely coarse grained with a seemingly chaotic arrangement of large boulders and little matrix material. Some of the diamictite, such as shown in Fig. 12, may be sediment originating from mass wasting, slumping, and avalanching onto the ice surface that was transported as supraglacial sediment and later let down as the ice melted. Some of the diamictite, such as shown in Fig. 12, may have formed during deglaciation, as the processes of mass wasting, slumping, and avalanching likely continued as material slumped, slid, and avalanched off the steep-sided valley as it was being freed from ice. Catastrophic sedimentation events (debris flows triggered by rock falls) may also have occurred during deglaciation and contributed large amounts of sediment to the sequence as seen in Fig. 12.

Diamictite constitutes a significant portion, probably more than 20 % of the fill in the Olta paleovalley. Additional diamictite units with tillite characteristics may be present in the valley, but because much of the diamictite forms steep valley walls, it is largely inaccessible for detailed logging and field description.

Description of Sandstone and Diamictite at the Campsite

Extensive areas of alternating sandstone and siltstone with three prominent interbeds of diamictite are exposed along the south valley wall at the Campsite (Fig. 7). Thin-bedded (10- to 20-cm thick), massive, very fine sandstone (Sm) is interbedded with thin-bedded (10–30-cm thick), ripple cross-bedded, very fine sandstone (Sr) and laminated siltstone (Fl). The massive sandstone beds are more resistant in outcrop (Figs. 16, 17, and 18). The total thickness of the section is about 16 m. The thickness of the diamictite layers ranges from about 0.7 m to about 2 m. The uppermost diamictite layer is about 0.7 m thick, the middle diamictite layer is about 1 m thick, and the lowest diamictite layer in the section is about 1.5 m thick. The lowest diamictite layer is continuous along the south valley wall in the approximate east–west direction for about 120 m and is described here in detail.

The diamictite is mostly matrix-supported massive (Dmm) and stratified (Dms) and is composed predominantly of cobbles and boulders of weathered granite and phyllite (Figs. 16 and 17). The clasts vary from rounded to very angular (tabular). Topping the diamictite is a bed of fairly well-sorted fine gravel and granules (Gm) that is about 30 cm thick (Fig. 16). The gravel bed (Gm) is composed predominantly of feldspar, quartz, biotite, and muscovite. The top of the gravel bed (Gm) strikes about 70° and dips about 4°. The top of the gravel bed (Gm) is smooth and somewhat undulating, whereas the bottom of the diamictite bed is highly irregular (Fig. 16), with large clasts projecting down into the contoured laminated silt and sand beds below (Fig. 18). In some areas, the underlying sandstone beds have discrete folds (Fig. 17). The two predominant orientations of the approximately parallel fold axes are about 320° and 50–65°. Some of the sandstone is ripple cross-bedded. The orientation of the ripples indicates a paleocurrent direction toward about 240°.

Fig. 16 Gravel (Gm), boulder-rich diamictite (Dmm), and deformed sandstone and siltstone layers (Sm, Sr, Fl), overlying alternating sandstone and siltstone beds at the Campsite location. Hammer in lower center of photo is 26 cm long

Fig. 17 Boulder-rich diamictite overlying deformed sandstone layers at the Campsite location. Hat for scale

Fig. 18 Base of diamictite at the Campsite location. Scale is 15 cm long, and placed at the contact of the diamictite with the underlying, highly contoured laminated siltstone

Depositional Models for Diamictites at Campsite

Several depositional mechanisms may have acted to produce the three diamictite beds which lie within the sequence of alternating sandstone and siltstone beds at the Campsite location (Figs. 16 and 17). This discussion focuses on the characteristics of the 2-m-thick lowermost diamictite in the section. The main depositional models considered are:

1. Subaqueous debris flow originating from sediment flow or mass wasting off the glacier front
2. Subaqueous debris flow originating from sediment flow or mass wasting off the valley walls
3. Subglacial deposition from ice advancing into the flooded valley

Saturated conditions are inferred for all models, and it is likely that soft sediment deformation occurred with concurrent dewatering, as is indicated by the convolute bedding in the fine-grained sediment underlying the diamictite (Figs. 16 and 18).

Debris Flow Models

The debris flow models both entail sediment or a mixture of sediment and water entering the body of water in the flooded valley, moving along the valley floor due to density differences, and then being deposited rapidly by immobilization of a mass of sediment. Debris flows stop moving when the driving force of the flow becomes less than the viscosity, shear resistance, and friction (Lowe 1982; Mulder and Alexander 2001).

Debris flows can be lacking cohesion, with frictional strength due to interlocking grains, or cohesive with yield strength due to a cohesive component such as clay (Lowe 1982; Mulder and Alexander 2001). Debrites may be recognized by the following sedimentary features: poor sorting, relatively higher mud content (than the associated sandstones), shear fabric at the base and sides, clasts protruding from the top of the bed, clasts floating in a matrix, lack of grading and traction structures (such as cross bedding or parallel laminations), fluid-escape structures, and sharp upper contacts (Middleton and Hampton 1976; Lowe 1982; Mulder and Alexander 2001).

The diamictite bed at the Campsite location has several sedimentary features that are common to debrites including poor sorting, relatively higher mud content than the associated sandstones, shear fabric at the base, clasts floating in a matrix, lack of grading and traction structures, fluid-escape structures, and a sharp upper contact.

The lowermost diamictite bed may be a debrite, but the distribution of the diamictite is more consistent with origination of the debrite off the ice front rather than the valley walls. Debris flow or mass wastage off the valley wall is expected to produce a fan- or wedge-shaped deposit along the valley wall, rather than a bed with fairly uniform thickness over a distance of over 100 m along the valley wall, as is found at the Campsite location. Therefore, the diamictite bed may represent ice advance into the flooded valley. The distribution of the diamictite is also consistent with a subglacial origin as a bed of sediment deposited from an advancing glacier.

Some striking features of the diamictite bed, not common to debrites, include a smooth, relatively even, or slightly undulating upper surface (clasts do not protrude from the top of the bed) (Figs. 16 and 17) and deformation structures that penetrate the underlying beds to depths similar to the full thickness of the diamictite bed (Fig. 17). The apparent smooth upper surface of the diamictite bed is to some degree due to the gravel (Gm) capping of the diamictite (Figs. 16 and 17). The gravel (Gm) is fairly well sorted and may serve to drape the underlying very poorly sorted diamictite, obscuring irregularities in the top of the diamictite. The apparent smooth upper surface of the diamictite may also in part be due to the large clasts settling into the underlying sediments following deposition.

Subglacial Model

With either a debris flow or a subglacial origin of the diamictite, it is likely that some postdepositional deformation occurred as boulders settled through the matrix and deformed the underlying sediment. But the large-scale folds that penetrate the underlying strata are not likely entirely due to gravitational settling. The folds, whose axes have a predominant orientation transverse to the axis of the valley, and to the presumed ice-flow direction, may be due to the shear stress from the overriding ice. Sedimentary features observed in the diamictite (including poor sorting, relatively higher mud content than the associated sandstones, shear fabric at the base, clasts floating in a matrix, lack of grading and traction structures, fluid-escape structures, and a sharp upper contact) which are common to debrites are also common to tillites. The diamictite could have originated as a bed of sediment moving in traction with the advancing ice front. The advancing ice and layer of sediment moving in traction with the ice overrode soft layered substrate and produced folds in the sediment layers. Alternatively, the diamictite could have been a debrite that originated off the advancing ice front, which was overridden as the ice advanced. Folds in the underlying layered sediment would reflect ice-flow direction in both cases.

The surface of the gravel layer (Gm) topping the diamictite (Fig. 16) dips up-valley. An up-valley dip is also presumably an up-glacier dip, which is consistent with a subglacial origin. Also the composition of the gravel bed is very similar to the composition of the tillite matrix.

The gravel (Gm) may have originated from a turbid plume of englacial and/or subglacial meltwater discharging from the glacier. Turbidites are formed by sediments gradually settling out suspension and progressively aggrading a bed (Lowe 1982). Recognition criteria for sandy turbidites include good sorting, relatively low mud content, normal grading and traction structures, and inverse grading at their bases (Middleton and Hampton 1976; Lowe 1982). The gravel (Gm) is well sorted and has a low mud content but lacks grading and traction structures. If the gravel (Gm) originated as a turbidite, it is suggested that the sediment concentration was likely very high but that deposition was still particle by particle in order to drape the underlying diamictite.

The lower diamictite likely represents an ice-front advance; either it is a debrite originating off the ice front, or it is a subglacial deposit, i.e., a tillite, or a combination of both. The three diamictite beds in the section may record three minor ice-front advances into the flooded valley.

Paleoenvironmental Setting

The paleoenvironmental setting for the Olta paleovalley is likely similar to that reported from the southwestern part of the Paganzo basin in the Jejenes subbasin near the city of San Juan (Kneller et al. 2004). Here a paleovalley, interpreted to be a paleofjord, contains deposits that record marine flooding and infilling with glacial and associated sediments (Kneller et al. 2004). Diamictites, including tillites, are found locally in the paleovalley, but the majority of the valley fill consists of mudstone generated by turbid plumes of glacial meltwater and sandy turbidites including mass-transport sediments (Kneller et al. 2004).

Deglaciation may also have been concurrent with or was triggered by marine transgression on the eastern side of the Paganzo basin, as marine transgression into the Malanzán–Olta paleovalley is indicated by marine microplankton (*Cymathiosphaera* sp., *Greinervillites* sp., and *Navifusa variabilis*) of Namurian–Westphalian age, discovered in mudstone and shales of the Malanzán Formation (Limarino et al. 2002). The strictly lacustrine interpretation of the Malanzán Formation (Limarino and Césari 1988; Andreis et al. 1986; Buatois et al. 1994; Sterren and Martínez 1996) is reinterpreted as sedimentation at the inner part of a paleofjord by Limarino et al. (2002).

Clay mineralogy (an illite + chlorite assemblage), in a sequence of mudstones, shale, and fine- and medium-grained sandstones, indicates marine flooding of fluvial valleys in the Olta–Malanzán area (Net et al. 2002). These marine-flooded valleys were possibly fjords where turbidites and Gilbert-type delta sedimentation occurred (Net et al. 2002). Marine fossils, found near the bottom of the delta sequence by Gutiérrez and Limarino (2001), support this interpretation of a marine-flooded valley. Sedimentology (ripple cross-laminated sandstone with foresets dipping in opposite directions) indicates the influence of tidal currents in the Olta paleovalley (Limarino et al. 2002).

The Preglacial Ancient Weathering Front (Late Devonian?–Earliest Carboniferous?)

The abundance of large boulders and granite sand observed in the sedimentary sequence and the overall coarse texture of the tillite and other diamictites are likely due to the erosion and local incorporation of intensely weathered basement rock (i.e., Ordovician) (Carignano et al. 2008, 2009). See below for discussion of the origin of these weathered boulders.

Similar intensely weathered rock underlying glacial deposits has been described from Canada (Ryan et al. 2005) and Europe (Lidmar-Bergstrom et al. 1997; Migón and Lidmar-Bergstrom 2001). The weathering of the Ordovician basement rock occurred prior to the onset of glaciation in the Paganzo basin, perhaps sometime between the Late Devonian and the Earliest Carboniferous, when much warmer–wetter climates dominated. An age previous to the Late Devonian for this deep weathering period could not be ruled out, but there is no evidence yet to suggest an older firm chronology. Deep zones of corestones and saprolites comprised a paleosurface or weathered horizon that may have extended vertically along joints in the basement rock for tens or hundreds of meters. Within the weathered horizon, pods of relatively unweathered material or corestones, in the central position of the blocks, were surrounded by coarser granitic saprolitic material (basically, a gruss) that was poorly consolidated and easily eroded by fluvial, alluvial, and ultimately glacial processes, as the climate changed to wetter and colder conditions. Glacial erosion and other associated erosional processes likely stripped the pre-weathered rock surfaces and deposited large amounts of sediment locally. See Fig. 19a–e.

Fig. 19 The ancient preglacial weathered surface (Late Devonian?–Earliest Carboniferous?). (**a**) Large granitic corestones in situ, underlying the Carboniferous glaciogenic sedimentary rocks. (**b**) Basal contact of the Carboniferous glaciogenic sedimentary rocks overlying the ancient weathering front. See a large rounded granitic corestones next to the person and a second one, both in situ, immediately underneath the glacial sediments. (**c**) A clear basal contact of the glaciogenic sediments overlying the in situ weathered granites. (**d**) A detail of the basal contact of the tillites and outwash sediments on top of the weathered granites. (**e**) In situ granitic, rounded corestones, surrounded by weathered materials and gruss. See the joints that separate the granite blocks, of which the corestones represent the unweathered portion

The occurrence of these corestones and the ancient paleoweathering front that includes them are a significant fact that confirms the existence of preglacial (i.e., pre-Middle Carboniferous) deep chemical weathering under perhaps hyper-tropical climates (see Rabassa 2010, 2014; Rabassa et al. 2010, 2014; Compagnucci 2014).

These findings describe paleoclimatic and paleoenvironmental conditions that have not been clearly described before. This part of the Sierras Pampeanas and the Paganzo basin were already a positive (i.e., continental) area in very ancient times, maybe at least since the Late Devonian (?) or even before. In fact, sometime between the uplifting of the Ordovician granites and metamorphic rocks, and the large-scale climate change (ice house) that installed the Carboniferous glaciations, a hyper-tropical climate was dominant in the region which forced the development of very deep chemical weathering conditions. The available information is so far not enough to define other characteristics of the ancient weathering front, but it may have been even several hundred meters thick, considering the irregular topography found underneath the glaciogenic deposits and the significant elevation difference between unweathered granites and schists located at higher elevations and the weathered gruss and corestones which are found at much lower elevations, in just very short distances. To our knowledge, this is the first time these paleoclimatic and paleoenvironmental conditions are described for this period, Late Devonian to Earliest Carboniferous, in cratonic areas of Argentina.

Sequence of Events

A major control on the nature of the sedimentary deposits found in the paleovalley is the weathering of the Ordovician granitic basement rock that perhaps occurred during the Late Devonian or other pre-Middle Carboniferous period of warmer–wetter climate, as it has been stated above. The deep zones of saprolite and corestones that were thus formed were easily eroded and partially denudated by the Carboniferous valley glacier that advanced into the area and eroded a characteristic U-shaped valley. Saprolite and corestones are still found in situ in the ancient weathering front or have been eroded and incorporated, mostly as basal glacial load, into the basal till of the Carboniferous sequence.

Locally ice advanced over soft sediment and produced deformation. Deformation and incorporation of local material indicates the ice was in contact with the bed. Grounded ice advanced into a valley which was flooded by marine waters at least part of the time. The ice may have readvanced several times during a period of general ice recession, because the thick sequence of diamictite found in the paleovalley may include one or more units that were deposited by or from grounded ice as subglacial till. The thick sequence of diamictites exposed at La Chimenea may be the remnant of end moraine(s), marking the farthest extent of ice advance to the southwest, to the mouth of the Olta paleovalley.

As sea level rose, the ice receded, and deltaic, lacustrine, and marine sediments, fine-grained sandstones and mudstones with dropstones, and fanglomerates infilled the Olta paleovalley. The thick sequence of diamictites exposing Mid-Valley in the paleovalley may be the remnant of recessional moraine(s), marking a stable position of the ice front, or a minor readvance of ice into the Olta paleovalley. As the ice margin retreated farther up the valley to the northeast, alternating beds of sandstone and siltstone were deposited in the valley filled by glacial meltwater and marine flooding. The three diamictite beds, lying within the sequence of alternating sandstone and siltstone, may record three minor ice-front readvances into the flooded valley. Eventually the valley was deglaciated. Fluvial processes dominated during the Permian, and deposits exposed in the northeast portion of the paleovalley are mostly red, cross-bedded sandstones. Weathering and erosion resulted in lowering of the land surface. Predominantly fluvial processes are exhuming the paleovalley or paleofjord.

Summary and Conclusions

This study presents a record of glacier advance and retreat preserved in Carboniferous strata found in the Olta paleovalley on the eastern side of the Paganzo basin. Although a glacial origin was inferred for the paleovalley, a direct glaciogenic origin had not been inferred for any of the diamictites in the paleovalley (Andreis et al. 1986; Buatois and Mángano 1995; Sterren and Martínez 1996). Diamictites at La Chimenea, and Mid-Valley, interpreted by this study to be subglacial tillite, provide direct evidence of valley glaciation. Deformation in the sandstone underlying the diamictite indicates warm-based conditions as the glacier advanced over soft deformable sediment. The composition of the bouldery diamictite (mostly locally derived from weathered granitic basement rock and granite corestones of an ancient paleoweathering front) is attributed to high rates of deposition and the susceptibility of pre-weathered granitic basement rock to glacial and other erosional processes. Concurrent with and following glaciation, other processes including mass-transport (slumping, rafting, sliding, avalanching, and debris flow) likely operated in the steep-sided valley and contributed large amounts of diamictite and other sediment to the valley fill. Corestones, weathered from the basement rock during the Late Devonian or other pre-Middle Carboniferous period of intense weathering, constitute the large clasts in the diamictite and associated deposits.

The lower diamictite bed at the Campsite represents an ice-front advance, either as a debrite originating away from the ice front, or as a subglacial deposit, i.e., a tillite, or a combination of both. The three diamictite beds within the sequence of alternating sandstone and siltstone at the Campsite location may record three minor ice-front advances into the flooded valley. The valley was likely a transitional (fjord) environment, as micropaleontological material (Gutiérrez and Limarino 2001) and clay mineral assemblages (Net et al. 2002) indicate a marine transgression into the area during the mid-Carboniferous.

References

Andreis R, Bossi G (1981) Algunos ciclos lacustres de la Formación Malanzán (Carbónico superior) en la región de Malanzán, Sierra de los Llanos, provincia de La Rioja. 13° Congreso Geológico Argentino, Actas, 4, 639–655

Andreis R, Leguizamón R, Archangelsky S (1986) El paleovalle de Malanzán: nuevos criterios para estratigrafia del Neopaleozoico de la Sierra de los Llanos, La Rioja, República Argentina. Bol Acad Nac Cienc Cordoba 57:3–119

Benn DI (1995) Fabric signature of subglacial till deformation, Bredamerkurjokull, Iceland. Sedimentology 42:735–747

Benn DI, Evans DJA (1996) The interpretation and classification of subglacially deformed materials. Quaternary Sci Rev 15:23–52

Boulton GS (1971) Till genesis and fabric in Svalbard, Spitsbergen. In: Goldthwait RP (ed) Till: a symposium. Ohio State University Press, Columbus, pp 41–72

Boulton GS (1972) Modern arctic glaciers as depositional models for former ice sheets. Q J Geol Soc Lond 128:351–393

Boulton GS (1978) Boulder shapes and grain-size distribution of debris as indicators of transport paths through a glacier and till genesis. Sedimentology 25:773–799

Boulton GS (1996) Theory of glacial erosion, transport and deposition as a consequence of subglacial deformation. J Glaciol 42:43–62

Brown NE, Hallet B, Booth DB (1987) Rapid soft bed sliding of the Puget Glacier lobe. J Geophys Res 92(B9):8985–8997

Buatois LA, Mángano MG (1995) Post glacial lacustrine event sedimentation in an ancient mountain setting: Carboniferous Lake Malanzán (western Argentina). J Paleolimnol 14:1–22

Buatois LA, Limarino CA, Césari SN (1994) Carboniferous lacustrine deposits from the Paganzo basin, Argentina. In: Gierlowski-Kordesch E, Kelts K (eds) Global geological record of lake basins. Cambridge University Press, Cambridge, pp 135–140

Carignano C, Rabassa J, Socha B, Mickelson D (2008) Nueva Interpretación de la Secuencia sedimentaria del Paleovalle Glacial Exhumado de Olta (Provincia de La Rioja) y Hallazgo de perfiles de meteorización indicadores de ambientes previos a la Glaciación de Montaña del Carbonífero Medio en la Cuenca de Paganzo. XII Reunión Argentina de Sedimentología, Abstracts volume, p 48

Carignano C, Cioccale M, Rabassa J, Socha B (2009) Paleogeomorfología de la Sierra de Olta, Sierras Pampeanas, La Rioja: Superficies de planación del Silúrico?-Carbónico Inferior? y paleogeoformas de Glaciación de Montaña del Carbónico. Simposio Paleosuperficies en regiones cratónicas y Evolución del paisaje en el largo plazo. IV Congreso Argentino de Cuaternario y Geomorfología, La Plata. Abstracts volume, p 5

Clayton L, Mickelson DM, Attig JW (1989) Evidence against pervasively deformed bed material beneath rapidly moving lobes of the southern Laurentide Ice Sheet. Sediment Geol 62:203–208

Compagnucci RH (2014) Modeling the atmospheric circulation and climatic conditions over Southern South America during the late history of the Gondwana supercontinent. In: Rabassa J, Ollier C (eds) Gondwana landscapes in southern South America. Springer, Dordrecht, pp 113–134

Cutler PM, MacAyeal DR, Mickelson DM, Parizek BR, Colgan PM (2000) A numerical investigation of the ice-lobe-permafrost interaction around the southern Laurentide ice sheet. J Glaciol 46:311–325

Dreimanis A (1976) Tills: their origin and properties. In: Leggett RF (ed) Glacial till: an interdisciplinary study, Special publication 12. Royal Society of Canada, National Research Council of Canada, Ottawa, pp 11–49

Dreimanis A (1988) Tills, their genetic terminology and classification. In: Goldthwait RP, Matsch CL (eds) Genetic classification of glacigenic deposits. Balkema, Rotterdam, pp 17–84

Dykstra M, Kneller BC, Milana JP (2006) Deglacial and postglacial sedimentary architecture in a deeply incised paleovalley–paleofjord– The Pennsylvanian (Late Carboniferous) Jejenes Formation, San Juan, Argentina. Geol Soc Am Bull 118:913–937

Elson JA (1988) Comment on glacitectonite, deformation till, and comminution till. In: Goldthwait RP, Matsch CL (eds) Genetic classification of glacigenic deposits. Balkema, Rotterdam, pp 85–88

Engelhardt HF, Kamb B (1998) Basal sliding of Ice Stream B, West Antarctica. J Glaciol 44:223–230

Evans DJA, Phillips ER, Hiemstra JF, Auton CA (2006) Subglacial till: formation, sedimentary characteristics and classification. Earth Sci Rev 78:115–176

Eyles N, Eyles CH, Miall AD (1983) Lithofacies types and vertical profile models; an alternative approach to the description and environmental interpretation of glacial diamict and diamictite sequences. Sedimentology 30:393–410

Gutiérrez PR, Limarino CO (2001) Palinología de la Formación Malanzán (Carbonífero Superior) La Rioja, Argentina: nuevos elementos y consideraciones paleoambientales. Ameghiniana 38:99–118

Ham NR, Mickelson DM (1994) Basal till fabric and deposition at Burroughs Glacier, Glacier Bay, Alaska. Geol Soc Am Bull 106:1552–1559

Hart JK (1998) The deforming bed/debris-rich ice continuum and its implications for the formation of glacial landforms (flutes) and sediments (meltout till). Quaternary Sci Rev 17:737–754

Hooyer TS, Iverson NR (2000) Diffusive mixing between shearing granular layers: constraints on bed deformation from till contacts. J Glaciol 46:641–651

Kneller BC, Milana JP, Buckee C, Ja'aidi O (2004) A depositional record of deglaciation in a paleofjord (Late Carboniferous [Pennsylvanian] of San Juan Province, Argentina): the role of catastrophic sedimentation. Geol Soc Am Bull 116:348–367

Lawson DE (1979) A comparison of the pebble fabric orientation in ice and deposits of the Matanuska Glacier, Alaska. J Geol 87:629–645

Lawson DE (1981) Distinguishing characteristics of diamictons at the margin of the Matanuska Glacier, Alaska. Ann Glaciol 2:78–84

Lawson DE, Kemmis TJ (1983) Genetic differentiation of Wisconsin age glacigenic diamictons in north-central Iowa, USA. Abstracts with Program, Geological Society of America, North-Central Section, 15, 223

Lidmar-Bergstrom K, Olsson S, Olvmo M (1997) Paleosurfaces and associated saprolites in southern Sweden. In: Widdowson M (ed) Paleosurfaces, recognition and reconstruction and paleoenvironmental interpretation, Geological Society of London special publication 120. Geological Society, London, pp 95–124

Limarino CO, Césari SN (1988) Paleoclimatic significance of the lacustrine Carboniferous deposits in northwestern Argentina. Palaeogeogr Palaeocl Palaeoecol 65:115–131

Limarino CO, Gutiérrez P (1990) Diamictites in the Agua Colorada Formation (northwestern Argentina): new evidences of Carboniferous glaciation in South America. J S Am Earth Sci 3:9–20

Limarino CO, Césari SN, Net LI, Marenssi SA, Gutiérrez RP, Tripaldi A (2002) The Upper Carboniferous postglacial transgression in the Paganzo and Rio Blanco basins (northwestern Argentina): facies and stratigraphic significance. J S Am Earth Sci 15:445–460

López Gamundi OR (1987) Depositional models for the glaciomarine sequences of Andean late Paleozoic basins of Argentina. Sediment Geol 52:109–126

López Gamundi OR (1997) Glacial-postglacial transgressions in late Paleozoic basins of southern South America. In: Martini IP (ed) Late glacial and post-glacial environmental changes – Quaternary, Carboniferous-Permian, and Proterozoic. Oxford University Press, New York, pp 147–168

López Gamundi OR, Limarino CO, Césari SN (1992) Late Paleozoic paleoclimatology of central west Argentina. Palaeogeogr Palaeocl Palaeoecol 91:305–329

Lowe DR (1982) Sediment gravity flow: II. Depositional models with special reference to the deposits of high-density turbidity currents. J Sediment Petrol 52:279–297
Mickelson DM, Ham NR Jr, Ronnert L (1992) Comment on "Striated clast pavements: products of deforming subglacial sediment?". Geology 20:285
Middleton GV, Hampton MA (1976) Subaqueous sediment transport and deposition. In: Stanley DJ, Swift PJP (eds) Marine sediment transport and environmental management. Wiley, New York, pp 197–218
Migón P, Lidmar-Bergstrom K (2001) Weathering mantles and their significance for geomorphological evolution of central and northern Europe since the Mesozoic. Earth Sci Rev 56:285–324
Mulder T, Alexander J (2001) The physical character of subaqueous sedimentary density flows and their deposits. Sedimentology 48:269–299
Munro-Stasiuk MJ (2000) Rhythmic till sedimentation: evidence for repeated hydraulic lifting of a stagnant ice mass. J Sediment Res 70:94–106
Net LI, Alonso MS, Limarino CO (2002) Source rock and environmental control on clay mineral associations, lower section of Paganzo Group (Carboniferous), northwest Argentina. Sediment Geol 152:183–199
Page S, Litvak VD, Limarino CO (2002) Nueva edad en el basamento de la sierra de Los Llanos, La Rioja: Intrusividad pampeana en el arco famatiniano. Revista Asociación Geológica Argentina 57:345–348
Paterson WSB (1994) The physics of glaciers, 3rd edn. Pergamon Press, Trowbridge, 480 p
Piotrowski JA, Tulaczyk S (1999) Subglacial conditions under the last ice sheet in northwest Germany: ice-bed separation and enhanced basal sliding? Quaternary Sci Rev 18:737–751
Piotrowski JA, Mickelson DM, Tulaczyk S, Krzyszkowski D, Junge FW (2001) Were deforming subglacial beds beneath past ice sheets really widespread? Quatern Int 86:139–150
Piotrowski JA, Larsen NK, Junge FW (2004) Reflections on soft subglacial beds as a mosaic of deforming and stable spots. Quaternary Sci Rev 23:993–1000
Piotrowski JA, Larsen NK, Menzies J, Wysota W (2006) Formation of subglacial till under transient bed conditions: deposition, deformation, and basal decoupling under a Weichselian ice sheet lobe, central Poland. Sedimentology 53:83–106
Porter PR, Murray T (2001) Mechanical and hydrological properties of till beneath Bakaninbreen, Svalbard. J Glaciol 47:167–175
Rabassa J (2010) Gondwana paleolandscapes: long-term landscape evolution, genesis, distribution, and age. Geociências 29(4):541–570. São Paulo, UNESP, Brazil
Rabassa J (2014) Some concepts on Gondwana landscapes: long-term landscape evolution, genesis, distribution and age. In: Rabassa J, Ollier C (eds) Gondwana landscapes in southern South America. Springer, Dordrecht, pp 9–46
Rabassa J, Socha BJ, Carignano C, Mickelson D (2007) Sedimentological record of Carboniferous mountain glaciation in an exhumed paleovalley, Eastern Paganzo Basin, La Rioja Province, Argentina. Uruguay National Geological Congress, Abstracts
Rabassa J, Carignano C, Cioccale M (2010) Gondwanic Paleosurfaces in Argentina: an introduction. Geociências 29(4):439–466. São Paulo, UNESP, Brazil
Rabassa J, Carignano C, Cioccale M (2014) A general overview of Gondwana landscapes in Argentina. In: Rabassa J, Ollier C (eds) Gondwana landscapes in southern South America. Springer, Dordrecht, pp 201–245
Ramos VA (1988) The tectonics of central Andes; 30°–33°S latitude. In: Clark S, Burchfiel D (eds) Processes in continental lithosphere deformation, Geological Society of America special paper 218. Geological Society of America, Boulder, pp 31–54
Ryan RJ, O'Beirne-Ryan AM, Zentilli M (2005) Rounded cobbles that have not traveled far: incorporation of corestones from saprolite in the South Mountain area of southern Nova Scotia, Canada. Sedimentology 52:1109–1121
Socha BJ (2007) Basal till and tillite: sedimentary characteristics and inferred sediment sources, mode of transport, depositional environment and general conditions of the glacier bed. Unpublished Ph.D. dissertation, Department of Geological Sciences, University of Wisconsin-Madison, USA, 247 pp. University Microfilms International Microform 3294146

Socha BJ, Carignano C, Rabassa J, Mickelson DM (2006) Sedimentological record of Carboniferous mountain glaciation in an exhumed paleovalley, Eastern Paganzo Basin, La Rioja Province, Argentina, paper no. 22–17, Geological Society of America Annual Meeting Abstracts with Programs, 38, 7, 62

Sterren AF, Martínez M (1996) El Paleovalle de Olta (Carbonífero): Paleoambiente y Paleogeografía. 13° Congreso Geológico Argentino and 3° Congreso de Exploración de Hidrocarburos, Actas, 2:89–103

Truffer M, Harrison WD (2006) In situ measurements of till deformation and water pressure. J Glaciol 52:175–182

Truffer M, Moytyka RJ, Harrison WD, Echelmeyer KA, Fisk B, Tulaczyk S (1999) Subglacial drilling at Black Rapids Glacier, Alaska, U.S.A.: drilling method and sample description. J Glaciol 45:495–505

Truffer M, Harrison WD, Echelmeyer KA (2000) Glacier motion dominated by processes deep in underlying till. J Glaciol 46:213–221

Wysota W (2007) Successive subglacial depositional processes as interpreted from basal tills in the Lower Vistula valley (N. Poland). Sediment Geol 193:21–31

Planation Surfaces of Central Western Argentina

Marcelo Zárate and Alicia Folguera

Abstract The region of central western Argentina is situated in a complex geological and tectonic setting with transitional features between major environmental, tectonic, and geomorphological domains. Preserved remnants of planation surfaces, which form the summit of major geological units, occur as discontinuous and isolated exposures with a remarkable geographical distribution along 600 km (33°–39° S). The main goal of this chapter is to summarize the available information and describe the main features and geological context of the planation surfaces preserved in the areas of the San Rafael block, Payenia, the Chical-có plain, and the Chadileuvú block situated in the southern part of Mendoza and western La Pampa provinces, central western Argentina. The common feature of the areas under analysis is the occurrence of remnants of planation surfaces at different altitudes which cut across bedrock of different composition (igneous, metamorphic, and sedimentary units) and varied age ranging from the Middle Proterozoic (Grenvillian age) to Middle Triassic continental sedimentary rocks. The development may have begun in the Triassic, before the Gondwana breakup and more than one planation surface might be present. During the Late Miocene synorogenic deposits covered the planation surface at the San Rafael block, very likely Payenia, and the Chical-có plain; no conclusive evidence is available in the area of the Chadileuvú block. As a result of tectonic activity (circa 7–6 Ma?), the region was fractured, differentially uplifted, and started a general exhumation process that partially removed the Late Miocene sedimentary cover from the San Rafael block and the eastern margin of Chical-có plain.

M. Zárate (✉)
Instituto de Ciencias de La Tierra y Ambientales de La Pampa (CONICET-UNLPAM) – Facultad de Ciencias Exactas y Naturales, UNLPAM, Avenida Uruguay 151, 6300 Santa Rosa, La Pampa, Argentina
e-mail: marcelozarate55@yahoo.com.ar

A. Folguera
Instituto de Geología y Recursos Minerales, SEGEMAR, Av. Julio A. Roca 651 piso 10 oficina 16, 1322 Buenos Aires, Argentina

J. Rabassa and C. Ollier (eds.), *Gondwana Landscapes in southern South America*, Springer Earth System Sciences, DOI 10.1007/978-94-007-7702-6_13,

Keywords Gondwana • Central Western Argentina • Cratonic areas • Etchplains • Pediplains

Introduction

The occurrence of ancient erosion surfaces has been reported in different mountain ranges of central Argentina since the beginning of the geological investigations of the territory (Stelzner 1885; Keidel 1916; Nágera 1939, among others). Except for general comments, the geomorphological, sedimentological, and tectonic significance of erosion surfaces remained barely unexplored in Argentina until recently when these landforms became the focus of renewed interest (e.g., Carignano et al. 1999; Demoulin et al. 2005; Rabassa 2010; Rabassa et al. 2014, this volume).

In central western Argentina, ancient erosion surfaces have been reported at various areas and localities between 33° and 38° S across southern Mendoza and western La Pampa provinces (Fig. 1). West of San Rafael (Mendoza), Feruglio (1946) and Dessanti (1956) made the first general descriptions followed by Polanski (1963) who provided a broad account of what he called a peneplain surface. Later, González Díaz (1972a, b, c) described the characteristics of the surfaces in several other areas south of San Rafael. Further south Calmels (1996) described the western part of La Pampa province as an ancient landscape.

Ongoing geological survey on the Neogene and Quaternary evolution of central Argentina (Folguera and Zárate 2009, 2011a) reveals the large geographical extension of planation surfaces across the Andean piedmont of southern Mendoza province along with an ample part of western La Pampa province. The preserved remnants of planation surfaces, which form the summit of major geological units, make up discontinuous and isolated exposures of variable areal extension ranging from few kilometers to hundreds of kilometers.

The region under study, located in between the Andean orogenic belt and the Andean foreland, exhibits transitional features of major environmental, tectonic, and geomorphological domains. Thus, it represents a key area to understand the role and significance played by planation surfaces during the geological evolution of Gondwana and its subsequent breakup. The remarkable geographical distribution of planation surfaces rises several questions on their genesis, age, and implications to interpret the tectonic behavior and geological evolution of the region. Do the remnants represent a single or several erosional surfaces? What is the tectonic significance of the landforms? What has been the role played by the Andean orogeny in their present configuration?

The main goal of this chapter is to summarize the available information and describe the main features and geological context of the planation surfaces preserved in central western Argentina. Their geomorphological and geological significance is discussed within the framework of the Middle–Late Mesozoic and Cenozoic tectonic dynamic. At a regional scale, a general comparison and correlation with planation surfaces described in the Sierras Pampeanas (or Pampean ranges) of Córdoba and San Luis provinces and the ranges of southern Buenos Aires province is presented.

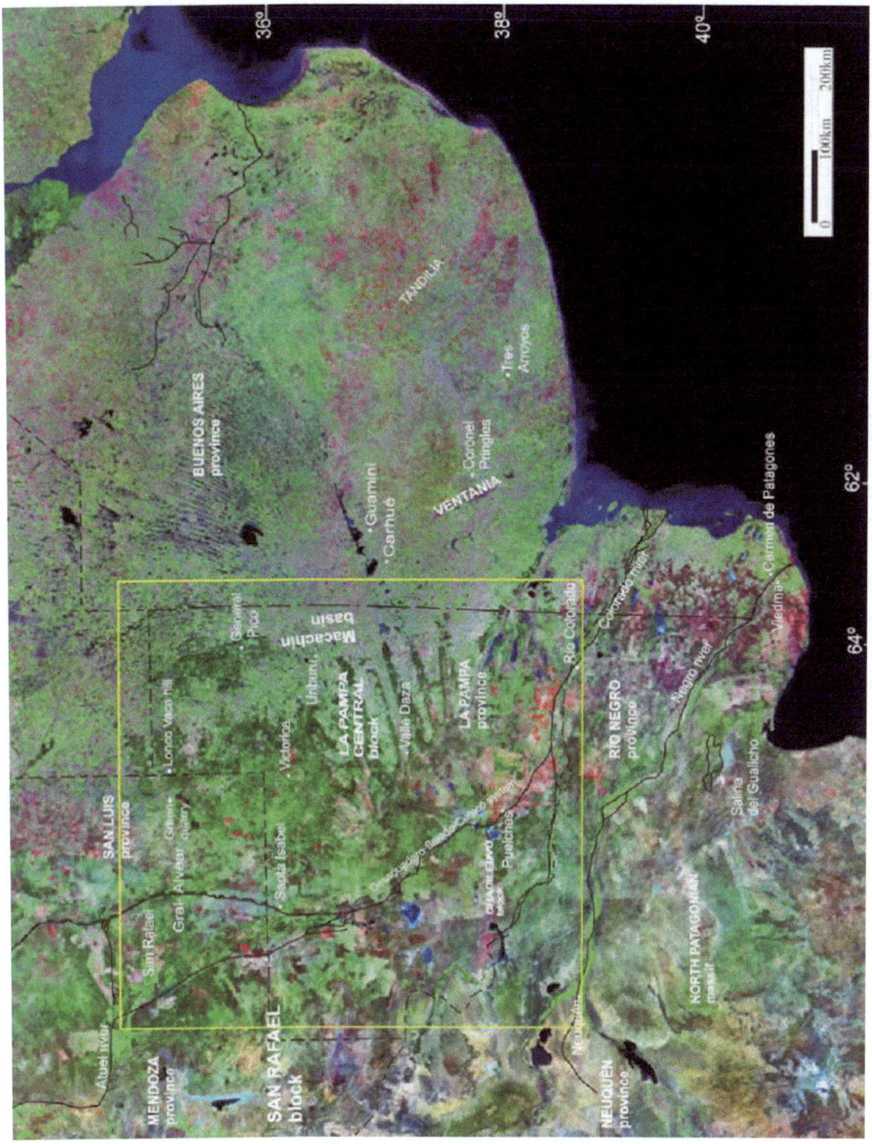

Fig. 1 Geographical location of the region under study

The Studied Region

The region under study encompasses the southeastern part of Mendoza province and the western sector of La Pampa province (Fig. 1) and extends along nearly 600 km from north to south between the northern tip of the San Rafael block (~34° S) and the Colorado River (38° S). It shows a variable width from east to west, ranging from 80 to 100 km in the northern and central areas to around 200 km in the southern part. From a geomorphological and environmental viewpoint, it is a transitional region between the Andean domain, the Pampean plain of central Argentina, and Patagonia. The valley of the Desaguadero–Salado–Curacó (Fig. 1) is taken as the boundary between the Andean piedmont and the westernmost part of the Pampean plain (Zárate 2009); the Colorado River is usually considered the geographical boundary between the Pampean plain and Patagonia (Coronato et al. 2008). The climate is dominantly arid in the west grading to semiarid conditions eastwards. The drainage system includes numerous ephemeral streams, either integrating endorheic basins in some areas or being tributaries of allochthonous streams such as the Atuel, the Diamante, and the Desaguadero–Salado–Curacó rivers. The Atuel and the Diamante rivers are perennial streams of seasonal regime generated by the melting of snowfalls at their headwaters located further west in the high Andes. These two rivers are tributaries of the Desaguadero–Salado–Curacó fluvial system, a major stream around 1,200 km long, presently misfit which, north of the studied region, is fed by other major allochthonous rivers (such as the Mendoza, Tunuyán, San Juan, and Bermejo rivers). In turn, the Desaguadero–Salado–Curacó fluvial system is a tributary of the Colorado River (Fig. 1), another allochthonous stream that drains a segment of the Andes Cordillera in southernmost Mendoza province and Neuquén province.

Geological and Structural Setting

The region under study has been originally considered as a single geological unit named as the "Provincia Geológica Sanrafaelino–Pampeana" (Sanrafaelian–Pampean geological province) by Criado Roqué and Ibáñez (1979). To date, according to the most recent geological subdivision of the Argentine territory (Ramos 1999), the region is comprised into the geological provinces of the San Rafael block, Payenia, and Chadileuvú block (also called Las Mahuidas; Ramos 1999) (Fig. 2). It is a tectonically complex setting composed of autochthonous and allochthonous terrains accreted to the Rio de la Plata craton during the Paleozoic evolution of the southwestern margin of Gondwana (Ramos 2010). The stratigraphic record consists mainly of Middle–Late Proterozoic basement covered by Paleozoic sedimentary rocks and a Permian–Triassic igneous complex; after a prolonged stratigraphic hiatus, Neogene continental deposits were accumulated accompanied by significant volcanic activity during the Pliocene and the Quaternary in the westernmost part of the study area (Fig. 1).

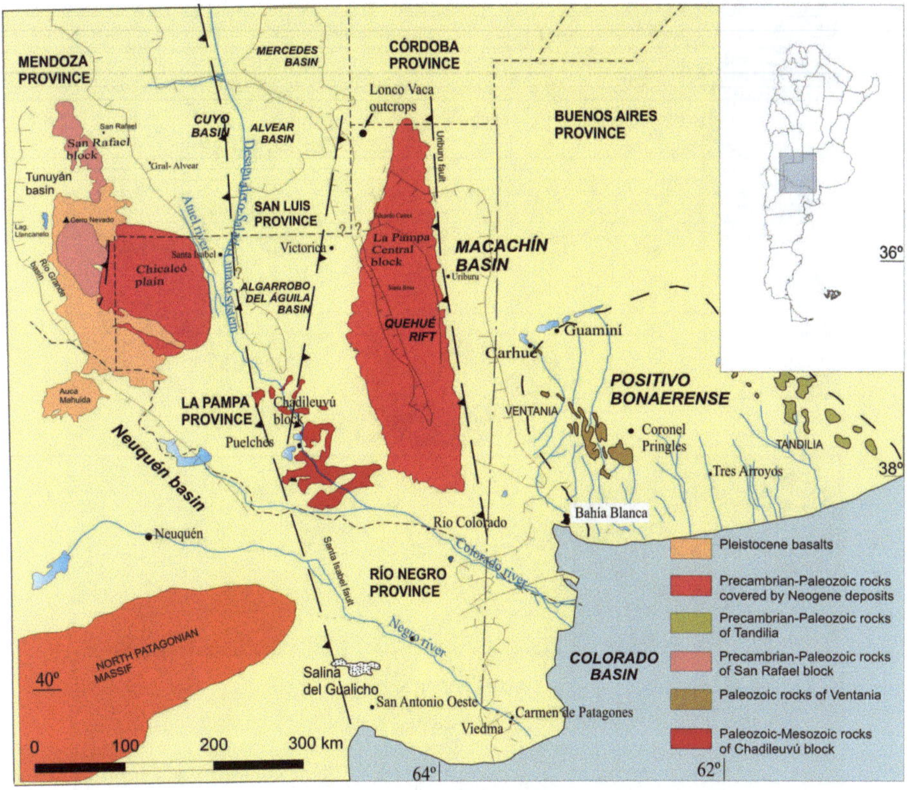

Fig. 2 Morphostructural units of the region under study; sedimentary basins and blocks

According to Criado Roqué and Ibáñez (1979), the region is characterized by major E–W trending fault lines which dominated during the Precambrian that changed to a prevailing north–south fracture system pattern around the Precambrian–Paleozoic boundary. The region was affected by block movement along fractures lines since the Carboniferous, while the faults were later reactivated since the Triassic to the Quaternary (Criado Roqué and Ibáñez 1979). The Atuel River and the southern reach of the Desaguadero–Salado–Curacó system are structurally controlled by a NNW–SSE fault zone (Fig. 2) named Santa Isabel thrust (Folguera and Zárate 2011b).

The San Rafael block (Figs. 3 and 4), segmented into a series of minor blocks, is a fault-bounded structural unit (Polanski 1963; Ramos and Folguera 2005). The block is bounded by longitudinal tectonic basins including Triassic sedimentary basins eastwards (Alvear subbasin, Criado Roqué 1979; Algarrobo del Aguila basin, Kostadinoff and Llambías 2002), Triassic–Cenozoic basin of the "Cuenca Neuquina" (Neuquén basin) southwards (Criado Roqué and Ibáñez 1979), and Cenozoic sedimentary basins (Rio Grande basin, Tunuyán basin) westwards (Ramos and Folguera 2005).

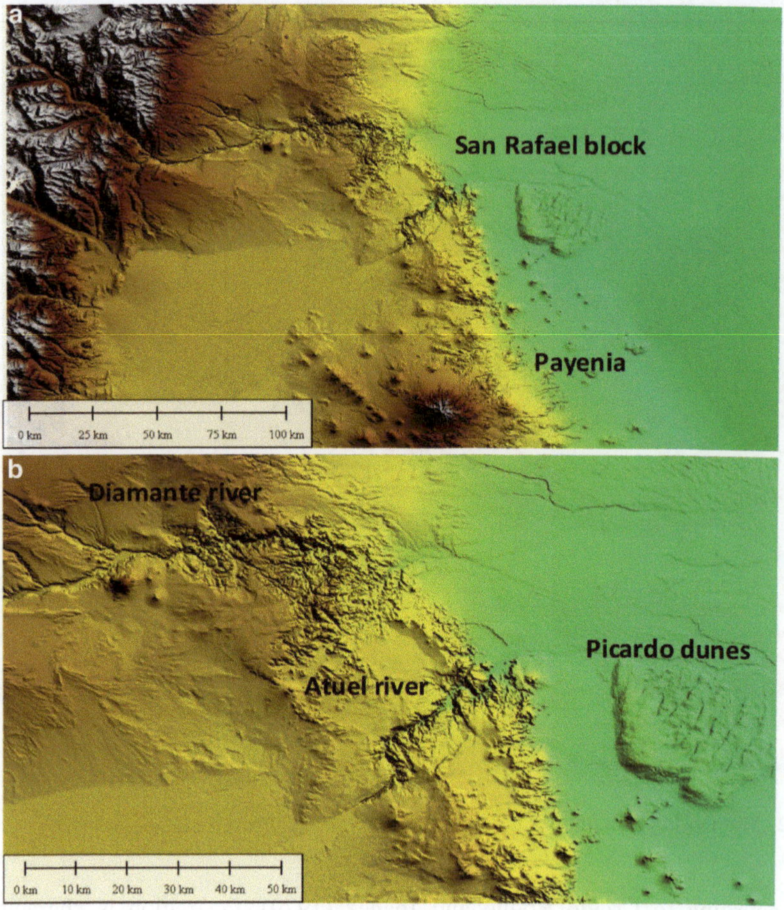

Fig. 3 San Rafael block and Payenia geological provinces. (**a**) SRTM image of the San Rafael block and northern part of Payenia. (**b**) SRTM image of the San Rafael block between the valleys of the Atuel and Diamante rivers

Following the synthesis by Ramos (1999), the block is composed of a Middle Proterozoic metamorphic basement, exposed north of Cerro Nevado volcano (Figs. 3a and 4a) and covered by Early to Middle Paleozoic deformed sedimentary rocks. The sequence is unconformably overlain by Carboniferous marine and continental deposits followed by a thick succession of Permian–Triassic volcanic rocks (Choiyoi Group); continental deposits accumulated in a rift basin (Llantenes) during the Middle–Late Triassic. After a long stratigraphic hiatus, sedimentation was reinitiated with the accumulation of Late Miocene and Pliocene continental deposits that partially covered the San Rafael block (Ramos 1999). These deposits, grouped into the Aisol Formation (González Díaz 1972a), represent a fluvial

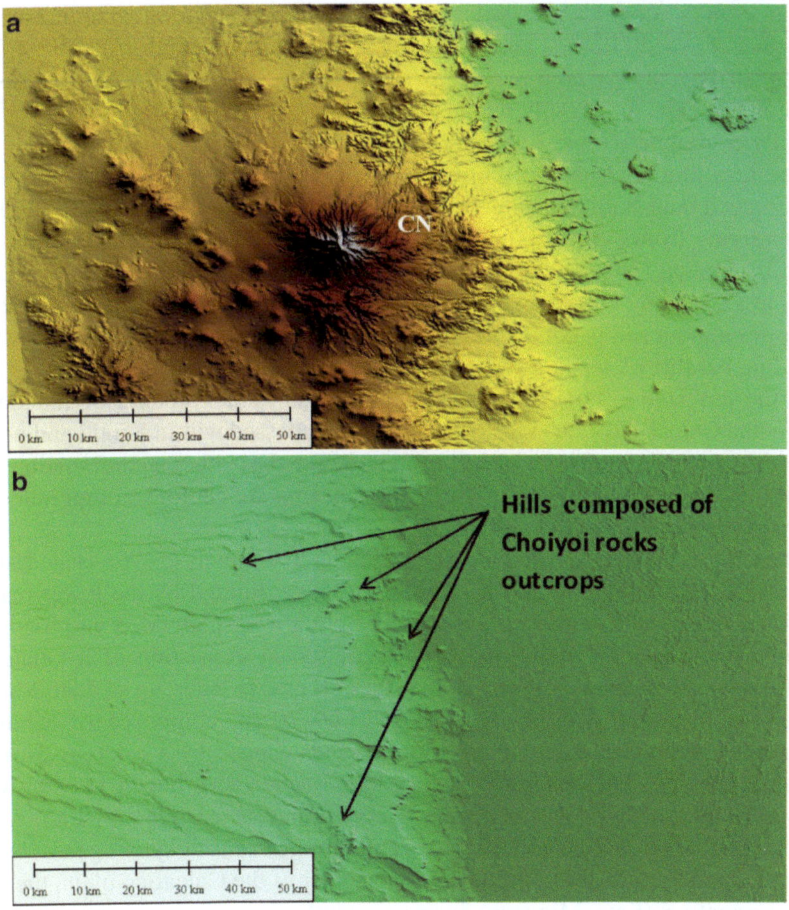

Fig. 4 Payenia and Chical-có plain. (**a**) SRTM image of volcanic cones of northern Payenia in the surroundings of Cerro Nevado (*CN*); (**b**) SRTM image of the eastern part of Chical-có plain showing the escarpment composed of Late Miocene synorogenic deposits (Cerro Azul Formation)

environment of Middle–Late Miocene to Pliocene (?) age according to the vertebrate fossil content (Forasiepi et al. 2011). One of the most extensive exposures is found along the escarpment of Arroyo Seco de la Frazada (Figs. 8 and 9b, c).

Payenia is an extensive volcanic field (Figs. 3a and 4a) situated in the Andean back arc (Ramos and Folguera 2005), resulting from volcanic activity developed at the margins of the San Rafael block (Polanski 1954, in Polanski 1963), since the Late Miocene to the Quaternary with a peak of activity between 7.8 and 4.8 Ma (Kay 2002). It is composed of extensive lava flows mostly of basaltic composition with andesitic, rhyolitic, and dacitic components.

Numerous stratovolcanoes, volcanic calderas domes, and monogenetic volcanoes were then formed (Ramos and Folguera 2005) (Figs. 3a and 4a). The volcanic activity of Payenia covered numerous areas of the San Rafael block, particularly the southern part. As a result, the areal extensions of both geological provinces are partially overlapped (Figs. 2 and 3a).

The resulting landscape of both the San Rafael block and northern Payenia is an elongated fault-bounded mountain block with a general N–S trend and a relative relief varying from ~500–700 m west of San Rafael city to around 120 m west of Agua Escondida (36° 09′ 19.56″ S; 68° 18′ 09.80″ W) situated at an altitude of 1,100 m a.s.l. Many volcanic cones of different sizes (3,800 m a.s.l., Cerro Nevado) occur throughout and become more common southwards where volcanic landforms including basaltic plateaux are dominant.

In the surroundings of Agua Escondida, the San Rafael block is bounded to the east by the Chical-có plain (Figs. 2 and 4b) encompassed within a rigid and positive morphostructural unit. This area has been interpreted as either a continuation of the San Rafael block (Núñez 1976) or as part of a separate geological unit named the Las Matras block (Sato et al. 2000).

The Chical-có plain, interpreted as a pediplain by Calmels (1996), is an eastward sloping plain around 200 km long and 90 km wide, limited by the lower reach of the Atuel River valley (Figs. 2 and 4b). The plain was subdivided into a northern sector, named oriental plain, and a southern sector, consisting of a Quaternary basaltic plateau (Calmels 1996). The northern sector is made up of Late Miocene continental deposits with a thickness of ~60–70 m, composed of sandy silt sediments, unconformably overlain by fine sands capped by a calcrete crust with volcanic pebbles. The described sedimentary sequence, grouped into the Cerro Azul Formation (Melchor and Llambías 2000), may be correlated with the Aisol Formation (Folguera 2011), both units representing Andean synorogenic deposits (Folguera and Zárate 2009). In the northern sector of the Chical-có plain, the Cerro Azul Formation (Late Miocene) forms a mantle that unconformably overlies a Proterozoic to Early Mesozoic bedrock of varied composition and age (Fig. 5). Middle Proterozoic (Grenvillian age) igneous rocks together with outcrops of Paleozoic (Cambrian to Permian) sedimentary rocks are exposed along the western margin of the Atuel River (36° 46′ S, 67° 07′ W) (Sato et al. 2000). The stratigraphic record is completed by magmatic rocks (igneous and volcanic) of Middle Permian–Middle Triassic age (Choiyoi Group) which form isolated exposures rising from the Chical-có plain, along with some other outcrops in the surroundings of the western margin of the Atuel River (36° 18′ S, 67° 15′ W).

The Chadileuvú block (Llambías and Leverato, 1975, in Sruoga and Llambías 1992) (Las Mahuidas sensu Ramos 1999) is located in the southwest of La Pampa province (Figs. 1 and 2). The block is bounded by the Macachín and Colorado basin eastwards and southwards, the Neuquén basin westwards, and La Pampa Central block northwards (Fig. 2). It is made up of a Late Proterozoic basement intruded by Early Paleozoic plutonic rocks, interpreted as the southern extension of the igneous and metamorphic complex that composed the Pampean ranges of Córdoba and San Luis (Linares et al. 1980; Kostadinoff et al. 2001).

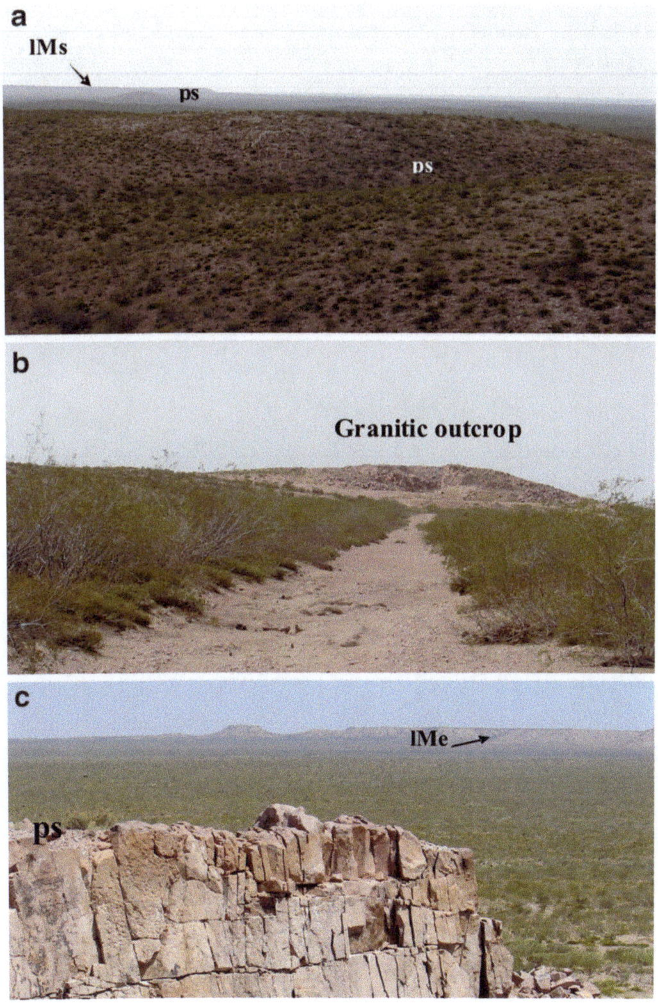

Fig. 5 Eastern margin of Chical-có plain. (**a**) Remnants of planation surface (*ps*) as flat-topped hills of Permian–Triassic (Choiyoi Group) granite, Late Miocene surface (*lMs*) at the background. (**b**) Lomas del Guanaco, hills composed of granitic outcrop. (**c**) Close-up of the granitic outcrop shown in (**b**); flat-topped surface (planation surface, ps); escarpment composed of Late Miocene deposits (*lMe*) of the Cerro Azul Formation

Continental Permian deposits are exposed in the western part. Numerous volcanic and plutonic exposures of the Choiyoi igneous complex (Permian–Early Triassic) are present. It is worth noting that Casadío et al. (1999) reported the occurrence of continental Cretaceous deposits (the Colorado Formation) (Fig. 6) in an area (38° 12′S, 64° 29′ W) that they considered to be located in the Colorado tectonic basin. However, the geological and geomorphological survey performed in this study suggests that the Cretaceous deposits are situated within the eastern

Fig. 6 Cretaceous outcrops of the Chadileuvú block. (**a**) Cretaceous deposits composed of conglomeradic sandy siltstones. (**b**) General view of the surface capped by calcrete crust

margin of the Chadileuvú block. The stratigraphic record is completed by a calcrete crust (~4 m thick) overlain by Late Miocene continental deposits (the Cerro Azul Formation) that crop out along the western, northern, and eastern borders of the Chadileuvú block. The block is mantled by a thin veneer (1–3 m thick on average) of Late Pleistocene–Holocene eolian deposits. It is worth noting that the La Pampa central block, situated northwards and eastwards, has been differentiated into an independent morphostructural unit on the basis of the large extension of the Late Miocene sedimentary cover which overlies the Precambrian–Mesozoic basement that crop outs at the Chadileuvú block.

The Chadileuvú block landscape consists of a series of low-relief ranges in the western part that grades, eastwards and northwards, into a plateau-like landform of gentle topography. The ranges, named as the western Pampean ranges by Calmels

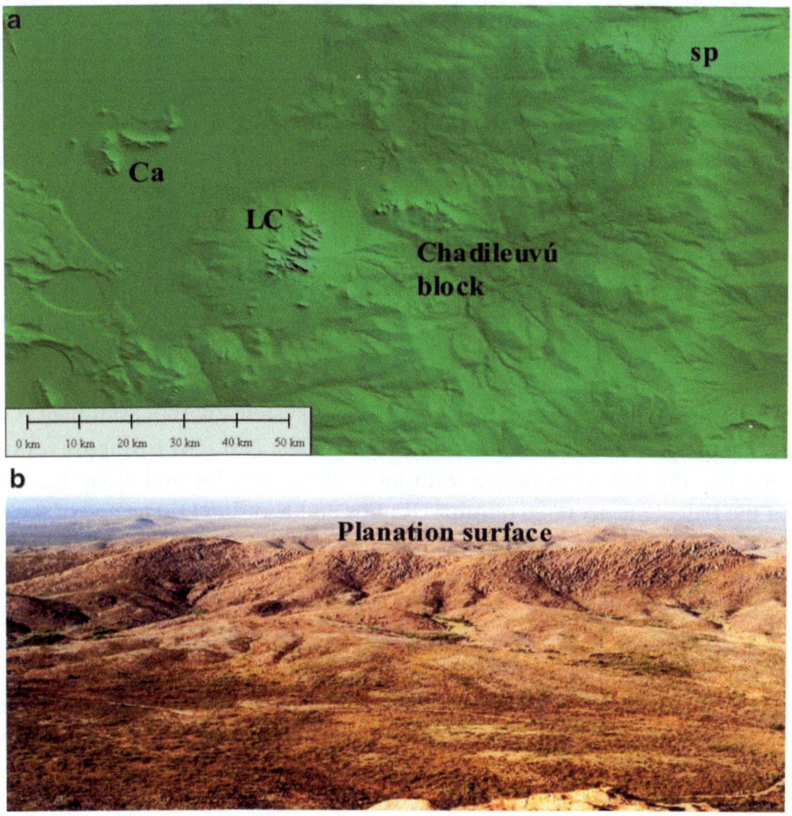

Fig. 7 Western margin of Chadileuvú block. (**a**) SRTM image of Lihuel Calel (*LC*), Carapacha range (*Ca*), structural plain (*sp*) composed of Late Miocene deposits. (**b**) NW view of the flat-topped summit of Sierra de Lihuel Calel from Cerro Sociedad

(1996), consist of Paleozoic–Early Mesozoic outcrops of moderate relief, on average 80–100 m above the surroundings. They include the Sierra de Lihuel Calel, Sierra Chata, Sierra Chica, and Sierra de Carapacha (Fig. 7).

To the north the Chadileuvú block grades into the La Pampa Central block (Fig. 2), a N–S trending morphostructural unit, around 300 km long and 150 km wide on average, bounded by inverse faults (Folguera and Zárate 2011a). It is composed of a 150–200 thick mantle of Late Miocene continental sediments capped by a calcrete crust (Cerro Azul Formation), in turn buried by a thin veneer of Late Pleistocene–Holocene eolian deposits. The Cerro Azul Formation unconformably overlies a Late Precambrian to Permian–Triassic bedrock exposed at some localities (e.g., Valle Daza, Lonco Vaca) of the westernmost margin of the La Pampa Central block. It is the same bedrock that composes the Chadileuvú block. The resulting landscape is an elongated block of around 70–100 m of relative relief; its summit is a gentle plain, sloping 2° eastwards, interpreted as a structural plain (Calmels 1996).

Tectonic Framework and History

The tectonic history of the region under analysis can be divided into four major stages. The first stage is associated to the development of the proto-margin of Gondwana in the Early and Middle Paleozoic that comprises the collision of different terrains against the Proterozoic margin of Gondwana; the second stage occurred in the Late Paleozoic and is related to the formation of the first Andean chain and the formation of Pangea; the third stage is connected to a generalized extension during the Triassic–Cretaceous associated with the Pangea breakup and the opening of the South Atlantic Ocean; the last stage occurred in the Cenozoic and is basically related to the formation of the present Andean chain and the central region of Argentina.

The Triassic extension is characterized by rift systems with a dominant NW–SE trend (Charrier 1979; Uliana and Biddle 1988). The NW–SE rifting direction was controlled by the basement fabric, with most of the rifts located along the hanging wall of the Paleozoic accreted terrains (Ramos and Kay 1991). The age of the rifting progressed from N to S (Mpodozis and Ramos 1990). In turn, the formation of the Colorado and Macachín basins is related to the opening of the South Atlantic Ocean during the general extension of the third stage. Synrift deposits of Late Jurassic–Early Cretaceous age are covered by marine Late Cretaceous and Cenozoic deposits (Yrigoyen 1999; Zambrano 1972; Lesta et al. 1978, 1980).

In the Late Cenozoic, between 15 and 13 Ma, an important flexural subsidence and a high sea level (Ramos and Alonso 1995) produced a sea transgression which diachronically invaded the Andean foothills (Malumián et al. 1998).

Today, the latitudinal segment situated south of the present flat slab zone (~34°–40° S) is formed by two distinct mountain systems separated by Neogene foreland basins. The first system includes the Malargüe fold and thrust belt associated to the Río Grande basin and its southern continuation and the Chos Malal and Agrio fold and thrust belts which were formed during the lower Cretaceous and in the Middle–Late Miocene. The second system comprises the San Rafael block along with the Chadileuvú and the Pampa Central blocks.

The eastern Andean sector has been constructed by the stacking of Late Triassic half-grabens and locally by thin-skinned structures that deformed Late Mesozoic and Cenozoic successions from 15 to 8 Ma (Giambiagi et al. 2008). At these latitudes, the Andean uplift was recorded in the Río Grande foreland basin, with more than 2,500 m of deposits developed between 34° and 37° S that have been partially cannibalized because of the uplift of the San Rafael block (Ramos and Folguera 2005). This deformation pulse coincides with the migration of the calc-alkaline arc magmatism described by Kay (2002) between 11.7 and 4.8 Ma, resulting from of the steepening of the subducted slab since the Middle Miocene until 4.8 Ma, when this arc volcanism finished in the Chachahuén hill south of the study area (37° 4′ 60″ N; 68° 52′ W) (Kay et al. 2006).

The event associated with the uplift of the San Rafael block that was called "main neotectonic phase" by Polanski (1963) occurred in the Middle–Late Miocene according to chronological calibration by continental fossil vertebrate remains recovered from synorogenic deposits (Soria 1983). In addition, distal foreland sedimentation was under progress in the extra-Andean central region during the Late Miocene. The sedimentation process ceased together with the Late Miocene shallowing event of the subducted slab at these latitudes (Ramos and Folguera 2005). The immediate consequences of the shallow subduction regime would have been the inversion of Late Triassic structures in the outer foreland zone, producing the uplift of both the San Rafael block (Ramos and Folguera 2005) and La Pampa Central–Chadileuvú blocks (Folguera and Zárate 2012).

During the Pliocene and Quaternary, extensional structures were reported in the Río Grande and Tunuyán basin (Polanski 1963; González Díaz 1964). This extensional period has been recorded in the back-arc inner part (Folguera et al. 2005). The Late Pliocene–Quaternary development of extensional N–NW depressions was pointed out by Folguera et al. (2003; i.e., Las Loicas and Loncopué basins).

In the Early Pleistocene, the uplifted San Rafael block collapsed as a consequence of the steepening of the subducted slab, whereas the injection of hot asthenosphere formed the Quaternary Payenia volcanic province. During the last million years, minor transpressional stressed affected the eastern front of the San Rafael block (Bastías et al. 1993; Costa et al. 2004, 2006; Lucero 2002) that deformed Pleistocene volcanic sequences. These deformations in the orogenic front are associated with evidence of crustal seismicity (Tello 1994).

Remnants of Planation Surfaces

A relative large remnant of a planation surface is exposed in the northern part of the San Rafael block (Fig. 8). It is the best preserved of the region under analysis forming the summit of the block across an area nearly 100 km long and 50 km wide between 34° and 35° S. Further south (Payenia), it is represented by isolated and discontinuous remnants apparently because of partial burial by Late Cenozoic basaltic flows (González Díaz and Fauqué 1993).

North of the Diamante River, the planation surface was described as an old flattened landscape named "exhumed peneplain of the San Rafael block" ("peneplanicie exhumada del bloque de San Rafael") by Polanski (1963). It is a moderately undulating, low-relief plain, with smooth and rounded hills of around 30–50 m of relative height and gentle slopes (Fig. 9). The planation surface exhibits a highest elevation of around 1,700 m a.s.l. descending to 1,500 m a.s.l. at the northern tip of the San Rafael block (Polanski 1963) (Figs. 8a and 9a).

An "extensive destruction surface" was later recognized south of the Atuel River (Fig. 8b) which represents the continuation of the San Rafael peneplain described by Polanski (1963) north of the Diamante River (González Díaz 1972a). The surface exhibits low relative relief and is incised by short and deep canyons along the Atuel River margin; it is situated at an altitude of around 1,150–1,200 m a.s.l. that progressively decreases southwards. Some low hills (Cerro Carrizalito, Loma Los

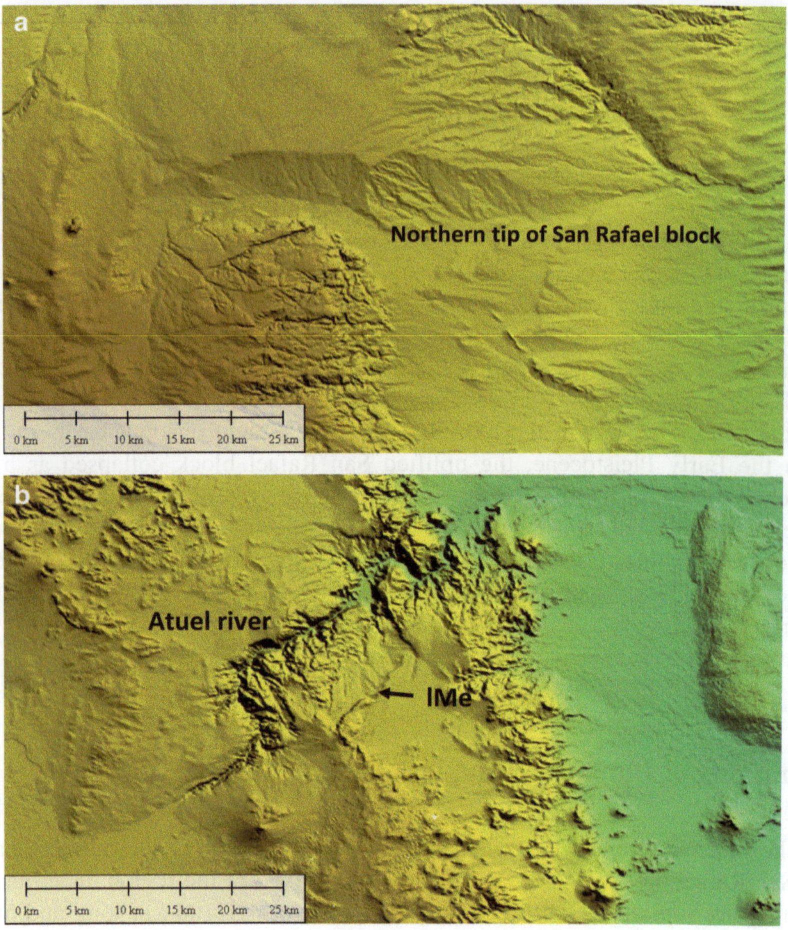

Fig. 8 San Rafael block (**a**) SRTM of the northern part of San Rafael block (**b**) SRTM of the San Rafael block by the Atuel river showing the escarpment (lMe) composed of late Miocene-Pliocene? deposits (Aisol Formation)

Molles) may represent monadnocks (González Díaz 1972a). The planation surface is clearly identified when it is cut across Permian–Triassic igneous rocks (Choiyoi Group) (Fig. 9c); instead it shows a more irregular topography forming structural terraces and small mesas on Triassic sedimentary rocks (González Díaz 1972a, p. 99).

In the area between the Cerro Nevado volcano and the surroundings of Agua Escondida and La Matancilla, situated in the southern part of the San Rafael block and Payenia, isolated exposures of a planation surface have been reported (Holmberg 1973; González Díaz 1972b, c). However, a much larger extension is assumed, likely masked by the burial of Quaternary lava flows (González

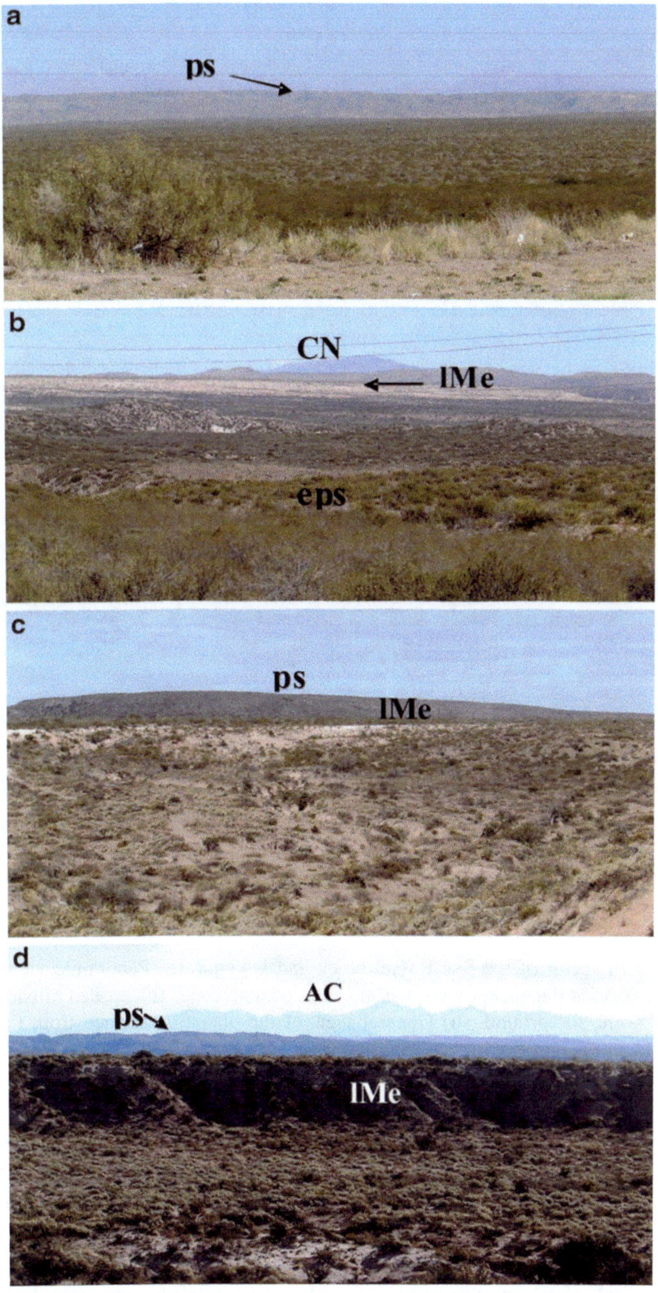

Fig. 9 San Rafael block. (**a**) West view of planation surface (*ps*) at the northern tip of the San Rafael block. (**b**) Arroyo Seco de la Frazada, panoramic view to the SW. Cerro Nevado (*CN*) at the background, Late Miocene escarpment (*lMe*) at the central part with exhumed planation surface (*eps*) on the foreground. (**c**) Close view of the Late Miocene escarpment (*lMe*) and planation surface (*ps*) on Choiyoi Group exposures at the background. (**d**) Late Miocene escarpment (*lMe*), planation surface (*ps*); the Andes Cordillera (*AC*) at the background

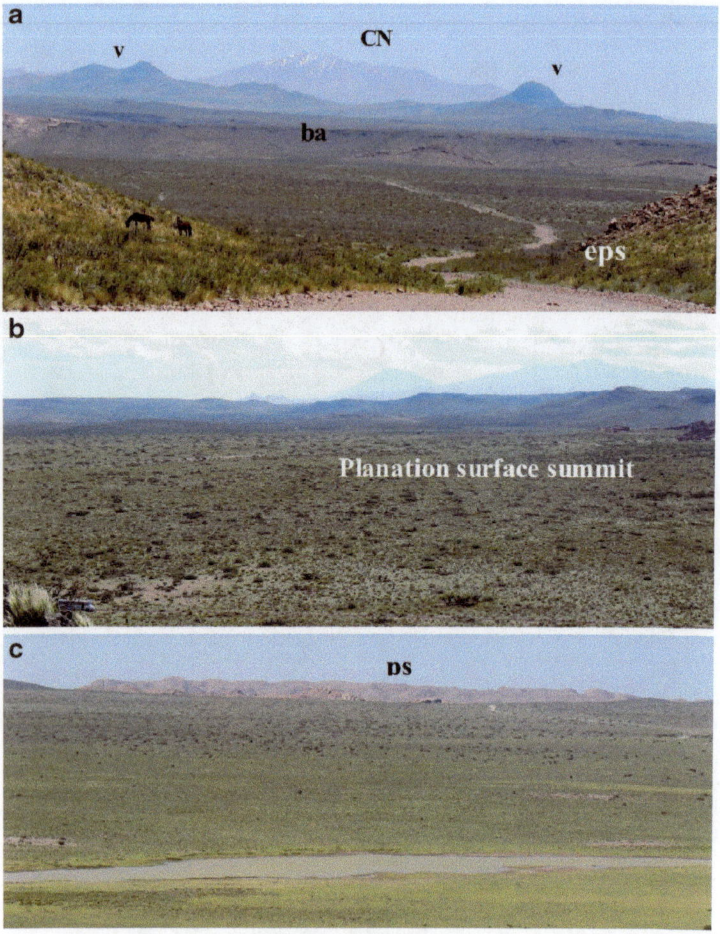

Fig. 10 Southern sector of the San Rafael block and Payenia. (**a**) Panoramic view to the north, Cerro Nevado (*CN*) at the background, basalt (*ba*), volcanic cones (*v*), eroded hillside of planation surface (*eps*) at the foreground. (**b**) General view of the planation surface from the summit. (**c**) Planation surface (*ps*) on Choiyoi outcrops at the background

Díaz 1972c). The planation surface cuts across Permian and Carboniferous rocks. Its remnants, situated at an absolute elevation varying from 1,900 to 2,000 m a.s.l., consist of relatively reduced exposures of flat-topped hills which exhibit leveled summits (Holmberg 1973), and a low relative relief varying from 50 to 80 m above the surrounding volcanic plain (Fig. 10).

In the Chical-có plain, remnants of a planation surface are traced along several flat-topped hills of low relative relief. These elevations, located along the western margin of the Atuel River, are made up of Middle Proterozoic outcrops (Las Matras hills), Paleozoic sedimentary rocks (west of Las Matras hills), and Permian–Triassic igneous rocks (Lomas del Guanaco, among others) (Fig. 11). Las Matras hills are

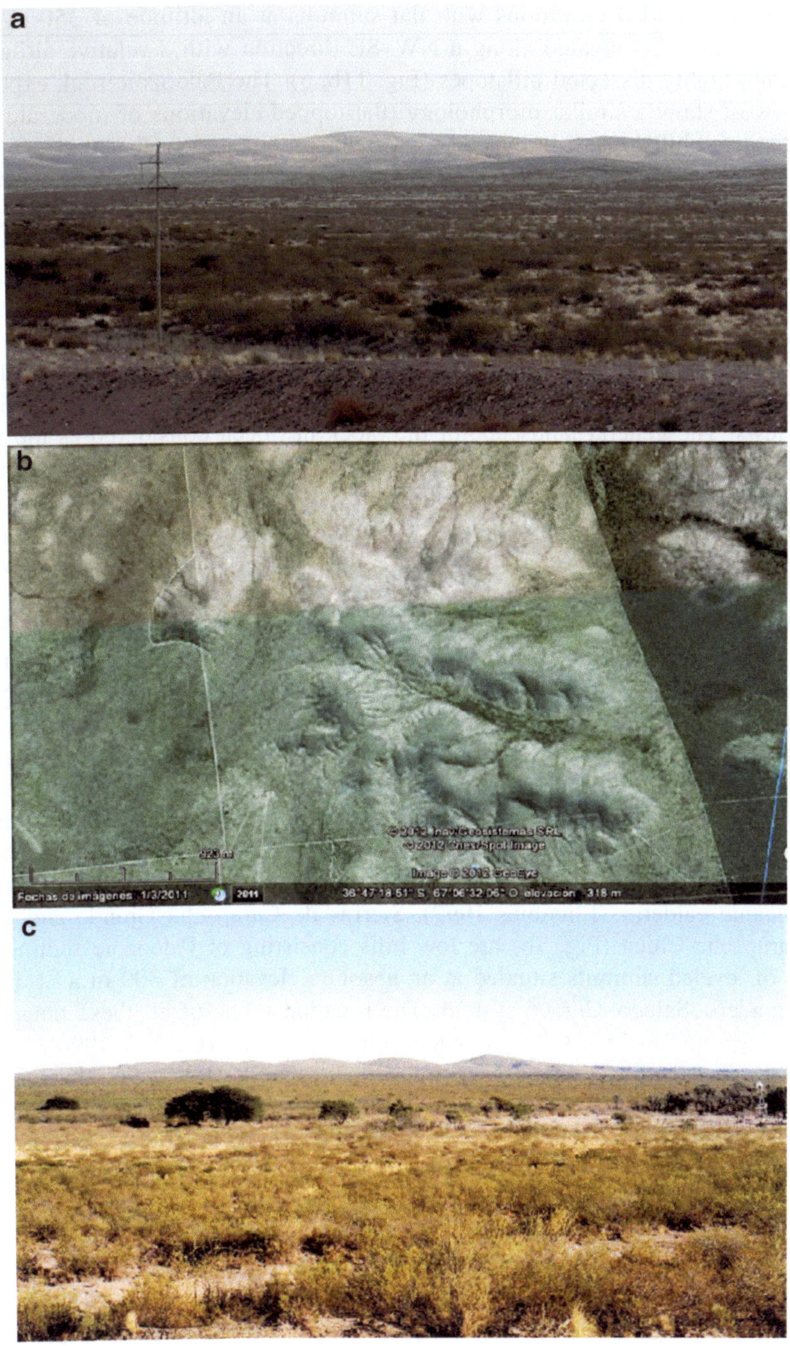

Fig. 11 Chical-có plain. (**a**) Flat-topped summit and rounded topography of Lomas del Centinela composed of Choiyoi volcanic rocks (Permian–Triassic). (**b**) Google Earth image of Las Matras hills made up of Precambrian igneous rocks. (**c**) Las Matras hills. Panoramic view to the northeast from the Atuel River floodplain

smooth and rounded elevations with flat summits at an altitude of 350 m a.s.l.; they are roughly elongated along a NW–SE direction with a relative altitude of 20 m and highly dissected hillslopes (Fig. 11b, c). The Paleozoic rock exposures to the west show a similar morphology (flat-topped elevations of moderate relief and dissected hillslopes). Lomas de los Guanacos, situated further north (Fig. 5), consists of several Permian–Triassic outcrops at an altitude of 360 m a.s.l. with flat and leveled summits, a relative altitude of around 30 m above the surrounding plain, and highly dissected hillslopes. These groups of hills are surrounded by a low-lying erosion surface carved in the Late Miocene Cerro Azul Formation, exposed along the escarpment situated to the west that presently undergoes an active process of erosional retreat.

The landforms of the Chadileuvú block are the least known among the different areas comprised in the region under study; no descriptions are available but simple overviews. The general landscape of the block has been considered as the result of a prolonged process of peneplanation (Ramos 1999). The plateau-like aspect of the block consists of a moderately undulating and hilly plain at some sectors. It cuts across bedrock of varied lithology ranging from Late Proterozoic to Cretaceous units and covered by the very thin veneer of Late Quaternary eolian deposits which mask and make smoother the general landscape.

The relative relief of the block is around 100 m from west to east across a 150 km long transect. The maximum heights are situated in the western margin where the "Sierras Pampeanas Occidentales" (Western Pampean ranges) are located. Sierra de Lihuel Calel (Fig. 7a), the best known of all the ranges, represents one of the topographic highs (500 m a.s.l. at Cerro Sociedad) of a large ignimbritic plateau of rhyolitic and minor dacitic composition related to magmatic activity that occurred during the Permian and the Early Triassic (Sruoga and Llambías 1992; Aguilera et al. 2014, this volume). Together with Sierra Chica, Sierra de Choique Mahuida, and other two hills (Las Piletas and Colón Mahuida), they are probably remnants of volcanic calderas (Llambías 1999). Sierras de Carapacha Grande and Sierra de Carapacha Chica (Fig. 7b) are low hills consisting of Paleozoic sedimentary rocks of leveled summits situated at an absolute elevation of 300 m a.s.l. by the Desaguadero–Salado–Curacó system. The rounded shape of all these ranges has been interpreted as the result of a long weathering process (Calmels 1996).

To the east, the general landscape of the block is moderately undulating resulting from fluvial dissection by several ephemeral systems. The resulting fluvial divides are cut across the block bedrock and exhibit flat surfaces at altitudes varying from 270 to 300 m a.s.l.

Bedrock exposures also show flat summits along the eastern margin (Fig. 12). Cerro de los Viejos (38° 28′S–64° 26′ W) is the most illustrative example with an absolute altitude of around 200 m a.s.l. Its leveled summit (Fig. 13b, c) is cut across Paleozoic igneous rocks deformed by metamorphism (Tickyj et al. 1997). North of Cerro de los Viejos, the outcrops of the Cretaceous deposits (the Colorado Formation) reported by Casadío et al. (1999) are located. The deposits form a leveled surface at an altitude of 190–200 m a.s.l. (Fig. 6b) deeply dissected

Fig. 12 Eastern margin of the Chadileuvú block. (**a**) SRTM image of the eastern border showing the remnants of the structural plain (*sp*) composed of Late Miocene deposits of the Cerro Azul Formation. (**b**) Google Earth view of Cerro de los Viejos

by ephemeral fluvial streams flowing northwards. Remnants of a structural plain surface composed of Late Miocene continental deposits (the Cerro Azul Formation) are exposed northwards of the Cretaceous outcrops and east of Cerro de los Viejos, forming flat-topped elongated elevations (mesas) with a relative relief of around 100 m and highly dissected lateral hillsides (Figs. 12a and 13a). Hence, the morphological characteristics of the bedrock exposures at the fluvial divides and the eastern margin of the Chadileuvú block permit to infer the occurrence of planation

Fig. 13 Eastern margin of the Chadileuvú block. (**a**) Remnant of the Late Miocene structural surface. (**b**) General view of Cerro de los Viejos from the east. (**c**) Leveled summit of Cerro de los Viejos

surfaces. It is cut across sedimentary rocks and the metamorphic–igneous complex ranging in age from the Late Proterozoic to the Permian–Early Triassic (the Choiyoi Group); the Cretaceous outcrops of the eastern margin also show a summit that exhibits the morphology of a planation surface.

Discussion

The common feature of the different areas examined in the San Rafael block and Payenia is the occurrence of preserved remnants of planation surfaces at different altitudes which cut across a bedrock of different composition (igneous, metamorphic, and sedimentary units) and varied age ranging from the Middle Proterozoic (Grenvillian age) to Middle Triassic continental sedimentary rocks at the northern part of the San Rafael block. Several remnants are carved on the outcrops of the Choiyoi Group (Permian–Early Triassic). A comparable geomorphological setting is apparent across the Chadileuvú block where planation surfaces cut across bedrock of varied lithology.

Age of the Planation Surfaces

According to Ollier (1991), the age of an erosion surface must be younger than the rock it cuts across and older than any rock that covers it. Consequently, the age of the planation surface in the San Rafael block–Payenia is bracketed between the Early–Middle Triassic and the Late Miocene. Polanski (1963) interpreted that the planation surface was an ancient plain of subaerial destruction formed during the Mesozoic and the beginning of the Tertiary. Considering the stratigraphic record of the Chadileuvú block, the planation surface may have started to develop after the Early Triassic (the Choiyoi igneous complex), as in the San Rafael block–Payenia units. Melchor and Casadío (1999) assigned a pre-Tertiary age to the "peneplain landscape" ("paisaje de peneplanicie") represented by low hills such as Sierra de Carapacha composed of Paleozoic and Triassic outcrops. In this respect Ramos (1999) pointed out that the Chadileuvú block has been relatively stable during most of the Mesozoic and the Cenozoic, undergoing an intense peneplanation process that continues until the present.

Number of Planation Surfaces and Correlation

Do all the remnants represent the same surface at every geological unit? In the current state of the research, it is difficult to elucidate how many planation surfaces are present in the region because of the discontinuous nature of the preserved remnants, the tectonic dynamic of the region, and the lack of detailed information. The remnants found in the San Rafael block and Payenia are considered to represent the same planation surface (González Díaz 1972a, b, c; González Díaz and Fauqué 1993). The altitudinal variation from north to south has been attributed to differential tectonic uplift during the Neogene and the Quaternary (González Díaz and Fauqué 1993). In the Chadileuvú block, the best preserved remnants are also scattered and discontinuous. As a working hypothesis, it is proposed that the leveled summits of the ranges along with the fluvial divides of the block interior are remnants of a same major planation surface which may be correlated with the planation surface of the San Rafael block–Payenia. Cerro Sociedad at Sierra de Lihuel Calel is probably a monadnock similar to those described by Polanski (1963) and González Díaz (1972a) in the northern segment of the San Rafael block. The occurrence of a surface that cuts across the Cretaceous sediments at a lower elevation in relation to the surrounding bedrock outcrops might indicate a younger erosion surface than the surface exposed westwards.

Subsurface geological information indicates that the Precambrian–Paleozoic bedrock of La Pampa Central block is also an extensive planation surface (Vogt et al. 2010; Folguera 2011) covered by the Late Miocene mantle of the Cerro Azul Formation. This interpretation is indirectly supported considering that the basement complex of the La Pampa Central block and the Chadileuvú block is the southern continuation of the Pampean ranges of Córdoba and San Luis where remnants of planation surfaces have been reported (Carignano et al. 1999).

Exhumation

Polanski (1963) interpreted that the planation surface of the San Rafael block was covered by Late Cenozoic continental deposits and later (Early Pleistocene) exhumed by erosion. Hence Polanski called it "peneplanicie exhumada" (exhumed peneplain). The presence of Late Miocene continental deposits partially covering the planation surfaces is observed at the northern segment of the San Rafael block (the Aisol Formation deposits), the Chical-có plain (the Cerro Azul Formation deposits), along with the western and eastern margins of the Chadileuvú block (the Cerro Azul Formation deposits). The geomorphological characteristics of the eastern border of the Chical-có plain and the margins of the Chadileuvú block are also evidence of exhumation comparable to those reported at the northern segment of the San Rafael block. In the Chical-có plain the general geomorphological and geological context indicates that the planation surface represented by the leveled summits of several hills (Lomas de los Guanacos, Lomas de las Matras) has been exhumed by the erosional retreat of the Late Miocene sedimentary cover (the Cerro Azul Formation) along the fault line of the Atuel River (Santa Isabel fault; Folguera and Zárate 2011b). In the eastern margin of the Chadileuvú block, the outcrop of Cerro de los Viejos is an exhumed exposure of a planation surface surrounded to the east by remnants of the Late Miocene sedimentary cover of the Cerro Azul Formation. At this locality, the deposits have been removed by fluvial erosion of minor drainage systems flowing eastwards. Similar relationships are observed in the western margins of the Chadileuvú block.

So far, there are no reports of Late Miocene deposits in the central area of the Chadileuvú block where detailed fieldwork is needed. In addition, the localities with remnants of planation surfaces in the southern part of the San Rafael block–Payenia units (localities of Agua Escondida, Sierras Porfíricas, among others) do not exhibit any evidence of having been covered by Late Miocene deposits.

Polanski (1963) interpreted that the exhumed peneplain of the San Rafael block was fractured and underwent a long differential descending movement in the Early Tertiary. This initiated a cycle of continental sedimentation that ended in the Pliocene burying almost completely the northern tip of the San Rafael block. A later episode of piedmont uplift (placed by Polanski in the Early Pleistocene) generated a fluvial cycle which started the exhumation of the planation surface. Recent data and information, however, suggest an earlier uplift and exhumation of the San Rafael block associated with the Miocene tectonic dynamic of the Andes. The erosional exhumation of both the San Rafael block together with the erosional retreat of the fault scarp at the Chical-có plain (Santa Isabel thrust) and the Chadileuvú block margins began approximately 7–6 Ma (?). In the La Pampa Central block, the uplift triggered the incision of the transversal valleys. As a result a fluvial erosion cycle initiated with the Late Miocene uplift and gave way to incision of the major fluvial valleys and its tributaries. The overlying Late Miocene cover of the much softer continental deposits (the Cerro Azul Formation) began to be removed by erosion and generated the exhumation of the planation surfaces and the dissection by fluvial erosion. According to the geomorphological characteristics displayed in the examined areas, the exhumation process is still under way as suggested by the present active erosional retreat of the escarpments composed of Late Miocene deposits.

Regional Correlation

At a regional scale, other ranges of central Argentina, east and west of the region under study, are characterized by smooth and rounded shapes and the presence of several planation surfaces varying in number from 4 to 5, with the oldest tentatively considered of Middle Triassic to Jurassic age. The genesis has been interpreted in the context of the Mesozoic tectonic evolution of the Gondwana supercontinent.

Erosion surfaces have been reported from the Pampean ranges of Córdoba and San Luis along with the ranges (Sierra de la Ventana, Sierras de Tandil) of southern Buenos Aires province. In the Pampean ranges of Córdoba and San Luis, Carignano et al. (1999) identified relict landforms, interpreted as planation surfaces; a Jurassic age is assigned to the oldest, with two others thought to be formed in the Late Jurassic–Late Cretaceous interval; a fourth planation surface is assigned to the Late Cretaceous–Paleocene and the fifth to the Miocene. The genesis of the different planation surfaces is linked with the tectonic evolution of the Gondwana supercontinent.

Demoulin et al. (2005) proposed a morphogenetic model for the evolution of Sierra de la Ventana and Sierras de Tandil in southern Buenos Aires province consisting of a pre-Cretaceous surface preserved at the highest elevation of both ranges which started to evolve after the Early Triassic. Likely the surface was formed during approximately 80 My of tectonic quiescence ending with the Late Jurassic–Early Cretaceous rifting process that opened the South Atlantic Ocean. A second surface began to form after the opening of the South Atlantic Ocean and continued to develop during approximately 100 My until the Oligocene. The third surface is believed to be associated with an Early Miocene compressive phase of the Andean domain (Demoulin et al. 2005). The highest surface at both ranges is thought to be formed prior to the Gondwana breakup in the Late Jurassic. A second regional surface is identified at lower elevation and thought to be associated with the rifting process that led to the opening of the South Atlantic Ocean.

The discontinuity and scattered nature of the planation surface remnants across the study region and the lack of chronological control are the main constraints to propose a correlation scheme among the examined areas and the neighboring ranges of Buenos Aires, Córdoba, and San Luis provinces. As the oldest rocks cut across by the erosion surface are of Late Permian–Early Triassic age, it is possible to hypothesize that the development of the erosion surface in the San Rafael block–Payenia, Chical-có plain, the Pampa Central block, and the Chadileuvú block might have started by then in the Middle–Late Triassic prior to the Gondwana breakup. If this were the case, then, the surface might be correlated with the pre-rift planation surface identified in the ranges of Buenos Aires province and Sierras Pampeanas.

No definitive evidence, however, is available to confirm the existence of a regional planation surface after the rifting process as it was suggested in Sierras Pampeanas and the ranges of southern Buenos Aires. In this respect, considering the presence of Cretaceous deposits in the northeastern part of the Chadileuvú block which were associated with the rifting process of the Colorado basin (Casadío et al. 1999), we might suspect the occurrence of a younger surface, at least of Cretaceous age in broader terms and hence younger than the surface exposed elsewhere in the block.

Final Remarks

The areas of the Chadileuvú block, the Pampa Central block, and the San Rafael block–Payenia unit exhibit conclusive geomorphological evidence of the development of planation surfaces. Their stratigraphic records are characterized by a long stratigraphic hiatus that suggests that they were major emerged landforms since the Early–Middle Mesozoic, providing sediments to the neighboring sedimentary basins. The regional analysis performed reveals the notorious similarity on the geomorphological evolution of cratonic areas across central western Argentina. Tectonic conditions of relative stability prevailed in the foreland since at least the Middle–Late Triassic (i.e., prior to the Gondwana breakup) and extended through the Middle–Late Mesozoic and Paleogene. The denudation process must have been controlled by the base level of the neighboring tectonic basins.

In light of the general geological characteristics and configuration of the landscape in the region under analysis, it is hypothesized that planation surfaces both in the San Rafael–Payenia units and the Chadileuvú block might have started their development at least sometimes after the Early Triassic (youngest age of the Choiyoi igneous complex) and likely prior to the rifting that led to the Gondwana breakup and the opening of the South Atlantic Ocean in the Late Jurassic.

The lack of morphological uniformity of the planation surfaces across the region is the result of the Late Miocene episode of Andean deformation which fragmented the surfaces into segments at different absolute elevations, modified the gradient, and constrained the correlation of isolated exposures. The landforms were buried by Late Miocene synorogenic deposits (the Aisol and Cerro Azul formations) at some areas that began to be exhumed around 7 Ma as a result of the steepening of the Nazca plate at this latitude.

Future work will be oriented to adjust the chronology and describe in detail the morphology of planation surfaces. Also, fieldwork in the barely explored central area of the Chadileuvú block will be the focus of forthcoming studies. If this area remained uncovered, the planation surface was continuously exposed until to date representing the basin margin of the extensive Late Miocene sedimentary basin of the synorogenic deposits that compose the Cerro Azul Formation.

Acknowledgments We thank Jorge Rabassa for inviting us to participate with this contribution. The research was supported by the Servicio Geológico Minero Argentino (SEGEMAR) and Project 234 FCEN-UNLPam. Walter Bertotto, Hugo Tickyj (HT), and Claudia Montalvo are gratefully acknowledged for providing information and assistance in the field (HT). Special thanks to Vivi Martínez for helping with the figures.

References

Aguilera EY, Sato AM, Llambías E, Tickyj H (2014) Erosion surface and granitic morphology in the Sierras de Lihuel Calel, province of La Pampa, Argentina. In: Rabassa J, Ollier C (eds) Gondwana landscapes in southern South America. Springer, Dordrecht, pp 393–422

Bastías H, Tello GE, Perucca LP, Paredes JD (1993) Peligro sísmico y Neotectónica. In: Ramos VA (ed) Geología y Recursos Naturales de Mendoza. 12° Congreso Geológico Argentino and 2° Congreso de Exploración de Hidrocarburos, Relatorio, pp 645–658

Calmels AP (1996) Bosquejo Geomorfológico de la Provincia de La Pampa. Universidad Nacional de La Pampa, Santa Rosa, 110 pp

Carignano C, Cioccale M, Rabassa J (1999) Landscape antiquity of the central eastern Sierras Pampeanas (Argentina): geomorphological evolution since Gondwanic times. Z Geomorphol NF Suppl-Bd 118:1–24

Casadío S, Manera T, Parras A, Montalvo C, Cornachione G (1999) Primer registro en superficie de sedimentitas continentales del Cretácico superior en la cuenca del Colorado, sureste de La Pampa. Revista de la Asociación Geológica Argentina 55(1–2):128–133

Charrier R (1979) El Triásico en Chile y regiones adyacentes de Argentina. Comunicaciones 26:1–37

Coronato A, Coronato F, Mazzoni E, Vázquez M (2008) The physical geography of Patagonia and Tierra del Fuego. In: Rabassa J (ed) The late Cenozoic of Patagonia and Tierra del Fuego, Developments in quaternary sciences, 11. Elsevier, Amsterdam, pp 13–55

Costa C, Cisneros H, Salvarredi J, Gallucci A (2004) Nuevos datos y reconsideraciones sobre la neotectónica del margen oriental del bloque de San Rafael. 12° Reunión sobre Microtectónica y Geología Estructural, Cafayate, Argentina. Abstracts, 7

Costa C, Gardini C, Diederix H, Cisneros H, Ahumada E (2006) The active Andean orogenic front at the Southernmost Pampean flat-slab. Backbone of the Americas. Abstract with Programs 15–1.

Criado Roqué P (1979) Subcuenca de Alvear (Provincia de Mendoza). In: Geología Regional Argentina. Academia Nacional de Ciencias Córdoba, 1, pp 811–836

Criado Roqué P, Ibáñez G (1979) Provincia Geológica Sanrafaelino-Pampeana. In: Geología Regional Argentina, Academia Nacional de Ciencias Córdoba, 1, pp 837–869

Demoulin A, Zárate M, Rabassa J (2005) Long-term landscape development: a perspective from the southern Buenos Aires ranges of east central Argentina. J S Am Earth Sci 19:193–204

Dessanti RN (1956) Descripción geológica de la Hoja 27c, Cerro Diamante. Boletín 85, Dirección Nacional Geología y Minería, Buenos Aires, 79 p

Feruglio E (1946) Sistemas orográficos de la Argentina. Geografía de la República Argentina, Gaea, IV. Buenos Aires

Folguera A (2011) La reactivación neógena de la Pampa Central. Unpublished doctoral thesis. Universidad de Buenos Aires, Buenos Aires, 198 pp

Folguera A, Zárate M (2009) La sedimentación neógena continental en el sector extrandino de Argentina central. Revista de la Asociación Geológica Argentina 64(4):692–712

Folguera A, Zárate M (2011a) Neogene sedimentation in the Argentine foreland between 34°30′ S and 41° S and its relation to the Andes evolution. In: Salfity JA, Marquillas RA (eds) Cenozoic geology of the Central Andes of Argentina. SCS Publisher, Salta, pp 123–134

Folguera A, Zárate M (2011b) El corrimiento Santa Isabel, posible frente orogénico del Bloque de San Rafael. 18º Congreso Geológico Argentino, Neuquén, p 101

Folguera A, Zárate M (2012) The Neogene reactivation of the Extra-Andean Central Pampean Plain (latitude 34°–37° S), La Pampa Province, Argentina. 13º Congreso Geológico Chileno. Simposio de Geodinámica y deformación cortical, Antofagasta, Chile, pp 126–128

Folguera A, Ramos VA, Melnick D (2003) Recurrencia en el desarrollo de cuencas de intraarco. Cordillera Neuquina (37°30′ S). Revista de la Asociación Geológica Argentina 58(1):3–19

Folguera A, Ramos VA, Zapata T, Spagnuolo M, Miranda F (2005) Pliocene to Quaternary retro-arc extension in the Andes at 35°–37°30′ S. 6°International Symposium of Andean Geodynamics, Barcelona

Forasiepi A, Martinelli A, de la Fuente M, Diéguez S, Bond M (2011) Paleontology and stratigraphy of the Aisol Formation: Neogene, San Rafael, Mendoza province: new evidences. In: Salfity JA, Marquillas RA (eds) Cenozoic geology of the Central Andes of Argentina. SCS Publisher, Salta, pp 135–154

Giambiagi LB, Bechis F, García V, Clark A (2008) Temporal and spatial relationships of thick- and thin-skinned deformation: a case study from the Malargüe. Tectonophysics 459:123–139

González Díaz EF (1964) Rasgos geológicos y evolución geomorfológica de la Hoja 27d (San Rafael) y zona occidental vecina (Provincia de Mendoza). Revista de la Asociación Geológica Argentina 19:151–188

González Díaz EF (1972a) Descripción geológica de la Hoja 27d San Rafael, Provincia de Mendoza. Servicio Nacional Minero Geológico, Boletín 132:1–127, Buenos Aires

González Díaz EF (1972b) Descripción geológica de la Hoja 30e Agua Escondida, Provincias de Mendoza y La Pampa. Servicio Nacional Minero Geológico, Boletín 135:1–79, Buenos Aires

González Díaz EF (1972c) Descripción geológica de la Hoja 30d Payún-Matrú, Provincia de Mendoza. Dirección Nacional de Geología y Minería, Boletín 130:1–88, Buenos Aires
González Díaz EF, Fauqué LE (1993) Geomorfología. In: Ramos VA (ed) Geología y Recursos Naturales de Mendoza. XII° Congreso Geológico Argentino and II° Congreso de Exploración de Hidrocarburos 1(17):217–234. Buenos Aires
Holmberg E (1973) Descripción geológica de la hoja 29d – Cerro Nevado, provincia de Mendoza. Servicio Nacional Minero-Geológico, Boletín 144:1–71
Kay SM (2002) Tertiary to recent transient shallow subduction zones in the Central and Southern Andes. 15° Congreso Geológico Argentino (Calafate), Actas 3:282–283
Kay SM, Mancilla O, Copeland P (2006) Evolution of the Backarc Chachahuen volcanic complex at 37° S latitude over a transient Miocene shallow subduction zone under the Neuquén Basin. In: Kay SM, Ramos VA (eds) Evolution of an Andean margin: a tectonic and magmatic view from the Andes to the Neuquén basin (35° -39° S lat.), Special paper 407. Geological Society of America, Boulder, pp 215–246
Keidel J (1916) La geología de las Sierras de la provincia de Buenos Aires y sus relaciones con las montañas del Cabo y los Andes. Anales de la Dirección Nacional de Geología y Minería 9:3. Buenos Aires
Kostadinoff J, Llambías E (2002) Cuencas sedimentarias en el subsuelo de la provincia de La Pampa. 5º Congreso de Exploración y Desarrollo de Hidrocarburos, Instituto Argentino del Petróleo, archivos electrónicos, Mar del Plata
Kostadinoff J, Llambías E, Raniolo LA, Alvarez GT (2001) Interpretación geológica de la información geofísica de la provincia de La Pampa. Revista de la Asociación Geológica Argentina 56(4):481–493
Lesta PJ, Turic MA, Mainardi E (1978) Actualización de la información estratigráfica en la Cuenca del Colorado. 7° Congreso Geológico Argentino 1:701–713. Buenos Aires
Lesta PJ, Mainardi E, Stubelj R (1980) Plataforma Continental Argentina. In: Turner JC (coord) 2º Simposio de Geología Regional Argentina, 2, pp 1577–1601. Academia Nacional de Ciencias, Córdoba
Linares E, Llambías EJ, Latorre CO (1980) Geología de la provincia de La Pampa, República Argentina, y geocronología de sus rocas metamórficas y eruptivas. Revista de la Asociación Geológica Argentina 35(1):87–146
Llambías E (1999) Las rocas ígneas gondwánicas. El magmatismo gondwánico durante el Paleozoico superior-Triásico. Geología Argentina. Instituto de Geología y Recursos Minerales, Anales 29(14):349–376
Lucero C (2002) Sistema estructural Aisol-Tordillo, bloque de San Rafael. 15° Congreso Geológico Argentino, Actas 1:363–368. El Calafate
Malumián N, Suriano JM, Cobos JC (1998) La Formación Barranca Final en su localidad tipo, Mioceno, Cuenca del Colorado. 10º Congreso Latinoamericano de Geología and 6º Congreso Nacional de Geología Económica, Actas 1:125–130
Melchor R, Casadío S (1999) Hoja Geológica 3766-III, La Reforma: Boletín Secretaría de Minería de la Nación, Servicio Geológico Minero Argentino, Boletín 295:1–63
Melchor R, Llambías EJ (2000) Hoja Geológica 3766-I, Santa Isabel. Secretaría de Minería de la Nación, Servicio Geológico Minero Argentino, Boletín 344:1–43
Mpodozis C, Ramos VA (1990) The Andes of Chile and Argentina. In: Ericksen GE, Cañas Pinochet MT, Reinemud JA (eds) Geology of the Andes and its relation to hydrocarbon and mineral resources, Earth sciences series, 11. Circumpacific Council for Energy and Mineral Resources, Houston, pp 59–90
Nágera JJ (1939) Geografía Física de la República Argentina. In: Manito O, Nágera JJ (eds) Geografía física de las Américas y de la República Argentina. Editorial Kapelusz, Buenos Aires, 232 p
Núñez E (1976) Descripción geológica de la Hoja 31e, Chical-có, Provincias de Mendoza y La Pampa. Servicio Geológico Nacional, Unpublished report, 92 pp
Ollier C (1991) Ancient landforms. Belhaven Press, London/New York, 233 pp

Polanski J (1963) Estratigrafía, neotectónica y geomorfología del Pleistoceno pedemontano entre los ríos Diamante y Mendoza. Revista de la Asociación Geológica Argentina 17(3–4):127–349
Rabassa J (2010) Gondwana paleolandscapes: long-term landscape evolution, genesis, distribution and age. Geociencias 29(4):541–570, São Paulo, UNESP
Rabassa J, Carignano C, Cioccale M (2014) A general overview of Gondwana landscapes in Argentina. In: Rabassa J, Ollier C (eds) Gondwana landscapes in southern South America. Springer, Dordrecht, pp 201–245
Ramos VA (1999) Las provincias geológicas del territorio argentino. In: Caminos R (ed) Geología Argentina. Instituto de Geología y Recursos Minerales, Buenos Aires, 29, 3, 41–96
Ramos VA (2010) The Grenville-age basement of the Andes. J S Am Earth Sci 29:77–91
Ramos VA, Alonso RN (1995) El Mar Paranense en la Provincia de Jujuy. Revista del Instituto de Geología y Minería 10:73–80, San Salvador de Jujuy, Argentina
Ramos VA, Folguera A (2005) El origen de la Payenia (Mendoza sur y norte de Neuquén): Evolución estructural y tectomagmática de los Andes a estas latitudes. 16º Congreso Geológico Argentino, Actas 2:115–122
Ramos VA, Kay S (1991) Triassic rifting and associated basalts in the Cuyo basin, central Argentina. In: Harmon R, Rapela C (eds) Andean magmatism and its tectonic settings, Special paper 265. Geological Society of America, Boulder, pp 79–91
Sato AM, Tickyj H, Llambías E, Sato K (2000) The Las Matras tonalitic-trondhjemitic pluton, central Argentina: Grenvillian age constraints, geochemical characteristics, and regional implications. J S Am Earth Sci 13:587–610
Soria M (1983) Vertebrados fósiles y edad de la Formación Aisol, provincia de Mendoza. Revista de la Asociación Geológica Argentina 38(3–4):299–306
Sruoga P, Llambías E (1992) Permo-Triassic leucorhyolitic ignimbrites at Sierra de Lihue Calel, La Pampa province, Argentina. J S Am Earth Sci 5(2):141–152
Stelzner A (1885) Beitraege zur Geologie und Paleontologie der argentinischen Republik – translation. In: Actas de la Academia Nacional de Ciencias, Córdoba, 8 (1924)
Tello G (1994) Fallamiento cuaternario y sismicidad en el piedemonte cordillerano de la provincia de Mendoza. Argentina. 7° Congreso Geológico Chileno, Actas 1:380–384. Concepción
Tickyj H, Dimieri L, Llambías E, Sato A (1997) Cerro de los Viejos (38º 28′ S–64º 26′ W): cizallamiento dúctil en el sudeste de La Pampa. Revista de la Asociación Geológica Argentina 52(3):311–321
Uliana MA, Biddle KT (1988) Mesozoic-Cenozoic paleogeographic and geodynamic evolution of southern South America. Revista Brasileira Geociências 18:172–190
Vogt H, Vogt T, Calmels AP (2010) Influence of the post-Miocene tectonic activity on the geomorphology between Andes and Pampa Deprimida in the area of Provincia de La Pampa, Argentina. Geomorphology 121(3–4):152–166
Yrigoyen M (1999) Los depósitos cretácicos y terciarios de las cuencas del Salado y del Colorado. In: Caminos R (ed) Geología Argentina. Instituto de Geología y Recursos Minerales, Anales 29, 21, 645–649, Buenos Aires
Zambrano JJ (1972) La cuenca del Colorado. In: Leanza A (ed) Geología Regional Argentina. Academia Nacional de Ciencias, Córdoba, pp 419–438
Zárate M (2009) El paisaje pampeano a través del tiempo. In: Berón M, Luna L, Bonomo M, Montalvo C, Aranda C, Carrera Aizpitarte M (eds) V Congreso de Arqueología de la Región Pampeana. Mamul Mapu. Pasado y Presente desde la Arqueología Pampeana. Editorial Libros del Espinillo, 19–32

Erosion Surface and Granitic Morphology in the Sierra de Lihuel Calel, Province of La Pampa, Argentina

Emilia Y. Aguilera, Ana Maria Sato, Eduardo Llambías, and Hugo Tickyj

Abstract This chapter aims to characterize the landforms of the ignimbrite landscape in the Sierra de Lihuel Calel, province of La Pampa, Argentina. DEM data from SRTM were used for the analysis of morphometric characteristics and other parameters, so as to obtain longitudinal profiles and identify the boundaries of the paleosurface. By means of the interpretation of satellite imagery and field observations, the different features of this ancient landscape were identified as well as the lithology and fabric of the volcanic and pyroclastic rocks, which were petrographically studied. Textural transformations occurring in ignimbrites provided properties to these rocks which are similar to those common in granitic rocks. The studied landforms might be regarded as pseudo-granitic landforms, and thus major and minor landforms are described. Fracturing has been found as a significant factor in landform evolution.

Keywords Gondwana • Argentina • La Pampa • Ignimbrites • Granitic geomorphology

E.Y. Aguilera (✉)
Facultad de Ciencias Naturales y Museo, Universidad Nacional de La Plata (UNLP), Calles 122 y 60, 1900-La Plata, Argentina

DAIS (Dirección de Aplicación de Imágenes Satelitarias), Calle 7 N° 1267-2°P, 1900-La Plata, Argentina
e-mail: eaguilera@fcnym.unlp.edu.ar

A.M. Sato • E. Llambías
Centro de Investigaciones Geológicas (CIG), CONICET-UNLP, 1900 La Plata, Argentina

H. Tickyj
Departamento de Ciencias Naturales, Facultad de Ciencias Exactas y Naturales, Universidad Nacional de La Pampa, Avda. Uruguay 151, L6300CLB, Santa Rosa, La Pampa, Argentina

J. Rabassa and C. Ollier (eds.), *Gondwana Landscapes in southern South America*, Springer Earth System Sciences, DOI 10.1007/978-94-007-7702-6_14,

Introduction

Paleosurfaces are relict topographic surfaces corresponding to ancient surfaces of regional extent. As a consequence of their evolution, they record the effects of surficial alteration which have been the result of very long periods in which weathering, erosion, or no-deposition alternated (Widdowson 1997). Their importance is based upon the fact that they are key geomorphological elements in the long-term landscape evolutionary reconstruction.

The Sierra de Lihuel Calel (i.e., the Lihuel Calel hilly ranges) raise above the ignimbritic plateau which composes (45 km^2) pat of the Chadileuvú Block (see Zárate and Folguera 2014, this volume), in the province of La Pampa, Central Argentina. According to Ramos (1999), this region would have maintained a relatively stable position during most of the Mesozoic and Cenozoic, thus being subject of "prolonged peneplainization." The ignimbritic plateau is characterized by a smoothly undulating topography, with the exception of the Sierra de Lihuel Calel which rise above the surrounding plains as if it was a large-scale inselberg. It is composed of a thick sequence of ignimbrite beds, coarsely stratified, of an approximate N-S strike, and inclinations varying between 20 and 25°. The erosion patterns reflect erosion processes in the long term, probably periods of up to 200 Ma, possibly as a response to extensive exposition to atmospheric and climatic conditions. This investigation aims to characterize the landforms of the ignimbrite landscape found in these Sierras, where the nature, texture, and structure of ignimbrites have conditioned and affected the resulting landforms.

The methodological criteria proposed by Twidale (1982, 1983) and Vidal Romaní and Twidale (1998) were followed in the different steps leading to the analysis of this landscape developed exclusively on ignimbrites, paying special attention to the varying size and scale of the different landforms found in these landscapes. This study has started at the scale of the paleosurface developed on top of the ignimbrite plateau, identifying different sets of major and minor landforms.

Study Area

The Sierra de Lihuel Calel are located in the province of La Pampa, Central Argentina, centered around 38° 00′ lat. S and 65° 36′ long. W (Fig. 1). The sierras are crossed by National Route 152. This range is the highest topographic feature found in the immense, semidesert plains of this region, characterized by large salt lakes in a very arid country. The Sierra de Lihuel Calel are part of the Lihuel Calel State Park, which has abundant autochthonous flora and fauna and cultural heritage, such as noted rock art.

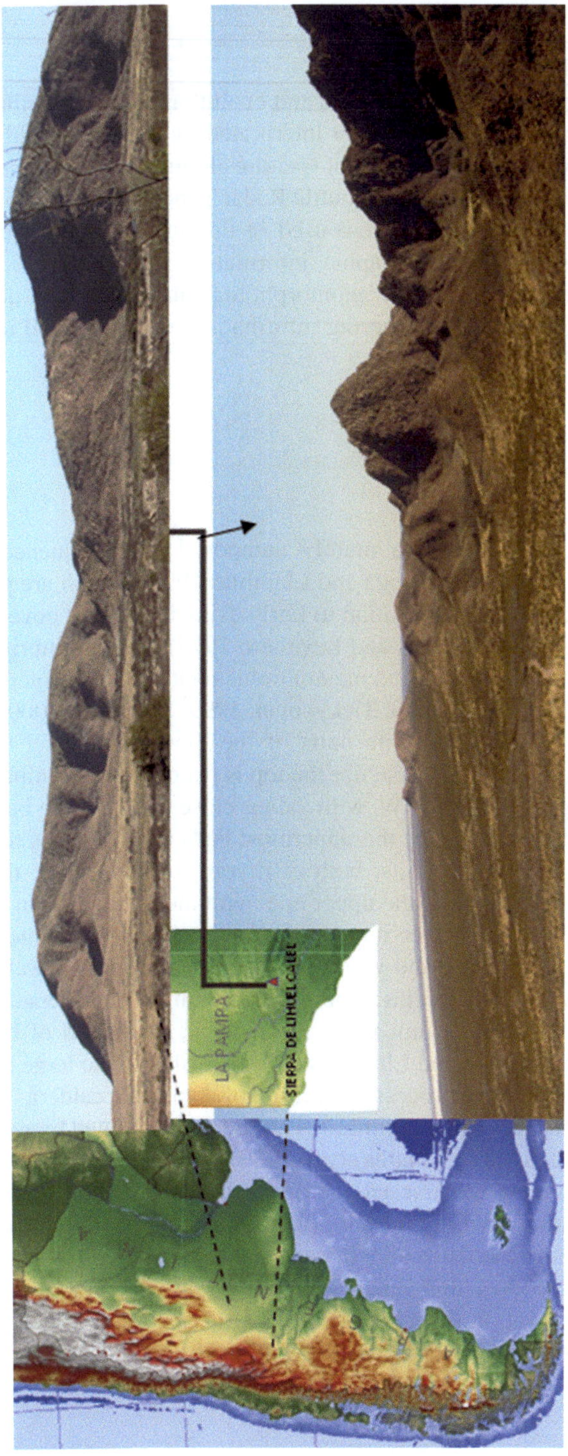

Fig. 1 Sierra de Lihuel Calel, location map

Methodology

Those factors intervening in the genesis and evolution of the ignimbrite landforms were identified and described and later interpreted by means of field observations, joint density analysis, and orientation of fracture alignments using satellite imagery. Landsat, Google Earth, and SRTM (Shuttle Radar Topography Mission) images and DEM (digital elevation models) were used in this investigation. These materials were operated using GIS (geographic information systems), which were used as a base for the geological and geomorphological studies, the qualitative and quantitative analysis, and the gathering information to be included in topographic sections.

Geological Setting

The Sierra de Lihuel Calel are mainly composed of a sequence of rhyolitic ignimbrites (Llambías 1975; Sruoga and Llambías 1992), which are part of a more extended rhyolitic plateau, of Permian to Early Triassic age that covers most of the province of La Pampa (Llambías and Leveratto 1975). The ignimbrites overlie the basement integrated by igneous and metamorphic rocks of Late Proterozoic to Early Paleozoic age (Linares et al. 1980; Tickyj et al. 1999; Sato et al. 2000).

The thickness of the ignimbrite units of the Sierra de Lihuel Calel exceeds 950 m (Fig. 2), but neither the base nor the top is exposed. The ignimbrite beds are homoclinally dipping to the WNW, with values close to 25° at the basal levels and values closer to 15° observed in the uppermost beds (Fig. 3). This thick sequence is composed of two cooling units, both of rhyolitic composition: the lower one, with a thickness of 440 m, and the upper one, with more than 450 m thick (Sruoga and Llambías 1992). The lower one overlies very thin flows, which as a whole are no more than 20 m thick and which are composed of highly welded, rhyolitic ignimbrites and volcanic breccias. The upper cooling unit is separated from the lower one by a thin cooling unit up to 50 m thick, composed of highly welded dacitic ignimbrites. Sruoga and Llambías (1992) attributed the large thicknesses of the two cooling units to the process of infilling of a volcanic caldera.

The age of the Lihuel Calel ignimbrites has been determined by means of Rb-Sr isochrones at 238 ± 5 Ma by Linares et al. (1980) and at 240 ± 2 Ma by Rapela et al. (1996). After the intense volcanic activity during the Permian and Triassic, there was no further magmatic activity in the area. Moreover, these rocks were never covered by marine strata since the Paleozoic. It was only in the Late Miocene that the whole complex was covered by a thin veneer of continental sedimentary rocks, whose top shows a highest elevation of only 285 m a.s.l., a reason for which the Sierra de Lihuel Calel were never entirely covered by these sediments.

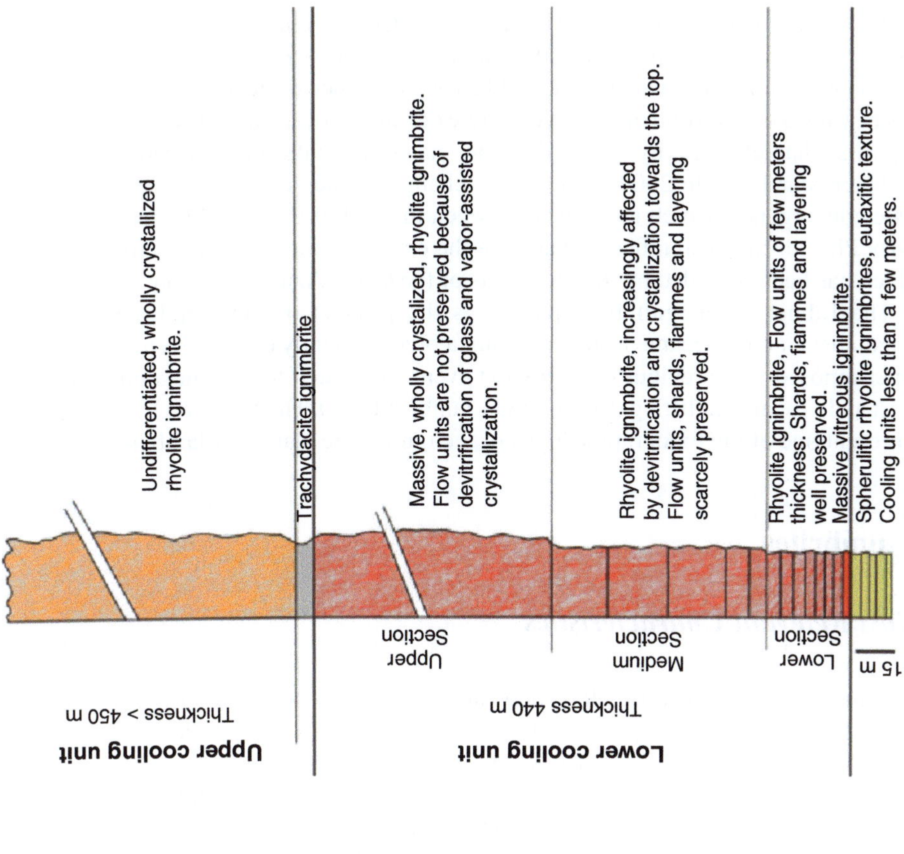

Fig. 2 Geological section of the ignimbrites of Sierra de Lihuel Calel

Fig. 3 View of the Sierra de Lihuel Calel from north to south, indicating in the cooling unit the sectors in which the fiammes, shards and layering are preserved

These continental sediments belong to the Cerro Azul Formation (Llambías 1975). This unit is composed of silts, sandy silts, and very fine, silty sands, with abundant calcium carbonate nodules and strong evidence of pedogenetic processes. It is bearing a varied suite of fossil vertebrate of Late Miocene age. These are aeolian deposits characterized by loess-like fine materials, with interbedded paleosols (Folgueras and Zárate 2009). Due to the change of the subduction angle and the plate convergence, which took place between the end of the Late Miocene and the Early Pleistocene, a significant change in the environmental conditions took place, which has been recorded by the sediments of the Cerro Azul Formation.

The Lihuel Calel ignimbrite plateau is the product of the Gondwana silicic magmatism, and it became installed within a predominantly extensive environment, superimposed to a previous compression event associated to the thickening of the crust and orogenic uplifting (Sruoga and Llambías 1992). The Sierra de Lihuel Calel represent one of the topographic highs of the ignimbrite plateau (Llambías 1973).

Ignimbrites

Most Relevant Characteristics

Ignimbrites are pyroclastic flow deposits, which are composed of vitroclasts, crystalloclasts, lithic components, pumice, and gas. The ignimbrite bodies have variable thickness, both in vertical and lateral directions, according to several conditions, such as underlying topography, crystalloclast, pumice, lithic clasts, and rock density. The large volume of ignimbrite eruptions may produce flooding of the landscape forming extensive plateaus. The pyroclastic flows are complex volcanic units, with a large variety of intervening processes in their formation, and they are chaotically deposited due to the violence and speed of the eruptive style. These characteristics are recorded in the large spectrum of observed facies occurring in the pyroclastic deposits.

When referring these rocks as ignimbrites, reference is made to their texture and the kind of depositional process but not to their chemical or petrographic composition. Thus, the composition must be included in the description, for instance, rhyolitic, dacitic, or andesitic ignimbrite (Llambias 2003).

The processes accompanying the deposition such as compaction, cooling, and crystallization during the vapor phase may introduce major textural changes. The high temperature (over 550 °C) of these flows frequently produces the welding of these deposits, though sometimes these units do not show evidence of welding. Welding takes place by agglomeration and deformation of the vitroclasts. The welded ignimbrites have eutaxitic texture, which is a product of pumice deformation.

Welded ignimbrites cool out as a homogeneous material and usually they exhibit columnar jointing, and less frequently, joints occur in the unwelded zones. Jointing takes place due to thermal contraction. Generally, the joint separation varies from a few centimeters to one meter. This variation is directly related to the degree of welding, whereas the wider jointing is related to lesser degrees of welding, whereas a closer jointing is associated with intense welding. Another factor that has influence on the joint separation is the cooling rate, where a wider spacing would reflect a slower cooling rate (Wilson 1993). For instance, in the volcanic province of Cappadocia, Turkey, which is characterized by large ignimbrite deposits of varied thickness and welding degree, columnar jointing is found even in the non-welded zone, which extends towards the uppermost zone, where strange erosion landforms, such as "fairy chimneys," villages, monasteries, and churches, are developed (Ollier 1988). Jointing has also been recognized in welded ignimbrites such as the Bishop, Tuffs, or Bandelier ignimbrites, where joints are frequent in the upper non-welded zone and the vapor phase has been channelized along these fractures, altering sectors of the units. However, in the basal portions of these ash flows the joints are absent. For example, in the Taupo ignimbrite plateau, New Zealand, where pumice deposits developed a landscape characterized by rounded hills, when the ignimbrite-welded zone is exposed, it generates bare rock vertical cliffs, following the joint systems. It is common that the upper pumice zone is removed by erosion. In this case, the welded ignimbrite acts as the cap rock, developing mesas, tables, and buttes. These landforms evolve towards ignimbrite tors by size reduction, which are typical of the North Island of New Zealand. The Mamaku Plateau of New Zealand covers an area of 1,500 km^2, approximately. The tor development may be the result of different degree of welding, which may vary in short distances and favor the differential erosion (Ollier 1988), being the joint spacing in the ignimbrite flows determined by the degree of welding as well.

Petrography

The landforms herein described belong to the middle and high portions of the welded ignimbrite, where the original eutaxitic texture is modified by devitrification and/or crystallization processes assisted by vapor phase, which produced a porphyritic texture, with noticeable crystalloclasts as "phenocrysts," surrounded by a groundmass varying between microcrystalline (grain size from 4 to 63 μm) and very fine-grained (from 63 to 125 μm) to fine-sized crystalline (from 125 to 250 μm). It is possible to follow the continuous upward variation in the recrystallization texture of the ignimbrite.

In the basal portion of the ignimbrite (a few tens of meters), the eutaxitic texture that results from compaction and flattening of pumice fragments is significant, with elongated "fiammes" of up to 2 × 10 cm (Fig. 4a, b). Between 20 and 25 % of the crystalloclast, characterize this section with quartz, acid plagioclase, and sanidine, in that order of abundance, in addition to scarce, altered lithoclasts of acid volcanic rocks. The fiammes show spherulitic devitrification, with interlocking open spherulites of fan to bow-tie types (following the terminology by McArthur et al. 1998), with fiber length of up to 100 μm, and sweeping extinction. Sectors with very fine crystal aggregates of clean quartz feldspar would correspond to crystallization starting in vapor phase in the original vesicles. Welded glass shards in the vitroclastic matrix are aligned in bands and slightly deformed at the crystal margins. They present devitrification to a cryptocrystalline to fine-grained crystalline aggregates, of less than 10 μm in grain diameter (Fig. 4c, d).

In the following 50–100 m, the fiammes are fewer and less visible at the microscopic scale, although lenses of texture different to that of the matrix are microscopically distinguished as lenses of texture are clearly different from that of the matrix. They are elongated and undulating dominium, with wispy to ragged terminations, with inner spherulitic texture and axiolithic margins, and with coarser grain size upwards, which are transformed into granophyric to graphic aggregates (Fig. 4e, f). The quantity of crystals and lithoclasts is larger than 35 %, and they are composed of quartz, sanidine, plagioclase, and, to a lesser extent, opaque or mafic minerals. The volcanic lithoclasts are scarce, showing sporadic distribution. The matrix is crystallized into a felsitic, very fine-grained crystalline aggregate, in which minute nuclei of biotite and fluorite are observed. The shape of the preexisting vitroclasts may be recognized at plain polarized light.

In a stratigraphical position above the previously described units, the middle sections of this cooling unit which are relevant to this chapter are placed. The fiammes are not visible in the outcrops of this section, with the exception of very sporadic, centimeter-long lenses distinguishable by a darker coloration. In the matrix, it is not possible to identify the dominium corresponding to fiammes, because the texture is inhomogeneous and with irregular sectors bearing very fine- to fine-grained crystalline aggregates, passing to quartz, feldspar, and biotite clean aggregates, in which fluorite, epidote, and opaque minerals participate in a lesser proportion. In the sectors with finer aggregates, the matrix presents micro-poikilitic growth of quartz, feldspar, biotite, and scarce amphibole, including microcrystalline to very fine-crystalline grains, forming micro-poikilitic mosaics of sizes between 100 and 200 μm (Fig. 4f). Likewise, the crystals of quartz, sanidine, and plagioclase show overgrowth along their margins, at the expense of the finer matrix (Fig. 4g, h), which stay as inclusions that still depict the preexisting crystalline margins. The modal counting of the crystals, without their poikilitic growth rims, indicates around 40 % adding quartz, sanidine, and plagioclase, with lesser proportion of opaque minerals and allanite. Along the entire profile, the quartz crystals appear as

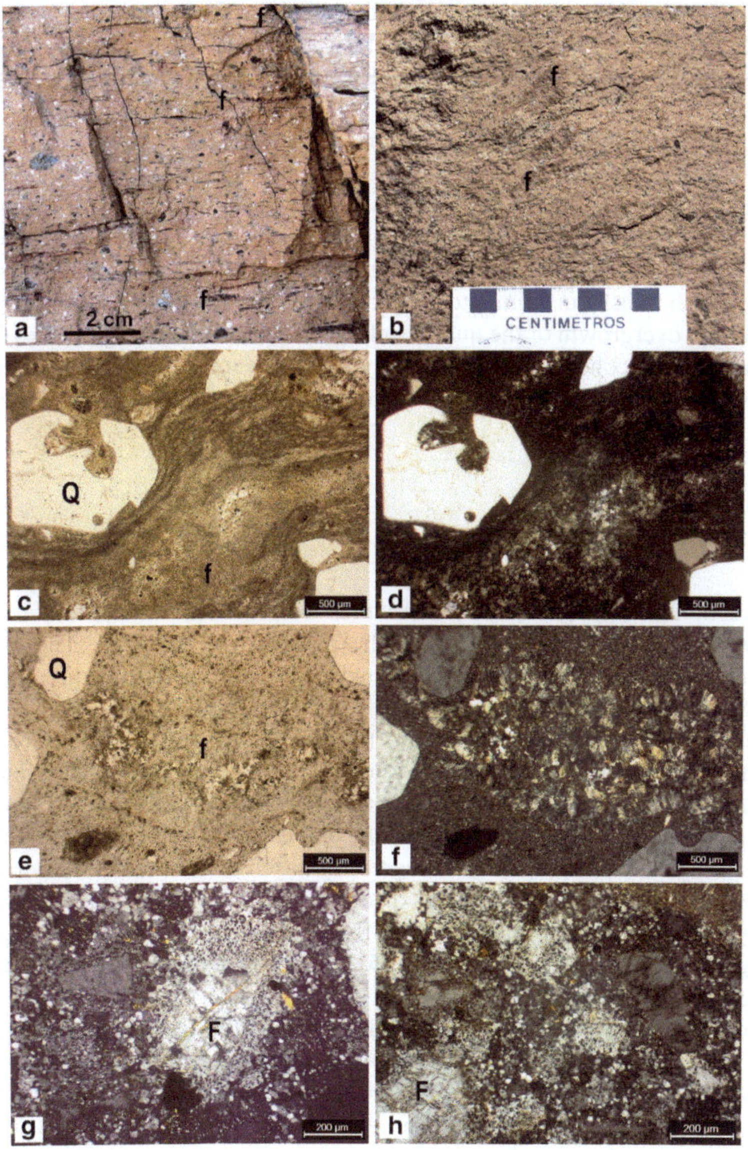

Fig. 4 (**a**) Eutaxitic texture defined by thin lenses of fiamme, with mm-scale quartz, feldspar, and lithic clasts. Ignimbrite from the basal layers. (**b**) Thicker lenses of fiamme, few meters above the former ignimbrite. (**c**) Fiamme with ragged end and flattened glass shards wrapping around quartz crystals. Plane polarized light. (**d**) Same view as (**c**), with crossed nicols. Fiamme with incipient spherulitic devitrification and vitroclastic to cryptocrystalline matrix. (**e**) Fiamme is recognized as a lens-shaped domain containing clean quartz-feldspar aggregates. Plane polarized light. (**f**) Same view as (**e**), with crossed nicols. Spherulitic textures of the fiamme are transformed into coarser granophyric aggregates. Matrix is very finely crystalline felsitic aggregate. (**g**) Poikilitic overgrowth of feldspar crystal enclosing surrounding smaller grains of the matrix. Crossed nicols. (**h**) Micro-poikilitic texture formed by patches of quartz and feldspar in the matrix, enclosing smaller grains of the former felsitic aggregates. Crossed nicols. *f* fiamme, *Q* quartz, *F* feldspar

fragments, with subhedral or hexagonal trending shapes, with conspicuous engulfed margins, and also of skeletal pattern. Feldspar crystals are fragmentary, euhedral to subhedral, and with documented engulfments, which present a varied degree in albitization.

In this manner, the volcaniclastic origin of this cooling unit, very well preserved at the basal ignimbrite levels, remains erased upwards by devitrification and/or crystallization assisted by vapor phase, as suggested by the late crystallization of biotite and fluorite. Both the fiammes and the vitroclasts have been transformed in very fine- to fine-grained crystalline aggregates, with additional micro-poikilitic growth, which transform them into a mosaic of high inner cohesion. Adding to this, the poikilitic overgrowth of the quartz and feldspar crystalloclasts provides the rock an even stronger cohesion and a "porphyritic" appearance, typical of a coherent igneous rocks (see terminology in McPhie et al. 1993), as could be a sub-volcanic or plutonic intrusive body.

Fractures

The directions and length of major fractures were measured on satellite imagery and their influence on the erosion of the plateau was analyzed. The preferred regional direction of fractures is NW/SE and ENE/WSW, and some directions such as NE/SW are locally important.

Concerning the joints and according to their genesis, those directly related to regional tectonics are recognized in systems, which show a geometric pattern in which two joint systems are intercrossed (Fig. 6a). Other systems are composed of sets of parallel joints which are linked to the emplacement and cooling of the rocks.

Fracturing took place in various degrees in different sectors of the Sierras. The ignimbrite beds are stratified in layers of strike NNE/SSW, with inclinations of 20–25°. Bedding is partially obliterated by the development of dense perpendicular jointing. In the Cerro Lihuel Calel (Fig. 5) columnar jointing is found, where their spacing is of centimeter scale.

In other sectors, the deposits occur with very intense fracturing, with development of various systems of subvertical joints superimposed to columnar jointing. The upper level is characterized by poorly preserved columnar jointing, since it occurs in well-developed rounded shapes, whose bases exceed 2 m, occasionally with horizontal partition in the upper and basal portions of these blocks, which may be related to flow plane jointing.

Unloading joints are not ruled out, because more than 450 m of ignimbrites accumulated on top of the studied levels, within the ancient caldera (intra-caldera ignimbrites). The upper portion was eroded after it was tilted up to 20°. The age of the tilting is unknown. During the tilting the NW/SE and ENE/WSW fractures were generated and also others following some other directions such as NE/SW. During erosion processes, the subcircular pattern joints may have been generated (Fig. 6b). This type of fracturing could have controlled later on the development of "nubbins" and tors.

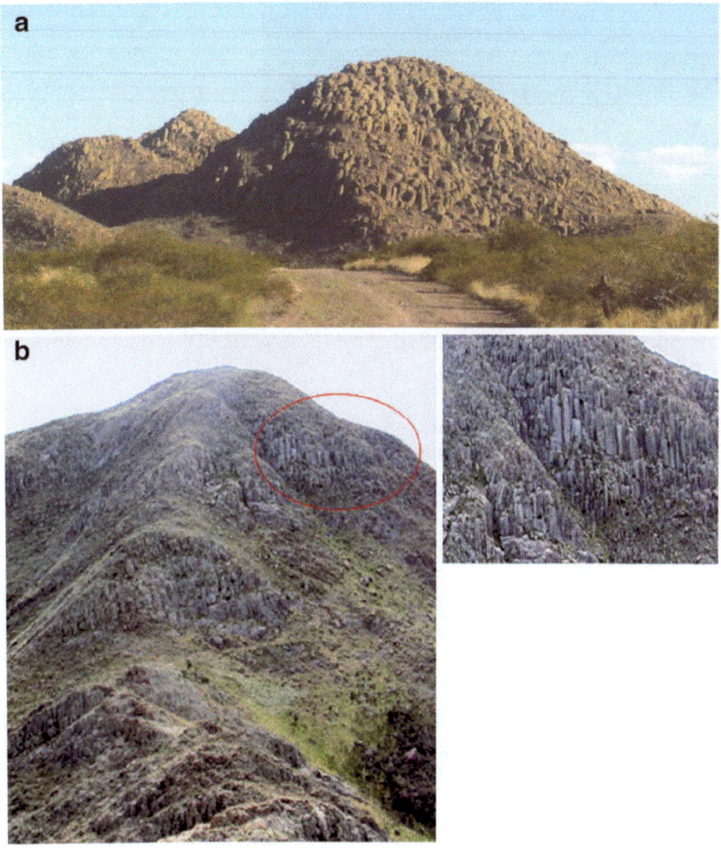

Fig. 5 (**a**) View from the north of Cerro Lihuel Calel, where the stratification with inclination to the west is observed, cut by perpendicular jointing. (**b**) Another mountain showing columnar jointing in a good degree of preservation

As a consequence of the elimination of the upper cover by erosion, changes in the stress and deformation conditions would have taken place. Towards the upper levels, the rocks would have developed a denser fracturing pattern, a product of the load removal and discharge expansion. The rocky bodies were thus fragmented by this fracturing network, which became later on penetration pathways for the weathering/denudation agents affecting the ignimbrites.

Ignimbrite Inselberg Landscapes

The isolated ranges of the Sierra de Lihuel Calel constitute an inselberg landscape, in which noted inselbergs are rising abruptly from the monotonous Pampean plains (Fig. 7a). These inselbergs are surrounded by pediments which may be differentiated

Fig. 6 Fractures of varying order of magnitude, seen in GeoEye/Google Earth image and field photographs. (**a**) In *yellow*, the crossing of two sets of major joints related to basement tectonics. (**b**) Enlargement of the same image in which numerous minor fractures with subcircular geometry may be observed

Fig. 7 (**a**) Plain, surrounding the Sierra de Lihuel Calel (inselberg). This landform is characterized by the alternation of highs and depressions (see the undulating road). (**b**) View of a sector of the plain, crossed by an ephemeral stream

from the plain itself. These pediments are usually covered by regolith. The regolith is a mantle-like deposit, with rounded to subangular, ignimbrite clasts, with a reddish sandy matrix. The clasts have a maximum size range between 5 and 10 cm. Underlying these materials and in contact with the fresh ignimbrite, a reddish weathered zone is observed, which is easily disintegrated when hit with the geological hammer.

The inselbergs themselves are domes of bare rock. Their morphology is varied, from elliptically elongated and low whaleback type to true domes whose dimensions in length and height are practically equal. Some inselbergs occur in groups. Individual domes are developed from huge joint blocks. The evolution of inselbergs is accompanied or followed by formation of other landforms, such as corestones, nubbins, and castle koppies.

The question arises whether the inselbergs were formed by intense chemical weathering and stripping or by fluvial erosion and slope retreat-forming pediments and pediplains.

The ignimbrite mantles are easily weathered when they come in contact with circulating waters. This is favored by its mineralogical composition in which feldspar is one of the more abundant components. They present a fracturing pattern similar to that of granitic rocks, that is, an orthogonal joint system, and other randomly oriented fractures as well as numerous micro-fissures (Vidal Romaní and Twidale 1998). The rocks of the Sierra de Lihuel Calel have been exposed to atmospheric and underground waters over a very long time period. Deep weathering and erosion have lowered the relief of the area, with the development of a regolith cover and immediately below it a weathering front. As a consequence of the denudation of the regolith cover, the weathering front has become exposed at the surface, and the plain can be regarded as an etchplain.

Landform Identification

The landforms, of all sizes, are interpreted as having had a sub-superficial origin, since they are developed at the weathering front (Mabbutt 1961). Among the larger landforms, the following features may be recognized: the regional plain and inselbergs, domes, nubbins, castle koppies and tors.

The regional plain is located surrounding the Sierra de Lihuel Calel. This landform is characterized by the alternation of elevations and depressions, a very smooth relief, and a discontinuous regolith cover. Towards the east, the plain gradually loses elevation, following the regional slope, and provides certain monotony to the landscape. It constitutes a planation surface or an erosion surface, in the sense of Ollier (1991). See Figs. 7a, b, 8, and 9.

In the representation of the topography of the study area (Fig. 8), a local base level is observed in between the 202 and 242 m contour lines. It is a topographic level with smooth undulations and frequent presence of lakes, ponds, and salt lakes. The fluvial channels of ephemeral nature discharge towards the La Amarga, Urre

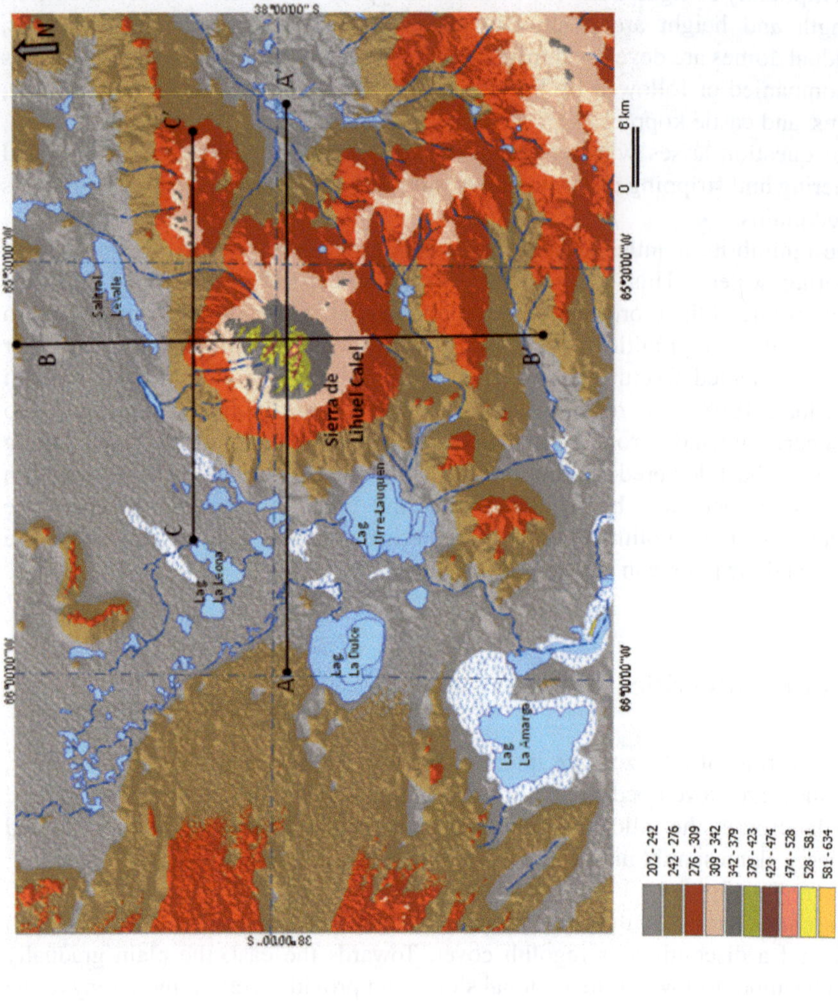

Fig. 8 An SRTM digital elevation model, classified following certain contour line intervals which are indicated in the list of symbols. The alignments correspond to the base lines of the topographic profiles A-A′, B-B′ and C-C′

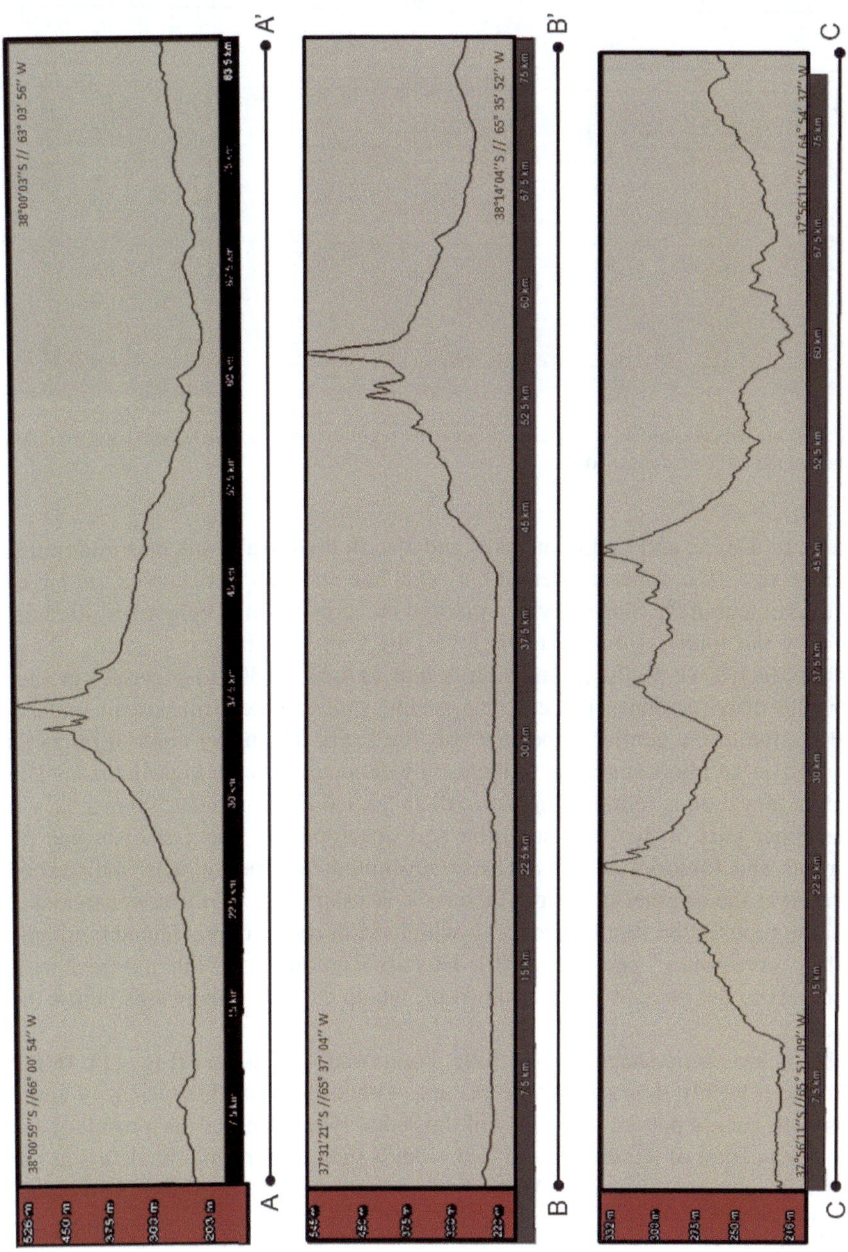

Fig. 9 Sections along the A-A′, B-B′, and C-C′ profiles of Fig. 8. The diagrams show the shape of the terrain surface, where the Sierra de Lihuel Calel is clearly seen, in a W–E direction in A-A′ and C-C′. The Sierra is seen from S to N in the B-B′ profile, where it appears surrounded by the regional plain

Fig. 10 All the unconsolidated materials compose the regolith cover in which fresh rock, rounded boulders named as "corestones," occur

Lauquen, La Dulce, and La Leona lakes and also in the Lavalle salt lake and others of smaller size. The contour interval between 242 and 276 m occupies the larger superficial extent in the Sierras, with a marked incision by small ephemeral streams, favored by the joint systems. At this level, the Cerro Cortado is located, whose summit scarcely exceeds the 300 m contour line. In the 276–309 m interval, a greater topographic dissection takes place, with a tendency towards dendritic stream pattern, as is the same in the gentler slopes towards the E/SE. The upper contour intervals clearly reflect an important areal reduction by denudation, with high incision. The maximum elevation is found at the Cerro de la Sociedad Científica (590 m a.s.l.).

The upper part of this unit has little soil development, but it is disturbed by plant roots and fossorial animals. The zone immediately below it is undisturbed and preserves the original structure of the rock as a saprolite. All the unconsolidated material composes the regolith cover in which fresh rock, round-shaped boulders, named as "corestones," occur (Fig. 10). This association of regolith and fresh rock corresponds to the ancient weathering front, which extends downwards below the surface.

Nubbins are dome-shaped, small hills, composed of boulders (Fig. 11). In the foreground a highly degraded dome occurs, with orthogonal fracturing, in some sectors with curved pattern and partly of the radial type. Nubbins (as a result of the in situ degradation of the dome) and rocky fields that are forming the debris at the foot of the dome are observed. At the background, the Cerro Cortado is located, where primary structures of the ignimbrite beds, with development of bedding, are observed (Fig. 11).

Inselbergs are relatively high and steep-sized hills, with dominant vertical jointing and subordinate curved jointing and the development of vertical walls which were cut by the curved fracture system (Fig. 12).

Fig. 11 *Left image*: curve jointing in dome landforms, associated to a vertical system with progressive development of nubbins at the foreground. At the background is the Cerro Cortado where primary structures of the ignimbritic beds are observed. *Right image*: development of nubbins in the sides of a dome

Fig. 12 Inselbergs with dominant vertical jointing and subordinate curved jointing

Fig. 13 Dome landforms of the whaleback type. Rocky surfaces shown are part of the flanks and tops of domes, either partially exhumed or in process of exhumation

Dome landforms of the whaleback type are developed at Cerro Alto, with curved jointing and degradation in nubbins and rocky fields that determine flat surfaces with subhorizontal jointing. Rocky surfaces are part of the flanks and tops of partially exhumed (or in process of exhumation) domes, which limit flat surfaces with horizontal to subhorizontal jointing. Favored by the slope inclination, the sliding of slabs or boulders feeds incipient stone accumulations (Fig. 13).

Castle koppies are domes whose morphology is controlled by fracturing systems. In them, orthogonal jointing is observed, limiting angular blocks. This type of dome

Fig. 14 Castle koppies limited by orthogonal joints, located in the Valle de las Pinturas (Arroyo de las Sierras)

Fig. 15 Tors with varied morphology along the sides of Cerro Alto, where many slumped blocks are found, some of them could have been "rocking stones" in the past

has been given its name due to their likeness to castle shapes (Fig. 14). Horizontal and vertical joints are noted. The upper blocks show certain rounding with respect to the blocks at the base of the features, which are more angular.

Tors is the name given to rocky blocks limited by fractures, which occur disseminated in different areas, in various topographic levels and exhibiting multiple morphologies. Their basal portions are usually planar, oblique, or rounded (Fig. 15).

Fig. 16 "Rocking stone" ("Piedra Movediza") or balancing rock: block of several cubic meters. This rock is located W of the Cerro Alto and it is not balancing anymore, since its two pedestals were destroyed

They are normally found in groups, but isolated landforms also occur. There are examples of tors in unstable equilibrium, which are known as "balancing rocks" (Fig. 16).

Regolith zones are observed in very ample surfaces where nuclei of fresh rock may be detected as if they were submerged into regolith. It should be noted that the weathering and the dismantling of the regolith has been incomplete. The landscape is characterized by flat sections where it is interrupted by the outcropping of blocks, groups of boulders, nubbins, and castle koppies. The regolith is preserved in the lower zones where residual corestones are still outcropping. Strip-shaped zones are coincident with the boundaries between different flow units. The limits between units affected differentially the local weathering rate, resulting in a steplike topography (Fig. 17).

Small-Scale and Minor Landforms

Alveoli correspond from rounded to ellipsoidal holes, of centimeter size, which occur in groups. They merge altogether and in depth, they form larger cavities, sometimes separated by just thin walls. They appear in surfaces of medium to steep inclination. They may be found in the inner parts of rock cavities named as tafoni (Fig. 18).

Fig. 17 Depressions bounding flow units, a product of regolith mobilization

Fig. 18 Alveoli developed in the inner part of tafoni; some of them are separated by thin rock walls

Fig. 19 (**a**) Tafoni appear as arranged in certain orientations, according to weakness surfaces within the ignimbrite mantles, following the boundaries in between flow units. (**b**) In the inner part, saline chemical precipitation is observed, in this case depicted by a *red circle*. (**c**) The inner part of the cavity is covered by alveoli, mamelonar shapes, and depressions of the scallop type. (**d**) Tafoni developed from fracture plane convergence

Tafoni are also integrated by cavities, which are larger than the hollows and big enough to develop caves. They occur grouped in circular or elliptical patterns, some of them with their floor covered by debris. In some opportunities, they appear arranged in certain orientations, according to weakness surfaces within the ignimbrite mantles, following the boundaries in between flow units (Fig. 19a). At the inner part of some tafoni, saline precipitation is observed, in this case depicted by a red circle (Fig. 19b). In other cases, the inner part of the cavity is covered by alveoli, with mamelonar shape and depressions of the scallop type (Fig. 19c). Some tafoni are developed following the convergence of fracture planes (Fig. 19d).

Gnammas, also known as weathering pits or rock basins, are depressions developed on surfaces with little inclination. They occur isolated or in groups. Their more frequent diameters are in the decimeter scale, though some others have been also identified at the centimeter scale or even surpassing a 1 m diameter. Their shapes are either circular or elliptical. Some of these gnammas are coalescent and their limiting borders are diffused (Fig. 20a). Others have a differential location along lines of weakness following joints (Fig. 20b), although in some cases they do not follow any visible structural control. It is common to observe the presence of outflow channels in some of these gnammas (Fig. 20c). These landforms, known in

Fig. 20 (**a**) Convergent gnammas and their limiting borders are diffused. (**b**) Location along lines of weakness following joints. (**c**) Gnammas with extruding channel that discharges in the rills concentrating superficial runoff. (**d**) Cascading gnammas, which are produced by one gnamma located in an upper topographic level of the massif which becomes coalescent with gnammas at lower levels. The sediment yield follows the water discharge; note the sediments accumulated in the lower gnamma

the German language as Opferkessel and Blutrinnen, have different sizes even when developed in the same rocky surface, and this is interpreted as different stages of growth with evolution and continuous generation of new gnammas (Fig. 20d).

The landscape of the Sierra de Lihuel Calel allowed the recognition of all the types of gnammas: (a) pit gnammas, deeper than wider; (b) pan gnammas, rather shallow in relation to their diameter; in these features, overhanging slopes are frequent; (c) cylindrical basins; and (d) armchair gnammas, located in gentle slopes, with a triangular transversal section, which represent an intermediate step towards the tafoni morphology (Fig. 21).

Exfoliation includes flaking, spalling, and spheroidal weathering (Ollier 1965, 1967). Spalling is widely present, as granular disintegration and lichen colonization are. These surfaces show laminae of a few millimeter in thickness, which massively cover the entire outcrop (Fig. 22a, b). They are also observed in the inner portion of tafoni. Some boulders show not only sapling but spheroidal exfoliation (Figs. 22c and 23).

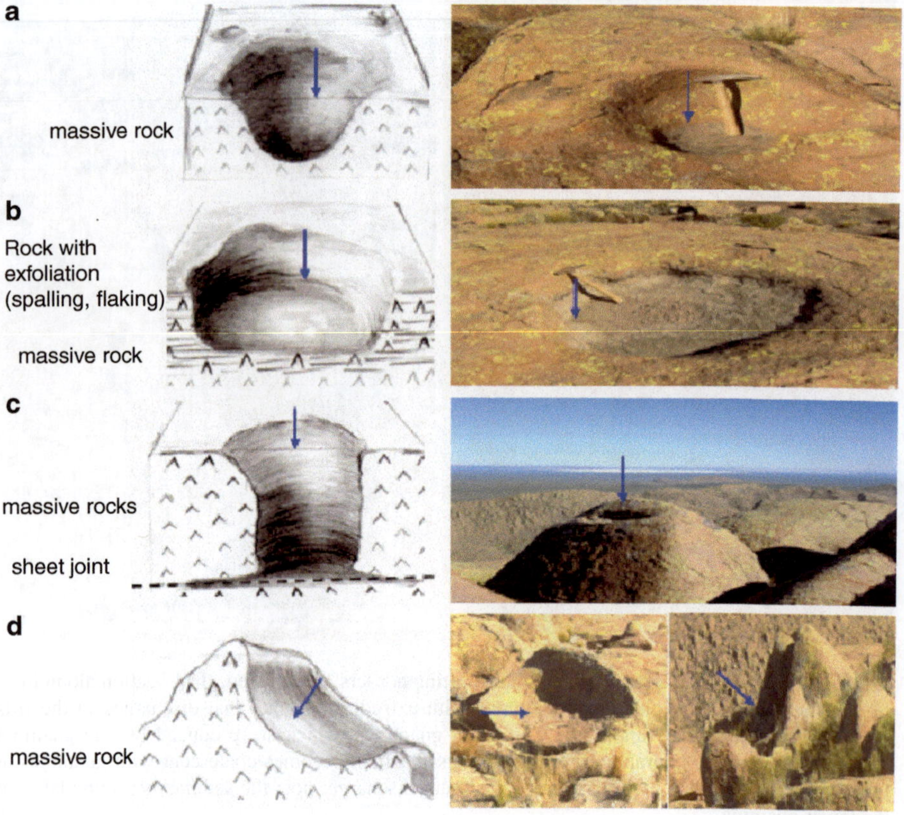

Fig. 21 The types of gnammas recognized in Lihuel Calel: (**a**) pit gnammas. (**b**) pan gnammas. (**c**) cylindrical basins, and (**d**) armchair gnammas

Bedrock drainage channels (also known as fluting, gutters, rille, and karren forms) are drainage channels that develop in the roofs and sides of domes, exhibiting incision of the dome due to these features. Some of them are controlled by structures or discontinuity surfaces. Their size is varied with an average depth of 10 cm, width of 30–50 cm, and a longitude of several meters (Fig. 24a, b).

Flared Slopes

Along the margins of the rocky massifs and also along the sides of residual blocks and boulders, it has been observed that the development of slopes with steep inclination and concave cross section has been termed as "flared slopes." These features have also been called inverted slopes, excavated walls, and concave

Fig. 22 (**a**) and (**b**) Flaking in rounded boulders, where numerous laminae occur as flakes. (**c**) Rock nucleus with flaking cover and incipient spheroidal exfoliation

Fig. 23 (**a**) Rounded blocks with spheroidal exfoliation and granular disintegration. (**b**) Small-scale fracturing

slopes. At least three levels of flared slopes have been recognized in the study area (Fig. 25a). Those of larger dimensions reach the metric scale and correspond to the differentiated levels of the summit surface. Those flared slopes developed in residual boulders and blocks reach decimeter scale. Some of them show a lateral transition to cavities of the tafoni type.

Fig. 24 Bedrock drainage channels; (**a**) and (**c**) channels in steep slopes; (**b**) and (**d**) parallel channels controlled by structure

Discussion

The study of these landforms has posed several questions:

(a) Are inselbergs formed by deep weathering and stripping or by fluvial erosion and slope retreat-forming pediments and pediplains?
(b) Are there common patterns between the landforms developed in ignimbrite rocks and those formed in granitic landscapes?

In the first case, concerning the formation of inselbergs, some authors attribute them to slope recession; another hypothesis includes two processes: deep weathering and planation (Twidale 1982); and a third one suggested mantle planation (Mabbutt 1961; Ollier 1978).

To a larger extent, these hypotheses not only explain the genesis of inselbergs, but they could be also applied to the formation of nubbins, castle koppies, tors, and other minor features.

Some tors are evidently the result of deep weathering, a product of a period of intense chemical weathering following joints and continued by a period in which the degraded material is mobilized. However, some tors lack these features indicating deep weathering, such as spheroidal weathering. These landforms suggest such

Fig. 25 Flared slope; note the development of the concave slope. At the base, a *red arrow* indicates the occurrence of several tafoni

genetic conditions in which weathering and denudation acted at the same time on rocks of different resistance, for instance in this case, welded ignimbrites and poorly welded ignimbrites.

With respect to flared slopes, the sub-superficial alterations of the rocky massif at the contact with the alteration mantle play a significant role in their formation (Twidale 1962). In those areas with scarce or null inclination, the corrosion plain is the product of a homogeneous alteration mantle (Twidale 1990). In the other extreme of the system, the exhumed reliefs are placed, where corrosion works in the margins of the rocky massif, progressively making even more vertical its walls. Chemical weathering is concentrated and reduced to zones favored by underground phreatic level. If the phreatic level is constant for a long period of time, undermining of the walls produced flare slope formation. The condition needed for its development is a stabilization of the alteration mantle, so as chemical weathering etched the morphology. In the second phase, the landform became exposed and degradation and stripping took place. Ollier and Bourman (2002) suggested that concave slopes may also be formed by subaerial weathering and that the retreat of these slopes may generate rectilinear footslopes. These rectilinear slopes have been related to flared slopes. The combination of concave and rectilinear slopes may produce convergent relief features that may be formed by different processes.

Gnammas (rock basins) are genetically related to weathering and mechanical erosion, because these cavities may retain water and clastic, loose materials as products of the cavity growth. This material is removed by water currents and it erodes and enlarges the cavity.

(c) The common patterns developed in the Lihuel Calel ignimbrites are similar to those found in typical granitic landscapes.

In the granitic landscapes, weathering acts along joint systems that are bounding blocks. Such weathering is not homogeneous, since it is more intense in the margins and much less effective in the remaining parts of the blocks. For this reason, the blocks which were originally subangular and angular are gradually advancing in their degree of roundness. Likewise, the weathered rock is usually removed and denudated, finally exposing the rounded rock nuclei, such as corestones and other boulders.

In the Sierra de Lihuel Calel example, the development of unloading joints, parallel to the topographic surface (sheeting) and others of subcircular pattern, would be able to control the development of nubbins and tors.

Another intervening similitude parameter is the massive character (of textural nature) of the rock. In ignimbrites, the poikilitic overgrowth of the quartz and feldspar crystalloclasts provided the rock an even stronger cohesion and a "porphyritic" appearance, typical of coherent igneous rocks, as could be a sub-volcanic or plutonic intrusive body.

Conclusions

With respect to the genesis of the larger- and smaller-scale landforms recognized in the Sierra de Lihuel Calel, the joint action of several agents such as weathering (either sub-surficial or subaerial), slope retreat and, to a lesser proportion, fluvial action would be responsible of their formation.

The exposure of the weathering front, the exhumation of corestones included in a regolith mantle and showing spheroidal weathering suggests the participation of sub-superficial alteration with a later removal of the regolith.

Weathering intervenes in the disaggregation of the materials following the joint systems, as the depressions are deepened and slope retreat achieves a larger significance, acting these processes in a combined manner (mixed genesis), perhaps in response to small variations of the climatic factors, considering also that certain forms of subsurface weathering are superimposed on subaerial weathering forms, as in the case of spheroidal weathering in corestones with flaking exfoliation, which sometimes may mask other types.

This landscape of inselbergs, nubbins, castle koppies, tors, and other landforms is developed on ignimbrite rocks, with textural properties and joint systems similar to those occurring in granitic rocks.

Acknowledgments This investigation was financed by the PIP Project 112-200801-00119 of CONICET. The authors are greatly indebted to Cliff Ollier for his permanent support and careful revision of the first draft of the manuscript and to Jorge Rabassa for his kind invitation to present this chapter.

References

Folgueras A, Zárate M (2009) La sedimentación neógena continental en el sector extrandino de Argentina central. Asociación Geológica Argentina, Revista 64(4):692–712

Linares E, Llambías EJ, Latorre C (1980) Geología de la provincia de La Pampa, República Argentina y geocronología de sus rocas metamórficas y eruptivas. Asociación Geológica Argentina, Revista 34(1):87–146

Llambías EJ (1973) Las ignimbritas de la sierra de Lihuel Calel, provincia de La Pampa. Actas 5 Congreso Geológico Argentino 4:55–67

Llambías EJ (1975) Geología de la provincia de La Pampa y su aspecto minero. Dirección de Minas de la provincia de la Pampa, Informe inédito, Santa Rosa, 38 p

Llambias EJ (2003) Geología de los cuerpos ígneos. Instituto Superior de Correlación Geológica y Asociación Geológica Argentina Serie de Correlación Geológica 15:1–182

Llambías EJ, Leveratto MA (1975) El plateau riolítico de la provincia de La Pampa, República Argentina. Actas II Congreso Iberoamericano de Geología Económica 1:99–114

Mabbutt JA (1961) Basal surface or weathering front. Proc Geol Assoc Lond 72:357–358

McArthur AN, Cas RAF, Orton GJ (1998) Distribution and significance of crystalline, perlitic and vesicular textures in the Ordovician Garth Tuff (Wales). Bull Volcanol 60:260–285

McPhie J, Doyle M, Allen R (1993) Volcanic textures: a guide to the interpretation of textures in volcanic rocks. CODES Key Centre, Hobart, 197 pp

Ollier CD (1965) Some features of granite weathering in Australia. Zeitschrift für Geomorphologie 9:285–304

Ollier CD (1967) Spheroidal weathering, exfoliation and constant volume alteration. Zeitschrift für Geomorphologie 11(1):103–108

Ollier CD (1978) Inselbergs of the Namib Desert, processes and history. Zeitschrift fir Geomorphologie, Supplement 31:161–176

Ollier CD (1988) Volcanoes. Blackwell, Oxford, 228 p

Ollier CD (1991) Ancient landforms. Belhaven, London, p 233

Ollier CD, Bourman RP (2002) Flared slopes, foot-slopes, and the retreat of overhanging slopes: examples of convergent landform development. Phys Geogr 23:321–334

Ramos VA (1999) Las Provincias Geológicas del Territorio Argentino. Geología Argentina, Instituto de Geología y Recursos Minerales, Anales 29(3):41–96, ISSN:0328–2325

Rapela CW, Pankhurst RJ, Llambías EJ, Labudía C, Artabe A (1996) "Gondwana" magmatism of Patagonia: inner Cordilleran calc-alkaline batholiths and bimodal volcanic provinces. In: Third international symposium on Andean geodynamics, Saint-Malo, France, pp 791–794

Sato AM, Tickyj H, Llambías EJ, Sato K (2000) The Las Matras tonalitic-trondhjemitic pluton, central Argentina: Grenvillian-age constraints, geochemical characteristics, and regional implications. J S Am Earth Sci 13:587–610

Sruoga P, Llambías EJ (1992) Permo-Triassic leucorhyolitic ignimbrites at Sierra de Lihuel Calel, La Pampa province, Argentina. J S Am Earth Sci 5(2):141–152

Tickyj H, Basei MAS, Sato AM, Llambías EJ (1999) U-Pb and K-Ar ages of Pichi Mahuida Group, crystalline basement of south-eastern La Pampa province, Argentina. In: Actas II South American symposium on isotope geology, Carlos Paz, Argentina, pp 139–144

Twidale CR (1962) Steepened margins of the inselberg from north-western Eyre Peninsula, South Australia. Zeitschrift fur Geomorphologie 6:51–69

Twidale CR (1982) Granite landforms. Elsevier, Amsterdam, 372 pp

Twidale CR (1983) Pediments, peneplains and ultiplains. Revue de Geomorphologie Dynamique 32:1–35

Twidale CR (1990) The origin and implications of some erosional landforms. J Geol 98:343–364

Vidal Romaní JR, Twidale CR (1998) Formas y paisajes graníticos. Universidade da Coruña, A Coruña, 411 pp

Widdowson M (ed) (1997) Palaeosurfaces: recognition, reconstruction and palaeoenvironmental interpretation, Special publication, 120. Geological Society of London, London

Wilson CJN (1993) Stratigraphy, chronology, styles, and dynamics of late Quaternary eruptions from Taupo volcano, New Zealand. Philos Trans R Soc Lond A 343:205–306

Zárate MA, Folguera A (2014) Planation surfaces of Central Western Argentina. In: Rabassa J, Ollier C (eds) Gondwana landscapes in southern South America. Springer, Dordrecht, pp 365–392

Paleo-landscapes of the Northern Patagonian Massif, Argentina

Emilia Y. Aguilera, Jorge Rabassa, and Eugenio Aragón

Abstract The dominant geomorphological unit of the Northern Patagonian Massif landscape is a regional planation surface, eroded across the crystalline basement (plutonic and metamorphic rocks), eruptive rocks of the Gondwana cycle (Early to Middle Carboniferous), and Jurassic volcanic rocks.

The most important active climate during the genesis of this surface had a very significant role, developing intense chemical weathering extending to variable depths with the corresponding degradation of the rocky material exposed at the surface. Remnants of the weathering profiles, both outcropping and fossilized by burial, are identified and described. Such a particular mega-landform was developed in a cratonic environment, mainly as a product of deep weathering, and it is interpreted as a denuded surface, an etchplain formed by corrosion followed by erosion.

The analysis of the relationships between relief, saprolite, and rock cover throughout time suggests that the most important factor for the classification of the present landscape is the duration of exposure of the crystalline basement at the surface, from the end of the Paleozoic and during the entire Mesozoic. This conclusion has essential relevance for the evaluation of the effects of Mesozoic tectonics and the powerful weathering under certain climatic conditions.

E.Y. Aguilera (✉)
Facultad de Ciencias Naturales y Museo, Universidad Nacional de La Plata (UNLP), Calles 122 y 60, 1900-La Plata, Argentina

DAIS (Dirección de Aplicación de Imágenes Satelitarias), Calle 7 N° 1267-2° P, 1900-La Plata, Argentina
e-mail: eaguilera@fcnym.unlp.edu.ar

J. Rabassa
Laboratorio de Geomorfología y Cuaternario, CADIC-CONICET, Ushuaia, Tierra del Fuego, Argentina

Universidad Nacional de Tierra del Fuego, Ushuaia, Tierra del Fuego, Argentina
e-mail: jrabassa@gmail.com

E. Aragón
Centro de Investigaciones Geológicas (CIG), Calle 1 N° 644, 1900, La Plata, Argentina

J. Rabassa and C. Ollier (eds.), *Gondwana Landscapes in southern South America*, Springer Earth System Sciences, DOI 10.1007/978-94-007-7702-6_15,

It is estimated that this paleosurface would have initiated its development towards the end of the Paleozoic, but later modifying also the Jurassic volcanic rocks that preceded the rifting processes that lead to the opening of the Southern Atlantic Ocean. Finally, the tectonic activity during the early Tertiary produced the exhumation of the planation surface, which was buried by its own regolith, reactivating erosion surfaces and small drainage basins. However, it is possible that some areas of the planation surface had never been covered by other rocks, other than its own overlying weathering products.

Our results suggest that the landscape features should not be assigned to Quaternary morphogenesis, but instead, they have evolved over a very long time, perhaps 100 Ma or even more. These observations refer to Mesozoic times, and therefore the time scale used for the discussion of the geomorphology of the Northern Patagonian Massif should be enlarged to properly analyze the evolution of the ancient landscapes of this cratonic region.

This chapter contributes to the analysis of comparative studies of global geomorphology of cratonic areas, where planation surfaces record very long periods in which the speed of crustal deformation is highly compensated by planation processes.

Keywords Gondwana • Argentina • Northern Patagonian Massif • Etchplains • Granitic geomorphology

Introduction

The development and evolution of the landscape of the Northern Patagonian Massif is clearly related to the peculiar morpho-structural environment of this cratonic geological province and with the morpho-climatic systems from tropical to temperate and from humid to dry that were concomitant with the tectonic processes of uplift and structural deformation and later continental tear apart. Therefore, many of the observed landforms may have had their origins as early as 300 Ma, before the process of Gondwana breakup started.

The dominant mega-landform of this landscape is a very large planation surface, a geomorphological unit of regional magnitude associated with residual relief at various scales.

The different factors which determined the impact of ancient weathering rates, such as lithology, fracturing and jointing of the rocky blocks, their spatial variation, and climate and drainage, are very significant elements for this model of landscape evolution.

A varied morpho-structural context shapes up the landforms of the basement landscape, where the primary characteristics of the rocks such as mineralogy, texture, and structure determine their weathering and erosion, and their later transformation into regolith or saprolite (Aguilera et al. 2010; Aragón et al. 2005; Rabassa et al. 2014). In this chapter, the term "saprolite" is restricted to the whole volume of in situ weathered rock, whereas the term "regolith" applies to all overlying materials (including, loess, alluvium), that is, every mineral and rocky materials between fresh rock and fresh air, and it may involve also the saprolite, if it is still in situ (C. Ollier, 2012, personal communication). A fundamental aspect in the evolution of this landscape is the fracture systems in the rocks, both faults and joints, which fragment the granitic and metamorphic bodies generating appropriate ways of fluid circulation which selectively attack the minerals. The scale of observation of the rock fracturing includes those fractures of regional magnitude which articulate large landscape units, faults that divide plutons with relative block displacement, and joint systems generated by the emplacement of igneous bodies and micro-fractures.

The landforms determined by joint networks separate the basement rocks in parallelepipeds, affected by subaerial erosion which made progress through the joints, widening them and progressively rounding the edges of the blocks.

The variation of these factors that determine the weathering rate regulates the shape of the weathering front, which is generally markedly irregular and whose extension in depth does not necessarily preserve relationships with the surficial forms. During periods of climatic and tectonic stability, the weathering and denudation rates are almost balanced in the long term, equating the thickness and the depth advance of the weathering profile.

Modifications in the climatic régime or in the crustal uplift may change this dynamic equilibrium state, increasing erosion rates. The adjustments of the geomorphological systems have strong incidence in the stability of the weathering mantle, which may be partially or completely mobilized.

Location of the Studied Areas

The Northern Patagonian Massif, also known as the Somuncurá Massif (Fig. 1), is located in the provinces of Río Negro and Chubut, Patagonia, Argentina, with a minor extension into the province of Neuquén, where it is locally known as the Sañicó Block. It is included in the geographical region known as extra-Andean Patagonia, with a prevailing tableland relief, with low ranges and sierras, depressions, hollows, and very broad terraced fluvial valleys, with a dominant west-east direction and carved by allochthonous streams with sources in the Northern Patagonian Andes.

The climate is semiarid to arid and cold, with strong winds coming from the west. The aridity of the present climate is a consequence of the Late Cenozoic uplift of the Patagonian Andes, which formed a barrier in the pathway of the humid winds coming from the Southern Pacific Ocean. The principal vegetation formation is the steppe, due to the scarcity of precipitation.

Three areas of the Massif are analyzed in this chapter: (a) one is located in the province of Río Negro, between 67°30′ and 70°30′ west longitude and 40°00′and 42°00′ south latitude; (b) another is found in the province of Chubut between the

Fig. 1 Regional location of Patagonia, Argentina, showing the Northern Patagonian Massif and the geotectonic setting

sites of Paso del Sapo and Piedra Parada, along the margins of the Río Chubut, between 42°30′ and 42°45′ south latitude and 70°00′ and 70°45′ west longitude, respectively; and (c) the western portion of the Massif, which has been marginally affected by the Andean tectonics.

Methodology

The analysis of the paleosurface was done using field data, superposition of geological and geomorphological maps and topographic profiles, aerial photographs, digital elevation models, satellite imagery, and petrographic analysis of thin sections.

Regional Geological Framework

The crystalline basement of the Northern Patagonian Massif is formed by metamorphic rocks and syntectonic granitoid rocks, extending from the Late Precambrian to the Early Paleozoic. In the eastern portion of the Massif, marine deposits are overlying these rocks in angular unconformity, in an environment of clastic marine platform of Silurian to Early Devonian age. The basement is intruded by granitic plutons of the Gondwana cycle, that is, from the Late Paleozoic to the Early Triassic. These plutons are overlain by ignimbritic-tuffaceous complexes belonging to the same cycle and whose radiometric ages extend from the Late Paleozoic to the Middle Jurassic inclusive (Llambías and Rapela 1984).

These are two superimposed magmatic cycles. The first one is formed solely by rocks of plutonic nature and of Carboniferous age. The second is composed of eruptive volcanic rocks, both lavas and ignimbrite flows, which formed an extensive ignimbrite plateau of Neopaleozoic to Late Triassic-Early Jurassic age. This latter event is associated with Early Jurassic plutonic processes of extensional nature (Rapela and Alonso 1991).

The crystalline basement formed a positive area at least during the Triassic and perhaps even before. This positive region was covered by fossil flora bearing sedimentary rocks, which occupied the periphery of the craton. The Jurassic sediments developed in continental environments with pyroclastic participation were covered by continental deposits of Early to latest Cretaceous age and even from the Tertiary, in which it is possible to identify the marine transgressions of the Maastrichtian-Danian, Eocene, and Neogene (Miocene), respectively. A transgression from the Pacific Ocean has been recorded in a small portion of the western slope of the Massif, bearing Early Cretaceous marine sedimentary rocks.

The Northern Patagonian Massif then became a sub-positive element, which divided towards the north and south two important depositional centers starting in the Early Cretaceous: the Neuquén Group Basin and the Chubut Group Basin, respectively. Overlying the Massif rocks, only a thin and scattered Cretaceous sedimentary cover proves the physical connections between both basins in those times. Another important feature of the Massif is the existence of extensive basaltic volcanism of Cenozoic age. It started feebly during the Eocene, to continue with prominent alkaline basalt flows during the Oligocene that covered most of the Massif surface, particularly in the Meseta (or high-plain) de Somuncurá.

The Massif is characterized by the development of basaltic plateau flows and alkaline and per alkaline rocks coming from a reduced number of eruption centers. This cratonic unit is related to a tractive régime, associated with an ephemeral hot spot, located in the extra-Andean region and connected to a process of crustal thinning which, in just a few millions of years, gushed out large volumes of basalts (Kay et al. 1993).

The volcanic activity developed later on with important alkaline-acid events. The Miocene basaltic volcanism is restricted to the western portion, whereas the most modern flows occur in peripheral positions of the Massif (Ramos 1999).

The Tertiary continental deposits are interfingered with the basaltic volcanism, achieving only a modest thickness. Their development in the Massif is poor; meanwhile, through this whole period, the Massif behaved as a relatively stable, positive cratonic unit. The Quaternary deposits are represented by fluvial and piedmont deposits, with limited glacial activity along the western margin.

The structure of the Massif is characterized by large-block tectonics, in its eastern half controlled by the extensional régime related to the opening of the South Atlantic Ocean and to the north with the development of the aulacogen basin of the Colorado-Negro rivers. The basement blocks defined tilted semi-graben systems, some of them with oblique development referred to the margin of the craton, and controlled by ancient structures (Cicciarelli 1989).

In the tectonic evolution during the Middle to Late Triassic in the north-central sector of the Northern Patagonian Massif, a marked transcurrent tectonic activity of dextral faults of east-west strike and kilometer-scale displacements is documented (Giacosa et al. 2007). The faults affect the Early Paleozoic metasediments, the Gondwana granitoids, and the rocks of the Los Menucos Complex, in a similar manner.

The present configuration of the regional structure is a consequence of the Famatinan, Gondwanan, and Patagonian orogenies and, to a lesser degree, the Andean orogeny, with the development of basement blocks, tilted by compression and transpression with dominance of northwest-oriented fracturing (Ramos and Cortés 1984).

During the Andean cycle, a minor tectonic inversion took place in the western sector of this geological province, and ancient alignments of northwestern strike were reactivated in the central and eastern areas. The effects of this orogeny have been feeble and, in some portions, even absent.

Paleoclimates

Paleontological evidence indicates that during the Mesozoic, the climate of the region was warm to hot and humid. From the paleo-floristic point of view, no plants are known in the Middle to Late Paleozoic, whereas in those formations attributed to the Mesozoic, the paleo-flora is important due to the existence of rich Triassic and some Jurassic tapho-floras (Arrondo et al. 1984). Likewise, paleo-palynological records of great biodiversity reflect abundant plant life which characterized the Northern Patagonian region during the Mesozoic (Volkheimer 1984). Concerning the Cenozoic fossil floras, the groups of tapho-floras studied by Romero and Dibbern (1984) are indicators of gradual changes in the environmental conditions, from a temperate-hot and very humid climate to a cooler and drier one. The micro-floristic associations that occurred in the Late Eocene-Early Oligocene suggest that there were areas localized in lower topographic sites with humid climate and protected environments, as well as others corresponding to conditions of higher and drier sites of undulating relief, exposed to maritime wind régimes with abundant seasonal rainfall (Pothe de Baldis 1984).

Studies of mixed paleo-floras during the Cenozoic of Chile and Argentina suggest that their development took place under relatively warm temperatures and high annual rainfall, with little seasonal variability (Hinojosa and Villagrán 2005).

Landscape Evolution

To understand the genesis of the landscape features and its physical pattern on a long time scale, the period between the end of the Paleozoic and the entire Mesozoic is taken into consideration. This very long period is coincident with a global change in the Earth tectonic régime, moving from compressive conditions to the initiation of the Mesozoic distension.

A fundamental importance is attributed to the role of the climatic and tectonic controlling factors and also to the analysis of weathering profiles which may be considered as firm bases on which the study of the long-term development of these landscape. Specific key factors determining the conditions in which these landforms were generated are herein analyzed.

The model of landscape evolution used in this chapter is that proposed by Wayland (1934), who coined the term "etchplanation" and explained its meaning. When the alteration material is eroded by fluvial action, the weathering front is exposed as an "etchplain" (Wayland 1934), named by other authors as a "chemical corrosion plain" (Twidale 1982, 1985, 1987, 1989, 1990; Vidal Romaní 1989; Vidal Romaní and Twidale 1998). The spatial variation of the factors that determine the weathering rates of the rocks, particularly lithology, fracturing and jointing of the rocky blocks, climate, and drainage comprise important elements for this landscape evolution model. Numerous names have been proposed for the different types of etchplains. For instance, Thomas (1978) has developed a classification, taking into account the lithology and structure of the rocks, the degree of dissection, denudation of the regolith, and the detailed morphology of the weathering front.

Surveying the Gondwana Paleosurface

The Gondwana paleosurface covers a large area exceeding 20,000 sq. km in the study area. Its morphology is that of a plain, but no perfect planation has in fact taken place. It is actually an undulating plain, with low, rounded summits, of little local relief, which nevertheless has well-defined, drainage divides (Fig. 2a–c).

Regional geological profiles show the low-relief, topographic surface developed on top of plutonic and volcanic rocks, on which the Late Cretaceous and Paleogene/Neogene sedimentary rocks overlie in total concordance. Younger units are also the bedrock for the development of younger planation surfaces (Aguilera 2006a, b, 2007; Aguilera and Rabassa 2010) (Fig. 3).

In the central-northern portion of the Massif, the smoothly undulating landscape is monotonous, with flat sections and sporadic rounded hills, whereas in the western area, a low hill relief is present with a few higher crests.

In the western area, next to the Limay and Collón Curá rivers, close to latitudes 40°00′ to 40°45′ S, low-range profiles are defined. The paleosurface is noted for its smooth topography, composed of hills of concordant summits and isolated

Fig. 2 (**a**) Panoramic view of the planation surface, in the way to Laguna Blanca, (**b**) Panoramic view of the planation surface. In the foreground, remnants of the Cretaceous sedimentary rocks. The road follows the planation surface. At the background, the Sierra de Quepuniyeu, (**c**) Panoramic view of the planation surface. Rounded landforms limited by an orthogonal fracture network may be observed. Note the absolute flatness of the paleosurface

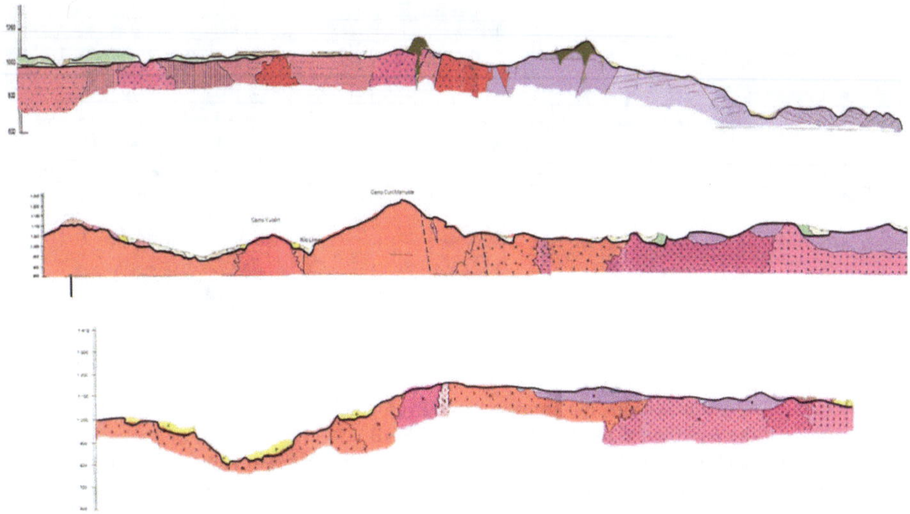

Fig. 3 Geological sections showing the planation surface cutting across diverse geological structures. Regionally, the surface is essentially flat, in spite of a few local topographic expressions (Source: SEGEMAR, Geological Sheets 4169-I and 4169-II). In *reddish colors*, plutonic rocks; in *violet*, *purple*, and *pinkish* tones, volcanic rocks

wetlands. In its central part, frequent small depressions are found, which during rainy seasons become temporary ponds and shallow lakes. The drainage pattern of the westernmost higher block is of the radial type, oriented towards the periphery, which is conterminous with the larger streams that have their local base level in the Río Limay.

In this portion of the Massif, the paleosurface reaches its maximum elevations over 1,250 meters above sea level (m a.s.l.). This zone would have been uplifted in a block pattern, inferring the existence of a large fault but with no erosion scarp, due to intensive fluvial erosion. The dissection degree is very marked with the development of deep valleys, favored by lines of weakness along fractures, faults, and joints. Drainage patterns of the dendritic and/or parallel types are noted for these structural controls (Fig. 4).

The few areas with a major topographic expression locally show steeper slopes. It is the case of landforms composed of convex hills, interrupted by the development of rocky crests with higher resistance to weathering and erosion (differential weathering). These crests are usually symmetrical and are partially buried by weathered materials, mostly mobilized. The granitic bodies appear segmented by faults and joints, the latter following two directions in an almost orthogonal pattern, which improve the water infiltration penetration to deeper levels. In these sectors, the maximum intensity and extension of the fragmentation and disintegration processes take place, which facilitate the decomposition of the rocks in blocks and boulders of varied sizes. Thus, residual landforms corresponding to varied lithology occur in the paleosurface, such as inselbergs, domes, castle koppies, nubbins, bornhardts,

Fig. 4 In pinkish-reddish colors, the granitic rocks, over which the paleosurface is developed, may be observed, at the *left side* of the Landsat image 7 RGB 435. The network of darker tones due to the fracturing and the high degree of dissection is noted. In the lower right sector of the image, numerous depressions and hollows occur aligned, suggesting an ancient drainage system

and rocky crests, accompanied by middle- to small-scale landforms as blocks, balls, gnammas, taffonis, and corestones (Aguilera et al. 2010).

To the east, the Massif gradually loses altitude and the paleosurface is found at elevations of 1,000, 800 and 750 m a.s.l. The differential nature of the weathering and the erosion processes operate at all scales. At the level of the mineral composition, it acts in terms of mineral stability and some textural parameters, such as grain size. A slow but steady transformation of the primary to secondary minerals takes place.

At the scale of outcrops, the surface covers large areas of similar lithological composition with relatively uniform resistance to erosion. The rocks of the crystalline basement show that the patterns of joints and faults play a dominant role in differential weathering. Some sectors have been identified with an almost horizontal topography over granitic and ignimbrite lithology. In certain areas, a pattern of paleo-drainage, formed by numerous aligned depressions, may be observed (Fig. 4).

This sector, with a markedly flat landscape, appears to be interrupted in some areas by topographic landforms which generated positive and negative relief (Fig. 5). Near La Esperanza, groups of aplite and lamprophyre dykes intruded granitoid rocks. These injected masses of varied lithological composition show a larger (aplites) or lesser (lamprophyres) resistance to erosion than the bedrock, thus forming alternating highs or depressions.

In the region of Los Menucos, the homogeneity of the paleosurface is interrupted by a structural depression which is bounded by a fault system over 100 km long. Additionally, structural highs of dome shape are observed there, with a radial-type drainage pattern, with development of gnammas or "pias" (i.e., rock kettles; Fig. 6).

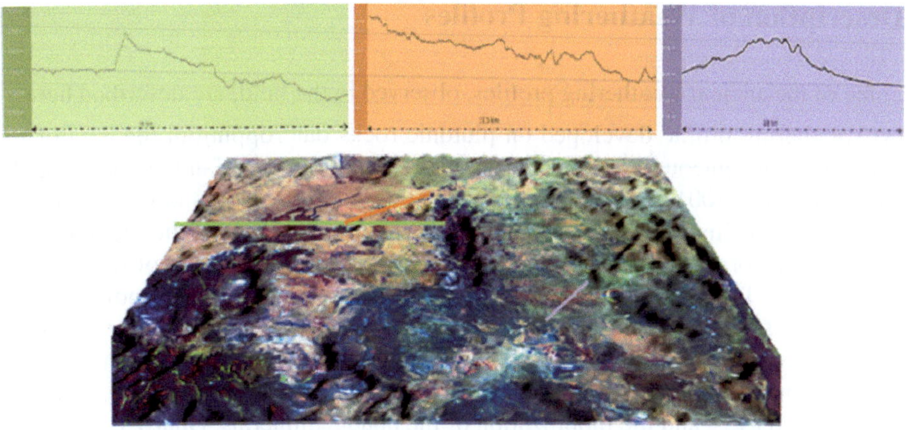

Fig. 5 Digital elevation model with superimposed satellite image Landsat 7 RGB 541. The vertical exaggeration is ×4. The *colored lines* indicate the position of the topographic profiles shown, with no scale exaggeration, at the central sector of the Massif

Fig. 6 La Esperanza-Los Menucos sector, satellite image Landsat 7 RGB 435. The plutonic body of the Calvo Granite has been segmented in two halves by the Loma Blanca fault. The plutonic bodies have topographic expression, with development of gnammas pias

Description of Weathering Profiles

Some of the ancient weathering profiles, observed in the field, are described here.

(a) Weathering profile developed on plutonic rocks outcropping in the area of La Esperanza, an igneous body named as the Calvo Granite (Llambías and Rapela 1984; Llambías 2001), composed of potash feldspar, oligoclase, quartz, and biotite. The weathering impact is recognized by the reddish appearance of the rocky surface and the transformation, though in a varied degree, total or partial, of the mineral components. The observation of changes at the mineralogical level indicates that the potash feldspar has usually offered a larger resistance to weathering than plagioclase, since this has been transformed in a powdery material in which kaolinite is dominant, whereas biotite is altered to vermiculite – with the corresponding volume increase – and the mobilization of the opaque minerals. Quartz is abundant and resistant to alteration. Its grain size, from medium to coarse, makes it susceptible to alveolar erosion; therefore, the pressure exerted by expansion on each crystalline union increased with the growth of grain size (Derrau 1970). The quartz texture is granulated, hypidiomorphic, with anhedral and subhedral crystalline faces, which provides a lesser cohesion than if all crystalline faces were anhedral. Towards the upper levels of the profile, the mineralogy and primary texture of the original rock occurs obliterated by the dominance of clayey materials and iron oxide minerals. The original textures may be distinguished towards the lower zones, but there is a loss of coherence of the grains and an increase in porosity.

The upper portion of the profile is slightly disturbed by scarce roots of xerophile vegetation and slope erosion. The part immediately below does not show perturbation, and it retains structures of the original rock, "in situ" contour of dykes and unweathered corestones. It is observed that the primary minerals have not changed their composition; thus, volumetric changes are not generated. Note that iso-volumetric weathering is characteristic of saprolite.

Dykes have quartz-feldspar composition and anhedral micro-grained texture, showing high resistance to weathering. They occur in the topography as positive features, and they maintain their extent without tectonic perturbation (not mobilized, "in situ"), together with the whole granitic mass or partially altered to regolith. According to the deepening level of the profile, stripes, fringes, zones, and rounded arenization masses occur together with other areas of fresh granite. Arenization, in the sense of Derrau (1970), indicates an altered but not mobilized, granitic material, thus being a saprolite that behaves as a slightly compacted sand, and the unequal fracturing and fissuring of the material permits the persistence of unweathered corestones and shows that the bottom of the weathering profile is not totally flat (Fig. 7).

(b) A remnant of the planation surface developed upon the Jurassic volcanic rocks may be observed between the localities of Paso del Sapo and Piedra Parada, sites located along the Río Chubut valley. On such surface, a regolith mantle is preserved, a product of alteration of the underlying volcanic units, and it is overlain unconformably by the quartzitic sandstones of the Paso del Sapo Formation, of Late Cretaceous age.

Fig. 7 Arenization in altered granitic material, not mobilized yet. The saprolite acts as slightly compacted sandstone, with uneven fracturing and fissuring of the material allows the persistency of unweathered corestones, showing that the bottom of the weathering front is not absolutely flat

A bluish-gray-colored regolith, with purple to reddish sectors, has developed over the volcanic rocks, in which a whitish mottled zone gradually evolves, as a result of transformation of plagioclase into clayey materials, mainly kaolinite. The regolith shows a clast-supported texture of approximately 1 m in thickness, with angular clasts of volcanic rocks and a scarce epiclastic matrix of sandstones. The clasts of andesitic rocks show whitish lumps (pseudomorphs of glomeroporphyritic texture) due to plagioclase alteration into clayey materials, basically kaolinite. This process of alteration to kaolinite generated an aggregate of pseudomorphic alteration in both the clasts and the matrix.

The horizontality of the paleosurface is observed over all the Jurassic outcrops. Towards the east the Cretaceous sediments have covered the paleosurface and fossilized its regolith, though this paleosurface is now being gradually exhumed by erosion.

At a regional scale, this planation is observed where the flat surfaces with scarce topographic expressions dominate, generating rounded landforms of the whaleback type, as the larger forms. The advance of peeling out of the surfaces and the selective atmospheric action as a function of the structural characteristics leads towards the conservation of more resistant compartments, under the shape of residual hills such as inselbergs, which survive in the landscape as erosion remnants (Fig. 8).

The areas of multiple or superimposed lava produce structural surfaces that are horizontal. Water can percolate between flows and through joints in the basalt, resulting in chemical attack and weathering.

Fig. 8 Residual hills (inselbergs) stand up in the landscape as erosion remnants. The flat surfaces with scarce topographic expressions generate rounded landforms of the "whaleback" type, near Paso del Sapo

Fig. 9 Residual hills (inselbergs). The *dotted red line* represents the reconstruction of the paleosurface developed in Jurassic volcanic rocks. In the *yellow rectangle*, semicircular structures due to acting weathering processes may be observed, with rounding of their corners and a cover of loose materials of reddish color, adopting a spherical weathering type of structure

Spheroidal weathering is observed in jointed basalt. Weathering starts at the joints and works inward. As weathering proceeds, the central zone remains unaltered, with rounding of the corners producing a corestone, The corestone may be surrounded by concentric shells of partially weathered basalt producing a type of weathering structure called "spheroidal" which, according to Ollier (1984, 1991), would be related to element migration within the rocky mass (Fig. 9). Due to the extent of this paleosurfaces, the two phases in which these chemical corrosion landforms evolve, in the sense of Twidale (1989), may actually occur. Where the model of the weathering profile that takes place in underground conditions has been preserved, due to the fossilization of the regolith, the second phase concerns the denudation of the regolith and the exposure of the weathering front (Twidale 1989).

General Considerations About the Stratigraphy and the Late Mesozoic and Cenozoic Paleo-geomorphology in the Western Portion of the Northern Patagonian Massif

The western sector of the Northern Patagonian Massif in the province of Río Negro deserves a particular description, since it had a clear surface drainage direction to the west, towards the Pacific Ocean. This region has been marginally affected by the Andean tectonics, with major changes in the drainage direction which switched to the north-northeast when the Andean ranges emerged. In this area, three superposed paleo-landscapes occur, which developed sometime between the Early to Middle Cretaceous (?) and the latest Cretaceous-Paleocene, the Oligocene-Early Miocene, and the Early to Middle Pliocene. Some of the various resulting landforms have been exhumed and partially modified by the erosion agents during the Quaternary, but others have been preserved in the surface and never covered again by thick marine or continental sedimentary sequences.

The reconstruction of the paleo-landscapes has been possible thanks to several volcano-sedimentary units in terrestrial environment that had partially buried these landscapes, notably preserving the respective paleo-landforms. The oldest paleo-landscape is a partially exhumed etchplain, equivalent to that described above, developed during the Early to Middle Cretaceous on the granites and metamorphic rocks of the crystalline basement (Figs. 10 and 11) and also on tilted, volcano-sedimentary Triassic-Jurassic units. This etchplain was formed under tropical climate and deep chemical weathering processes. The development of the following paleo-landscape partially destroyed this Cretaceous etchplain, which was initially covered by the Late Cretaceous continental red beds of the Angostura Colorada Formation and then by the Early Tertiary sequences. The regional drainage was then in direction to the Pacific Ocean, since the Andean Cordillera had not been uplifted yet. The landscape was buried by the Late Paleocene-Eocene, volcano-sedimentary beds of the Ventana Formation (formerly known as the "Serie Andesítica"). The orogenic movements of the Late Oligocene pushed up the Patagonian Andes Cordillera, thus forcing an inversion of the drainage, now towards the Atlantic Ocean. Erosion was essentially of fluvial origin, and it entrenched a deep landscape on top of these Early Tertiary volcanic units, as well as on the etchplain developed on top of the crystalline basement and the Mesozoic sequences, the Comallo paleosurface (Fig. 12). This was a very well-integrated relief, which had reached the maturity stage (in the classical Davisian sense) when the deposition of the ignimbrites and tuffs of the Collón Curá Formation (Middle to Late Miocene)

Fig. 10 Deeply weathered crystalline basement, near Comallo, western Río Negro province. Professor Cliff Ollier for scale (Photo J. Rabassa 2001)

Fig. 11 Deeply weathered Paleozoic metamorphic rocks, near Comallo

Fig. 12 In the background, the Comallo paleosurface, the Late Mesozoic etchplain that dominates the landscape of the Northern Patagonian Massif

took place. This volcanic formation rapidly buried the existing depressions, with emerging low hills as the unique positive elements of the landscape. At the Miocene-Pliocene boundary, new orogenic movements finally emplaced the Patagonian Andes at their present position and forced changes in the climate of the region, gradually leading to the present cold, semiarid steppes.

The information presented here is a synthesis of the basic aspects of the paleo-geomorphology of the region (Rabassa 1974, 1975, 1978a), reviewed according to new concepts related to long-term landscape evolution (Rabassa et al. 2010). The study revealed abundant information about the existence of several superposed paleo-landscapes, some of them being discussed in detail below. In this chapter, it is herein analyzed the fundamental characteristics of these paleo-landscapes.

The more relevant topics of the studied region have been discussed elsewhere (papers mentioned above). The stratigraphy of the cited area has been described by Rabassa (1975, 1978b), and it was compiled after extensive fieldwork over 250 sections and selected localities. Table 1 presents a summary of the identified litho-stratigraphic units.

Considering the cited paleo-landscapes, the most relevant ones to the purpose of this chapter are the Cretaceous and Early Tertiary landscapes. A large portion of the ancient landscapes is still being denudated and exhumed. These ancient landscapes were eroded in several stages, which probably started in the Early to Middle Cretaceous. The area located between the localities of Pilcaniyeu Viejo and Neneo

Table 1 Stratigraphy of the western margin of the Northern Patagonian Massif (Slightly modified from Rabassa 1974, 1975, 1978b)

Litho-stratigraphic units	Lithology	Geochronological units	Radiometric ages
Alluvium	Alluvial and colluvial sands and gravels and other sediments	Holocene	<10 ka 14C BP
Pichi Leufu Drift (Great Patagonian Glaciation)	Till and other glacigenic deposits	Early Pleistocene	ca. 1 Ma
Piedmont deposits	Gravels and sands	Pliocene to recent	
Late Miocene to Pleistocene basalts	Basalts and scoria cones	Late Miocene to Pleistocene	8–1.5 Ma
Collon Cura Fm.	Ash-flow and ashfall tuffs, fluvial sedimentary rocks	Middle to Late Miocene	15.5–11 Ma
Ventana Fm.	Basalts, trachytes, rhyolites, tuffs, volcano-sedimentary rocks	Late Paleocene to Eocene	60–50 Ma
Angostura Colorada Fm.	Sandstones, siltstones, and mudstones, tuffs, continental red beds	Late Cretaceous	
"Comallo beds"	Volcano-sedimentary complex	Triassic to Late Jurassic	
Crystalline basement	Granitic, granitoid, and metamorphic rocks	Middle to Late Paleozoic	

Rucá shows the existence of several remnants of an extended high plain, carved by fluvial erosion. It is a smoothly undulating plain with very rounded summits, which lie around 1,200–1,300 m a.s.l., developed on top of the crystalline basement (Fig. 13) and the faulted-tilted, Triassic-Jurassic "Comallo beds" (Fig. 14). The occurrence of these Jurassic beds provides a maximum age for this planation surface. This paleo-landform shows towards the west and southwest the evidence of erosion action along its margins. The crystalline rocks are deeply weathered, indicating an ancient weathering front. This weathering front is unconformably covered by the red beds of the Angostura Colorada Fm. (Late Cretaceous) (Fig. 15). Many small paleo-depressions were also formed, probably as irregularities in the ancient weathering front after extensive denudation. These depressions were much later filled up by the tuffaceous, volcanic, and sedimentary rocks of the Ventana Formation (Paleocene-Eocene). The reconstruction of the ancient relief indicates that these depressions, probably related to the Cretaceous weathering front, and wide, extensive fluvial valleys of low gradient were draining westwards, as the crystalline basement is gradually disappearing in this direction. The pre-Tertiary drainage towards the Pacific Ocean had been already noted by Groeber (1929, p. 66). The volcanic rocks of the Ventana Fm. extended well beyond the margins of their marine depositional basin along the shallow Pacific Ocean coast and covered the crystalline basement and the Cretaceous rocks, preserving the pre-Late Paleocene landforms. This may be clearly observed in the surroundings of the town of

Fig. 13 Inselberg developed on Permian granites, south of Pilcaniyeu, province of Río Negro

Fig. 14 The "Comallo beds" volcano-sedimentary complex (Triassic-Jurassic) exposed in the foreground. The Paleozoic granites and metamorphic rocks appear at the background. The Late Mesozoic planation surface equally cuts through all these ancient units

Fig. 15 Continental sedimentary rocks of the Angostura Colorada Fm., Late Cretaceous, lying on top of the Comallo paleosurface, the Late Mesozoic planation surface, probably an etchplain

Pilcaniyeu and other hills of the region, where isolated granitic and metamorphic hills have been buried by radially dipping tuffs of the Ventana Fm., presently being exhumed.

In summary, the older lithological complex integrated by the crystalline basement and the Jurassic rocks was affected by regional faulting, thus generating large uplands and depressions, which later controlled the Cenozoic landscape. The whole set was actively weathered under very warm and wet climate, which accounts for the remains of the ancient weathering front. Later, this etchplain was covered by Cretaceous sediments and partially exhumed during the Early Tertiary, with prolonged denudation of the saprolite debris and in-filling of the preexisting depressions by the Early Tertiary pyroclastic and volcanic rocks. Extensive denudation during the Late Cenozoic has exhumed the very large remnants of the ancient etchplain, presently found in the landscape.

Conclusions

Field evidence indicates that the rocky materials have been exposed to an intense and extensive erosion processes. As a consequence, a very large planation surface developed. The landforms and patterns of this landscape show, both at the small and large scales, the influence of the geological structure, generating pathways for the weathering processes. The interaction between a series of different climatic events over extensive time periods, the alteration processes of the rocky masses, and the erosion and transport of the regolith are responsible for the surficial morphogenesis.

The analysis of the relationship between relief, saprolite, and rock cover through time suggests that the most important factor for the differentiation of the present relief is the time of exposure of the land surface.

This paleosurface would have basically been developed in two phases. The first one would have started, though locally, to evolve in the Late Permian or Triassic, partially affecting the granitoid rocks of the Mamil Choique Formation (Early Permian), the Los Menucos Group of Late Triassic age, and the dioritoid bodies of the middle Jurassic. This period corresponds to the start of weathering attack by atmospheric action, in which the exposed rock was transformed and disaggregated, a process that developed an important alteration mantle, which buried the fresh rock under its own saprolite. The weathering was directly dependent upon the climatic conditions. Paleoclimatic studies reveal that the climate during the latest Paleozoic and the Mesozoic was very appropriate for the alteration of the granitic masses, due to very high temperature and moisture. These two factors are key issues in these processes, and the planation process culminates in the Late Mesozoic, a period almost 100 Ma long, from the Early Jurassic to the Early Cretaceous.

Finally, a period of remobilization occurred, and the surface was exposed by denudation of the regolith. The most plausible explanation for the mobilization of the regolith would be a crustal uplifting produced during the Andean tectonics, where the Tertiary tectonics would have elevated the base level and reactivated the erosion processes.

As this surface is not totally flat, but presenting instead irregularities of a varied degree, the alteration material has not been entirely removed, but still persists in the landscape, which contains altered and fresher rocks, separated by the weathering front.

Thus, the origin of this paleosurface is attributed to deep chemical weathering, with later removal of the regolith, producing an etchplain (Wayland 1934).

The precisely determined age of these elements of ancient landscapes and their stratigraphic position reinforce the hypothesis that a large part of the Northern Patagonia Massif was a positive element of the landscape for a very long time. Due to the action of different erosion agents, including deep weathering, the paleosurface has persisted with little change over very extensive time periods. The characteristics of these ancient landscapes are not compatible with models proposed for the evolution of the Patagonian landscape that start from the uplift of the Andean Cordillera. It is necessary to extend the time scale needed for the genesis of this landscape and provide a new model for the evolution of the ancient landscape of the North Patagonian Massif. The chronology of the series of events that this model implies clearly widens our understanding of the present landscape and puts it in the context of global tectonics and possibly enables correlation with planation surfaces in other continents.

Acknowledgments The authors want to express their deepest gratitude to Professor Cliff Ollier for his valued comments and observations, both in the field in 2001 and on a preliminary version of this manuscript.

References

Aguilera EY (2006a) Identificación y distribución de distintas superficies geomorfológicas de escaso relieve local, por medio de sensores remotos en el área del Macizo Nordpatagónico, Provincia de Río Negro. Unpublished doctoral Thesis, Facultad de Ciencias Naturales y Museo de La Plata, Universidad Nacional de La Plata, Argentina, 285 pp

Aguilera EY (2006b) Superficie de planación Gondwánica en el Macizo Nordpatagónico. III Congreso Argentino de Cuaternario y Geomorfología, Abstracts, Córdoba, Argentina

Aguilera EY (2007) Teledetección y SIG aplicados a la regionalización geomorfológica en el Macizo Nordpatagónico. In: SIMPÓSIO BRASILEIRO DE SENSORIAMENTO REMOTO, 13, (SBSR), 2007, Florianópolis. Anais. São José dos Campos, INPE, pp 1935–1942. CD-ROM, On-line. ISBN 978-85-17-00031-7

Aguilera EY, Rabassa J (2010) Origin of the Northern Patagonian Massif regional paleosurface. Geociências 29(4):467–478. São Paulo, UNESP

Aguilera EY, Aragón E, Carretero S (2010) The paleosurface on the Paso del Sapo volcanic rocks, Chubut, Argentina. Geociências 29:479–486. São Paulo, UNESP, Brazil

Aragón E, Carroll AR, Aguilera EY, Cavarozzi CE (2005) Ancient Gondwana landforms in Patagonia. Gondwana, 12, "Geological and Biological Heritage of Gondwana" Abstracts, Academia Nacional de Ciencias, Mendoza, p 46

Arrondo OG, Artabe AE, Morel EM, Petriella B (1984) Floras fósiles Mesozoicas. IX Congreso Geológico Argentino (San Carlos de Bariloche). Relatorio 2(1):367–372

Cicciarelli MI (1989) Análisis estructural del sector oriental del Macizo Nordpatagónico y su significado metalogénico. Unpublished doctoral thesis, Facultad de Ciencias Naturales y Museo de La Plata, Universidad Nacional de La Plata, Argentina, 179 pp

Derrau M (1970) Geomorfología. Editorial Ariel, Barcelona, 442 pp

Giacosa R, Lema H, Bosteros A, Zubia M, Cucchi R, Di Tomasso (2007) Estructura del Triásico de la región norte del Macizo Nordpatagónico (40°–41° S, 67°30′–69°45′O), Río Negro. Revista Asociación Geológica Argentina [online] 62(3):355–365. ISSN 1851-8249

Groeber PF (1929) Líneas fundamentales de la geología del Neuquén, Dir. Gral. Minas, Geol., Hidrol., Buenos Aires, Publ. 58. Min. de agric. de la Nación, Buenos Aires

Hinojosa LF, Villagrán C (2005) Did South American Mixed Paleofloras evolve under thermal equability or in the absence of an effective Andean barrier during the Cenozoic? Palaeogeogr Palaeoclimatol Palaeoecol 217(1–2):1–23

Kay SM, Ardolino A, Franchi M, Ramos V (1993) Origen de la Meseta de Somún Cura: distribución y geoquímica de sus rocas volcánicas máficas. 12º Congreso Geológico Argentino and 2º Congreso de Exploración de Hidrocarburos, Actas, 4, pp 236–248

Llambías EJ (2001) Geología de los Cuerpos Ígneos. In: Serie de Correlación Geológica 15, Instituto Superior de Correlación Geológica, INSUGEO, San Miguel de Tucumán, 232 pp

Llambías EJ, Rapela CW (1984) Geología de los complejos eruptivos de La Esperanza, provincia de Río Negro. Revista de la Asociación Geológica Argentina 39(3–4):220–243

Ollier C (1984) Weathering, 2nd edn. Longman, Essex

Ollier CD (1991) Ancient landforms. Belhaven, London

Pothe de Baldis ED (1984) Microfloras fósiles Mesozoicas. 9° Congreso Geológico Argentino (San Carlos de Bariloche). Relatorio 2(4):393–411

Rabassa J (1974) Geología superficial en la región de Pilcaniyeu-Comallo, provincia de Río Negro. Unpublished doctoral thesis, Universidad Nacional de La Plata

Rabassa J (1975) Geología de la región de Pilcaniyeu-Comallo, provincia de Río Negro, Argentina. Fundación Bariloche, Publicaciones del Departamento de Recursos Naturales y Energía, N° 17, San Carlos de Bariloche

Rabassa J (1978a) Paleorrelieves cenozoicos en la región de Pilcaniyeu-Comallo, provincia de Río Negro, Argentina. Actas 7° Congreso Geológico Argentino, Neuquén, pp 77–87

Rabassa J (1978b) Estratigrafía de la región de Pilcaniyeu-Comallo, provincia de Río Negro, Argentina. Actas 7° Congreso Geológico Argentino, Neuquén

Rabassa J, Carignano C, Cioccale M (2010) Gondwana paleolandscapes of Argentina: an introduction. Geociências 29(4):439–466. São Paulo, UNESP

Rabassa J, Carignano C, Cioccale M (2014) A general overview of Gondwana landscapes in Argentina. In: Rabassa J, Ollier C (eds) Gondwana landscapes in southern South America. Springer, Dordrecht, pp 201–245

Ramos V (1999) Las Provincias Geológicas del Territorio Argentino. Instituto de Geología y Recursos Minerales. Geología Argentina. Anales 29(3):41–96. Buenos Aires

Ramos V, Cortés JM (1984) Estructura e interpretación tectónica. IX Congreso Geológico Argentino (San Carlos de Bariloche). Relatorio 1(12):317–346

Rapela CW, Alonso G (1991) Composición litológica y geoquímica del batolito de la Patagonia central. VI Congreso Geológico Chileno, Actas, 1, Viña del Mar, pp 236–240

Romero EJ, Dibbern MC (1984) Floras fósiles Cenozoicas. 9° Congreso Geológico Argentino (San Carlos de Bariloche). Relatorio 2(2):373–382

Thomas MF (1978) The study of inselbergs. Zeitschrift für Geomorphologie. N. F., Suppl. Bd. 31:1–41. Berlin

Twidale CR (1982) Granite landforms. Elsevier, Amsterdam. Pub. Com, 372 p

Twidale CR (1985) Old land surfaces and their implications for models of landscape evolution. Revue Géomorphologie Dynamique 34:131–147

Twidale CR (1987) Etch and intracutaneous forms and their implications. Aust J Earth Sci 34: 367–386

Twidale CR (1989) La iniciación subsuperficial de las formas graníticas y sus implicaciones en las teorías generales de la evolución del paisaje. Cuadernos Laboratorio Xeologico de Laxe, 13, Coruña, Spain, pp 49–68

Twidale CR (1990) The origin and implications of some erosional landforms. J Geol 98:343–364

Vidal Romaní JR (1989) Granite geomorphology in Galicia, (NW Spain). Cuadernos Laboratio Xeologico de Laxe, 13, Coruña, Spain, pp 89–163

Vidal Romaní JR, Twidale CR (1998) Formas y paisajes graníticos. Universidade da Coruña, A Coruña, Spain, 411 pp

Volkheimer W (1984) Microfloras fósiles Mesozoicas. IX Congreso Geológico Argentino (San Carlos de Bariloche). Relatorio 2(3):383–391

Wayland EJ (1934) Peneplains and some erosional platforms, Geological survey of Uganda, Annual report and bulletin. Protectorate of Uganda Geological Survey, Department of Mines, Kampala, pp 77–79

The Rhyolitic Plateau of the Marifil Formation (Jurassic): A Gondwana Paleosurface in the Southeastern Portion of the Northern Patagonia Massif

Oscar A. Martínez and Jorge Rabassa

Abstract Along the southeastern border of the Northern Patagonian Massif of the provinces of Río Negro and Chubut, an extensive surface is presently called the "Rhyolitic" or "Ignimbritic Plateau." This large geomorphological unit has a geographical extension which exceeds 50,000 km^2 and it is located between 40°30′ and 44° lat. S and between the Atlantic Ocean coast and 67°30′ long. W. It is characterized by a smooth topography of low and rounded hills, shallow endorheic basins, and a poorly integrated drainage network. The drainage network is mostly nonfunctional and roughly coincident with the bedrock fracture system. Bedrock is almost exclusively composed of the acid volcanic and pyroclastic rocks of the Marifil Formation of Early to Middle Jurassic age. A significant proportion of the identified positive landforms present form and nature very similar to that of "bornhardts," as defined by Twidale (Revista de la Asociación Geológica Argentina 62(1):139–153, 2007), basically for granites. Bornhardts are uncovered dome hills (Twidale, Revista de la Asociación Geológica Argentina 62(1):139–153, 2007) which are usually frequent in Gondwana landscapes (Fairbridge, Encyclopedia of geomorphology. Ronald, New York, 1968). Furthermore, the ubiquitous presence of "corestones" (isolated, large, rounded boulders), which are taken as indicators of an ancient, deep weathering front, supports the hypothesis that these paleosurfaces were generated by long-term, intense chemical weathering processes. The deep weathering would have occurred over at least 25 Ma, between the Middle and Late Jurassic, under a hot and moist paleoenvironment and under extremely stable

O.A. Martínez (✉)
Universidad Nacional de la Patagonia-San Juan Bosco, Sede Esquel, Esquel, Chubut, Argentina
e-mail: oscarm@unpata.edu.ar

J. Rabassa
Laboratorio de Geomorfología y Cuaternario, CADIC-CONICET, Ushuaia, Tierra del Fuego, Argentina

Universidad Nacional de Tierra del Fuego, Ushuaia, Tierra del Fuego, Argentina
e-mail: jrabassa@gmail.com

J. Rabassa and C. Ollier (eds.), *Gondwana Landscapes in southern South America*, Springer Earth System Sciences, DOI 10.1007/978-94-007-7702-6_16,

tectonic conditions. The mobilization, denudation, and later sedimentation of the regolith/saprolite formed under such conditions would have taken place during several erosion episodes, mostly under tectonic forcing, between the Late Jurassic and the Late Cretaceous. The important clay and other secondary mineral accumulations (some of them significant sources of uranium) in the region would have a direct genetic relationship with the development of these paleosurfaces. From the Late Miocene onwards, the colder and drier conditions that were imposed in the region by the uprising Andes and the establishment of mountain glaciers and ice caps during numerous glaciations allowed the modification of this landscape by hydro-eolian processes which generated the widely distributed endorheic depressions (locally known as "bajos sin salida") by deflation and occasionally reworked the surviving rocky hills by abrasion.

Keywords Gondwana • Argentina • Ignimbrite plateau • Etchplains • Granitic geomorphology

Introduction

Extensive paleosurfaces remnants, ranging in age from the Jurassic to the Early Tertiary, are found in cratonic areas of Argentina (Rabassa et al. 2010), such as the Sierras Pampeanas of central and northern Argentina, the Tandilia and Ventania ranges of Buenos Aires province, the Northern Patagonian Massif, and the Deseado Massif. They have been found even in the Malvinas/Falklands islands (Clapperton 1993). These paleosurface remnants are developed over igneous and metamorphic rocks and their origin has been related to intense, deep chemical weathering during long periods of tectonic stability (Rabassa et al. 1996, 2010). These landscapes are characterized and identified by the presence of peneplains, etchplains, and pediplains and smaller scale landforms such as dome hills (bornhardts), inselbergs, corestones, rocking stones, weathering front and weathering profile remnants, and duricrusts, such as ferricretes, silcretes, and calcretes. Some of these features may be found also in similar landscapes of South America and South Africa, in times when both continents were still united or very close to each other in the Gondwana supercontinent (Rabassa 2010). For this reason, these paleolandscapes may be referred to as Gondwana surfaces, as they were described by Fairbridge (1968).

The results of some preliminary observations in a remote area of Argentine Patagonia are presented in this chapter. Work has been developed along the Jurassic Ignimbritic Plateau (Malvicini and Llambías 1974), an extensive geological unit that is part of the basement of the Northern Patagonian Massif (Fig. 1) corresponding to the outcropping volcanics of the Marifil Formation. In the framework of this chapter, these outcrops have been grouped in two main areas. The northern area (Fig. 2) extends between the town of Sierra Grande in the NE, the Somuncurá basaltic plain in the NW and W, the Bajo de la Tierra Colorada in the SW, and the Sierra Chata in the S (approximately at 42°50′ S). The southern area is roughly coincident with a section of the Río Chubut valley (Fig. 3) and it extends between the towns of Las Plumas and Dolavon, in the surroundings of the Florentino Ameghino Dam.

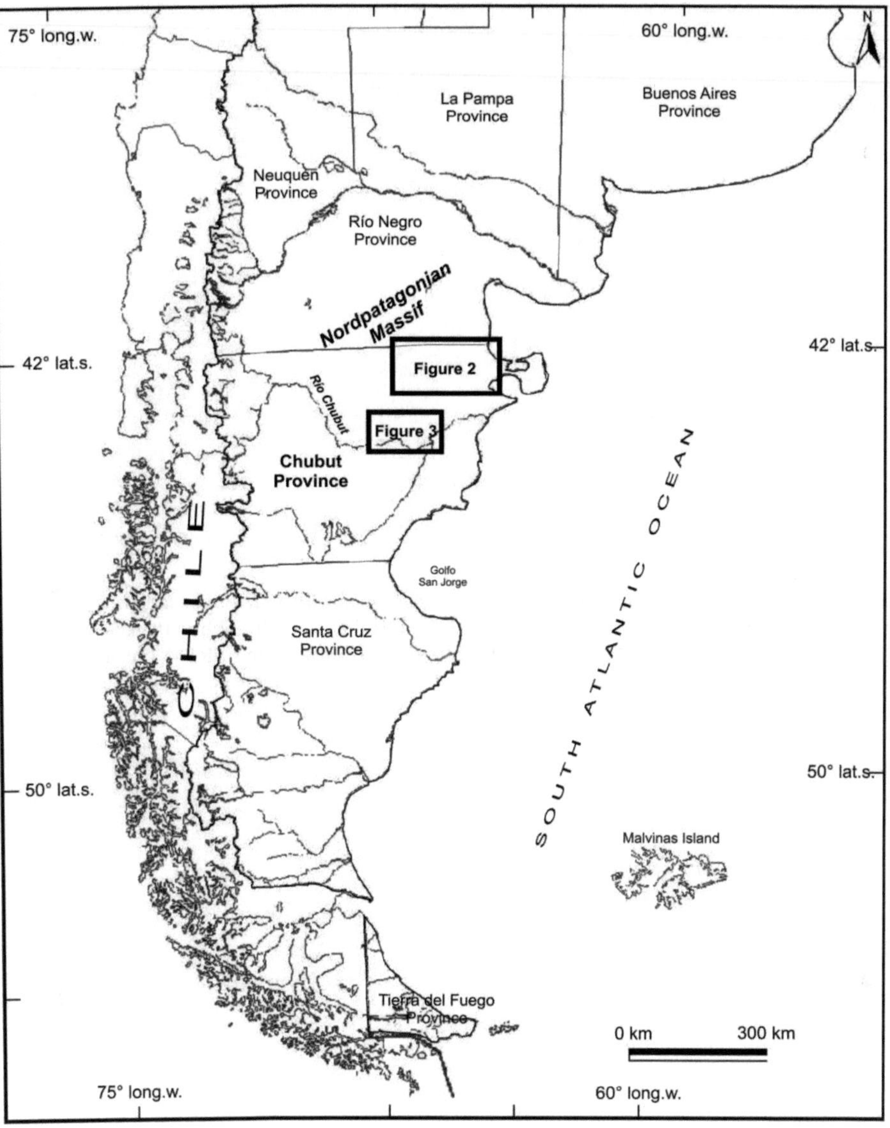

Fig. 1 Map of Patagonia, in which the position of the Northern Patagonian Massif is indicated. The two study areas are depicted in boxes corresponding to Figs. 2 and 3

There are many important previous contributions related to the nature of the geological units found in the area. See, for instance, Stipanicic et al. (1968), Malvicini and Llambías (1974), Ramos (1975), Aliotta et al. (1977), Haller (1978), Cortés (1981a, b), Panza et al. (2002), and Sacomani and Panza (2007). Others have described the magmatic events (Malvicini and Llambías 1974; Llambías et al. 1984;

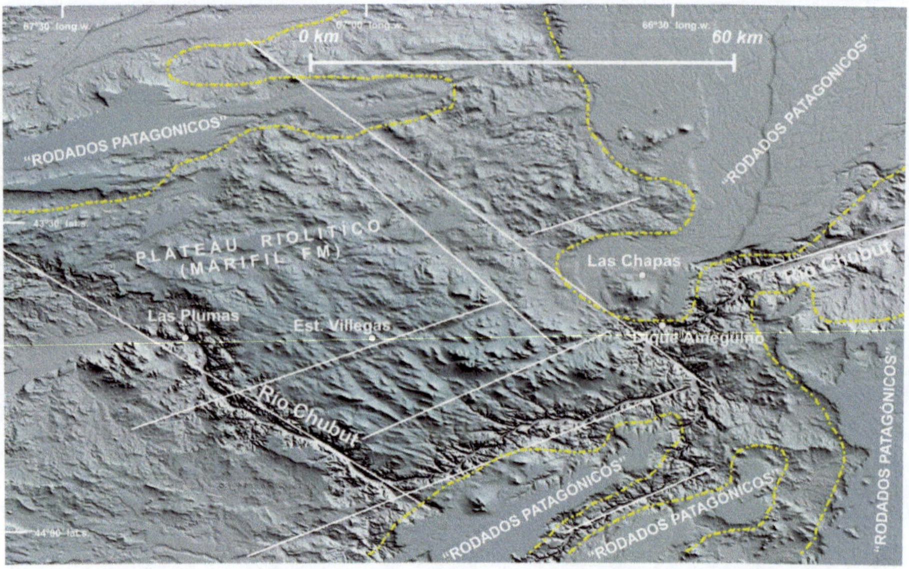

Fig. 2 Northern sector of the study area. The *dashed lines* indicate the boundaries of the Marifil Formation outcrops, which are named as the "Plateau Riolítico" or "Rhyolitic Plateau". The main tectonic alignments have been represented with *straight lines* on the SRTM image. "Rodados Patagónicos" are fluvial and pediment gravels and sands forming tablelands of Late Cenozoic age

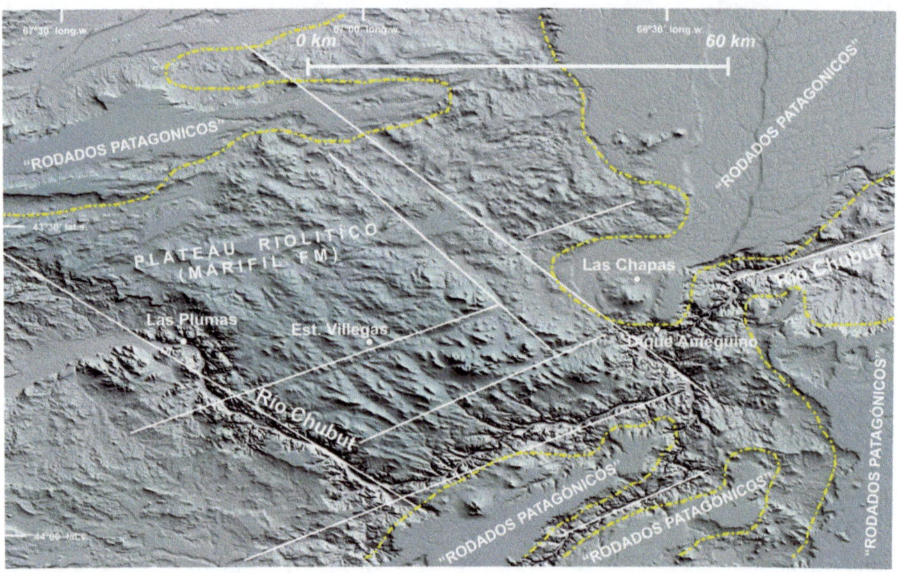

Fig. 3 Southern sector of the study area that corresponds to the lower valley of the Río Chubut. The approximate boundaries of the outer outcrops of the Marifil Formation (the southernmost portion of the Rhyolitic Plateau) have been indicated with *dashed lines* on the SRTM image. For the "Plateau Riolítico" and "Rodados Patagónicos" terms, see Fig. 2

Rapela and Pankhurst 1993; Aragón et al. 1996), the geochronology (Malvicini and Llambías 1974; Rapela and Pankhurst 1993; Pankhurst and Rapela 1995; Alric et al. 1996), and the structure and tectonics of the region (Cortés 1981b). Some other works relevant to the objectives of this chapter are those which have described and genetically interpreted the clay mineral deposits of the area (Angelelli and Stegman 1948; Olivieri and Terrero 1952; Hayase 1969; Hayase and Maiza 1969; Romero et al. 1974, 1975; Maiza and Hayase 1980; Cravero et al. 1991; Domínguez and Murray 1995) or the uranium mineral deposits that lie upon or near the Rhyolitic Plateau outcrops. The origin of these outstanding landscapes has not been studied previously and this chapter is the first scientific geomorphology paper on these topics, in this area.

As a conceptual framework, this chapter relates to the regional and extra-regional works of Rabassa (1978), Rabassa et al. (1996, 2010), and Aguilera and Rabassa (2010).

Geological Units of the Study Area

The eastern and southeastern slope of the Northern Patagonian Massif presents a relatively simple geology. The sequence starts with an igneous-metamorphic basement of Precambrian-Paleozoic age, represented by very few outcrops. The basement is overlain by a quite thick and extensive volcanic/pyroclastic cover, composed by the Marifil Formation (Malvicini and Llambías 1974), also known as the Marifil Complex (Cortés 1981a). This unit includes subordinate lava flows, with dominant pyroclastic and sub-volcanic units of acidic composition, mostly of Early to Middle Jurassic age, which outcrop in eastern Patagonia from southern Río Negro province (40° 30′ S) up to Bahía Bustamante (45° S) and which would probably extend under the sea into the Gulf of San Jorge (Fig. 1), covering an area of over 50,000 km^2 (Page et al. 1999). Due to the abundance of several ignimbritic varieties and the general mantle-like form, these enormously extensive outcrops are referred to as the Jurassic Ignimbritic Plateau or the Rhyolitic Plateau. The maximum thickness that has been mentioned for this unit is around 900 m (Malvicini and Llambías 1974; Page et al. 1999). It has been suggested that the emplacement mechanics of this large unit is related to a fissure-type model (Malvicini and Llambías 1974), to an intense volcanic activity through multiple volcanic vents (Llambías et al. 1984), and to the eruptive activity of wide calderas (Aragón et al. 1996). In all cases, the volcanic events would have taken place coevally during the initial stages of the opening of the southern Atlantic Ocean (Ramos 1983). Absolute radiometric dating obtained in several localities of the region (Cortés 1981a; Rapela and Pankhurst 1993; Pankhurst and Rapela 1995; Alric et al. 1996) has yielded minimum ages of 177 Ma for these rocks (Las Plumas locality, Fig. 3) and maximum ages around 183/187 Ma (in the area of Arroyo Verde, Fig. 2), which has defined an eruptive period in the region of approximately 10 Ma, with decreasing ages from N to S.

Fig. 4 Outcrops of the continental sandstones of the Chubut Group nearby the town of Telsen (Fig. 2)

Cortés (1981a) has proposed that this volcanism evolved from NW to SE. This same author has noted the intense fracturing of these outcrops, forming blocks of different dimensions and varied vertical and horizontal displacement. During the Middle and Late Cretaceous, the sedimentary rocks of the Chubut Group accumulated, corresponding to whitish ashfall tuffs, yellowish and reddish sandstones, light brown conglomerates, and grayish and greenish mudstones (Franchi et al. 2001, Fig. 4). According to Codignotto et al. (1978), these rocks would be distributed, with rather small thickness, from the west to the east with decreasing age. According to Page et al. (1999), they are just a very thin but extensive cover. On the other hand, Labudía et al. (2011) confirmed a thickness of 750 m for the same area (in a tectonic block south of the town of Telsen, Fig. 2), which increases towards the south and southeast, due to the inclination of bedrock in that direction (Cortiñas 1996). The geological section continues with the deposits corresponding to the La Colonia, Roca, and Salamanca formations, products of Eocene marine transgressions from the Atlantic Ocean during the Maastrichtian/Danian. The Atlantic transgressions during the Eocene and the Neogene are also represented in the region. A distinctive characteristic of the Northern Patagonian Massif is basaltic volcanism, ranging in age from the Eocene to the Miocene. Aragón et al. (2011a, b) believed this plateau (the basaltic plain or Somuncurá Meseta), located to the northeast of the study area with a mean altitude of around 1,200 m a.s.l., was formed before the extrusion of the Somuncurá basalts (30 Ma, Early Oligocene, Fig. 2).

Structure and Tectonics

A brief description of the post-Paleozoic geodynamics in the region is needed to allow the genetic interpretation of the regional landscape, taking into consideration, especially, that the present surficial aspect of the Rhyolitic Plateau is, to a large extent, the result of tectonic and epeirogenic differential movements of the various blocks that form it (Fig. 2). These displacements (uplifting, downwarping, tilting) have favored the erosion processes but have also contributed to the preservation of ancient rocks and landforms.

The structure of the area is the product of the action of different diastrophic phases corresponding to the Patagonian (Mesozoic) and Andean (Neogene) orogenic events. During the Early and Middle Jurassic, block fracturing, thinning, and fusion of the lower crust, associated to the initial phases of the southern Atlantic Ocean opening, took place in the region when these terrains were located in southwestern Gondwana (Ramos 1999). These tensional and transtensional movements permitted the extrusion of the ignimbrites and acid volcanic rocks that formed the Ignimbritic Plateau. Some of the regional fractures corresponding to this tectonic event, possibly of extra-regional scale (Coira et al. 1975; Panza et al. 2002; Sacomani and Panza 2007), were reactivated during the later tectonic or epeirogenic episodes and have a usually surficial expression as alignments and fracturing zones.

A significant tectonic episode for the region was the Araucanian phase (or intra-Malm movements) which took place in the Late Jurassic (Kimmeridgian). This extensional/transtensive deformation was caused by a fracturing process that alternatively elevated and downwarped the blocks of the pre-Jurassic basement and the Jurassic volcanic cover. These movements forced the Northern Patagonian Massif to become a positive element since the Early Cretaceous, which divided two important depositional centers of continental accumulation to the north and south, with red beds facies. The sediments that were deposited in the southern sector, that is, in the area surveyed in this chapter, are now the Chubut Group sedimentary rocks, deposited between 110 and 90 Ma, Middle to Late Cretaceous.

The Huantraicoan phase, that is, the inter-Senonian movements or main Patagonian phase that took place during the Campanian (74 Ma), was of high relevance in the formation of the physiography of the region. This compression reactivation locally folded the Cretaceous sedimentary cover (Panza et al. 2002). It also exhumed ample sectors of the volcanic plateau and thus generated the conditions for the erosion of a large portion of the Chubut Group sedimentary rocks, whereas other remnants were preserved in the downwarped blocks. Another consequence of these tectonic events is the inversion of the regional slope which, since that moment, tilted towards the Atlantic Ocean (Page et al. 1999). Next came deposits corresponding to the accumulation of the marine transgressions that took place during the Maastrichtian/Danian. The block located south of the Sierra Chata alignment (Fig. 2) is, according to Page (1987), a downwarped basement block covered by the Chubut Group, the Maastrichtian/Danian sedimentary rocks, and a thin veneer of the "Rodados Patagónicos" (i.e., the "Patagonian Gravels," Pliocene-Pleistocene). The main structural features of the area have been described by several authors (Windhausen 1918; Lapido and Page 1979; Haller 1981; Lapido 1981; Cortés 1981a, b; Page 1987; Franchi et al. 2001; Panza et al. 2002; Sacomani and Panza 2007). These features are horsts and grabens, limited by fractures, fracture belts, and alignments, some of them of regional scale (for instance, the fracturing belt of Cona Niyeu and the Sierra Chata, El Moro, and Telsen alignments, Fig. 2). The blocks present, almost without exception, positive thresholds and steps, which demonstrates the complexity of the mechanical response of these large rock units during the different diastrophic events. Thus, the distinct blocks and subblocks that form the Rhyolitic Plateau presently show different dip values and directions. This is the result of not only the rising or lowering processes of each block but also tilting.

Description and Interpretation of the Rhyolitic Plateau Surface

The Rhyolitic Plateau has a great physiographic homogeneity and it is characterized by the presence of rounded hills (Fig. 5), of low elevation since the topographic amplitude in the area rarely exceeds 250 m. Only occasionally, the intense jointing generates a rough relief with acute crests and high rocky walls (Haller 1981). The hills are bounded by a drainage network that, at the present time, has an ephemeral régime. Its texture is dense, although poorly integrated. In some sections of the existing channels, this fluvial system is practically inactive and shallow; endorheic closed basins have appeared (locally called "bajos sin salida"), with depths of only 5–6 m at most, and diameters of a few hundred meters to a few kilometers, which act as local base level for a large number of tributaries (Fig. 6). These frequent depressions, whose bottoms are some meters below the thalweg of the linked fluvial channels (Fig. 7), have an essentially eolian genesis and have developed after the dismantling of the fluvial network. The landscape shows a strong structural control with adjustment of the fluvial network to the general dense fracturing of the rocks (Fig. 6). This specific aspect of the Marifil Formation outcrops shows a large contrast with those in the region which are including other younger geological units, such as the Tertiary basaltic flows (Somuncurá basaltic plateau) or the wide plains of the Tertiary/Quaternary "Rodados Patagónicos" (Figs. 2 and 3).

Several authors have discussed the origin of these surfaces. Chebli et al. (1976) mentioned that the Chubut Group unconformably overlies a marked paleolandscape eroded on the Marifil Formation. According to Codignotto et al. (1978), the sedimentary rocks of the Chubut Group accumulated over a relative relief which it is assumed was smooth to moderate. Lapido (1981) described these outcrops as an exhumed regional erosion surface and assumed that the exhumation was in times younger than the Paleocene. In the surroundings of Sierra Chata (Fig. 2), over Jurassic acid porphyritic rocks, Haller (1981) identified rock pillars of more than 2 m of local relief (Fig. 8) which he assigned to differential weathering

Fig. 5 Overview of the Sierra Colorada (Fig. 2), a small hill range of domed shape composed of the characteristic rhyolites and acid porphyritic rocks of the Marifil Formation

controlled by jointing (corestones) and concluded that the overlying weathered debris would have been eliminated by deflation and fluvial erosion. Franchi et al. (2001) found evidence that the rigid basement, which includes the Marifil volcanic rocks, was thoroughly eroded up to the degree of regional peneplanation. Panza et al. (2002) and Sacomani and Panza (2007) described the outcrops of the Marifil Formation located near the Ameghino Dam (Fig. 3) and agreed with the interpretation of other authors (Lapido 1981; Ardolino and Franchi 1996) that these surfaces correspond to an exhumed or resurrected "peneplain," concluding that such exhumation still continues today. The concept of "exhumed peneplain" was coined by González Díaz and Malagnino (1984) as they described large sectors of the province of Río Negro located in the periphery of the Northern Patagonian Massif (Fig. 1), both in the areas near the Andean Cordillera and those closer to the Atlantic Ocean coast, thus including, obviously, a good portion of the Marifil Formation outcrops (mainly those located north of 42° S) studied in this chapter (Fig. 2). These authors proposed the name "Exhumed Peneplain of Río Negro" for "a landform (eroded on granitic and volcanic rocks) which corresponds with the final stage of the evolution of the Davisian cycle and which has been exhumed later on" (free translation by the present authors). These landforms would have been developed in the period between the end of the Triassic and approximately

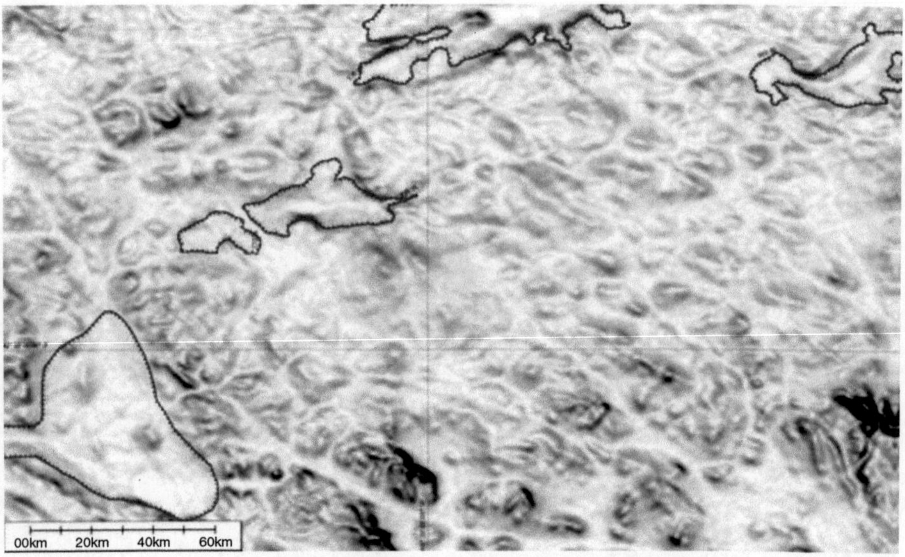

Fig. 6 Radar image (Shuttle Radar Topography Mission) of northwestern Sierra Colorada (Fig. 2), processed with the Global Mapper program. The rounded, domed shape of the Marifil Formation outcrops and the marked adjustment of the drainage network to the intense fracturing of the area are observed. Some of the larger endorheic depressions ("bajos sin salida") are indicated with a *line*, integrating the surficial drainage network

the Cretaceous-Tertiary boundary. They assigned their preservation to the large resistance of the rocks over which the erosional processes acted and the later burial by younger sediments. The geomorphology of the crystalline basement, studied by González Díaz and Malagnino (1984), in the northeastern part of the Northern Patagonian Massif, is reinterpreted by Aguilera and Rabassa (2010) who considered that this wavy landscape, with occasional rounded hills of very small relief, developed over an area of more than 20,000 km^2, is the product of deep weathering and later erosion of the regolith/saprolite levels during an extensive period of tectonic and climatic stability, which took place between the Permian-Triassic boundary and the Middle Cretaceous. Thus, this paleosurface would be an etchplain or deep weathering surface, not a peneplain in the sense of the Davisian cycle, and which would correspond to the Gondwana paleolandscape recognized in other localities of the Southern Hemisphere (for detailed references, see Rabassa (2010) and Rabassa et al. (2010)).

The preliminary character of this chapter includes a general survey of the huge extension of the Rhyolitic Plateau supported by detailed observations focused in a smaller, representative area, the Sierra Colorada (Figs. 2, 5, and 6). This range has an extension of 4 km by 1 km, with a general orientation of N 320° (Fig. 9). It presents a domed morphology (Figs. 5, 10, and 11), of thoroughly eroded aspect, although without erosion features which could be clearly correlated with the rounded shape

Fig. 7 Outcrops of the Marifil Formation rhyolites appear in the foreground, whose surface presents a typical scaly aspect and over which small basins have been developed, showing evidence of present frost weathering and deflation erosion. At the background, a hollow generated by deflation (essentially during the glacial periods) in the channel of a nonfunctional drainage line is observed

of the hills. Moreover, the cracks that act, in some cases very clearly (Fig. 11), as surficial drainage lines and are coincident with rock fractures and joints do not provide a convincing explanation the genesis of these landforms.

A relevant geometric feature of these hills is the absence of any defined pattern in orientation, shape, or inclination of their slopes (Figs. 6, 9, and 10). A consequence of this is that the observed morphologies cannot be related to the action of any particular agent, such as running water, wind, and glaciers, to establish the direction of displacement.

The occurrence of in situ weathered material represents unquestionable evidence of their origin by weathering of these landforms (Twidale 2007). In the study area, regolith accumulations in contact with these surfaces have not yet been found. However, considering the enormous extension of these outcrops and the reduced percentage of them which has actually been surveyed in detail, it may not be ruled out that such findings will be done in the future. On the other hand, the sedimentary deposits of clays and kaolin are very common in contact with the outcrops of the Marifil Formation rocks. These materials are typical products of chemical weathering. Anyhow, there are other elements that allow the suggestion

Fig. 8 Rock outcrops intensively modified by Quaternary eolian abrasion, though these ventifacts could have started even before their evolution, during the Neogene. Haller (1981) considered that these rocks correspond to a Jurassic volcanic unit, though older than the Marifil Formation, and that the landforms are in fact corestones produced by deep weathering

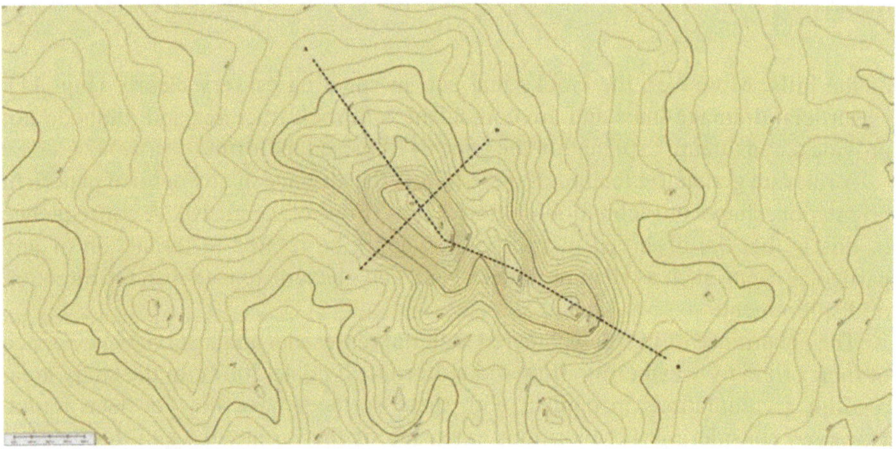

Fig. 9 Topographic map of the Sierra Colorada and longitudinal and transversal cross sections of this range, with exaggerated vertical scale

Fig. 10 Overview of the Sierra Colorada in which the domed morphology and the generally smooth surface that characterize most of the hills of the Rhyolitic Plateau are observed. Note the 3 fracture systems with vertical and subvertical inclination and the subhorizontal discontinuity planes. The latter seem to correspond with surfaces that separate the lava flow units

that the topography of the Rhyolitic Plateau actually corresponds to a paleosurface which represents an ancient deep weathering front. The domed morphologies are of high diagnostic value, many of which (including the Sierra Colorada) may be considered as similar to "bornhardts," being these landforms defined as "barren domed hills, with slopes which become steeper towards the base and which lack the development of debris cones, alluvial cones or soils, whose genesis has been controlled by the internal structure of the rock" (Thomas 1968) or as "rounded hills, usually isolated, generated under wet tropical conditions on massive rocks with fracture systems at large scale" (Allaby and Allaby 1999). These morphologies would be the result of a cycle in which two stages may be identified (Falconer 1911; Twidale 2007). A first stage took place when the compositionally homogeneous crystalline rocks (basically igneous and metamorphic rocks) underwent intense chemical weathering under conditions of extremely high moistures and moderate to high mean annual temperatures. The weathering front advances according to the cited climatic variables but also depending upon the texture, structure, and composition of the rock. The density and orientation of the fracture systems or any lithological discontinuities, such as joints, schistosity, or exfoliation, play a central role in the pattern that the weathering front developed in depth, because the front becomes progressively deeper in the sectors appearing more densely fractured. The depth achieved by the weathering front would be larger, proportionally to the

Fig. 11 Outcrops of the Marifil Formation in the southern end of the Sierra Colorada with their characteristic domed shape. A subhorizontal plane which seems to be the contact between two flow units may be observed. To the right, such a plane is intercepted by the topographic surface which indicates that the latter was generated clearly after the volcanic event

time along which the weathering agents acted. This means that a direct relationship exists between the thickness of the regolith/saprolite as a product of the epigenetic alteration and the tectonic and climatic stability of the area. The second stage of this cycle implies the mobilization of the weathered materials, thus finally leading to the exposition of the deeper unweathered rocks. This second stage starts immediately after a major tectonic and/or climatic change. The soils and vegetation cover existing up to that moment are stripped away by running water, mass movements, or wind starts as the saprolite is exposed to the agents. Finally, this stage ends with the full exhumation of the weathering front and the exposure of the etchplain. If the climatic/environmental conditions which took place after these events, usually of tectonic or epeirogenic nature, would be appropriate, as it was the case in extra-Andean Patagonia during the Late Cenozoic, these smoothly rounded landscapes would have been preserved until present times without major changes. The climatic conditions in Patagonia during the Cenozoic and particularly from the Early Oligocene onwards were never as warm and wet so as to reproduce the conditions that allowed the formation of the etchplain. Thus, all landforms and minor features associated with the etchplain must have been formed in the Late Mesozoic or, at most, in the Paleocene-Eocene, but they are never younger than this latter age.

A significant element that supports a weathering origin for the relief of the plateau is the presence of large rock boulders which clearly stand above the mean level of the surface. These boulders form, occasionally, pedestal rocks which seem to be disconnected from the bedrock but careful observations demonstrate that they are in situ and in contact with fresh rock (Fig. 8). As it was already proposed by Haller (1981), these conspicuous landforms are corestones generated by sub-superficial weathering that are later exposed when the overlying altered materials are eroded and denuded. The notable similarities existing between the landscape herein described here and those identified as Gondwana paleolandscapes studied in other regions of Patagonia and Argentina (see, for instance, Rabassa (1978), Rabassa et al. (1996, 2010), Aguilera and Rabassa (2010)) support an argument to assign a similar origin to the surface of the Rhyolitic Plateau of the provinces of Río Negro, Chubut, and a small area of eastern province of Neuquén, known as the "Sañicó Block."

The Gondwana Paleosurfaces of Patagonia and the Occurrence of Clay Deposits

The lower valley of Río Chubut (Fig. 3) is the most important kaolin-producing area of Argentina, with more than 250 kaolin quarries distributed over an area of approximately 1,000 km^2 (Panza et al. 2002). The kaolin deposits lie exclusively over the volcanic and pyroclastic rocks of the Marifil Formation. Although they occur with a very strong topographic control (Domínguez and Murray 1995), being preferably located in the more depressed parts of the Jurassic paleorelief (Romero et al. 1974; Panza et al. 2002), they are characterized by their higher horizontal development rather than vertical, forming conspicuous levels of important and continuous lateral extent. They underlie Tertiary sedimentary rocks belonging, for instance, to the Salamanca and Roca formations (Maastrichtian to Early Tertiary in age), and occasionally, they occur underneath the Cretaceous rocks of the Chubut Group, always separated by an erosional unconformity. In general, the base of the clay units is transitional, with the degree of clay formation processes diminishing downwards, until reaching the various facies of the Marifil Formation. The thickness of the layers of white kaolinite may vary between 1 and 6 m, although the total thickness of the alteration zone is of course much greater than these values (Panza et al. 2002). With the exception of some cases in which the clays are of clearly sedimentary origin, they do not show stratification. Structural control of the alteration is not observed, with the exception of very scarce and isolated situations, in which the clay formation process is coincident with faults or it is interrupted by them (Domínguez and Murray 1995). The evidence of tectonic displacement after the generation of these deposits is relevant because they help to establish their age. The more abundant mineral components are kaolinite, quartz, plagioclase, illite, montmorillonite, halloysite, sericite, and various types of iron oxide. Organic matter is frequently present.

Different opinions have been given about the origin of these accumulations. Angelelli and Stegman (1948), Olivieri and Terrero (1952), Aliotta et al. (1977), and Domínguez and Murray (1995) have postulated that they are the product of chemical weathering. These ideas were opposed by Hayase (1969) and Maiza and

Hayase (1975) who considered that they were formed (at least some of them) by hydrothermal alteration, although they recognized deposits of sedimentary type in the area. On the other hand, Romero et al. (1974) related the genesis of these accumulations to rock alteration due to the circulation of deep underground waters. Effectively, there are kaolin deposits of primary and sedimentary origin. The primary kaolin may be a product of weathering or hydrothermal origin. Kaolin of sedimentary origin, less abundant, is formed by erosion, transport, and deposition of the primary-type ones. They may present grain-size variations in a vertical direction (Maiza and Hayase 1980), cross-bedding (Romero et al. 1974), and other sedimentary structures that depict the water environment in which they were deposited. The base of these deposits is usually flat and clean, overlying without transition the Jurassic volcanic rocks. Another aspect that confirms their sedimentary origin is their location, without exception along the depressions or paleochannels of the Jurassic paleolandscape.

With respect to the primary deposits, either they are considered supergenetic or of hydrothermal origin, it is possible to recognize the textural features of the original rock which confirms that the clay formation processes took place in situ. Both processes are characterized for the development of zoning, which in the case of the supergenetic profiles shows a clear diminution of alteration with depth, although a similar situation occurs in some deposits considered to be of hydrothermal origin (Hayase 1969). The determination of the true origin using the mineralogical composition may be challenging because, for instance, kaolinite may be formed either by hydrothermal processes or by chemical weathering and that, in some circumstances, it is impossible to decide which of these processes is responsible for the clay formation processes (Grim 1968). In this respect, the works of Maiza and Hayase (1980) and Maiza et al. (2009), in some ore deposits in the province of Chubut, present a zoning of hydrothermal origin that is composed of a level of intense silicification, then a belt of alunite formation, continuing with a kaolinization zone that grades laterally to a sericite/chlorite sector that corresponds with the zone of lesser alteration. In contrast, the zoning due to weathering of the clays in the lower valley of the Río Chubut consists of a sequence that starts, at the upper part, with a level of white kaolinite that gradually changes downwards into a brownish-reddish kaolinite, occasionally with kaolinite veins, until reaching the lower levels of the unweathered volcanic rocks (Domínguez and Murray 1995).

In any case, the predominance of kaolinite with respect to any other alteration component is very characteristic for this region. This fact is depicted by Cravero et al. (1991). These authors concluded that the presence of kaolinite as the almost only neoformed mineral indicates excellent drainage conditions and intense lixiviation, characteristics of a humid climate. In the same sense, Besoain (1985) indicated that whatever the feldspar kaolinization process would be, a large water supply is always necessary. The composition of the parent material, the Marifil Formation, is also consistent with the almost exclusive generation of kaolinite. Romero et al. (1974) summarized the average mineralogical composition of these rocks in the study area which is composed of 33.2 % quartz, 33.8 % orthoclase, and 23.1 % acid plagioclase. The results of a regional study by Domínguez and Murray

(1995) indicate that the large majority of the kaolinite and clay accumulation were generated by weathering of the rhyolitic rocks. The hydrothermal origin is ruled out by these authors because of the absence of typical minerals found in these processes (sulfur, pyrite, pyrophyllite, alunite), because of the lack of structural control of the ore deposits (the usual ascending ways of the hydrothermal fluids), and because in none of the examples increase of the clay formation processes with depth was identified. Other significant arguments in favor of the supergenic origin are the results of the isotopic analyses of deuterium and ^{18}O presented by Cravero et al. (1991), which correspond to meteoric water action.

A profile of global weathering for present times was presented by Strakhov (1967) which included five main zones ordered in order to their diminishing depth/degree of weathering: (a) a deep saprolite zone in contact with the fresh rock; (b) a zone of illite/sericite, montmorillonite, and beidellite; (c) a zone of kaolinite; (d) a zone of ochre gibbsite; and the sequence ends with (e) a laterite layer. It is possible that this general alteration sequence would have never developed on the rhyolitic outcrops of the study area in the past. Nevertheless, there is an acceptable correlation between the three lower zones, with the possibility that the level of superficial duricrust (bauxite?) would have never developed here or that it had been eroded during the process of exhumation of the plateau, which would allow the inference that the original thickness of the alteration area was much larger.

In summary, the well-known presence of accumulations of kaolinite and other clays over the Marifil Formation outcrops would be indicating intense weathering of these rocks by lixiviation of rain water (Cravero et al. 1991), perhaps for a very long period, sometime between 180 Ma (the minimum approximate age of the Marifil Formation) and 150 Ma (the approximate age of the Araucanian tectonic phase in the Andean ranges which lead towards the exhumation of the Rhyolitic Plateau). The environmental conditions of this epoch (Middle to Late Jurassic) would have been very warm and humid in Patagonia, according to the paleontological records (Volkheimer 1967), or of temperate-moderate conditions (according to Archangelsky (1967)). Cravero et al. (1991) confirmed, by means of the isotopic content analyses of the kaolinites, the need of very high precipitation conditions to generate the appropriate environments. Besides, this method allowed these authors to calculate the values of mean paleotemperature which were found to be between 10 and 12 °C. These values should be considered as a minimum, because these authors do not rule out that the climate would have been highly seasonal, with a rainy season.

The Relationship Between the Paleosurfaces and the Uranium Deposits of the Region

A significant number of uranium mineral deposits of economic value are distributed in central Chubut province, near the study area. The Sierra de Pichiñales uranium district, located 30 km north of the town of Paso de Indios, and within it, the

Los Adobes, Cerro Cóndor, and Cerro Sólo mines should be mentioned. Due to the available reserves and strategic value, the latter has been the most studied in recent years, providing some clues to understanding the relationship between the occurrence of these valuable sedimentary deposits and the processes involved in the paleolandscape development.

The Cañadón Asfalto sedimentary basin, located immediately to the west of the study area (between 42° and 44°30′ S and 68°30′ and 70° W), developed during the Mesozoic and it is composed of meso-silicic volcanic rocks assigned to the Lonco Trapial Formation (genetically related to the Marifil Formation; Page et al. 1999). These rocks were overlain by the Late Jurassic continental sedimentary rocks of the Cañadón Asfalto Formation and later by the continental sequence corresponding to the Chubut Group, Middle to Late Cretaceous. At the base of this latter unit, already described, the Arroyo del Pajarito Member (of the Los Adobes Formation) is found. These are the rocks bearing the main uranium mineralization and are composed of a series of conglomerates and sandstones deposited in a high-energy, braided fluvial environment. The descriptions of a subunit found in the Cerro Sólo mine (Benítez et al. 1993) defined a clastic fraction with very high roundness where rhyolites, ignimbrites, and multicolored acid tuffs are dominant, including quartz grains of volcanic origin and sanidine. Montmorillonite and kaolin are very abundant as well. Organic matter is frequent and of essentially clastic origin, with frequent findings of large fossil trunks and other smaller plant remains. According to Benítez et al. (1993), the mineralization is dominantly epigenic, forming tabular to lenticular bodies, parallel to the stratification. The present mineral species, originated by uranium leaching, are mainly uraninite, uranopilite, and coffinite, generally associated with organic matter. Marveggio and Llorens (2011) determined that the Arroyo del Pajarito Member occurs as a grain-size fining-up sequence, with thicknesses of up to 150 m in the Cerro Sólo mine. These same authors concluded that the deposition of such a large volume of organic matter was concomitant with the almost exclusive contribution of rhyolitic volcanic clasts transported by the large Cretaceous fluvial networks which had their heads at the Marifil Formation Rhyolitic Plateau located to the east (Figs. 2 and 3). The observations of Allard et al. (2011) and Foix et al. (2011) have been oriented in the same direction, who suggested the development of paleo-valleys that were draining from the SE to the NE, locating the source area in the Marifil Formation outcrops.

It seems pertinent to propose that at least some of the uranium mines of central Chubut province are the end product of a process that was initiated under hyper-tropical conditions in the Late Jurassic, with the development of an important level of regolith product of deep chemical weathering of the acid volcanic rocks of the Marifil Formation, which bear uranium primary minerals. Apparently, these volcanic outcrops would be comparable to those found today in the surroundings of the Florentino Ameghino Dam (Fig. 3). The regolith would have been afterwards eroded and transported towards the W and NW, to the depositional center of the Cañadón Asfalto basin. The intense fluvial erosion would have exposed the unweathered and more competent levels of the Marifil Formation at the eastern border of the basin. These surfaces have survived almost without modification up to now, representing the ancient weathering fronts, with frequent corestones and bornhardts. The assumed hyper-tropical conditions that took place during the Jurassic may also be confirmed by the immense quantities of plant and organic remains accumulated at the base of the Chubut Group (Arroyo del Pajarito Member), which later became the basic materials for the natural concentration of the uranium minerals during diagenetic processes.

Cenozoic Morphogenesis

A characteristic of the Late Cenozoic that clearly differentiates it from the preceding times is its global climatic pattern, consisting in alternating glacial (cold) and interglacial (temperate/warm) periods, approximately in 100 ka cycles. Astronomic variables, related to the Earth's orbital parameters, combined with geological processes such as tectonics and volcanism, are essentially the factors controlling the occurrence, recurrence, and intensity of the climatic events. This global atmospheric context, which started in the Late Miocene, has introduced important geomorphological modifications over all continental regions and has been especially relevant in Patagonia, due to its latitude and the presence of the Andean Cordillera as its backbone in the western border. Thus, during glacial periods and at least since the Early Pleistocene, a mountain ice sheet was emplaced in the Andes (Clapperton 1993; Rabassa et al. 2005; Rabassa 2008), whereas in extensive extra-Andean regions, periglacial conditions were established (Clapperton 1993; Trombotto Liaudat 2008). These eastern territories were even more extensive due to the lowering of sea level during the glaciations, exposing the continental shelf, roughly doubling the present surface of Patagonia. Distinctive landscape features of Patagonia are the endorheic depressions or "hollows" ("bajos sin salida"). The origin of these depressions is a problem that has been discussed by many authors in the past (for references, see Martínez (2012)). At present, most of the Argentine geomorphologists accept that these hollows are not the product of just one single process and that there are many factors which, combined or not, have taken place in their development, although hydro-eolian processes are considered as dominant, particularly deflation (Martínez 2012). The Rhyolitic Plateau shows abundant shallow depressions of this nature, which exceptionally exceed a depth of 5 m (Fig. 7) and, generally, are of small size (less than 5 km along its longest axis). These depressions have irregular shapes, though always adjusted to the superficial drainage networks which are determined by the rock fracture patterns (Fig. 6). These frequent depressions were generated by surficial physical weathering (essentially frost processes) that acted over both bedrock and structural alignments (such as joints, faults, dykes, bedding, schistosity, and foliation), which then provided the fine materials that were later removed by deflation. However, it is possible that some of the today existing depressions have, at least partly, inherited their shape characteristics from irregularities of the original weathering front, formed and active under hyper-tropical climates in the Late Mesozoic. The irregularities were probably due to differential chemical weathering

Fig. 12 Ventifacts and wind-abraded rocks, carved on the Marifil Formation outcrops, located a few kilometers S of Sierra Colorada. These eolian erosion features are overprinted on the original morphology of the paleolandscape, during the glacial cycles of the Late Cenozoic. Simultaneously, during these cold and dry events, the numerous hollows found in this area were generated and later expanded

on varying rock types or fracturing patterns. Thus, at least some of these shallow and relatively small features are in fact relicts from the Late Jurassic, ancient etchplain and have been reactivated by Tertiary denudation and remodeled by a set of totally different processes.

Another unquestionable indication of the great influence of eolian erosion in the area is the size and abundance of ventifacts (Figs. 8, 12, and 13). Ventifacts are rock fragments, boulders, or gravels, and wind-abraded rocks are outcrops which show evidence of eolian abrasion, characterized by their shape and surficial features (Laity 1994). The shape of ventifacts and wind-abraded associated features, such as facets, keels, flutes, striation, grooves, scallops, and pits, depend upon the size of the rock, its density and hardness, primary texture, characteristics of the impacting particles (density, diameter, and roundness), wind characteristics and régime, and time length of exposition to abrasion. The fine-grained, hard rocks tend to generate facets, as more or less smooth, abraded surfaces, and a few surficial irregularities. Poorly homogeneous rocks, mostly coarse grained, are more limited in the possibilities for development of facets, with more surficial features. The sand particles are the main, if not the only agent with capacity to produce eolian abrasion, and other materials such as air, dust (clay/silt), ice, or snow should be definitively ruled out (Laity and Bridges 2009). This abrasion capability of sand with respect to other materials is due not only to the size but to the fact that the majority of these particles are made of quartz. The development of landforms due to eolian abrasion implies

Fig. 13 Overview of a surface modeled by eolian erosion, located a few kilometers S of Sierra Colorada. The grooving in these ventifacts and wind-abraded rocks has a general W-E orientation, thus indicating the dominant wind paleo-direction during glacial episodes

not only the presence of strong and frequent winds but also the availability of loose sand in the surroundings (Laity 1995). Another aspect that should be considered is the maximum elevation at which eolian abrasion loses efficiency. In general, it is assumed that on a flat topography, the sand particles lose their erosion capacity at 1 m high (Hobbs 1917). Some authors have argued that sand has no abrasion power above 40 cm (Schlyter 1994). Laity and Bridges (2009) demonstrated that several factors, such as the topographic accidents, may increase the erosion power by saltation, elevating the sand grains above 1 m high and producing erosion at upper levels. In the study area, near Sierra Chata (Fig. 2), outcrops of igneous rocks as corestones have been deeply eroded by the wind, with abraded features above 2 m high (Fig. 5). The ventifacts are excellent paleoenvironmental indicators because they are related to arid, in this case cold, conditions, with lack of vegetation and strong winds capable of transporting sand grains. The eolian abrasion is concentrated exclusively in the upwind side of the affected rocks, which is a valuable element to establish the dominant wind direction. In any case, ventifacts should not be interpreted as indicators of mean wind velocity, or when the wind pattern is or has been bidirectional or multidirectional, these erosion features do not allow the approximation to the strongest wind directions (Laity and Bridges 2009). The sense of the dominant wind in the study area, measured from the linear features on ventifacts (basically grooves), is coincident with that established by other authors (Haller 1981) and tends to be oriented according to a W/SW (upwind)-E/NE (downwind), roughly parallel to the present wind directions. However, although

Fig. 14 Weathering basin on outcrops of the Marifil Formation in the study area. Physical weathering processes and deflation are presently the main morphogenetic processes in the area and they were significantly more intense during Late Cenozoic glacial periods

Patagonia is well known for the frequency and intensity of the winds, the present environmental conditions do not match those needed for the development of such features.

The origin of the "bajos sin salida" and other wind erosion features of this region should be correlated with environmental conditions much drier and colder than today. Physical weathering (particularly, frost processes; Figs. 7 and 14) provides important quantities of sand (quartz, feldspar) to the wind, mostly obtained from the outcrops of the Marifil Formation but also from the Chubut Group exposures and other younger sedimentary rocks, which favors rock abrasion and simultaneously, associated to deflation, contributes to the development of the endorheic depressions. These rigorous climatic conditions should be placed in time during glacial periods, which were represented in extra-Andean Patagonia as periglacial environments, mostly with permafrost conditions south of 42° S.

Final Remarks

The Rhyolitic Plateau, located in the eastern portion of the Northern Patagonian Massif, is a very large geological/geomorphological unit, extending over more than 50,000 km^2, and composed of volcanic/pyroclastic rocks extruded during a roughly

10 Ma period, between the Early and Middle Jurassic ("a" in Fig. 15). Today, this plateau is noted for its homogeneous physiography with low, rounded, and domed hills, and a scarcely integrated drainage network, of ephemeral nature, on which the "bajos sin salida" have been developed, suggesting a prolonged inactivity of these fluvial systems. This mega-landform presents a general "erosive" aspect, with exposure of a rocky substrate with rounded hills and absence of depositional landforms. This has been interpreted by several authors as a "peneplain," corresponding to the final stage of the Davisian fluvial cycle. However, there are sufficient and solid arguments to consider a different genesis for these landforms, comparable to that described in other cratonic areas of Argentina and the Southern Hemisphere (see Rabassa (2010) for further references). These features would be related to (a) deep chemical weathering during a long period of tectonic and climatic stability, under warm and wet environmental conditions ("b" in Fig. 15), and (b) later remobilization of the regolith/saprolite due to erosion enforced by tectonic or epeirogenic reactivation ("c" and "e" in Fig. 15). The surface generated by these mechanisms is a deep-weathering plain or etchplain. The domed shape of the hills and the unweathered corestones derived from the volcanic bedrock are related to an ancient weathering front, exhumed today. Although ignimbrites are the most frequent rocks found in this plateau, it is also true that the rest of lithological types that integrate the Marifil Formation (rhyolites, porphyritic rhyolites, breccias) outcrop in all areas under study, thus suggesting that lithological composition has not played the most relevant role during the development of this homogeneous surface. It is more likely that structural features, such as fracture and joint density, have controlled the efficiency of weathering processes being more intense and, therefore, deeper where bedrock occurs with a higher degree of fracturing ("b" in Fig. 15). According to Twidale (2007), the combination of orthogonal fractures and laminar structures (sheet structures) favors the development of bornhardts and bornhardt-like features. A similar or even identical situation to that described by Twidale (2007) may be observed in many localities of the study area, including Sierra Colorada (Figs. 5, 10, and 11).

The abundant accumulations of epigenetic clays, which appear in contact and overlying rocks of the Marifil Formation, are a modest relict of the huge volumes of weathered rocks that covered a good portion of the plateau in the Jurassic ("b" in Fig. 15). A significant portion of the materials composing these weathering profiles, some of them several hundreds of meters thick, were removed and denudated when the Late Jurassic landscape was reactivated by the Tertiary Araucanian movements ("c" in Fig. 15), to become incorporated as clastic sediments in the Chubut Group basin ("d" in Fig. 15). It should not be ruled out that these weathered materials are also genetically linked to the formation of secondary uranium minerals which are abundant at the base of the lithostratigraphic unit. It is highly probable that, at least partially, the mostly nonfunctional, present drainage network found in the area would have started its development during such tectonic reactivation.

Apparently, the diastrophic events that forced the exhumation of the plateau were the inter-Senonian movements ("e" in Fig. 15), which not only provoked the differential dislocation of the rocky blocks but exposed them to erosion. The

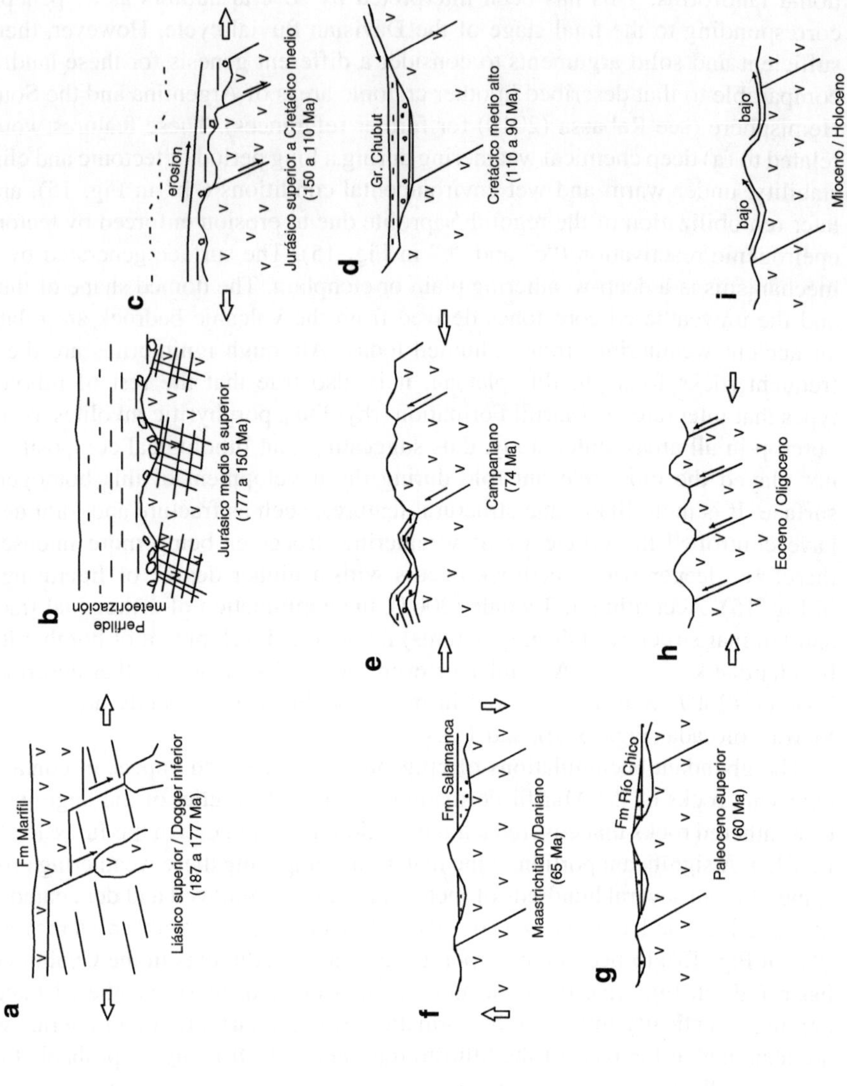
a
Fm Marifil
Liásico superior / Dogger inferior (187 a 177 Ma)
b
Perfil de meteorización
Jurásico medio a superior (177 a 150 Ma)
c
erosion
Jurásico superior a Cretácico medio (150 a 110 Ma)
d
Gr. Chubut
Cretácico medio alto (110 a 90 Ma)
e
Campaniano (74 Ma)
f
Fm Salamanca
Maastrichtiano/Daniano (65 Ma)
g
Fm Río Chico
Paleoceno superior (60 Ma)
h
Eoceno / Oligoceno
i
bajo
bajo
Mioceno / Holoceno

Fig. 15 **(a)** Extrusion of lavas and pyroclastic flows of the Marifil Formation during the initial phases of the South Atlantic Ocean rifting. **(b)** Phase of tectonic stability with intense chemical weathering in which most of the clay deposits of the region were formed. **(c)** The Araucanian movements initiated an erosion phase of essentially fluvial nature. **(d)** Associated with the uplifting of the Northern Patagonian Massif, the continental sedimentary rocks of the Chubut Group are irregularly accumulated over the ancient surfaces. **(e)** The inter-Senonian movements exhumed the Marifil Formation outcrops. **(f)** Eustatic and/or epeirogenic events favored the marine transgressions coming from the newly born South Atlantic Ocean. **(g)** The regional uplift forced the lowering of sea level and the accumulation of continental deposits. **(h)** The Andean orogeny reactivated the structures producing a new exhumation of some of the regional tectonic blocks and final denudation of the paleosurfaces. **(i)** The evolution of the landscape is controlled by the glacial cycles with a marked predominance of wind erosion, which completed the denudation and modeled the volcanic rock outcrops by abrasion and deepened the existing depressions by deflation and generating the endorheic depressions. At least some of these hollows probably correspond to original depressions of the weathering front, formed during the Late Mesozoic and controlled by differential erosion and/or fracturing. Thus, some of the present depressions are in fact landforms which inherited their basic shapes from the original Gondwana paleolandscape fluvial network was initiated in the Late Jurassic. Reactivation of the landscape took place in this region during the different phases of the Andean orogeny in the Early Tertiary ("h" in Fig. 15). This would have been an essentially erosive period, and possibly, the drainage network was very active. The tectonic or epeirogenic events after the Middle Jurassic which took place forced the different blocks of the plateau up, down, or tilted in different directions. This should have imposed a permanent adjustment on the fluvial network. Thus, the stream channels, strongly controlled by weakness zones and coincident with extensive and abundant alignments present in the area, slowly developed subsequent valleys. This is a relevant aspect because it could be accepted that erosion would have been concentrated through time in the same sectors, essentially those of more intense fracturing allowing a better conservation of the domed hills and bornhardts.

Finally, the present landscape of the study area has clear signals of intense and prolonged chemical weathering under environmental conditions of extremely high temperature and moisture, hyper-tropical climates which took place in the Mesozoic and which did not occur again later in geological history. Remobilization of alteration products and unweathered bedrock erosion was controlled by the degree of exhumation of each of the rocky blocks affected by different tectonic and/or epeirogenic events. During the Late Cenozoic, episodic, extreme environmental conditions also developed by these were exactly opposite to those which allowed the formation of the paleosurfaces. Especially during the Pliocene and the Pleistocene, during glacial periods, the region underwent very cold and dry conditions which favored denudation and wind erosion. This had a very marked geomorphological impact since deflation excavated sections of the nonfunctional fluvial channels, generating the characteristic Patagonian "bajos sin salida," endorheic depressions ("i" in Fig. 15). However, it should not be ruled out that at least some of the hollows are in fact relicts of the Late Jurassic etchplain, deeper portions of the weathering front denudated in the Tertiary, characterized by irregularities due to differential weathering based on different rock types or degree of fracturing.

Eolian abrasion features have been carved on the Marifil Formation outcrops as well as on isolated boulders such as corestones. These provide a good indicator of the morphodynamic impact of the wind during glacial epochs. These features may be generated only under intense and constant wind action associated with a large sand supply.

In summary, the Rhyolitic Plateau of the Marifil Formation in eastern Chubut province provides a perfect scenario to understand the deep chemical weathering environmental conditions under hyper-tropical climates in the Late Mesozoic, which produced an etchplain and a unique set of landforms at different scales on a Jurassic volcanic/pyroclastic complex. The chemical weathering of these rocks generated an enormous amount of regolith/saprolite which is related to the accumulation of kaolinite deposits and uranium secondary minerals of high economic interest. Future work in the region with detailed surveys in selected areas will undoubtedly provide a much clearer insight to understand the Gondwana paleolandscapes in Northern Patagonia.

Acknowledgments The authors are greatly indebted to Professor Cliff Ollier for his careful and most valued reviewing of a previous draft of this contribution. His comments and suggestions over the whole text and his personal discussion of some of the topics considered here are very much appreciated, as well as his colossal and indefatigable efforts to bring this manuscript into some sort of more readable English language.

References

Aguilera E, Rabassa J (2010) Origin of the Northern Patagonian Massif regional surface. Geociências, UNESP 29(4):467–478. São Paulo

Aliotta G, Domínguez E, Whewell R (1977) Sedimentología de las psammitas del Terciario inferior entre Dique Ameghino y Boca Toma, provincia del Chubut, Argentina. Revista de la Asociación Geológica Argentina 32(2):81–98

Allaby A, Allaby M (1999) A dictionary of earth sciences, Oxford paperback reference. Oxford University Press, Oxford

Allard JO, Giacosa R, Paredes JM (2011) Relaciones estratigráficas entre la Formación Los Adobes (Cretácico Inferior) y su sustrato Jurásico: implicancias en la evolución tectónica de la cuenca Cañadón Asfalto, Chubut, Argentina. XVIII Congreso Geológico Argentino, S15

Alric V, Haller M, Féraud G, Bertrand H, Zubia M (1996) Cronología $^{40}Ar/^{39}Ar$ del volcanismo jurásico de la Patagonia Extrandina. 13° Congreso Geológico Argentino and 3° Congreso de Exploración de Hidrocarburos, Actas 5:243–250. Buenos Aires

Angelelli V, Stegman E (1948) Estudio sobre arcillas y caolines de la República Argentina, Yacimientos Blaya Dougnac, Linares, Villanueva y Hércules. Revista de Industria Minera 47–54. Buenos Aires

Aragón E, Iñiguez Rodríguez A, Benialgo A (1996) A caldera field at the Marifil Formation, new volcanogenic interpretation, Norpatagonian Massif, Argentina. J S Am Earth Sci 9(5–6): 321–328

Aragón E, Castro A, Otamendi J, Brunelli D, Rabbia O, Aguilera E, Cavarozzi C, Ribot A (2011a) Isolated epeirogenic mountains in supra-subduction systems; the case of the North Patagonian Massif, Argentina. XVIII Congreso Geológico Argentino, Abstracts. Buenos Aires

Aragón E, Tassara A, Tocho C, Mendoza L, Spackman W, Rabbia O, Perdomo R, Bertotto W (2011b) Upper mantle geodynamic constrains beneath the North Patagonian Massif, Argentina. XVIII Congreso Geológico Argentino, Abstracts. Buenos Aires

Archangelsky S (1967) Estudio de la Formación Baqueró, Cretácico inferior de Santa Cruz, Argentina. Universidad Nacional de La Plata, Facultad de Ciencias Naturales y Museo

Ardolino AA, Franchi MR (1996) Hoja Geológica 4366-I Telsen. Dirección Nacional del Servicio Geológico Boletín 215:1–110. Buenos Aires

Benítez AF, Fuente A, Maloberti Landi V, Bianchi R, Marveggio de Bianchi N, Gayone M (1993). Evaluación de la Mina Nuclear Cerro Sólo, Provincia de Chubut. Parte I: Características geológicas y de la cuenca. XII Congreso Geológico Argentino. Actas:272–278

Besoain E (1985) Mineralogía de arcillas de suelos. Instituto Interamericano de Cooperación para la Agricultura, San José, Costa Rica. Serie de Libros y Materiales Educativos, 60, 1205 pp

Chebli GA, Nakayama C, Sciutto JC, Serraiotto AA (1976) Estratigrafía del Grupo Chubut en la región central de la provincia homónima. 7° Congreso Geológico Argentino, Actas 1:375–392

Clapperton C (1993) Quaternary geology and geomorphology of South America. Elsevier Science Publishers B.V, Amsterdam, 779 pp

Codignotto J, Nullo F, Panza J, Proserpio C (1978) Estratigrafía del Grupo Chubut entre Paso de Indios y Las Plumas, provincia de Chubut, Argentina. 7° Congreso Geológico Argentino, Actas 1:471–480

Coira B, Nullo F, Proserpio C, Ramos V (1975) Tectónica de basamento de la región occidental del Macizo Nordpatagónico (provincias de Río Negro y Chubut). Revista de la Asociación Geológica Argentina 30(1):361–383
Cortés JM (1981a) El sustrato precretácico del extremo noreste de la provincia de Chubut. Revista de la Asociación Geológica Argentina 36(3):217–235
Cortés JM (1981b) Estratigrafía cenozoica y estructura al oeste de la Península de Valdés, Chubut. Revista de la Asociación Geológica Argentina 34(4):424–445
Cortiñas J (1996) La Cuenca de Somuncurá-Cañadón Asfalto, sus límites, ciclos evolutivos del relleno sedimentario y posibilidades exploratorias. 13° Congreso Geológico Argentino and 3° Congreso de Exploración de Hidrocarburos, Actas 1:147–163
Cravero F, Domínguez E, Murray H (1991) Valores $\delta^{18}O$ y δD en caolinitas indicadores de un clima templado-húmedo para el Jurásico superior-Cretácico inferior de la Patagonia. Revista de la Asociación Geológica Argentina 46:20–25
Domínguez E, Murray H (1995) Genesis of the Chubut river valley kaolin deposits, Argentina, and their industrial applications. In: Churchmann GJ, Fitzpatrick RW, Eggleton RA (eds) Clays controlling the environment: proceedings of the 10th international clay conference Adelaide, Australia, 1993. CSIRO Publishing, Melbourne, pp 129–134
Fairbridge RW (1968) Encyclopedia of geomorphology. Ronald, New York
Falconer JD (1911) The geology and geography of northern Nigeria. Macmillan, London, 295 pp
Foix N, Allard JO, Paredes JM, Giacosa R (2011) Sistemas fluviales exhumados en la Formación Cerro Barcino (Cretácico Superior de la cuenca Cañadón Asfalto, Chubut): estilos fluviales y análogos actuales. XVIII Congreso Geológico Argentino, S15
Franchi M, Ardolino A, Remesal M (2001) Hoja Geológica 4166-III, Cona Niyeu. Provincia de Río Negro. Instituto de Geología y Recursos Minerales, Servicio Geológico Minero Argentino, Boletín 262, 83 pp. Buenos Aires
González Díaz EF, Malagnino EC (1984) Geomorfología. Relatorio 9° Congreso Geológico Argentino (San Carlos de Bariloche) 1(13):347–364. Buenos Aires
Grim RE (1968) Clay mineralogy. McGraw Hill Book Co., New York, pp 493–503
Haller MJ (1978) Estratigrafía de la región al poniente de Puerto Madryn, provincia de Chubut, República Argentina. 7° Congreso Geológico Argentino, Actas 1:285–297
Haller M (1981) Descripción Geológica de la Hoja 43h Puerto Madryn, provincia de Chubut. Servicio Geológico Nacional, Boletín 184:1–48
Hayase K (1969) Génesis del yacimiento de caolín de la Mina Villegas, Provincia de Chubut, República Argentina. Revista de la Asociación Geológica Argentina 24(1):55–71
Hayase K, Maiza P (1969) Génesis del yacimiento de caolín "Mina Equivocada", Los Menucos, Provincia de Río Negro, República Argentina. Revista de la Asociación Argentina de Mineralogía, Petrología y Sedimentología 1(1–2):33–47
Hobbs W (1917) The erosional and degradation processes of deserts with special reference to the origin of desert depressions. Ann Assoc Am Geogr 7:25–60
Labudía C, Alvarez G, Raniolo A (2011) Gravimetría del área de Telsen, sector oriental de la comarca nordpatagónica, provincia de Chubut. 18° Congreso Geológico Argentino, Abstracts, Buenos Aires
Laity J (1994) Landforms of aeolian erosion. In: Abrahams AD, Parsons AJ (eds) Geomorphology of desert environments. Chapman & Hall, London, pp 506–535
Laity J (1995) Wind abrasion and ventifact formation in California. In: Tchakerian VP (ed) Desert Aeolian processes. Chapman & Hall, London, pp 295–321
Laity J, Bridges N (2009) Ventifacts on Earth and Mars: analytical, field and laboratory supporting sand abrasion and windward feature development. Geomorphology 105:202–217
Lapido O (1981) Descripción Geológica de la Hoja 44g Cañadón Iglesias, provincia de Chubut. Servicio Geológico Nacional, Boletín 185:1–46
Lapido O, Page R (1979) Relaciones estratigráficas y estructura del Bajo de la Tierra Colorada (provincia de Chubut). 7° Congreso Geológico Argentino, Actas 1:299–313

Llambías E, Rapela C, Caminos R (1984) Las plutonitas y volcanitas del Ciclo eruptivo Gondwánico. In: Ramos V (ed) Geología y recursos naturales de la provincia de Río Negro. Relatorio 9° Congreso Geológico Argentino, pp 85–117

Maiza P, Hayase K (1975) Los yacimientos de caolín de la Patagonia. En República Argentina, 2° Congreso Ibero Americano de Geología Económica, Actas, pp 365–383

Maiza P, Hayase K (1980) Estudio comparativo de los yacimientos de caolín hidrotermal y sedimentarios del valle del Río Chubut. 6° Congreso Geológico Argentino, Actas 3:117–123

Maiza PJ, Marfil SA, Cardellach E, Zunino J (2009) Geoquímica de la zona caolinizada de Mina Estrella Gaucha, provincia de Chubut. Revista de la Asociación Geológica Argentina 63(3):226–432

Malvicini L, Llambías E (1974) Geología y génesis del depósito de manganeso Arroyo Verde, provincia del Chubut. 5° Congreso Geológico Argentino, Actas 2:185–202

Martínez O (2012) Los Bajos Sin Salida (BSS) del centro-norte de Patagonia y su relación con los ciclos climáticos del Cenozoico superior. 5° Congreso Argentino de Cuaternario y Geomorfología, Río Cuarto. Actas:175–185

Marveggio N, Llorens M (2011) Nueva edad de la roca hospedante de la mineralización de uranio, Yacimiento Cerro Sólo, Provincia de Chubut. XVIII Congreso Geológico Argentino, S14

Olivieri J, Terrero J (1952) El yacimiento de material caolínico "Don Emilio" (Departamento de Gaiman, Territorio del Chubut). Boletín Dirección Nacional de Minería 81:1–30. Buenos Aires

Page R (1987) Descripción Geológica de la Hoja 43g Bajo de la Tierra Colorada, provincia de Chubut. Servicio Geológico Nacional, Boletín 200:1–70

Page R, Ardolino A, de Barrio R, Franchi M, Lizuaín A, Page S, Silva Nieto D (1999) Estratigrafía del Jurásico y Cretácico del Macizo de Somún Cura, provincias de Río Negro y Chubut. In: El Jurásico Cretácico de la Patagonia y Antártida, Geología Argentina. Instituto de Geología y Recursos Minerales, Anales 29, 17, 3, pp 460–484

Pankhurst R, Rapela C (1995) Production of Jurassic rhyolite by anatexis of the lower crust of Patagonia. Earth Planet Sci Lett 134(1–2):23–36

Panza J, Sacomani L, Rodríguez M, Aragón E (2002) Hoja Geológica 4366–27 Campamento Villegas, escala 1.100.000. Boletín, 330, Servicio Geológico Minero Argentino, 154 pp

Rabassa J (1978) Paleorrelieves cenozoicos en la región Pilcaniyeu-Comallo, Provincia de Río Negro, Argentina. 7° Congreso Geológico Argentino, Actas 2:77–87

Rabassa J (2008) Late Cenozoic Glaciations in Patagonia and Tierra del Fuego. In: Rabassa J (ed) The late Cenozoic of Patagonia and Tierra del Fuego, vol 11, Developments in quaternary sciences. Elsevier, Amsterdam, pp 151–204

Rabassa J (2010) Gondwana paleolandscapes: long-term landscape evolution, genesis, distribution and age. Geociências 29(4):541–570. São Paulo, UNESP

Rabassa J, Zárate M, Cioccale M, Carignano C, Partridge T, Maud R (1996) Paisajes relictuales de edad gondwánica en areas cratónicas de Argentina. 13° Congreso Geológico Argentino and 3° Congreso de Exploración de Hidrocarburos, Actas 4, 219. Buenos Aires

Rabassa J, Coronato A, Salemme M (2005) Chronology of the Late Cenozoic Patagonian glaciations and their correlation with biostratigraphic units of the Pampean Region (Argentina). J S Am Earth Sci 20:81–103

Rabassa J, Carignano C, Cioccale M (2010) Gondwana paleosurfaces in Argentina: an introduction. Geociências 29(4):439–466. São Paulo, UNESP

Ramos V (1975) Geología del sector oriental del Macizo Norpatagónico entre Aguada Capitán y la Mina Gonzalito, provincia de Río Negro. Revista de la Asociación Geológica Argentina 30(3):274–285

Ramos V (1983) Evolución tectónica y metalogénesis de la Cordillera Patagónica. 2° Congreso Nacional de Geología Económica, Actas 2:107–124. Buenos Aires

Ramos V (1999) Las Provincias Geológicas del territorio argentino. In: Caminos R (ed) Geología Argentina, Instituto de Geología y Recursos Minerales, Buenos Aires, Anales 29(3):41–96

Rapela C, Pankhurst R (1993) El volcanismo riolítico del noreste de la Patagonia: un evento meso-Jurásico de corta duración y origen profundo. 12° Congreso Geológico Argentino and 2° Congreso de Exploración de Hidrocarburos, Actas 4:179–188. Buenos Aires

Romero A, Domínguez E, Whewell R (1974) El Area Caolinera del Departamento Gaiman, Provincia de Chubut, República Argentina. Fundación Bariloche, 80 pp. San Carlos de Bariloche

Romero A, Domínguez E, Whewell R (1975) Génesis de los yacimientos de caolín del río Chubut inferior. Segundo Congreso Iberoamericano de Geología Económica, 5, pp 423–444. Buenos Aires

Sacomani L, Panza J (2007) Hoja Geológica 4366-III Las Plumas, provincia del Chubut, escala 1:250.000. Servicio Geológico Minero Argentino, Boletín 291, 74 pp

Schlyter P (1994) Paleo-periglacial ventifact formation by suspended silt or snow-site studies in South Sweden. Geogr Ann 76:187–201

Strakhov NM (1967) Principles of lithogenesis, vol 1. Oliver and Boyd, 577 pp

Stipanicic P, Rodrigo F, Baulíes O, Martínez C (1968) Las formaciones pre-senonianas en el denominado Macizo Norpatagónico y regiones adyacentes. Revista de la Asociación Geológica Argentina 23(2):67–88

Thomas M (1968) Bornhardt. In: Fairbridge RW (ed) Encyclopedia of geomorphology. Reinhold, New York

Trombotto Liaudat D (2008) Geocryology of southern South America. In: Rabassa J (ed) The late Cenozoic of Patagonia and Tierra del Fuego, vol 11, Developments in quaternary sciences. Elsevier, Amsterdam, pp 255–268

Twidale RC (2007) Bornhardts and associated fracture patterns. Revista de la Asociación Geológica Argentina 62(1):139–153

Volkheimer W (1967) La paleoclimatología y los climas del Mesozoico Argentino. Revista Minería, Geología y Mineralogía 28(3):41–48

Windhausen A (1918) Rasgos de la historia geológica de la planicie costanera en la Patagonia septentrional. Boletín Academia Nacional de Ciencias de Córdoba 23(3–4):319–364

Meso-Cenozoic Paleotopographies and Paleolandscapes in the Deseado Massif (Santa Cruz Province, Argentina)

F. Bétard, J.P. Peulvast, Jorge Rabassa, and Emilia Y. Aguilera

Abstract The Deseado Massif is the southernmost part of a continent, outside of Antarctica, where Gondwana Landscapes may be observed and investigated. This chapter presents preliminary observations and field data about the Gondwana Landscapes of this cratonic area of Southern Argentina, one of the most remote, isolated, and less populated places on Earth. Under extreme cold–arid climate conditions, the region presents very scarce vegetation cover, which further enables the geomorphological observations. Remnants of planation surfaces of undisputable Late Mesozoic age, developed on Jurassic volcanic units and covered by Late Cretaceous and Paleogene sedimentary rocks, are exposed along tens of thousands square kilometers of this cratonic unit. In those remote times, the climate of this portion of Patagonia was very wet and warm, allowing the development of extensive chemical weathering.

Keywords Gondwana • Argentina • Santa Cruz • Paleosurfaces • Denudation rates

F. Bétard (✉)
Université Paris-Diderot, Sorbonne Paris Cité, UMR CNRS 8586 PRODIG, Paris, France
e-mail: francois.betard@univ-paris-diderot.fr

J.P. Peulvast
Université Paris-Sorbonne, UFR Géographie et Aménagement, Paris, France

J. Rabassa
Laboratorio de Geomorfología y Cuaternario, CADIC-CONICET, Ushuaia, Tierra del Fuego, Argentina

Universidad Nacional de Tierra del Fuego, Ushuaia, Tierra del Fuego, Argentina
e-mail: jrabassa@gmail.com

E.Y. Aguilera
Facultad de Ciencias Naturales y Museo, Universidad Nacional de La Plata (UNLP), Calles 122 y 60, 1900-La Plata, Argentina

DAIS (Dirección de Aplicación de Imágenes Satelitarias), Calle 7 N° 1267-2°P, 1900-La Plata, Argentina
e-mail: eaguilera@fcnym.unlp.edu.ar

J. Rabassa and C. Ollier (eds.), *Gondwana Landscapes in southern South America*, Springer Earth System Sciences, DOI 10.1007/978-94-007-7702-6_17,

Introduction

Paleolandforms and paleosurfaces are powerful indicators of long-term landscape evolution in various settings, particularly when studied in combination with stratigraphy or basin analysis (e.g., Widdowson 1997; Peulvast and Claudino Sales 2004; Demoulin et al. 2005; Lidmar-Bergström et al. 2013). Such paleotopographies might give evidence of former environments and landscapes, or paleolandscapes, as well as indications on uplift and erosional histories at various geological timescales (e.g., Peulvast et al. 2008; Bétard 2010). Studies of pre-Cenozoic paleolandforms and Gondwana paleolandscapes have been developed in several areas of Argentina (e.g., Carignano et al. 1999; Demoulin et al. 2005; Aguilera and Rabassa 2010; Rabassa 2010; Rabassa et al. 2010, 2014). In southern Extra-Andean Patagonia where Gondwana paleolandscapes were briefly mentioned by Rabassa et al. (1996, 2010), no detailed paleogeomorphological studies have been developed so far.

This chapter focuses on the Deseado Massif, a large platform area and volcanic province located in Southern Patagonia (Santa Cruz province, Argentina), in the foreland region of the Patagonian Andes. Based on the preliminary results of an extensive field survey achieved in March 2011, this chapter aims to provide a qualitative and quantitative insight into long-term landscape development during the Mesozoic and Cenozoic in the Deseado region. To achieve this, we propose a morphostratigraphic analysis of the regional landscape based on crosscutting relations between paleosurfaces, volcano–sedimentary stratigraphy, and continental paleoweathering.

In situ analyses and sampling of the alteration mantles developed on the Jurassic volcanics were performed for X-ray diffraction studies.

Owing to its richness in paleolandforms and stratigraphic markers of various types and ages, this region affords good opportunities for quantifying the magnitude of long-term uplift and denudation since the breakup of Gondwana, in a context of volcanic rifted margin located at the foreland region of an active orogen.

Study Area: Topography and Geological Setting

The Deseado Massif is located in Southern Patagonia (Santa Cruz province), separated from the Northern Patagonian Massif by the San Jorge Gulf Basin and from the Southern Patagonian and Fuegian Andes by the Austral Basin (De Giusto et al. 1980; Márquez et al. 2002; Andreis 2002a, b; Giacosa et al. 2002; Haller 2002; Panza and Haller 2002; Panza and Franchi 2002; Malumián 2002; Nullo and Combina 2002; Rabassa et al. 2010; see also the geological map of the province of Santa Cruz, Panza et al. 2003, and references cited in these papers). It constitutes the foreland region of the Patagonian Andes between 46° and 49°S latitude, and also belongs to the low-elevation rifted margin of Eastern Argentina bordering the Atlantic Ocean (Cavallotto et al. 2011). The main topographic features of the Deseado Massif are that of a very large tableland region, with general small local relief, located between 0 and 1,335 m a.s.l. (Fig. 1). All the topographic profiles on

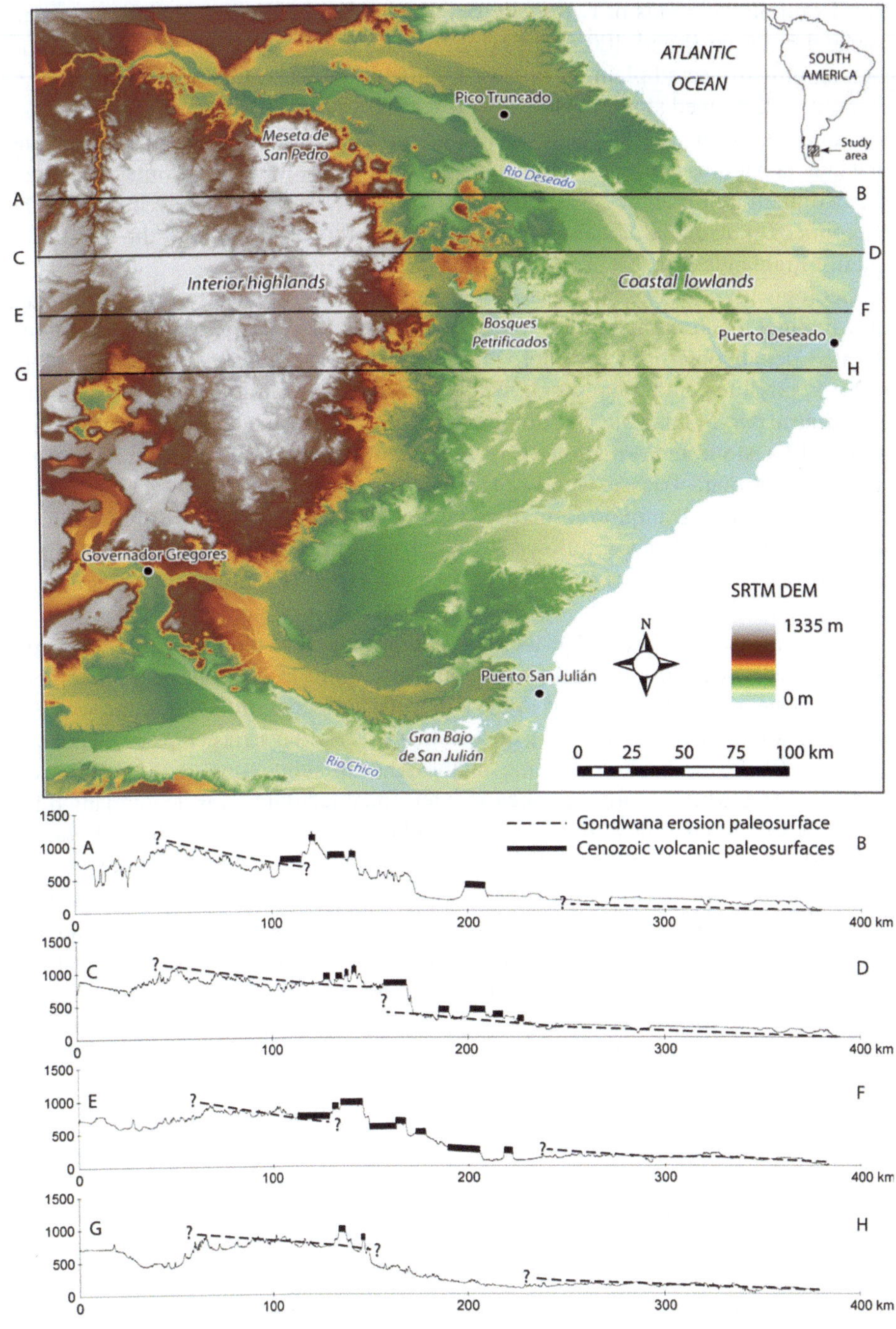

Fig. 1 Topography of the Deseado Massif (derived from SRTM data, version 04). *Lettered cross-sectional lines* refer to the four topographic profiles represented below

Fig. 1 outline the treads of two main topographic levels: the lower plain or coastal lowland between 0 and 400 m a.s.l., gently sloping seaward up to the coastline, and a highly dissected plain, or interior highlands, between 600 and 1,100 m a.s.l., without well-defined slope. Residual bedrock (inselbergs, bornhardts, and tors) and volcanic landforms (necks, buttes, and mesas) are scattered across the plains and plateaus at various altitudes. Higher local relief is found at the contact between the two main topographic levels, 150–200 km from the coast, where it defines the contours of an irregular, locally stepped, high escarpment delimiting the belt of the interior highlands (Fig. 1).

Geologically, the Deseado Massif belongs to a large platform area dominated by Middle–Late Jurassic silicic volcanic rocks (Guido et al. 2004; Fig. 2). Only isolated, small outcrops of Proterozoic–Paleozoic low-grade metamorphic rocks (schists, phyllites, and quartzites), locally intruded by granitoids, are found in the east-central part of this volcanic-dominated geological province (Río Deseado Complex and La Modesta Formation). Overlying these igneous–metamorphic basement rocks is a Permian–Triassic Gondwana siliciclastic cover sequence (La Golondrina, La Juanita, and El Tranquilo formations) which was deposited in a rift basin during an initial extensional phase, followed by block faulting and the development of half-grabens occupied by lakes and deltas (Homovc and Constantini 2001; Echavarría et al. 2005).

All along the Mesozoic, the tectonic and structural evolution of the Deseado Massif was related to the breakup of Gondwana and the opening of the southern Atlantic Ocean, on one side, and to the subduction process initiated on the western margin of southern South America, on the other side (Echavarría et al. 2005; Giacosa et al. 2010). In the back-arc setting of the Andean subduction zone, established as early as the Middle Jurassic, intense extensional tectonics was accompanied by fissure volcanism and lava flows of the Bajo Pobre Formation and, in the Middle–Late Jurassic, by more acidic volcanism that resulted in a large ignimbrite plateau of pyroclastic flows, laminated tuffs, and megabreccias of rhyolitic composition. These last volcanic and volcaniclastic rocks belong to the Chon Aike and La Matilde formations of the Bahía Laura Group (de Barrio 1989; Guido et al. 2004). The extensional regime which lead to the separation of the American and African continents resulted in the opening of small basins where clastic sediments accumulated from the Late Jurassic to the Late Cretaceous (Bajo Grande and Baqueró formations).

Several marine incursions occurred during the Tertiary, especially in the Early Paleocene (Salamanca Formation) and in the Late Oligocene–Early Miocene (Monte León Formation) (Malumián and Náñez 2011). Finally, during the Tertiary and Quaternary, olivine-rich basalt flows and tuffs were deposited over large areas of the Deseado Massif (Gorring et al. 1997) and today form dissected plateau remnants (mesas and buttes) at various elevations in the landscape. This is also the time of deposition of the "Rodados Patagónicos," or the "Patagonian Shingle Formation" as they were named by Darwin (1842), i.e., immense gravel accumulations of fluvial or glaciofluvial origin formed at various levels since the Late Miocene, between

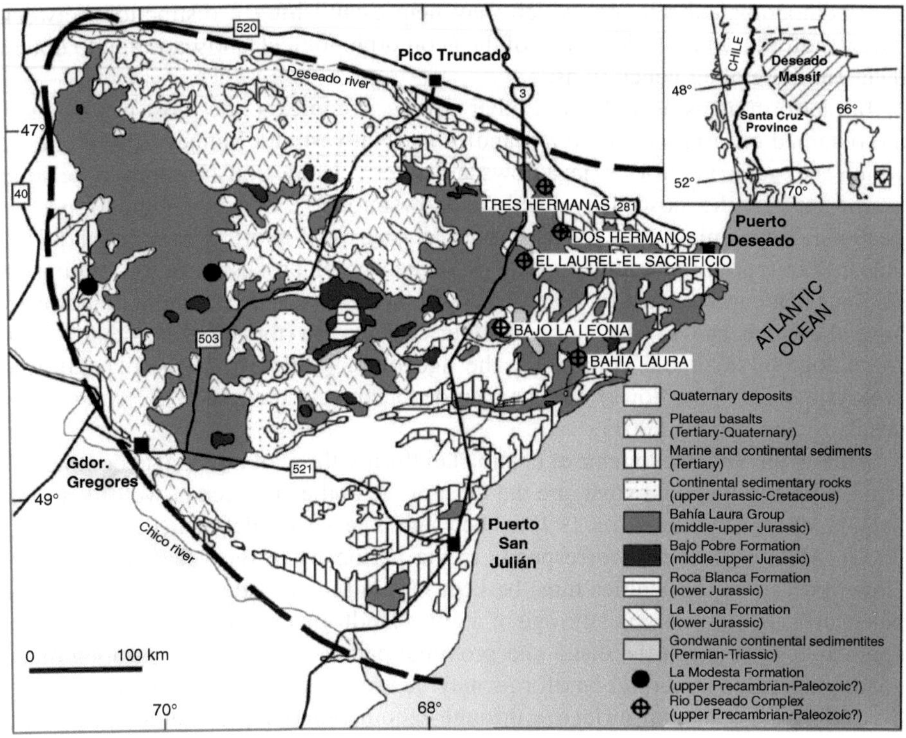

Fig. 2 Simplified geological map of the Deseado Massif (From Guido et al. 2004)

the Andean piedmont and the coast under the control of climatic, glacio-eustatic, and tectonic factors (Fidalgo and Riggi 1965, 1970; Martínez and Kutschker 2011). Most of these units are arranged in stepped systems of mesas, plateaus, and terraces. These surfaces appear widely excavated by wide and shallow valleys originating in the Andes and by multitudes of closed depressions of hectometric to plurikilometric scale, which were formed at all levels, down to −107 m for the deepest of them (Gran Bajo de San Julián), by hydro-eolian processes in the semiarid to arid conditions that have prevailed for 14–16 Ma (Blisniuk et al. 2005).

Conceptual and Methodological Remarks

This chapter is partly based upon the idea that ancient, inactive, or inherited landforms may survive in the landscapes over long periods and that their "age" implies minimal change by denudation processes since their creation (Ollier 1991; Summerfield et al. 1999). It also takes into account the following conceptual and methodological remarks which may help establishing a distinction between paleolandforms and transient parts of the geomorphological landscape and to obtain dates for these older generations.

In platform areas as well as in other structural contexts, the most precise ages are obtained on original structural landforms (Peulvast and Vanney 2001), among which endogenic landforms (lava flows and other volcanic constructions, coseismic scarps, etc.) are the most easily dated (Widdowson 1997). Corresponding to specific geological dynamics, they were formed during very short events (earthquake, volcanic eruption, meteorite impact) or sequences of processes (e.g., faulting or folding) that can be dated by relative or absolute methods. The original landforms have the same age as the corresponding structure or the ultimate stage of its formation. In favorable conditions, the age of the simplest of these landforms (volcanic cones, lava flows) may be also deduced from their state of preservation (Peulvast and Vanney 2001).

Among the sets of exogenic or erosion landforms, those of regional to continental scale, with large lateral extent, are the most prone to the presence of well-identified paleolandforms (Widdowson 1997). Their age is less well defined (Peulvast et al. 2009). At first glance, it corresponds to the period when they ceased to evolve. However, various difficulties must be kept in mind in any attempt to date erosion paleosurfaces. For example, the age of a given landform can differ from one part to another (e.g., between the distal and proximal parts of a pediment). Among other conditions, some structural landforms may be created by differential erosion after dissection of a planation surface or through exhumation of buried landscapes. Their age is well known when they relate to a well-dated generation of cyclic landforms or to a climatic sequence (e.g., the structural landforms of glacial origin identified in east Greenland: Peulvast and Vanney 2001). Therefore, inherited or fossil structural landforms (e.g., the paleo-hogbacks of southern Gaspésie, Quebec, still partly sealed by Visean conglomerates: Peulvast et al. 1996) give the best contribution to landform dating and geomorphic reconstructions. Some morphogenetic events which create new landforms may be easily identified thanks to stratigraphic markers with a well-known geometry and age. Primitive structural surfaces or aggradation surfaces corresponding to the ultimate stages of a sedimentary sequence are also useful, if well dated. They may be deformed without being immediately eroded, especially if the deformations correspond to faulting or regional vertical movements.

Initially flat and subhorizontal surfaces, especially planation surfaces, are considered to be the most convenient landmarks to reconstruct the local or regional tectonic and geomorphic history, provided that they are well identified, related with a former base level, and dated (Calvet and Gunnell 2008; Japsen et al. 2009). When well accomplished, planation processes have roughly the same obliterating effects ("resurfacing") on older topographies as aggradation processes. The problem most often encountered with these landforms (pediments, peneplains, pediplains, etchplains) concerns their age (Watchman and Twidale 2002), since most exposed landscapes are often subject to continued modification through dynamic factors and variation in rates controlled by climatic and tectonic environment (Widdowson 1997). This age may be found if a surface is preserved within a stratigraphy (unconformity), by dating the rocks immediately above and below, but in most cases, exposed surfaces prevail and only a maximum age is yielded by the bedrock upon which they developed. Unconformable sediment and characteristic weathering formations (i.e., saprolites, ferricretes, calcretes) may help dating planation surfaces or differentiated landscapes (Godard et al. 2001; Vasconcelos and Conroy 2003; Watchman and Twidale 2002). However, such dating is delicate because superficial formations are not always correlative of the planation process, which can be older. Moreover, formations which could be dated by radiochronological methods, as the extensive laterite covers in tropical regions, undergo a continuous geochemical evolution. They cannot be considered as closed systems and stable markers (Gunnell 2003; Nahon 2003), and some of them appear to be dependent on a later, slow incision of low-relief paleosurfaces (laterites: Thomas 1994). Finally, identification of synchronous remnants throughout a vast area is the most favorable case for evaluating tectonic deformations on a given period. Identification of diachronous surfaces or stepped sets of surfaces gives also clues on vertical movements and cyclic or noncyclic types of geomorphic evolution.

Although some authors assume that planation surfaces may be simultaneously developed at different altitudes in the same region in relation with separate base levels (African Surface of southern Africa: Partridge and Maud 1987), stepped systems of undeformed surfaces are generally considered as presenting a chronological meaning (Lageat and Robb 1984; Klein 1997). Elements of various ages and origins often coexist in apparently uniform topographies (Peulvast and Claudino Sales 2005). Uncertainties remain about their age, initial, terminal, before burying or dissection, or since exhumation (Dumont 1991). They often are diachronic, particularly if they result from scarp retreat. The residuals and scarps that bound them may correspond to lithological controls, without chronological meaning. Some surfaces cannot be dated since they are constantly reworked in the conditions of acyclic regime or slow degradation (Klein 1997). Possibly difficult on unequally resistant rocks, such as those which form the Jurassic volcanic sequence of the Deseado Massif, their preservation is better where factors of resistance are present (hard bedrock, hard residual cover deposits or saprolites such as laterites, silcretes, or bauxites) and in regions where vertical movements prevail over orogenic movements: this is the case of the Deseado Massif.

Results: Identification and Reconstruction of Paleosurfaces

Using the above-described criteria, several paleosurfaces of different natures were identified throughout the Deseado region:

1. Erosion paleosurfaces, the master of which is regionally represented by the extensive Gondwana paleosurface of Late Jurassic to Early Cretaceous age
2. Volcanic paleosurfaces of various ages, which were constructed as a direct result of basaltic lavas flow eruptions during the Tertiary and Quaternary

3. Sedimentary paleosurfaces corresponding to large gravel deposits of Neogene age, called "Rodados Patagónicos" ("Patagonian Shingle Formation"), mainly developed in distal position on the dissected platform, forming an extended Andean piedmont down to the coast

The Gondwana (Late Jurassic–Early Cretaceous) Erosion Paleosurface

In the Deseado Massif, the so-called Gondwana paleosurface (Rabassa et al. 2010) is the erosion surface that bevels the Middle–Late Jurassic volcanic rocks of the Bahía Laura Group. This surface is unconformably covered by the Cretaceous sediments of the Bajo Grande and Baqueró formations, indicating a Late Jurassic–Early Cretaceous age for the shaping of this erosional topography. Such Gondwana paleolandscapes were already mentioned in the Deseado Massif by Rabassa et al. (1996, 2010), who cited extensive erosion surfaces developed on Jurassic volcanics and volcaniclastics of the Chon Aike Formation and other units of the Bahía Laura Group.

Buried or exhumed elements of the Gondwana paleosurface were identified at various elevations in the landscape, in places where remnants of unconformable Neocomian to Aptian sediments were preserved or freshly exhumed. In the western highlands of the Deseado Massif, this surface is uplifted to more than 800 m a.s.l., where it records a general eastward slope (Fig. 1), although it is locally irregular due to younger volcanic processes (high volcanic necks such as the Cerros Madre e Hija; Fig. 3). To the NW of the "Bosques Petrificados" area (i.e., the National Monument of the Petrified Forests), it is locally downfaulted and buried by a large thickness of Cretaceous sediments, in half-grabens and small basins formed during the opening of the southern Atlantic Ocean in the Late Mesozoic. In the lower plain to the east, wider remnants of the Gondwana paleosurface were identified at low elevations, freshly exhumed or still buried by shallow thicknesses of Cretaceous and Tertiary sediments; at that place paleo-inselbergs, tors, and bornhardts shaped into Jurassic volcanic rocks regularly outcrop from the sedimentary cover (Fig. 4). A reconstruction of the paleosurface geometry in the coastal lowlands indicates the same eastward tilting as in the interior highlands, even if the downwarping appears much less pronounced in that area (Fig. 1).

Partly dissected or degraded, this surface displays numerous depressions, some of which will distinct from more recent depressions of hydro-eolian or other origin. They are interpreted as etch basins, carved between low-relief rock outcrops (tors and bornhardts, including necks and dykes), into irregularly weathered and kaolinized igneous basement rocks (Figs. 1 and 5). This general physiognomy of the bedrock surface topography of the Gondwana paleosurface makes it well recognizable in the regional landscape. The main question arises on the origin of the rock alteration often found at the surface (Fig. 5).

Fig. 3 Cerros Madre e Hija, seen from Bosques Petrificados. Note the silicified trunks of Araucariaceae at the foreground, partly exhumed from Jurassic ignimbrites (LaMatilde Formation) (Photo: J-P. Peulvast)

Weathering products represented by laterite superficial formations have been identified. In those areas where the alteration of the rocky block is very intense, mineral transformation bands are observed, which extend deeply into the rock-forming alteration mantles. In these mantles, sectors of varying color such as reddish, brownish, greenish, yellowish, and whitish are suggesting mineralogical and textural compositional changes. In this paleo-weathered surface, the in situ morphological description was completed and sampling was performed on residual deposits, detecting changes in the composition and fabric of the rocks, as gradation from massive to clastic stages, ferruginous crusts, and other mineralogical transformations such as neo-formed clays. By means of X-ray diffractometry performed on seven samples considered as representative of the different field-recognized zones, kaolinite, illite, halloysite, smectite, and quartz were identified (Aguilera et al. 2012) (Figs. 6 and 7).

The existence of residual products as a result of deep weathering suggests that they have been formed under intense regional weathering under wet-tropical climates sometime between the Late Jurassic and the Early Cretaceous, which were the dominant conditions in the Deseado Massif in these periods, as it has been shown by paleoclimatic studies (Aguilera et al. 2012).

The existence of residual deposits due to deep chemical weathering in tropical climates was already recognized by Cravero and Domínguez (1992, and references therein), who described kaolin deposits in Santa Cruz, at the southern portion of

Fig. 4 The Cerro Moro, a paleo-inselberg of Jurassic volcanic rocks (Chon Aike Formation) outcropping from the Late Cenozoic sedimentary cover (Photo: F. Bétard)

Fig. 5 Superficial morphology of the Gondwana paleosurface, displaying in situ altered Jurassic volcanic rocks in the plane of the irregular paleotopography (Photo: F. Bétard)

Fig. 6 Remnants of a saprolite mantle over Jurassic volcanic and pyroclastic rocks

the Deseado Massif. These kaolin-bearing units are of fluvial origin, developed within the Baqueró Formation (Middle to Late Cretaceous) over Middle Jurassic ash-flows (Chon Aike Formation) to Early Cretaceous ashfall tuffs (Bajo Grande Formation). This implies that the volcanic units would have been previously altered to kaolinite by deep chemical alteration. Therefore, the authors suggested that kaolin was formed by regional, chemical weathering under humid tropical climates in the Late Jurassic to the Early Cretaceous, being these the dominant environments on the Deseado Massif in those times. Afterwards, when the climate changed towards warm, temperate conditions (113–88 Ma; Nullo and Combina 2011), the weathering products were removed by subaerial fluvial processes to the accumulation areas in the Middle to Late Cretaceous. Nevertheless, previous or synchronous hydrothermal alteration of the igneous rocks cannot be excluded, in a regional context where volcanism was closely followed in time and space by widespread hydrothermal activity and vein formation (Giacosa et al. 2010; Guido and Campbell 2011, 2012).

The existence of younger erosion paleosurfaces is suspected in the region, as advocated by some authors (De Giusto et al. 1980; Rabassa et al. 2010). Highlighting the angular unconformity between the Bajo Grande Formation (Neocomian) and the Baqueró Formation (Aptian; Giacosa et al. 2010), preserved elements of an exhumed sub-Aptian paleosurface were locally identified in the landscape (Fig. 8). It might correspond to a local element of the probably diachronic Gondwana paleosurface, or to a distinct generation of planation surface. The possibility of younger, extensive erosion paleosurfaces, Mesozoic or Cenozoic in age, should be investigated further.

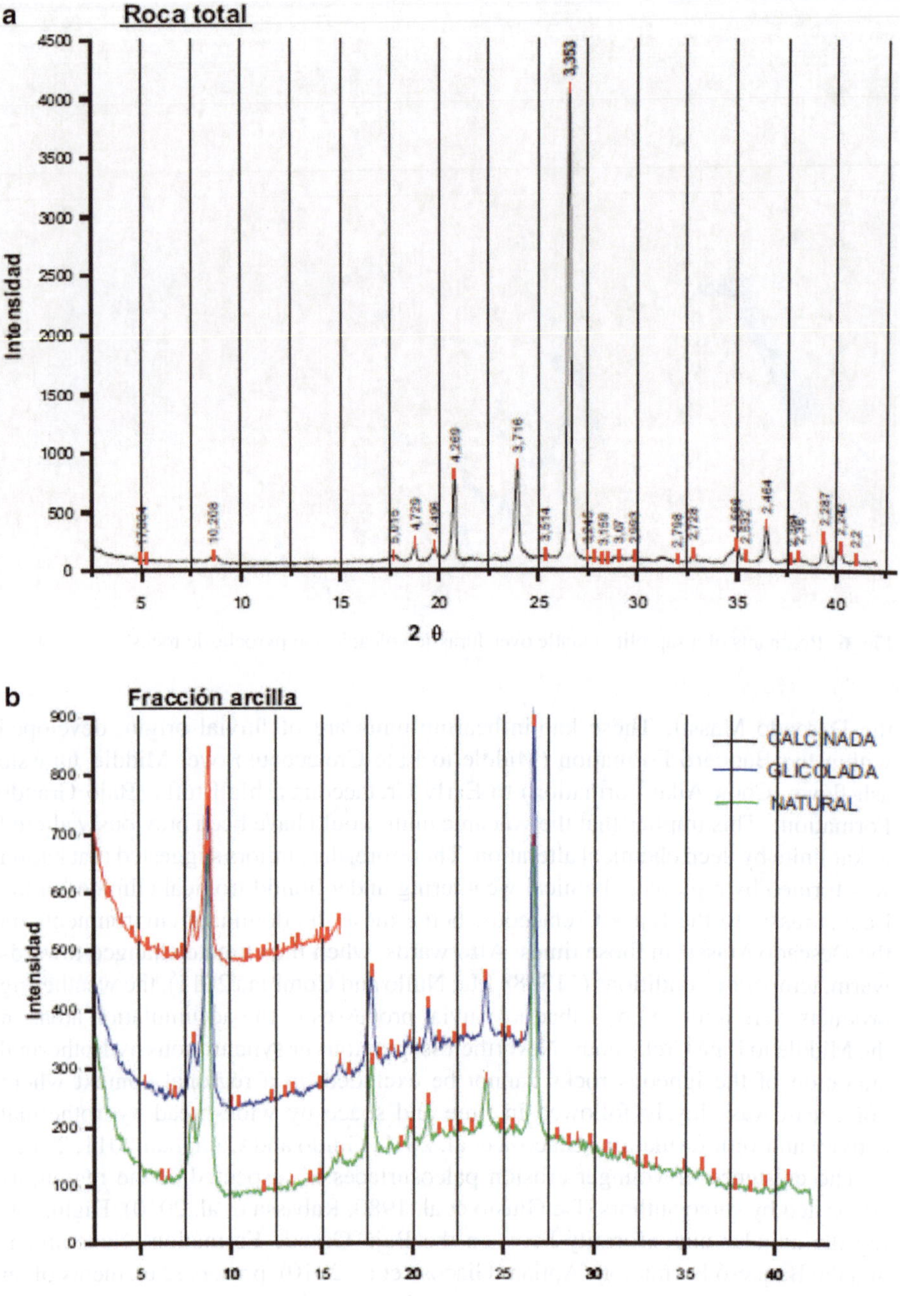

Fig. 7 X-ray diffractometry of total rock (**a**) and clay fraction (**b**), on seven samples considered as representative of the different zones identified in the field, in which kaolinite, illite, halloysite, smectite, and quartz are recognized

Fig. 8 Angular unconformity between the Bajo Grande Formation (Neocomian) and the Baqueró Formation (Aptian), with exhumed elements of a sub-Aptian paleosurface at the Bajo Grande area (Photo: F. Bétard)

Cenozoic Volcanic Paleosurfaces

Volcanically generated surfaces of various ages are ubiquitous in the Deseado region, where intense phases of volcanic activity have taken place since the Mesozoic. The volcanic paleosurfaces are currently represented as flat surfaces or plateau remnants which were originally created as a result of fluid basaltic eruptions in the region all along the Cenozoic. Two main generations of Cenozoic volcanic paleosurfaces may be distinguished in the present-day landscape:

1. Paleogene flat-lying and smooth volcanic surfaces, corresponding to the earliest basalt flows occurring since the breakup of Gondwana. In the region, Paleogene and later basalts mainly occur as generally well-preserved flows of melanocratic olivine basalts and basanites. The oldest of these lava flows is the Las Mercedes Basalt, dating from 64 to 63 Ma (Early-Middle Danian: Panza and Franchi 2002) (Fig. 8). The other main Paleogene lava flows are regionally represented by the Cerro del Doce Basalt, dating from between 60 and 40 Ma, and the Alma Gaucha Basalt, ranging from 30 to 23 Ma (Panza and Franchi 2002).
2. Neogene volcanic structural surfaces or mesas, corresponding to tholeiitic plateau lavas covered by less voluminous alkaline post-plateau flows, in relation to the opening of an asthenospheric "slab window" associated to ridge collision at the Chile Triple Junction (Gorring et al. 1997). Often intact but locally deeply dissected and reduced to narrow buttes and mesas, especially where thin layers overlay weakly resistant rocks (Fig. 9), these volcanic surfaces are locally topped

Fig. 9 "Las Pirámides" (SE of "Bosques Petrificados"): dissected Pliocene–Pleistocene basalt flow (La Angelita Formation) overlying weakly resistant, Jurassic volcaniclastic rocks (La Matilde Formation) (Photo: J-P. Peulvast)

Fig. 10 Cerro Lavatorio: Pliocene–Pleistocene cinder cone and basalt flow (La Angelita Formation), NE of Gobernador Gregores (Photo: J-P. Peulvast)

by Pleistocene cinder cones (Fig. 10). In the study area, the main plateau lavas range between 12 and 7 Ma in the western part of the back-arc region and between 5 and 2 Ma in the eastern part of the Deseado Massif (Gorring et al. 1997). The Pliocene–Pleistocene post-plateau basalts have ages from 3.4 to 0.125 Ma (Gorring et al. 2003), mainly corresponding to the basaltic lava flows of the La Angelita Formation.

Fig. 11 Stepped or inset lava flows: Meseta de Cali and dissection landforms E of Gobernador Gregores. Early Pliocene (*background, top*) and Late Pliocene–Pleistocene (*left side*) basalt flows. The intermediate level corresponds to Miocene gravels of the Pampa de la Compañía Formation (Photo: J-P. Peulvast)

Between the different phases of volcanic activity conducive to paleosurface construction, erosional processes became dominant and, consequently, active dissection and differential erosion led to the partial destruction and topographic inversion of most of the basaltic plateau lavas, whatever be their age. To the west, in the most elevated regions, the most recent flows are frequently inset below the older ones (Fig. 11). Therefore, the various ages and topographic positions of the volcanic paleosurfaces offer good opportunities and further constraints for deciphering the landscape evolution and denudation history of the Deseado region, especially for the Cenozoic times.

Late Cenozoic Sedimentary Paleosurfaces

The immense flat-lying surfaces corresponding to the "Rodados Patagónicos" (Late Miocene to Pleistocene gravels) as well as the wide terrace systems that are inset at lower levels along the major valleys may themselves be considered as paleosurfaces, since the conditions and processes of their sedimentation have disappeared, probably since the Late Glacial stage. They extend from the Andean piedmont down to the coast (Fig. 12), with a gentle WE regional dip which does not record the intense deformations of the older surface generations (Fig. 13). This piedmont mantle cover, corresponding to poorly consolidated conglomerates rich in

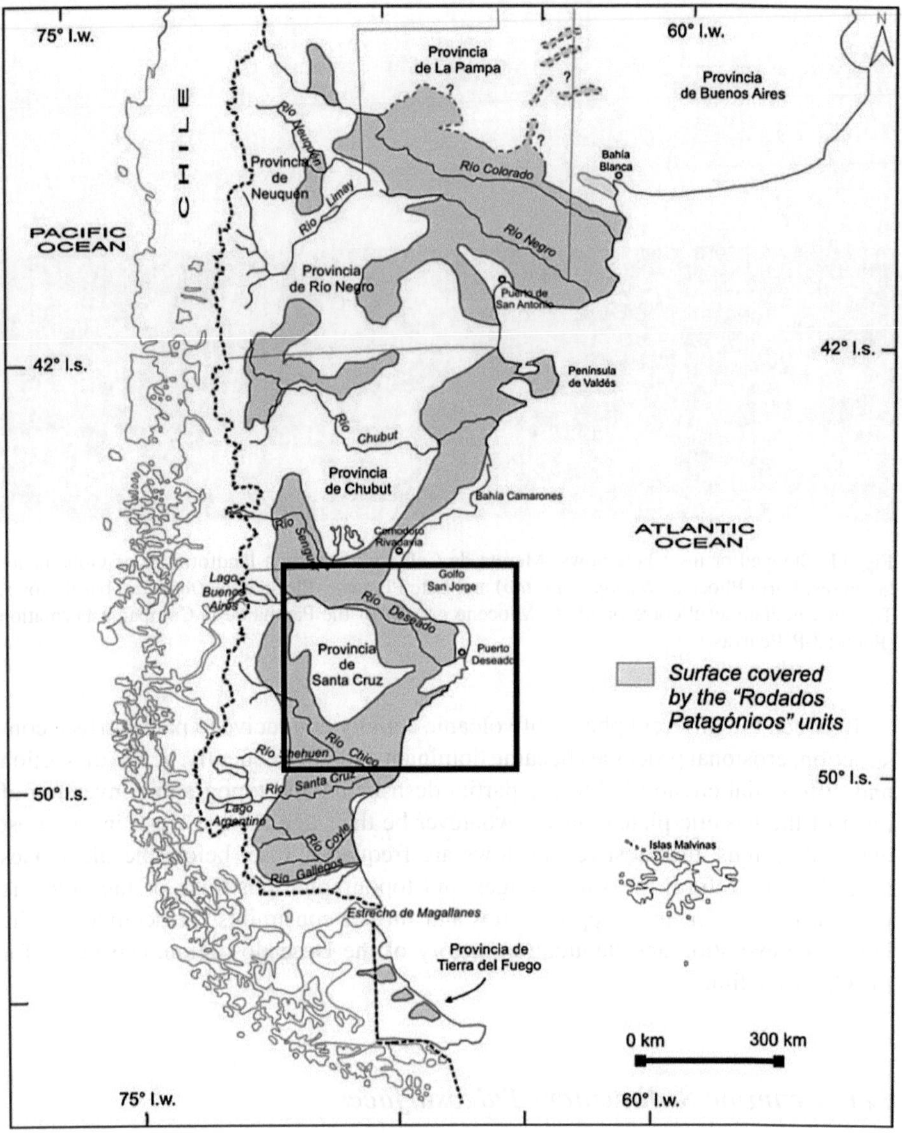

Fig. 12 Geographical distribution of the "Rodados Patagónicos" in southern South America (From Martínez and Kutschker 2011). The *rectangle* locates the study area

rounded pebbles and gently dipping eastward, would result from the coalescence of alluvial cones in very large fluvial systems (Martínez and Kutschker 2011). While these "Rodados Patagónicos" are likely attributable to the denudation response to accelerated crustal uplift since the Middle Miocene, the nature of the gravel sedimentation is also probably linked to a marked shift towards aridity at that time,

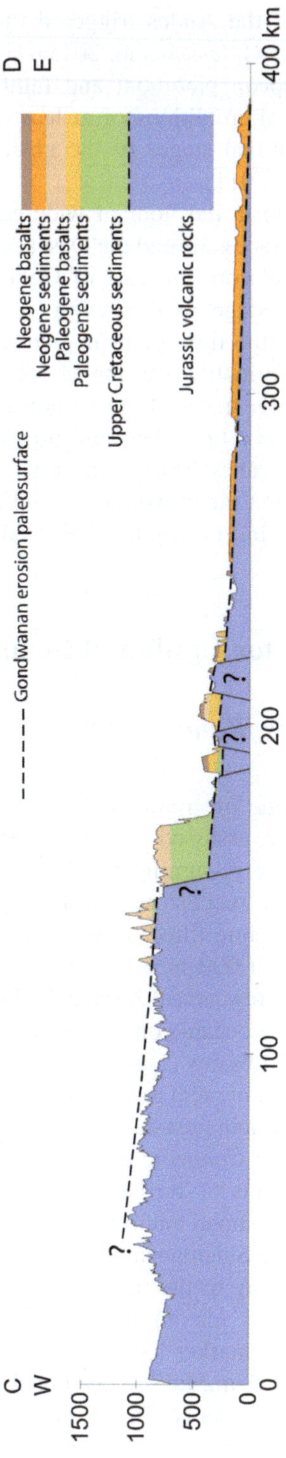

Fig. 13 Geological cross section throughout the Deseado region (profile C–D: see location on Fig. 1)

when the culminating uplift of the Andes triggered the appearance of semiarid conditions in southern Patagonia (Iglesias et al. 2011). However, relationships with marginal moraines on the Andean piedmont and faint topographies of braided channels reflect the fluvioglacial conditions of at least a part of their formation, during glacial stages or in terminal stages of the glaciations (outwash plains or sandur; Martínez and Kutschker 2011).

In the west, these deposits form the floor of wide corridors slightly or deeply inset in the higher and older surfaces, around the Deseado and Chico rivers, or even in deep abandoned valleys (EW corridor, east of Gobernador Gregores; Fig. 1), showing that a well-advanced stage of dissection of the plateaus was already achieved or still in progress at the time of their emplacement. To the east, their distal parts were spread unconformably over the Mesozoic to Oligocene–Miocene rocks, onlapping wider and wider tracts of the Gondwana paleosurface, owing to its lower position and its subdued relief. These uniform surfaces, only degraded by closed depressions and shallow valley heads, form a wide continuous fringe along the coast, where low sea cliffs cut the gravel layers and their beveled substrate, as well as the wide terrace systems inset along the major valleys.

Discussion: Implications for Uplift and Denudation Histories

Geometry of Deformations, Rates, and Causes of Tectonic Uplift

Geometry and relative amplitude of continental-scale tectonic movements since the end of the Jurassic can be assessed by the deformations recorded by the extensive Gondwana paleosurface. Assuming that the erosion paleosurface was near horizontal in the Late Jurassic, a reconstruction of the paleotopography indicates the geometry of a broad monocline tilted seaward, locally downfaulted, with a maximal differential uplift of ~1,000 m between the interior highlands and the coastal lowlands (Fig. 1). In a few areas, normal faulting related to extensional tectonics contemporaneous of the oceanic opening has been detected in the vicinity of Cretaceous grabens and small basins (Fig. 13). As a whole, both regional-scale up-doming and flexural deformation seem to have been the dominant style of crustal deformation in the Deseado region in post-Jurassic times.

Local amplitudes and rates of Cenozoic uplift were estimated in the Deseado Massif from the current elevations of Tertiary marine sediments, most of which were deposited in shallow, nearshore environments (Fig. 14). In the "Bosques Petrificados" area, the base of the Salamanca Formation of Paleocene age currently occurs at ~350 m a.s.l. According to the recalibrated Exxon curve (Miller et al. 2005), ancient sea level was at a maximum of +50 m in Paleocene times; so deformation of this stratigraphic marker locally records a Cenozoic crustal uplift of ~300 m, at a mean rate of 5 m.Ma^{-1} if averaged over the last 65 Myears, including possible periods of subsidence. In the same area, remnants of younger marine sediments (Monte Leon Formation, dating from Late Oligocene to Early

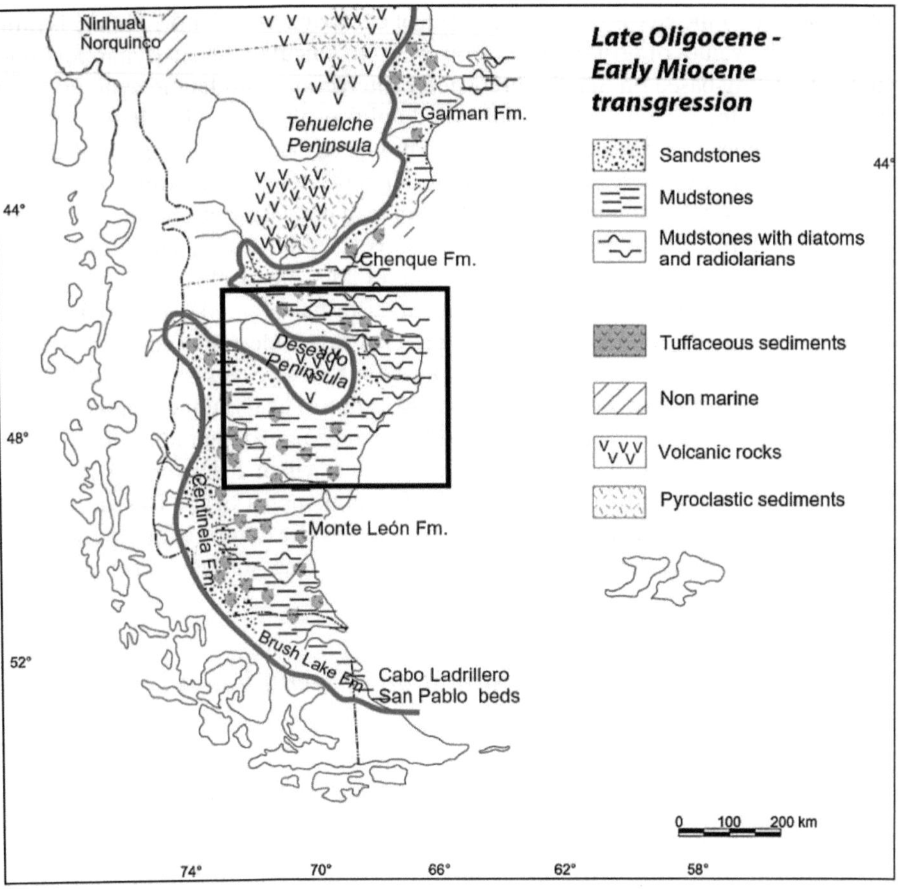

Fig. 14 Paleogeographical reconstruction of the Late Oligocene–Early Miocene transgression (26–20 Ma), responsible for the deposition of the Monte León Formation in the Deseado region (Adapted from Malumián and Náñez 2011). The *rectangle* locates the study area

Miocene) are found up to 600 m a.s.l. Assuming that sea level rose from −30 to +20 m in Late Oligocene–Early Miocene times (Miller et al. 2005), this indicates a Neogene crustal uplift of 500–600 m, at a rate of 20–25 $m.Ma^{-1}$ over the last 25 Myears. In fact, the apparent paradox in the record of uplift by the two sets of sedimentary strata might be explained by the active subsidence of that region until the Early Miocene (Guillaume et al. 2009), that was only recorded by the Paleocene Salamanca formation; it could just as well be due to differentiated vertical movements in different tectonic compartments delineated by normal faults. Anyway, maximal values of Neogene crustal uplift were recorded in the northern part of the Deseado Massif, where the base of the Monte Leon Formation today occurs at ~1,000 m a.s.l., just below the Pliocene basaltic paleosurface of the Meseta de San Pedro. Such an elevation for the marine sediments indicates a maximal crustal uplift of ~1,000 m for the Neogene, at a mean rate of ca. 40 $m.Ma^{-1}$. This estimate is quite similar to the values of post-Middle Miocene uplift calculated from dynamic topography modeling based on mantle–lithosphere interaction (Guillaume et al. 2009). It also confirms the regional-scale south-tilted uplift of the Deseado Massif detected from the geometry of post-Middle Miocene fluvial terraces (Guillaume et al. 2009).

All these data support the idea of an acceleration of tectonic uplift during the Neogene, in probable connection with the Andean convergence and the northward migration of the Chile Triple Junction (Gorring et al. 1997; Guillaume et al. 2009). This migration resulted in the opening of an asthenospheric "slab window" underneath Southern Patagonia, inducing a disturbance in the regional mantle convection. This slab window notably triggered a new widespread episode of back-arc volcanism in southern Patagonia, responsible for the establishment of basaltic plateau and post-plateau lavas in the Deseado region. The dynamic response of the lithosphere resulted in a switch from subsidence to generalized uplift in the Andean foreland that started in the Middle–Late Miocene, when the overall subduction dynamics changed (Guillaume et al. 2009). Dynamic uplift of the Deseado region continued during the Pleistocene, as demonstrated by the deformations recorded by marine terraces along the passive margin of Eastern Patagonia (Pedoja et al. 2011).

Topographic Inversion and Estimation of Denudation Depths and Rates

During the Cenozoic times, the volcanic province of the Deseado Massif has shown strong evidence of topographic inversion. Uplift and differential erosion resulted in an inversion of topography in places where flood basalts filling paleodepressions and paleovalleys became volcanic mesas or buttes. Estimates of denudation depths and rates on given durations can roughly be inferred from values of topographic inversion observed around Cenozoic volcanic paleosurfaces in different places of the Deseado Massif. A mean estimation of Cenozoic denudation rates can be obtained from the values of topographic inversion of Paleogene volcanic paleosurfaces. Maximum values of denudation depths are found in the central part of the Deseado Massif, where topographic inversion of Paleocene to Oligocene basaltic mesas reached 200–500 m. In all cases, the corresponding mean denudation rates are <10 $m.Ma^{-1}$ (most often <4–5 $m.Ma^{-1}$). Such low denudation rates for the Cenozoic are similar to those of vertical tectonic movements deduced from the altitudes of Paleocene marine strata. The same procedure can be applied from the values of topographic inversion of Neogene volcanic paleosurfaces. Maximum post-Miocene denudation depths are provided by the maximum value of topographic inversion observed around the Meseta de San Pedro, in the northern part of the Deseado Massif, i.e., ~800 m since 5 Ma. Close values of denudation depths are found in the central part of the massif, east of the "Bosques Petrificados" area, around Miocene basaltic mesas (~750 m since 15 Ma). In the last area, lower values of denudation depths are observed below Pliocene–Pleistocene basaltic plateau remnants, but the inversion occurred on shorter durations, also indicating mean denudation rates of 40–50 m.Ma^{-1} since the Miocene or later.

This set of data suggests an acceleration of denudation during the Neogene locally as high as 160 m.Ma^{-1}, as it has been observed in the northern part of the massif, around the Meseta de San Pedro. This perfectly matches the observations made on uplift rates accelerating during the Neogene (see above), especially in post-Middle Miocene times when started the generalized uplift of the Andean foreland in connection with ridge collision at the Chile Triple Junction (Guillaume et al. 2009). The sedimentary record of denudation for that period is contained in the Neogene continental deposits and piedmont mantle covers accumulated in the coastal low-lands, suggesting a definitive shift of depositional centers to the new Atlantic Ocean margin associated to an increasingly buoyant and eroding hinterland. Following the deposition of the marine and continental series of Late Oligocene–Early Miocene age (Monte Leon and Santa Cruz formations), the widespread accumulation of the Late Cenozoic "Rodados Patagónicos" (Martínez and Kutschker 2011) has taken place since the Middle–Late Miocene, in response to tectonic and/or climatic forcing. The fluvial terrace levels inset below the accumulation surface of the Late Cenozoic piedmont mantle cover results from the fluvial and fluvioglacial activity of rivers flowing from the Andes to the Atlantic Ocean, in the context of glacial–interglacial periods of the Pliocene–Pleistocene and glacio-eustatic fluctuations.

The presence of so many markers of the geomorphic, sedimentary, and tectonic history, and the well constrained reconstruction of a subsidence stage followed by uplift and relief inversion would make this region an ideal example for discussing the pertinence and the possible mechanisms of the controversial "episodic burial and exhumation model" recently advocated by some authors in other parts of the eastern margin of South America (Northeast Brazil: Japsen et al. 2012). Built on the basis of thermochronological analyses, and of extrapolations from scattered morphostratigraphic evidence, this model implies values and long-term rates of sedimentation and denudation, up to 200 m.Ma^{-1} in the Eocene and 150 m.Ma^{-1} since the Miocene, well beyond those generally observed in the Deseado region, in spite of a more stable tectonic environment, far from the Andes. Accompanied by thermochronological studies (fission track and U-Th/He on apatites), such a comparison would also permit a discussion of the calibration of these methods which often give results showing strong discrepancies with morphostratigraphic evidence (Peulvast et al. 2009; Ricordel-Prognon et al. 2010).

Conclusions and Perspectives

The current landscape of the Deseado Massif reflects a juxtaposition of stepped surfaces which correspond to paleolandforms and paleolandscapes of various types and ages, identified by the means of a morphostratigraphic analysis. Therefore, this preliminary study leads to original results concerning the long-term landscape development of this large platform area and volcanic province of Southern Patagonia. From the Late Jurassic to the Early Tertiary, the Deseado Massif has shown evidence of long-term geomorphological stability, as demonstrated by the good preservation of the Gondwana (Late Jurassic–Early Cretaceous) erosion paleosurface. This paleotopography, irregularly altered and/or weathered by kaolinization, was progressively buried below various thicknesses of Cretaceous–Tertiary sediments and basaltic lava flows, in a context of regional subsidence or relative tectonic quiescence. However, a generalized uplift occurring in the Neogene resulted in the topographic inversion of Cenozoic volcanic plateau lavas, the exhumation of buried paleosurfaces/stratigraphic unconformities, and the dissection of the regional landforms. All our data suggest an acceleration of uplift and denudation since the Miocene, in response to the tectonic forcing induced by changes in the Andean subduction dynamics at the time. The exact timing and causes of denudation history during Meso-Cenozoic times are still poorly constrained. The identification of other erosion paleosurfaces with their associated weathering signatures, and the detailed investigation of offshore sedimentary data, should help in deciphering the complex erosional history of that region. Because of the exploratory nature of this study, and of the lack of data on key problems such as the precise age of some volcanic formations and sedimentary deposits, many issues of long-term landscape evolution are not solved and consequently need further investigations.

References

Aguilera EY, Rabassa J (2010) Origin of the northern Patagonian massif regional paleosurface. Geociências 29(4):467–478. São Paulo, UNESP

Aguilera EY, Rabassa J, Imbelloni P (2012) Paleosuperficie de meteorización en el Macizo del Deseado. V Congreso Argentino de Geomorfología y Cuaternario, Abstracts. Río Cuarto, Córdoba

Andreis RR (2002a) Cuenca La Golondrina (depósitos del rift Pérmico y eventos magmáticos triásicos). In: Haller MJ (ed) Geología y Recursos Naturales de Santa Cruz, Relatorio, 15° Congreso Geológico Argentino, El Calafate, Buenos Aires, pp 71–82

Andreis RR (2002b) Cuenca Baqueró – secuencia volcaniclástica eocretácica del Macizo del Deseado. In: Haller MJ (ed) Geología y Recursos Naturales de Santa Cruz, Relatorio, 15° Congreso Geológico Argentino, El Calafate, Buenos Aires, pp 149–164

Bétard F (2010) Uplift and denudation history at low-elevation passive margins: insights from morphostratigraphic analysis in the SE Armorican Massif along the French Atlantic margin. C R Geosci 342:215–222

Blisniuk PM, Stern LA, Chamberlain CP, Idleman B, Zeitler PK (2005) Climatic and ecologic changes during Miocene surface uplift in the southern Patagonian Andes. Earth Planet Sci Lett 230:125–142

Calvet M, Gunnell Y (2008) Planar landforms as markers of denudation chronology: an inversion of East Pyrenean tectonics based on landscape and sedimentary basin analysis. In: Gallagher K, Jones SJ, Wainwright J (eds) Landscape evolution: denudation, climate and tectonics over different time and space scales, Special publications 296. Geological Society, London, pp 147–166

Carignano C, Cioccale M, Rabassa J (1999) Landscape antiquity of the Central-Eastern Sierras Pampeanas (Argentina): geomorphological evolution since Gondwanic times. Z. Geomorph. N.F., Suppl.-Bd 118:245–268

Cavallotto JL, Violante RA, Hernández-Molina FJ (2011) Geological aspects and evolution of the Patagonian continental margin. Biol J Linn Soc 103:346–362

Cravero MF, Domínguez EA (1992) Kaolin deposits in the Lower Cretaceous Baquer6 Formation (Santa Cruz province, Patagonia, Argentina). J S Am Earth Sci 6(4):223–235

Darwin C (1842) On the distribution of the erratic boulders and on the contemporaneous unstratified deposits of South America. Trans Geol Soc Lond 6:415–431

de Barrio RE (1989) Aspectos geológicos y geoquímicos de la Formación Chon Aike (Grupo Bahía Laura) en el noroeste de la provincia de Santa Cruz. Unpublished doctoral thesis, Facultad de Ciencias Naturales y Museo, Universidad Nacional de La Plata, La Plata, 175 pp

De Giusto JM, Di Persia CA, Pezzi E (1980) Nesocratón del Deseado. In: Turner JC (coord.) Geología Regional Argentina, vol 2. Academia Nacional de Ciencias, Córdoba, pp 1389–1430

Demoulin A, Zárate M, Rabassa J (2005) Long-term landscape development: a perspective from the southern Buenos Aires ranges of east central Argentina. J S Am Earth Sci 19:193–204

Dumont P (1991) Problèmes de datation des surfaces d'aplanissement au Zaïre. Bulletin de la Société géographique de Liège 27:175–185

Echavarría LE, Schalamuk IB, Etcheverry RO (2005) Geologic and tectonic setting of Deseado Massif epithermal deposits, Argentina, based on El Dorado-Monserrat. J S Am Earth Sci 19:415–432

Fidalgo F, Riggi JC (1965) Los rodados patagónicos en la Meseta del Guenguel y alrededores (provincia de Santa Cruz). Revista Asociación Geológica Argentina 20(3):430–443

Fidalgo F, Riggi JC (1970) Consideraciones geomórficas y sedimentológicas sobre los Rodados Patagónicos. Revista Asociación Geológica Argentina 25(4):272–325

Giacosa RE, Márquez M, Panza JL (2002) Basamento Paleozoico inferior del Macizo del Deseado. In: Haller MJ (ed) Geología y Recursos Naturales de Santa Cruz, Relatorio, 15° Congreso Geológico Argentino, El Calafate, Buenos Aires, pp 33–44

Giacosa R, Zubia M, Sánchez M, Allard J (2010) Meso-Cenozoic tectonics of the southern Patagonian foreland: structural evolution and implications for Au-Ag veins in the eastern Deseado Region (Santa Cruz, Argentina). J S Am Earth Sci 30:134–150

Godard A, Lagasquie JJ, Lageat Y (eds) (2001) Basement regions. Springer, Berlin/Heidelberg/New York, 306 p

Gorring ML, Kay SM, Zeitler PK, Ramos VA, Rubiolo D, Fernández I, Panza JL (1997) Neogene Patagonian plateau lavas: continental magmas associated with ridge collision at the Chile triple junction. Tectonics 16:1–17

Gorring ML, Singer B, Gowers J, Kay SM (2003) Plio-Pleistocene basalts from the Meseta del Lago Buenos Aires, Argentina: evidence for asthenosphere-lithosphere interactions during slab window magmatism. Chem Geol 193:215–235

Guido DM, Campbell KA (2011) Jurassic hot spring deposits of the Deseado Massif (Patagonia, Argentina): characteristics and controls on regional distribution. J Volcanol Geoth Res 203:35–47

Guido DM, Campbell KA (2012) Diverse subaerial and sublacustrine hot spring settings of the Cerro Negro epithermal system (Jurassic, Deseado Massif), Patagonia, Argentina. J Volcanol Geoth Res 229–230:1–12

Guido DM, Escayola MP, Schalamuk IB (2004) The basement of the Deseado Massif at Bahía Laura, Patagonia, Argentina: a proposal for its evolution. J S Am Earth Sci 16:567–577

Guillaume B, Martinod J, Husson L, Roddaz M, Riquelme R (2009) Neogene uplift of central eastern Patagonia: dynamic response to active spreading ridge subduction? Tectonics 28, TC2009. doi:10.1029/2008TC002324

Gunnell Y (2003) Radiometric ages of laterites and constraints on long-term denudation rates in West Africa. Geology 31(2):131–134

Haller MJ (2002) La Cuenca Triásica de El Tranquilo. In: Haller MJ (ed) Geología y Recursos Naturales de Santa Cruz, Relatorio, 15º Congreso Geológico Argentino, El Calafate, Buenos Aires, pp 83–86

Homovc JF, Constantini LA (2001) Hydrocarbon exploration potential within intraplate shear-related depocenters Deseado and San Julián Basins, Southern Argentina. AAPG Bull 85: 1795–1815

Iglesias A, Artabe AE, Morel EM (2011) The evolution of Patagonian climate and vegetation from the Mesozoic to the present. Biol J Linn Soc 103:409–422

Japsen P, Bonow J, Green P, Chalmers JA, Lidmar-Bergström K (2009) Formation, uplift and dissection of planation surfaces at passive continental margins – a new approach. Earth Surf Process Landforms 34(5):683–699

Japsen P, Bonow JM, Green PF, Cobbold PR, Chiossi D, Lilletveit R, Magnavita LP, Pedreira A (2012) Episodic burial and exhumation in NE Brazil after opening of the South Atlantic. Geol Soc Am Bull 124(5–6):800–816

Klein C (1997) Du polycyclisme à l'acyclisme en géomorphologie. Ophrys, Gap, 299 p

Lageat Y, Robb LJ (1984) The relationships between structural landforms, erosion surfaces and the geology of the Archean granite basement in the Barberton region, Eastern Transvaal. Trans Geol Soc S Afr 87:141–159

Lidmar-Bergström K, Bonow JM, Japsen P (2013) Stratigraphic Landscape Analysis and geomorphological paradigms: Scandinavia as an example of Phanerozoic uplift and subsidence. Glob Planet Change 100:153–171

Malumián N (2002) El Terciario marino: sus relaciones con el eustatismo. In: Haller MJ (ed) Geología y Recursos Naturales de Santa Cruz, Relatorio, 15º Congreso Geológico Argentino, El Calafate, Buenos Aires, pp 237–244

Malumián M, Náñez C (2011) The Late Cretaceous–Cenozoic transgressions in Patagonia and the Fuegian Andes: foraminifera, palaeoecology, and palaeogeography. Biol J Linn Soc 103: 269–288

Márquez MJ, Giacosa RE, Godeas M (2002) Los granitoides gondwánicos del borde oriental del Macizo del Deseado. In: Haller MJ (ed) Geología y Recursos Naturales de Santa Cruz, Relatorio, 15º Congreso Geológico Argentino, El Calafate, Buenos Aires, pp 57–70

Martínez OA, Kutschker A (2011) The 'Rodados Patagónicos' (Patagonian Shingle Formation) of eastern Patagonia: environmental conditions of gravel sedimentation. Biol J Linn Soc 103: 336–345

Miller KG, Kominz MA, Browning JV, Wright JD, Mountain GS, Katz ME et al (2005) The Phanerozoic record of global sea-level change. Science 310:1293–1298

Nahon D (2003) Altérations dans la zone tropicale. Signification à travers les mécanismes anciens et/ou encore actuels. C R Geosci 335:1109–1119

Nullo F, Combina A (2002) Sedimentitas terciarias continentales. In: Haller MJ (ed) Geología y Recursos Naturales de Santa Cruz, Relatorio, 15º Congreso Geológico Argentino, El Calafate, Buenos Aires, pp 245–258

Nullo F, Combina A (2011) Patagonian continental deposits (Cretaceous-Tertiary). Biol J Linn Soc 103:289–304

Ollier C (1991) Ancient landforms. Belhaven Press, London, 233 p

Panza J, Franchi M (2002) Magmatismo Basáltico Cenozoico Extraandino. In: Haller MJ (ed) Geología y Recursos Naturales de Santa Cruz. Relatorio, 15º Congreso Geológico Argentino, El Calafate, Buenos Aires, pp 259–284

Panza JL, Haller MJ (2002) El volcanismo jurásico. In: Haller MJ (ed) Geología y Recursos Naturales de Santa Cruz, Relatorio, 15º Congreso Geológico Argentino, El Calafate, Buenos Aires, pp 89–102

Panza JL, Sacomani E, Cobos JC (2003) Mapa Geológico de la Provincia de Santa Cruz, escala 1:750.000. Programa Nacional de Cartas Geológicas, SEGEMAR, Buenos Aires

Partridge T, Maud RR (1987) Geomorphic evolution of Southern Africa since the Mesozoic. S Afr J Geol 90:179–208

Pedoja K, Regard V, Husson L, Martinod J, Guillaume B, Fucks E, Iglesias M, Weill P (2011) Uplift of Quaternary shorelines in eastern Patagonia: Darwin revisited. Geomorphology 127:121–142

Peulvast J-P, Claudino Sales V (2004) Stepped surfaces and palaeolandforms in the northern Brazilian "Nordeste": constraints on models of morphotectonic evolution. Geomorphology 62:89–122

Peulvast J-P, Claudino Sales V (2005) Surfaces d'aplanissement et géodynamique. Géomorphol Relief Processus Environ 4:249–274

Peulvast J-P, Vanney JR (2001) Géomorphologie structurale (Terre, corps planétaires solides), vol 1: Relief et structure. Gordon & Breach (Paris) and BRGM (Orléans), 505 p

Peulvast J-P, Bouchard M, Jolicoeur S, Pierre G, Schroeder J (1996) Palaeotopographies and post-orogenic morphotectonic evolution around the Baie des Chaleurs (Eastern Canada). Geomorphology 16:5–32

Peulvast J-P, Claudino Sales V, Bétard F, Gunnell Y (2008) Low post-Cenomanian denudation depths across the Brazilian Northeast: implications for long-term landscape evolution at a transform continental margin. Glob Planet Change 62:39–60

Peulvast J-P, Bétard F, Lageat Y (2009) Long-term landscape evolution and denudation rates in shield and platform areas: a morphostratigraphic approach. Géomorphol Relief Processus Environ 2:95–108

Rabassa J (2010) Gondwana paleolandscapes: long-term landscape evolution, genesis, distribution and age. Geociências 29(4):541–570. São Paulo, UNESP

Rabassa J, Zárate M, Cioccale M, Carignano C, Partridge TC, Maud R (1996) Paisajes relictuales de edad Gondwánica en áreas cratónicas de Argentina. XIII Congreso Geológico Argentino and III Congreso de Exploración de Hidrocarburos, Actas, 4, Buenos Aires, p 219

Rabassa J, Carignano C, Cioccale M (2010) Gondwana paleosurfaces in Argentina: an introduction. Geociências 29(4):439–466. São Paulo, UNESP

Rabassa J, Carignano C, Cioccale M (2014) A general overview of Gondwana landscapes in Argentina. In: Rabassa J, Ollier C (eds) Gondwana landscapes in southern South America. Springer, Dordrecht, pp 201–245

Ricordel-Prognon C, Lagroix F, Moreau MG, Thiry M (2010) Lateritic paleoweathering profiles in French Massif Central: paleomagnetic datings. J Geophys Res 115:B10104. doi:10.1029/2010JB007419

Summerfield MA, Sugden DE, Denton GH, Marchant DR, Cockburn HAP, Stuart FM (1999) Cosmogenic isotope data support previous evidence of extremely low rates of denudation in the Dry Valleys region, southern Victoria Land, Antarctica. Geol Soc Spec Publ 162:255–267

Thomas MF (1994) Geomorphology in the tropics: a study of weathering and denudation in low latitudes. Wiley, Chichester/New York, 482 p

Vasconcelos PM, Conroy M (2003) Geochronology of weathering and landscape evolution, Dugald River valley, NW Queensland, Australia. Geochimica et Cosmochimica Acta 67:2913–2930

Watchman AL, Twidale CR (2002) Relative and "absolute" dating of land surfaces. Earth Sci Rev 58:1–49

Widdowson M (1997) The geomorphological and geological importance of paleosurfaces. In: Widdowson M (ed) Paleosurfaces: recognition, reconstruction and paleoenvironmental interpretation, Special publication 120. Geological Society of London, London, pp 1–12

Pseudokarst and Speleothems in the Chihuido Granite, Province of Mendoza, Argentina

Emilia Y. Aguilera, Silvina Carretero, and Jorge Rabassa

Abstract The core of the Chihuido Anticline is located precisely at Cerro Chihuido, Malargüe, southern Mendoza province, Argentina. This anticline represents the outcropping pre-Jurassic platform of the margin of the Neuquén Basin, which is composed of three volcano-sedimentary units separated by angular unconformities. The older unit corresponds to the volcano-sedimentary complex named as El Fortín, which concludes with the intrusion of a thick rhyolite-monzogranite dyke. The dyke, together with plutons corresponding to the Gondwana magmatism, is exposed on a paleosurface whose relief has been partly fossilized by much modern pyroclastic rocks. The thickness of the dyke varies from 0.2 km to almost 1 km. In this landscape, the dyke is the most remarkable topographic and geological characteristic, not only for its huge dimensions but also for its numerous weathering cavities that provide it with a quite peculiar aspect. Typical pseudokarst landforms are represented by tafoni cavities. The nature of pseudokarst is portrayed by selective erosion along joint planes and andesite composition blocks included in the dyke. In the wall of some

E.Y. Aguilera (✉)
Facultad de Ciencias Naturales y Museo, Universidad Nacional de La Plata (UNLP), Calles 122 y 60, 1900-La Plata, Argentina

DAIS (Dirección de Aplicación de Imágenes Satelitarias), Calle 7 N° 1267-2° P, 1900-La Plata, Argentina
e-mail: eaguilera@fcnym.unlp.edu.ar

S. Carretero
Facultad de Ciencias Naturales y Museo, Universidad Nacional de La Plata (UNLP), Calles 122 y 60, La Plata, Argentina

J. Rabassa
Laboratorio de Geomorfología y Cuaternario, CADIC-CONICET, Ushuaia, Tierra del Fuego, Argentina

Universidad Nacional de Tierra del Fuego, Ushuaia, Tierra del Fuego, Argentina
e-mail: jrabassa@gmail.com

J. Rabassa and C. Ollier (eds.), *Gondwana Landscapes in southern South America*, Springer Earth System Sciences, DOI 10.1007/978-94-007-7702-6_18,

cavities within the almost vertical dyke, opal speleothems have been formed from silica released by rock weathering. The speleothems were studied by means of optical microscopy, scanning electron microscopy (SEM), and X-ray diffraction. The study of these minor features of the granitic landscape, such as tafoni, and the associated speleothems, the description of their morphology, and the analysis of their composition are the main objectives of this chapter.

Keywords Weathering • Etchplain • Granitic geomorphology • Espelothemes • Geochemistry

Introduction

The geomorphological study of granite terrains shows a variety of features at the levels of macro- and microscale. Within the latter, tafoni (taffone, singular) are included, also known as weathering cavities. Although these features have been studied on other lithological types, tafoni are common as microlandforms in granitic rocks (Vidal Romaní 1984; Twidale and Vidal Romaní 2005), associated to other features of the granitic landscape. These microlandforms are limited by cavities or void spaces holding a concave profile with varied dimensions. There are classifications of the tafoni types, either in the evolutionary framework of the larger landforms (Vidal Romaní 1984; Vidal Romaní and Twidale 1998; Twidale and Vidal Romaní 2005), in which the cavities are distinguished as equidimensional or heterodimensional, according to their axial relationships, but they also receive various denominations concerning their localization in steep slopes, base of the boulders, etc.

Numerous works developed on the granitic microlandforms are coincident about a genesis in subaerial and/or superficial conditions, as a result of exogenous processes that include many physical and chemical processes. Likewise, these features could be the result of the interaction of exogenous and endogenous processes, such as negative exfoliation and deuterical alteration processes. Since these microlandforms have been identified in different morphogenetic environments, it has been assigned an origin related to process convergence. Finally, Vidal Romaní (1984) linked these microlandforms to rock geotechnical characteristics.

The different minerals that form granite, such as quartz, feldspar, and mica, show a differential behavior in relation to chemical weathering; therefore, a selective alteration of these rocks takes place, controlled by their texture, mineralogical composition, and micro-fissures. Thus, the dissolution acts also through the intracrystalline surfaces (Martini 1984; Urbani 1986; Galán 1988). Recent studies on rocks supposedly considered before as not affected by dissolution have exposed that dissolution is just the chemical aspect of a complex problem, including dissolution rate, duration of the chemical reactions, dynamics of water circulation, and the morphogenetic conditions under certain hydrological aspects (Galán 1991). Pseudokarst, in the sense of Vidal Romaní and Twidale (1998), is considered in this chapter in the context of features developed in granitic or volcanic acid rocks, which show intense similarities with equivalent features found in soluble rocks such as limestone. In Queensland, Australia, several caves have developed in granites,

Fig. 1 Localization of the dyke in Cerro Chihuido, south of Malargüe, Province of Mendoza, Argentina, 69° 34′ W and 35° 35′ – 35° 37 S

some of them such as South Bald Rock Cave and River Cave have opal speleothems of the coral-like type and flowstone on the walls (Webb and Finlayson 1984, 1987). In California, N–NE of San Diego, in the Cahuilla Cave Creek, Finlayson (1985) recognized speleothems of the coral-like type on the walls and flowstone on the floor; both speleothem types are basically of calcite (Webb and Finlayson 1987). Also in Galicia (NW Spain), Vidal Romaní and Vilaplana (1983), Vidal Romaní and Twidale (1998), and Vidal Romaní and Vaqueiro Rodríguez (2007) cited the development of speleothems in granitic rocks. Other authors studied the genesis of speleothems in granitic rocks under a variety of climatic conditions (Fernández Verdía 2000; Sanjurjo 2000). Later, Sanjurjo et al. (2006) studied these dissolution processes in granitic rocks. In Argentina, Cioccale et al. (2008) found siliceous speleothems in open cracks and tafoni walls, in the granitic rocks of the Achala Batholith (province of Córdoba).

Location of the Study Area

The Chihuido Anticline, located south of Malargüe (69° 34′ W; 35° 35′–35° 37 S), is close to National Route 40 and has been partially eroded by the Tronquimalal (in the south) and Loncoche (in the north) creeks (Fig. 1).

Geology

At the core of the Chihuido Anticline (Gerth 1928; Groeber 1947), igneous and sedimentary rocks ranging in age from the Permian to the Middle Jurassic are found. In this period, various cycles of igneous activity took place, following the active margin of the Gondwana supercontinent, with ample regional distribution and a very large volume of eruptive rocks, represented by three volcano-sedimentary units, separated by angular unconformities. The older unit corresponds to the El Fortín volcano-sedimentary complex, composed of breccias, lava flows and andesitic domes, and conglomerates and sandstones. The final episode of this unit corresponds to the intrusion of a thick dyke of rhyolite-monzogranite composition, which has been dated on 250 ± 5 Ma, by the U-Pb conventional technique. This age allows the correlation of El Fortín Complex with the Choiyoi Group, in the sense of Rolleri and Criado Roque (1970), found in the Cordillera Frontal and the San Rafael Block (Llambías et al. 2005).

The reconstruction of the geological evolution of this region has recorded the existence of angular unconformities in between the Permian and Late Triassic-Early Jurassic eruptive and sedimentary cycles, which are interpreted as tectonic events of varied intensity (Llambías et al. 2005). The sedimentological analyses of these units suggest the existence of an aggressive landscape as a result of the diastrophic movements. The unconformity between the Tronquimalal Group (the overlying unit) and the El Fortín Complex is correlated with the Huarpes diastrophic phase, sometime between 230 and 220 Ma in the Cordillera Frontal, although it could be diachronic. The Huarpes diastrophic phase had a great extension along the active margin of the Gondwana supercontinent. The cited authors remarked that the Triassic outcrops of the Chihuido Anticline correspond to the basin margin, a scenario mainly dominated by the volcanic landscape and not by basin subsidence.

Overlying the Choiyoi Group, continental and marine deposits pertaining to the "Precuyano" units and the Cuyo Group occur, being Early to Middle Jurassic in age. Then the continental and marine deposits of the Lotena Group follow, which culminate with the Oxfordian evaporites of the Auquilco Formation. Several units of Kimmeridgian to Barremian age of the Mendoza Group are found later on, composed of the Tordillo (sandstones), Vaca Muerta (bituminous shales), Chachao (limestones), and Agrio (black shales and limestones) formations. Afterwards, the Rayoso Group is found, of Aptian-Albian age, composed of the Huitrín and Rayoso formations (evaporites, sandstones, and limestones). Fluvial deposits of Late Cretaceous age are overlying, belonging to the Neuquén Group. Finally, Quaternary alluvial and colluvial deposits and basalt and ignimbrite flows complete the sequence.

The Rhyolite-Granite Dyke

In the Chihuido Anticline region, Backlund (1923) described for the first time a "granite or granitic porphyr," a name later changed to "Chihuido granite and porphyr" (Dessanti 1973). Later on, Llambías et al. (2005) analyzed the unusual thickness and the textural variations of this body, where the porphyritic and micrograined textures are remarkable. Based upon the granular textures, these authors classified the central sector of the dyke as fine-grained monzogranite, where the dyke reaches its maximum thickness. This thickness ranges up to almost 1 km N of the Tronquimalal creek. These authors observed that in the Loncoche creek, the thickness is of 500 m, occupying the monzogranite approximately 200 m of the central part, transitionally passing into rhyolites towards the margins of the dyke, where felsitic matrix dominates.

Methodology

Fieldwork for the reconnaissance and measuring the outcrops and their relief was performed, including identification of the different “in situ” textures of several sectors of the dyke, determination of the fracturing pattern, identification and measurement of the weathering cavities, lithological classification of xenoliths, and search and collection of speleothems. These materials were analyzed by optical microscopy, SEM, and X-ray diffractometry.

Results

In the field reconnaissance, the dike was found to have N trend and subvertical attitude. It is accompanied by a parallel system of joints and faults. At a regional scale, this outcrop is part of an exposed paleosurface, but only in the northern sector, where a paleolandscape has been preserved, composed of a fluvial valley buried by modern pyroclastic rocks (the Malargüe Ignimbrite, of Quaternary age; Fig. 2).

At a greater detail, many cavities are observed, most of them tafoni although cavities of the pia/gnamma type, or weathering depressions, are found as well.

Tafoni have sizes ranging from decimeters to meters, and they occur as dispersed in the walls of other granitic landforms. The tafoni host other features that are described here. The development of several shapes and varying morphology has been observed, with dominant equidimensional shapes. In lower zones of the body, their diameters are slightly larger than a few decimeters, whereas in the middle zone, they reach sizes of 60–70 cm, and in the upper part, they grow over 1 m. In the uppermost levels, the shapes of larger size are rocky shelters showing smaller tafoni in their inner portions (Fig. 3). Wall tafoni and basal tafoni have been identified. The larger number of tafoni developed in plunging walls. In very intensively jointed walls, spheroidal weathering features are identified, where rounded nuclei of fresher rock surrounded by thin concentric rock layers are observed (Fig. 4).

With a lower frequency, tafoni developed at the base of isolated boulders limited by orthogonal joint systems have been observed (Fig. 5).

Another group of tafoni of homogenous size, markedly smaller than the previous ones, is localized at the convergence of joint systems and aligned along joint planes (Fig. 6).

Fig. 2 (**a**) Panoramic view of the paleosurface. (**b**) Fluvial valley buried by the Malargüe Ignimbrite (Quaternary). (**c**) Details of the ignimbrite mantle covering the paleolandscape (the *arrow* shows the bottom of the valley)

Fig. 3 Summit sectors of the dyke where rock shelters are found showing development of tafoni of a smaller size

On horizontal surfaces on the top of the dyke weathering pits are present, generally associated with tafoni (also known as "pila", "pia", "vasque," or "gnammas"). They are depressions formed by excavation, cavities of variable dimensions that are slightly larger than 1 m and with depths around 0.60 m. These microlandforms keep rainwater temporarily, and they usually have sediment accumulation in their inner part. According to their morphology, gnammas of flat bottom have been recognized in which no discharge cannel has been observed. Some of these features show wall

Fig. 4 Features of spheroidal weathering, rounded nuclei of fresher rock surrounded by thin concentric layers

Fig. 5 Tafoni developed at the base of isolated boulders limited by orthogonal joint systems

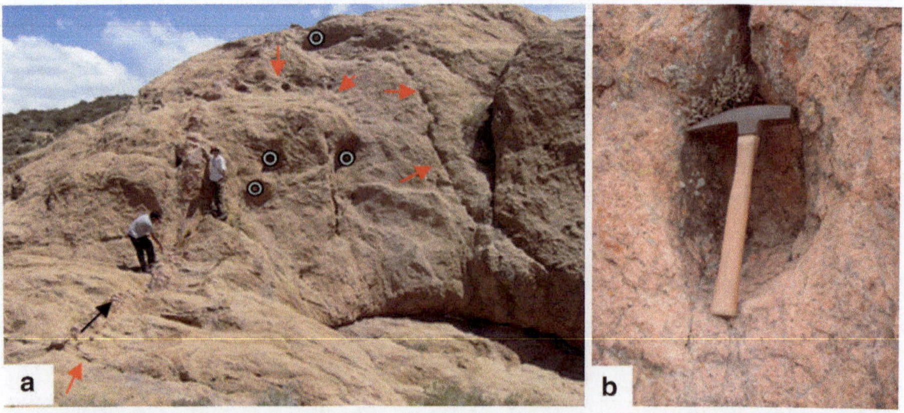

Fig. 6 (**a**) Alignment of smaller tafoni related to joint systems. *Red arrows* indicate aligned tafoni along joint planes. *Light blue circles* depict tafoni of greater size. The *black arrow* indicates an aplitic dyke. (**b**) Details of tafoni related to joints

Fig. 7 In the upper topographic levels of the dyke, weathering gnammas developed on horizontal surfaces are identified. (**a**) Cavities with diameter over 1 m and a depth of 0.60 m. (**b**) Details of the granitic debris in the inner part as products of rock degradation

degradation due to a longer exposition to weathering, granitic debris being common in their inner part as a product of rock disintegration (Fig. 7).

In the inner part of some tafoni and open joints, speleothems formed by very thin mineral crusts are found, composed of microcrystalline aggregates of light gray color, milky aspect, and mammillary surface (Fig. 8). At the petrographic microscope, rhythmite layering growth with banded opal is observed. Concerning

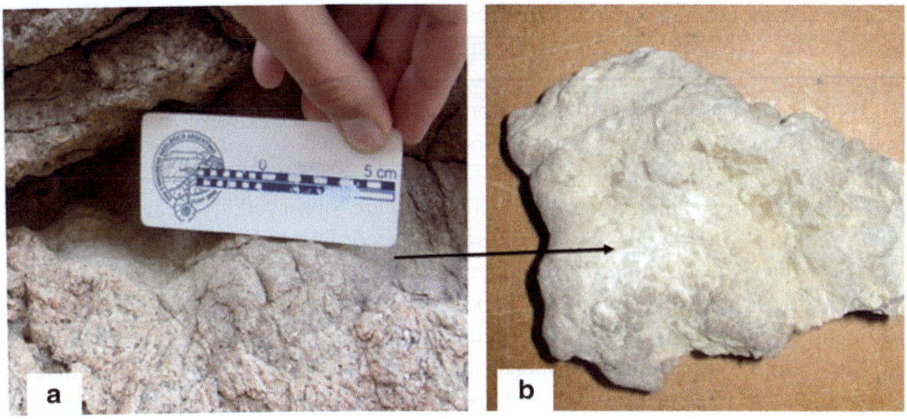

Fig. 8 (**a**) Speleothem along joints. (**b**) Macroscopic details of the speleothem composed of light-grayish, microcrystalline aggregates, of milky aspect and mammillary surface

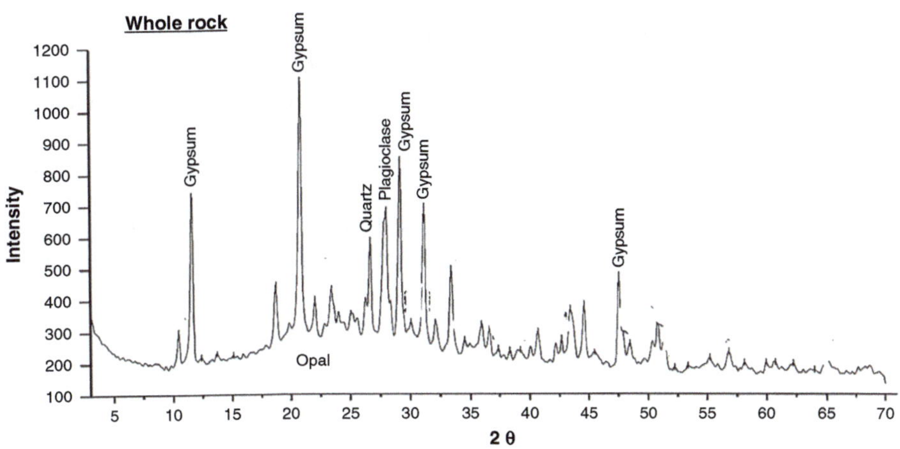

Fig. 9 The XRD records a diffuse peak of extended base with the center located around 26° (2θ~3.4 Å), which is typical of opal A. Peaks corresponding to crystalline silica superposed to opal and other angular values (2θ) are found. Other minerals present are feldspars, micas, and gypsum

their mineralogy, analysis by X-ray diffractometry determined that these deposits are basically composed of opal A. The samples show debris components (such as quartz, feldspars, and phyllosilicates) in smaller proportions. The results reveal the presence of amorphous silica, crystalline types of SiO_2, and other minerals typical of granitic rocks. The DRX of Fig. 9 shows a diffuse peak of extended base with the center located around 26° (2θ~3.4 Å) which is typical of opal A. Additionally, peaks corresponding to crystalline silica superimposed to the opal in other angular values (2θ) have been found. Other minerals present as contaminants are feldspars, mica and gypsum.

With respect to the SEM observations on speleothem samples, results in agreement with the different analyzed sectors were obtained, by contrasting the textures and porosity of the inner and outer surfaces, with approximations of 250× to 4,000×. Colloform and mammillary forms were observed, typical of solutions that precipitate in cavities. The textures vary according to the precipitation levels. Towards the external zone, rounded (mammillary) shapes are dominant, covered by gypsum crystals of threadlike shape. In the inner part, dish-shaped forms are noticeable, characterized by a continuous opal covering, with concoidal fracture and commonly with superficial perforations as little channels or tubes of diverse inclination (biological activity, perhaps?). These zones occur as interrupted by gypsum levels; these levels show a distinct compaction/porosity from the inner sector to the surface. In the latter case, they behave as thin threads or "whiskers," whereas in the inner zone, they form little rose-type shapes (Fig. 10a–d).

From a chemical point of view, the elementary analyses performed on each sector of the studied samples show that Si is the major component, accompanied by Al, Fe, K, Na, Mg, Ti, S, and Ca. Si keeps a constant relationship with Al (i.e., 4:1), whereas the relationship with Fe is 4:0.5, and a similar behavior with Mg, both elements of very low mobility. With respect to alkaline elements Na and K, elements of high mobility, although they are recorded in all samples, their contents are low, as expected. In a few samples, the content of Ti shares comparable values with K. Finally, S and Ca, abundant in the superficial crust under the mineralogical form of gypsum, are absent in the inner sectors where the opal patina massively covers the speleothem, with the exception of sectors in which thin, rose-type, gypsum layers are interbedded (Figs. 10, 1 and 2).

Conclusions

The granitic microlandforms analyzed in this chapter are the product of exogenous processes that included numerous physical and chemical processes, forming the so-called granitic pseudokarst. These processes were enhanced by endogenous properties related to the primary structures of the dyke, such as the presence of andesitic xenoliths. These xenoliths would have experimented differential responses to the cited physical and chemical processes, as reflected in the alteration/decomposition of their mineral components. Besides, also in the genesis of these microlandforms, the fissure joint system was involved as well. The pseudokarst process is enhanced by selective erosion along joint planes and andesitic composition xenoliths. Thus, the structural weaknesses of the rocky massif were determined by mineralogy, texture, and joint intersection.

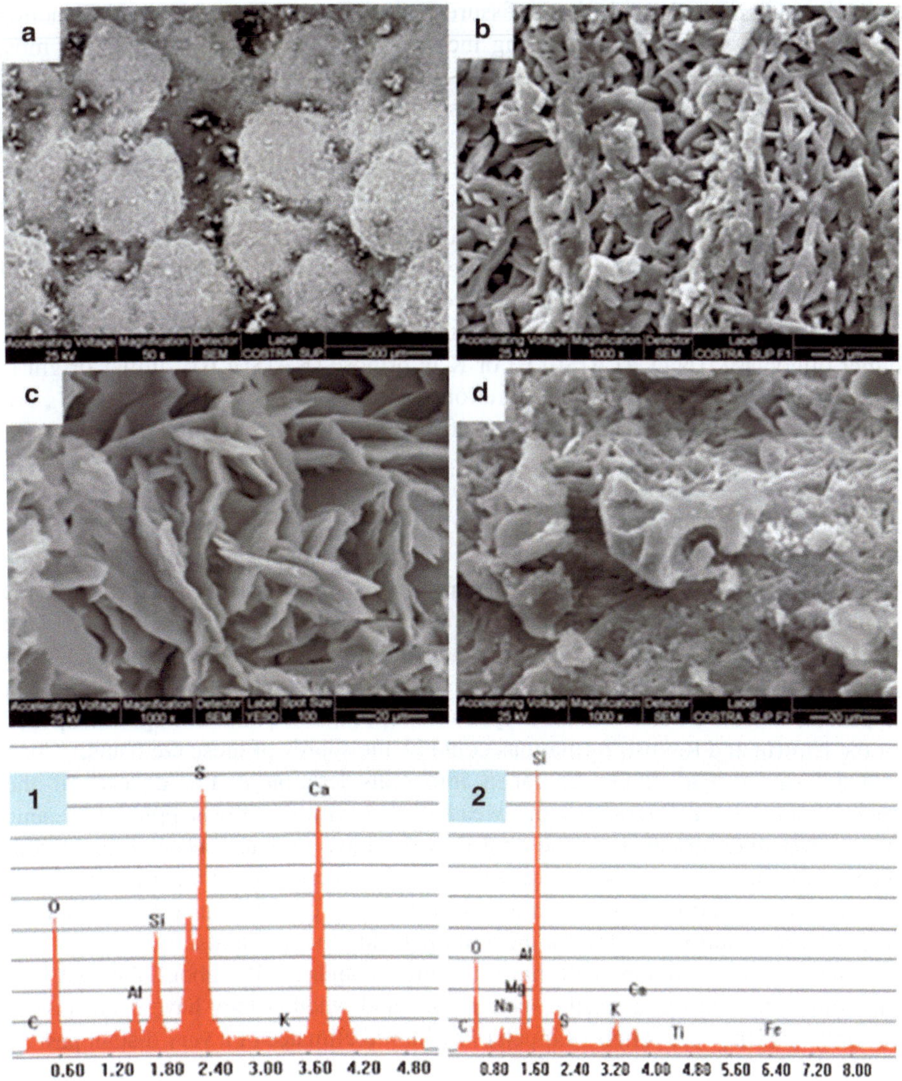

Fig. 10 (**a**) Superficial view of rounded landforms covered by gypsum crystals. (**b**) Details of a cavity developed over the *rounded shapes*. (**c**) Crystallization of gypsum with tendency to formation of *rose-like shaped* structures. (**d**) Inner sector of opal A with tubes/channels and compact gypsum. Semi-quantitative analysis: (*1*) an external and compact zone of the mammillary texture sample: the high values of S and Ca correspond to the gypsum crystals; (*2*) of the inner zone of the sample with continuous opal covering, with important presence of Si and, in lower proportion Al and Mg due to its lower mobility

The weathering of granite in the fissure systems was produced by the interaction of water and rock. Rock weathering includes several processes, being the most important the dissolution of the composing elements of the rock, and later, the precipitation of elements and soluble substances, originating speleothems.

With respect to the provenance of the elements recognized in the speleothems, Si, Al, Fe, K, Na, Mg, and Ti are considered as autochthonous, that is, direct products of weathering due to dissolution processes of the Chihuido granite, originally composed of orthose, oligoclase, quartz, and biotite. The provenance of Si is clearly linked to feldspars and crystalline quartz, although the intervention of biological activity cannot be ruled out, generating deep changes in P and even in biogenic opal. The alkaline elements Na and K are liberated from the decomposition of plagioclase, either oligoclase [(Na Ca) $AlSi_3O_8$] or K in orthose ($KAlSi_3O_8$); similar origin is assigned to Al. Meanwhile, the alteration of biotite $\{K_2(Mg, Fe^{2+})_{6-4}(Fe^{3+}, Al, Ti)_{0-2}[Si_{6-5}Al_{2-3}O_{20}](OH, F)_4\}$ is responsible for the concentration of Fe, Mg, Ti, K, and Al.

In relation to S and Ca, these elements are considered as foreign to granite, that is, allochtonous. Their provenance would be related to the Mesozoic formational units of the Neuquén Basin, characterized by important deposits of marine limestone of the La Manga Formation (Lotena Group), which culminates with Oxfordian evaporites (gypsum) of the Auquilco Formation, and also limestone belonging to the Chachao and Agrio formations (black shale and limestone) of the Mendoza Group (Kimmeridgian to Barremian). Likewise, extensive evaporate sheets with dominant gypsum and Ca-carbonate of the Rayoso Group, Aptian-Albian in age, composed of the Huitrín and Rayoso formations could be the source of these elements.

One aspect that should be taken into consideration is the genesis of the speleothems in the interior of the weathering cavities is the supply of organic matter and microorganisms. The opal deposits have probably been originated to a certain point by organic chemical reactions, but these studies have not proven it yet. The investigation should continue to solve some questions as the participation of microorganisms in the development of opal and other minerals.

Finally, future studies will have to inform and clarify about the age and environmental conditions of the speleothems and whether they have been formed recently or if they are relicts of much more humid climates of a very distant past.

References

Backlund H (1923) Der magmatische anteil der Cordillera von süd Mendoza. Meddeladen fran Abo Akademis Geologisk-Mineralogiska Institut 3:1–298, Abo

Cioccale MA, Pasquini AI, Depetris PJ (2008) Hallazgo de espeleotemas silíceas en rocas graníticas del batolito de Achala, Sierras Pampeanas de Córdoba. Revista de la Asociación Geológica Argentina 63(3):417–420

Dessanti RN (1973) Descripción geológica de la hoja 29b, Bardas Blancas (Provincia de Mendoza). Servicio Nacional Minero Geológico, Boletín 139:1–70, Buenos Aires

Fernández Verdía MA (2000) Estudio de la génesis de espeleotemas en granitoides en climas áridos (Salt River, Kellerberrin, Western Australia). Unpublished graduation thesis, Universidade da Coruña, A Coruña, Spain, 81 pp
Finlayson BL (1985) The formation of caves in granite. In: Sweeting MM, Paterson K (eds) New directions in karst. Geobooks, Norwich
Galán C (1988) Cavernas y formas de superficie en rocas precámbricas del Grupo Roraima (Guayana Venezolana). Bol SVE 23:1–12
Galán C (1991) Disolución y génesis del Karst en rocas carbonáticas y rocas silíceas: un estudio comparado. Munibe (CiencNat), SCAranzadi 43:43–72
Gerth E (1928) La estructura geológica de la Cordillera Argentina entre el Río Grande y el Río Diamante. Academia Nacional de Ciencias, Actas 10(2):125–172, Córdoba
Groeber P (1947) Observaciones geológicas a lo largo del Meridiano 70. Hojas Barda Blanca y Los Molles. Revista de la Asociación Geológica Argentina 2(4):409–433
Llambías EJ, Sato AM, Basei MAS (2005) El basamento pre-jurásico medio en el anticlinal Chihuido Malargüe: evolución magmática y tectónica. Revista de la Asociación Geológica Argentina 60(3):567–578
Martini J (1984) Rate of quartz dissolution and weathering of quartzite. Bull S Afr Speleol Assoc 25:7–10
Rolleri EO, Criado Roque P (1970) Geología de la provincia de Mendoza. IV Jornadas Geológicas Argentinas (Mendoza), Actas 2:1–60, Buenos Aires
Sanjurjo J (2000) Estudio comparado de la génesis de espeleotemas síliceos en granitoides en climas semiáridos y templado-húmedos. Unpublished graduation thesis, Universidade da Coruña, A Coruña, Spain, 72 pp
Sanjurjo J, Fernández Mosquera D, Vidal Romaní JR (2006) Procesos de disolución en ambientes seudocársticos graníticos. Actas de la IX Reunión Nacional de Geomorfología, Santiago de Compostela, pp 773–784, ISBN 84-9750-641-3
Twidale CR, Vidal Romaní JR (2005) Landforms and geology of granite terrains. Balkema, London, 351 pp
Urbani F (1986) Una revisión sobre las cavidades y demás formas relacionadas desarrolladas en las rocas cuarcíferas precámbricas del Grupo Roraima, Venezuela. Interciencia 11(6):298–300
Vidal Romaní JR (1984) Microformas graníticas tipo Tafoni (Cachola) y Gnamma (Pia). Un micromodelado sin relación con el clima o la estacionalidad. Caderno Laboratorio Xeolóxico de Laxe 7:273–277
Vidal Romaní JR, Twidale CR (1998) Formas y paisajes graníticos. La Coruña. Universidad da Coruña, 411 pp
Vidal Romaní JR, Vaqueiro Rodríguez M (2007) Types of granite cavities and associated speleothems: genesis and evolution. Nat Conserv 63:41–46
Vidal Romaní JR, Vilaplana JM (1983) Datos preliminares para el estudio de espeleotemas en cavidades graníticas. Cadernos Laboratorio Xeolóxico de Laxe 7:305–324
Webb JA, Finlayson BL (1984) Allophane and opal speleothems from granite caves in south-east Queensland. Aust J Earth Sci 31:341–349
Webb JA, Finlayson BL (1987) Incorporation of Al, Mg, and water in opal-A: evidence from speleothems. Am Mineral 72:1204–1210

The Exhumation of the Northern Patagonian Massif Gondwana Planation Surface Due to Uprising During the Oligocene

Eugenio Aragón, Emilia Y. Aguilera, Claudia E. Cavarozzi, and Alejandro Ribot

Abstract The altiplano (or high plain) of the Northern Patagonian Massif is a large, 100,000 km^2 geomorphological unit that rose from sea level to at least 1,200 metres above sea level (m a.s.l.) in Early Oligocene times, as a consequence of epeirogenic uplift. This uniform tableland feature is essentially a Cretaceous planation surface carved on Paleozoic igneous and metamorphic rocks of the Northern Patagonian Massif. This planation surface had been preserved by a thin and scattered cover of Maastrichtian-Danian marine sediments and Late Oligocene-Early Miocene basaltic flows. Erosion since Middle Miocene times at this tableland has exposed much of the Gondwana planation surface and developed numerous basaltic plateaus by relief inversion.

Keywords Altiplano • Patagonia • Plateau • Planation surface • Basalts

---

E. Aragón (✉) • C.E. Cavarozzi
Centro de Investigaciones Geológicas (Universidad Nacional de La Plata-CONICET), 1 Nº 644, (1900), La Plata, Argentina

Facultad de Ciencias Naturales y Museo, Universidad Nacional de La Plata, 122 y 60, s/n., (1900), La Plata, Argentina
e-mail: earagon@cig.museo.unlp.edu.ar

E.Y. Aguilera
Facultad de Ciencias Naturales y Museo, Universidad Nacional de La Plata (UNLP), Calles 122 y 60, 1900-La Plata, Argentina

DAIS (Dirección de Aplicación de Imágenes Satelitarias), Calle 7 N° 1267-2° P, 1900-La Plata, Argentina
e-mail: eaguilera@fcnym.unlp.edu.ar

A. Ribot
Facultad de Ciencias Naturales y Museo, Universidad Nacional de La Plata, 122 y 60, s/n., (1900), La Plata, Argentina

LEMIT – CIC, 52 entre 121 y 122, (1900), La Plata, Argentina

J. Rabassa and C. Ollier (eds.), *Gondwana Landscapes in southern South America*, Springer Earth System Sciences, DOI 10.1007/978-94-007-7702-6_19,

Introduction

A Gondwana planation surface is well preserved in the extra-Andean region of Northern Patagonia, in the Northern Patagonian Massif, a tectono-morphological, geological unit. The massif is composed of Proterozoic and Early Paleozoic metamorphic rocks, intruded by Ordovician, Carboniferous, and Permian igneous bodies, followed by Triassic and Jurassic volcanic complexes. This massif was strongly eroded from the Middle Jurassic to the Late Cretaceous (Rabassa 1978; González Díaz and Malagnino 1984; Franchi et al. 1984; Corbella 1984; Aguilera 2006; Aragón et al. 2003, 2005, 2008, 2009, 2010, 2011), feeding its cover sediments into the surrounding basins. These are the Neuquén Basin to the northwest, the Colorado Basin to the northeast, the Ñirihuau Basin to the west-southwest, and the Cañadón Asfalto Basin to the south (Fig. 1). A large Gondwana planation surface was then exposed as the sediments covering it were modified. This planation surface is finally covered and preserved by a thin sedimentary sequence of Late Cretaceous and Early Tertiary (Fig. 2a) continental-marine sediments (Malumián 1999) and basalts of Oligocene–Early Miocene age (Ardolino 1981; Corbella 1984; Franchi et al. 1984, and references therein).

In present times, the Northern Patagonian Massif shows a rather flat, topographic surface at an average elevation of 1,200 meters above sea level (m a.s.l.) (Coira 1979; Nullo 1978) that stands 500–700 m above the surrounding country and

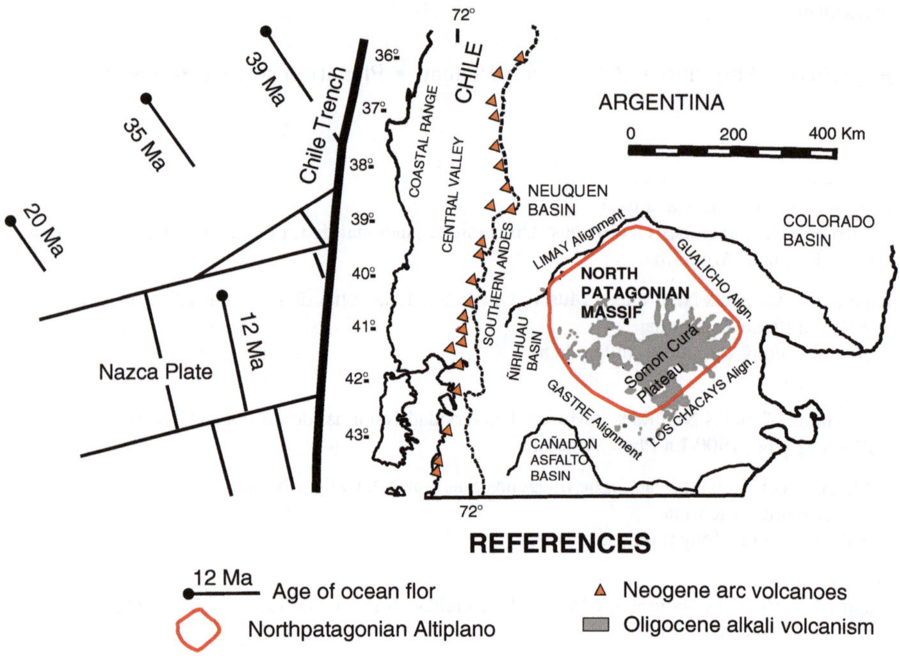

Fig. 1 Main Cenozoic geological features of Northern Patagonia

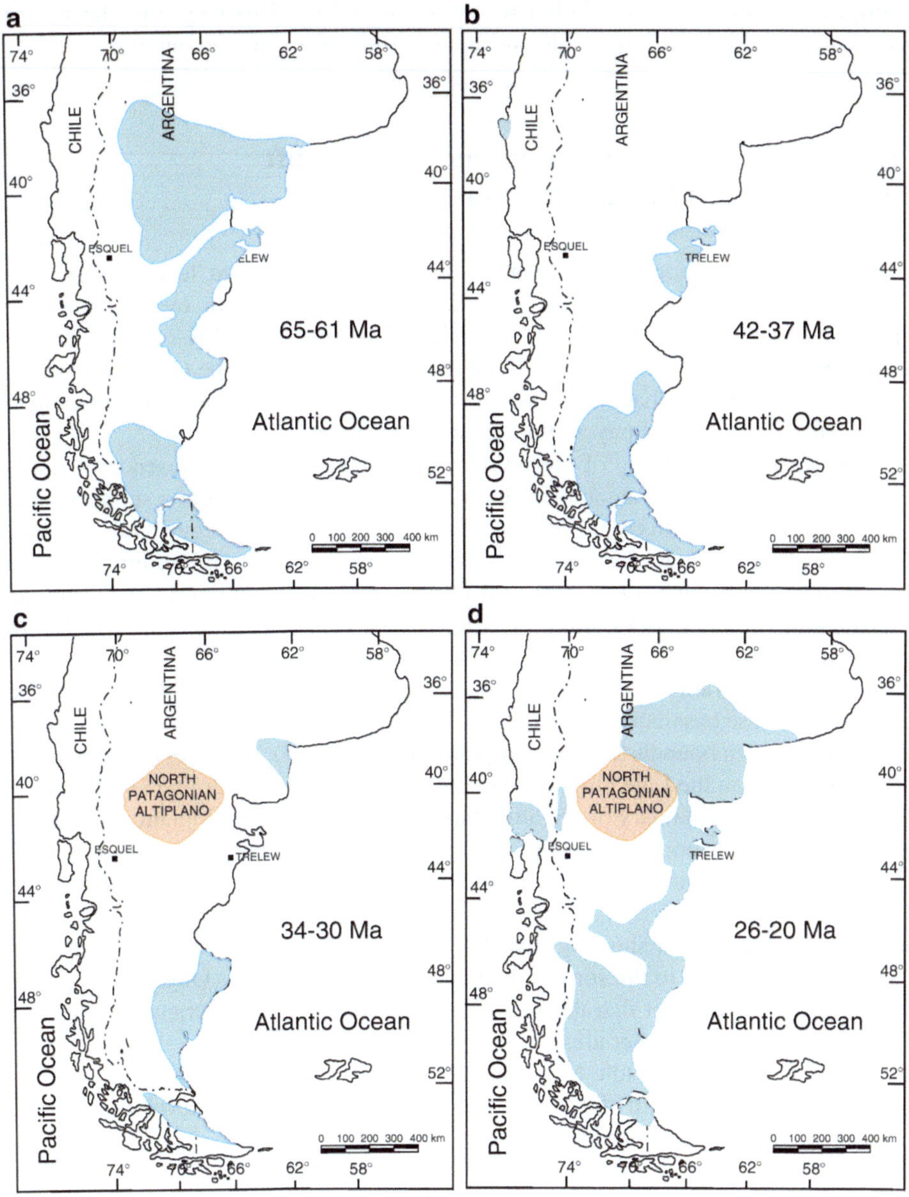

Fig. 2 Paleogeography of Patagonia, showing main Tertiary transgressions with respect to the Northern Patagonian Massif uprise: (**a**) Maastrichtian-Danian; (**b**) Eocene; (**c**) Oligocene; (**d**) Late Oligocene-Early Miocene

peripheral rivers (the Río Chubut to the south and the Río Negro to the north), showing that this high-lying tableland is an "altiplano" or high plateau.

Exhumation of the Altiplano Due to the Uplift of the Northern Patagonian Massif

The uplift of the Gondwana planation surface from near sea level positions to 1,200 m a.s.l. took place in the Early Oligocene (Fig. 2c), or even in the Late Eocene (Aragón et al. 2011), as a result of epeirogenic uplift of a surface of 100,000 km^2 (Fig. 1), constrained by four major alignments: the Gastre alignment to the southwest, the Chacays alignment to the southeast, the Gualicho alignment to the northeast, and the Limay alignment to the northwest.

The epeirogenic nature of the uplift is demonstrated by the presence of undeformed Maastrichtian-Danian marine sediments (Fig. 2a) that kept a subhorizontal position even at an average altitude of 1,100 m, scattered almost all over the surface of the altiplano. Instead, the Maastrichtian-Danian marine sediments stand at altitudes that range about 300–500 m and show deformation, at the altiplano surrounding country rocks along the Río Chubut to the south and the Río Negro to the north.

The age constrain to the uplift of the altiplano is set by the basalt flows above the tableland. These basalts cover more than 15,000 km^2 from the central to the eastern and southeastern boundaries of the altiplano (Figs. 1 and 3). The ages of these basalt flows range from the Late Oligocene to the Early Miocene (Ardolino 1981), and at the southeastern boundary of the altiplano, they spill from the altiplano edge down to the surrounding country as a cascade descending more than 500 m, showing that the altiplano had gained much of its altitude previous to the eruption of these lava flows.

A second argument showing that the North Patagonian altiplano has remained as such with respect to the surrounding country land since Late Oligocene times is suggested from the fact that the Late Oligocene-Early Miocene transgression was not able to flood the higher areas of the altiplano (Fig. 2d).

Erosion since the Middle Miocene at the altiplano of the Northern Patagonian Massif has exposed much of the Gondwana planation surface and developed plateau basalts by local relief inversion (Fig. 4).

Tectonic Setting

The Paleogene uplift history of Northern Patagonia is related to the major plate convergence changes caused by the collision of the South America plate with the Farallón-Aluk active ridge in the Late Paleocene (Cande and Leslie 1986).

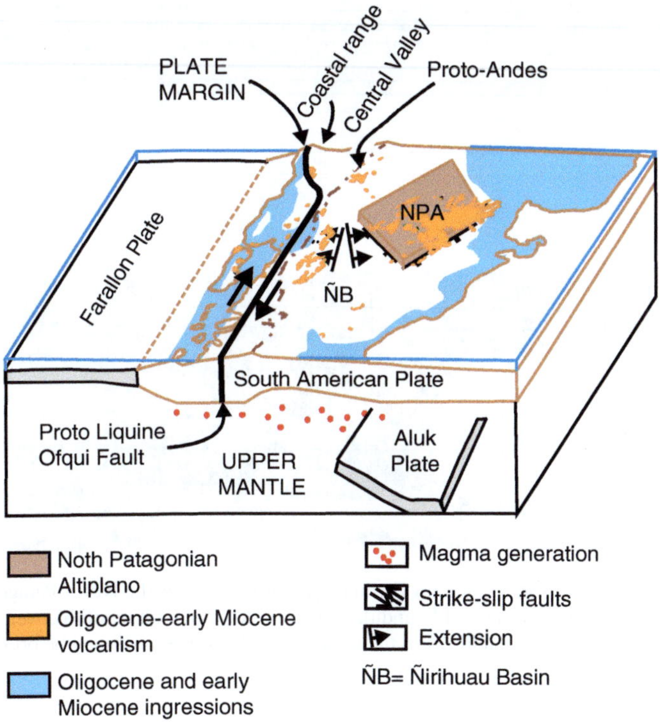

Fig. 3 Northern Patagonia 3D paleogeography and tectono-magmatic setting for the Late Oligocene-Early Miocene

The highly oblique convergence of the Farallón-SAM plates in the Paleogene was considered to have caused the development of oblique subduction (Rapela et al. 1987), but a more recent work suggests that the Aluk plate detached (Aragón et al. 2011), and the Farallón-SAM plates developed a transform margin (Fig. 3). During this time, extension has led to the development of the Ñirihuau Basin (González Bonorino 1973; Rabassa 1978; Cazau et al. 1989).

In this transform plate margin setting (Fig. 3), a large slab window will develop in time, leading first to regional mantle upwelling with the development of swelling, crustal uplift (Fig. 2b), and extension, with small basins that begin to develop at the former back-arc (Ñirihuau Basin). Finally, as the thermal swelling effect diminishes (Early Miocene), large transgressions from the Atlantic and Pacific oceans cover most of Northern Patagonia (Figs. 2d and 3). Then the tectonic scenario for the Northern Patagonian Massif uplift is extensional and coeval to the development of the Ñirihuau Basin at its southwestern side. The apex of uplift for this plateau must have been reached in the Early Miocene, as the Northern Patagonian Massif plateau was surrounded by the Atlantic and Pacific oceans transgressions (Figs. 5 and 6).

Fig. 4 Gondwana planation surface on Permian granitoid rocks at the North Patagonian Massif. A sector corresponding to the smoothly undulating, granitic plain with regolith cover. At the center of the photograph, a residual landform of the strongly eroded dome type may be observed

Fig. 5 A slightly convex sector of the paleosurface, corresponding to the "whale-back" type, with superficial development of severely eroded levels stressed by dominant horizontal fracturing over a much less developed vertical fracturing

Fig. 6 A detail of the severely eroded dome which occurs isolated in the granitic platform. Rounded boulders or corestones may be also observed

Concluding Remarks

The surface morphology of Northern Patagonia has at least three well-preserved nearly level surfaces: (a) the Gondwana planation surface caused by protracted erosion from the Middle Jurassic to the Late Cretaceous times, (b) the surface of the Maastrichtian-Danian marine sediments which flooded the Gondwana planation surface, and (c) the Late Oligocene-Early Miocene basalt lava flows.

Much of the Gondwana planation surface in Northern Patagonia is now exposed at the altiplano of the North Patagonian Massif. This is the reason for the nearly level surface morphology of this uplifted cratonic area. The exhumation of the Gondwana planation surface from sea level to the present altitude of 1,200 m a.s.l. could have started in the Late Eocene, but the main uplift of the unit may be related to the Oligocene extensional setting and mantle upwelling caused by the detachment of the Aluk plate and the development of a large slab window (Fig. 3) (Aragón et al. 2011). It was not until the altiplano of the Northern Patagonian Massif had reached a considerable altitude that the Late Oligocene-Early Miocene basalt flows (probably related to decompression melting) covered part of the altiplano and spilled down from its edges. Finally, Neogene erosion developed plateaus above the altiplano as a consequence of relief inversion. The large extent of some of these basaltic plateaus had erroneously led in the past to the conclusion that the altiplano of the North Patagonian Massif was caused by relief inversion. The debate continued until it was demonstrated that the basalt flows were younger than the uplift.

Acknowledgments This research was supported by the following grants: CONICET PIP 00/916 and UNLP 11/534 and UNLP 11/689.

References

Aguilera EY (2006) Identificación y distribución de distintas superficies geomorfológicas de escaso relieve local, por medio de sensores remotos en el área del Macizo Nordpatagónico, Provincia de Río Negro. Unpublished Doctoral Thesis, Facultad de Ciencias Naturales y Museo, Universidad Nacional de La Plata, La Plata, Argentina, 230 pp

Aragón E, González P, Aguilera EY, Cavarozzi C, Llambías E, Rivalenti G (2003) Thermal divide andesites-trachytes, petrologic evidence, and implications from Jurassic north Patagonian massif alkaline volcanism. J S Am Earth Sci 16(3):91–103

Aragón E, Carroll AR, Aguilera EY, Cavarozzi C (2005) Ancient Gondwana landforms in Patagonia. In: Gondwana 12, "Geological and Biological Heritage of Gondwana", Mendoza, Argentina, Abstracts, Academia Nacional de Ciencias, p 46

Aragón E, Aguilera EY, Cavarozzi C, Ubaldón MC, Ribot A (2008) La Caldera de Piedra Parada. Un volcán gigante de 50 millones de años, testimonio de cambios. In: Sitios de Interés Geológico, Ediciones Servicio Geológico Minero Argentino, Anales 46, 2, 669–682. Buenos Aires

Aragón E, Goin FJ, Aguilera EY, Woodburne MO, Carlini AA, Roggiero MF (2009) Paleogeography and paleoenvironments of Central Patagonia from the Late Cretaceous to the Miocene: the Paleogene Andean gap and the rise of the North Patagonian Altiplano, Abstracts, Workshop "Paleogeography and Paleoclimatology of Patagonia: effects on biodiversity", Universidad Nacional de La Plata-Dalhousie University-CONICET-ANPCyT- National Science Foundation: PIRE (Partnerships for International Research & Education), La Plata

Aragón E, Aguilera EY, Cavarozzi C, Ribot A (2010) The North Patagonian Altiplano and the Somón Cura basaltic plateau. Geociências 29(4):527–532, São Paulo, UNESP, Brazil

Aragón E, Dèramo F, Brunelli D, Castro A, Rivalenti G, Varela R, Rabbia O, Spackman W, Cavarozzi CE, Aguilera EY, Mazzucchelli M, Demartis M-L, Ribot A (2011) Tectono-magmatic response to major convergence changes in the North Patagonian suprasubduction system: the Paleogene subduction-transcurrent plate margin transition. Tectonophysics 509:218–237

Ardolino A (1981) El volcanismo Cenozoico del borde suroccidental de la Meseta de Somún Curá. Provincia del Chubut. In: 8º Congreso Geológico Argentino, Actas, 3, 7–23

Cande SC, Leslie RB (1986) Late Cenozoic tectonics of the Southern Chile trench. J Geophys Res 91(B1):471–496

Cazau L, Mancini D, Cangini J, Spalletti L (1989) Cuenca Ñirihuau. In: Chebli G, Spalletti L (eds) Cuencas Sedimentarias de Argentina, Correlación Geológica 6, 299–318

Coira BL (1979) Descripción geológica de la hoja 40d, Ingeniero Jacobacci: provincia de Río Negro. Servicio Nacional Minero Geológico, Carta geológico-económica de la República Argentina. Escala 1:200.000, Argentina. Ministerio de Industria y Minería, Argentina. Subsecretaría de Minería, Boletín 168, 5–99. Buenos Aires

Corbella H (1984) El vulcanismo de la altiplanicie de Somuncura. In: 9º Congreso Geológico Argentino, Relatorio, 1, 10, 267–300

Franchi MB, Nullo FE, Sepúlveda EG, Uliana MA (1984) Las sedimentitas Terciarias. In: 9º Congreso Geológico Argentino, Relatorio, 1, 9, 215–266

González Bonorino F (1973) Geología de la región de San Carlos de Bariloche, provincia de Río Ngero. Fundación Bariloche, Publicaciones del Departamento de Recursos Naturales y Energía, San Carlos de Bariloche

González Díaz EF, Malagnino EC (1984) Geomorfología de la Provincia de Río Negro. In: 9º Congreso Geológico Argentino, Relatorio, 2, 1–159

Malumián N (1999) La sedimentación y el volcanismo Terciarios en la Patagonia extraandina. In: Caminos R (ed) Geología Argentina. Servicio Geológico Nacional, Anales 29, 18, 557–578, Buenos Aires

Nullo FE (1978) Descripción geológica de la hoja 41d, Lipetrén: provincia de Río Negro, Carta geológico-económica de la República Argentina. Escala 1:200.000, Argentina. Ministerio de Industria y Minería, Argentina. Subsecretaría de Minería, Servicio Nacional Minero Geológico, Boletín 158, 5–89. Buenos Aires

Rabassa J (1978) Paleorrelieves Cenozoicos en la región de Pilcaniyeu-Comallo, Provincia de Río Negro. In: 7° Congreso Geológico Argentino, Actas, 2, 77–87

Rapela CW, Spalletti L, Merodio JC, Aragón E (1987) Temporal evolution and spatial variation of the lower Tertiary Andean volcanism (40–42° 30′ S). J S Am Earth Sci 1:1–14

Index

J. Rabassa and C. Ollier (eds.), *Gondwana Landscapes in southern South America*, Springer Earth System Sciences, DOI 10.1007/978-94-007-7702-6,

G

S

T

U

Z

MIX
Papier aus verantwortungsvollen Quellen
Paper from responsible sources
FSC® C105338

FSC
www.fsc.org

If you have any concerns about our products,
you can contact us on
ProductSafety@springernature.com

In case Publisher is established outside the EU,
the EU authorized representative is:
Springer Nature Customer Service Center GmbH
Europaplatz 3, 69115 Heidelberg, Germany

Printed by Libri Plureos GmbH
in Hamburg, Germany